Elementary Mathematical Modeling
A Dynamic Approach

James Sandefur
Georgetown University

THOMSON
BROOKS/COLE

Australia • Canada • Mexico • Singapore • Spain • United Kingdom • United States

Sponsoring Editors: Jennifer Huber/John-Paul Ramin
Assistant Editor: Stacy Green
Editorial Assistant: Lisa Chow
Marketing Manager: Leah Thomson
Project Manager, Editorial Production: Janet Hill
Print/Media Buyer: Vena Dyer
Production Service: Hearthside-Publishing Services—Anne Seitz

Text Designer: John Edeen
Illustrator: Hearthside Publishing Services
Cover Designer: Jennifer Mackres
Cover Image: Phototake
Cover and Interior Printer: Phoenix Color Corp
Compositor: ATLIS, Inc.

Printed in the United States of America

1 2 3 4 5 6 7 06 05 04 03 02

For more information about our products, contact us at:
Thomson Learning Academic Resource Center
1-800-423-0563

For permission to use material from this text, contact us by:
Phone: 1-800-730-2214
Fax: 1-800-730-2215
Web: http://www.thomsonrights.com

Library of Congress Control Number: 2001012345

Student Edition; ISBN 0-534-37803-X

Brooks/Cole–Thomson Learning
511 Forest Lodge Road
Pacific Grove, CA 93950
USA

Asia
Thomson Learning
5 Shenton Way #01-01
UIC Building
Singapore 068808

Australia
Nelson Thomson Learning
102 Dodds Street
South Melbourne, Victoria 3205
Australia

Canada
Nelson Thomson Learning
1120 Birchmount Road
Toronto, Ontario M1K 5G4
Canada

Europe/Middle East/Africa
Thomson Learning
High Holborn House
50/51 Bedford Row
London WC1R 4LR
United Kingdom

Latin America
Thomson Learning
Seneca, 53
Colonia Polanco
11560 Mexico D.F.
Mexico

Spain
Paraninfo Thomson Learning
Calle/Magallanes, 25
28015 Madrid, Spain

This text is dedicated to my sisters, Lynn Shepherd and Gayle Bright, who have always been there when I needed them.

Contents

Preface

This text introduces students to dynamic modeling; that is, modeling of situations that change over time. One purpose of mathematical modeling is to help solve problems encountered in the real world. The book begins by focusing on quantitative problems: finding the monthly payment for a mortgage or estimating the rate at which a body eliminates caffeine. Later, it focuses on qualitative problems: determining effective approaches to managing renewable resources or evaluating governmental approaches to stabilizing an economy.

The content and approach in my introductory modeling course, on which this text is based, has slowly evolved over the 12 plus years in which I have taught this course. One of my initial interests in developing such a course was to help students learn how to productively use the mathematics they have spent most of their lives studying. I also wanted the material to build throughout the semester, so that by the end of the course, students could develop and analyze complex models that led to a better understanding of real problems faced by society. Thus, by the end of the course, students will have renewed respect for mathematics as well as the confidence to know that they can use mathematics for their own needs—both personal and professional.

The modeling in this text requires the use of algebraic and geometric reasoning, combined with appropriate use of spreadsheets or calculators. While it is assumed the reader has a basic knowledge of algebra, most algebraic topics are briefly reviewed as needed. Solving most problems requires making appropriate choices about which tools to use and how to interpret the mathematical results within a real context.

To the Student

The only way to learn the material in this text is to be an active reader. This means that you should work through examples in the text as you read them, filling in any steps that were omitted. Unlike most math courses you have taken in the past, exact answers cannot be found for many of the problems in this text. This is often the case when using math to understand the world. In such cases, answers should be approximated to whatever precision makes sense, given the context of the problem. I have attempted to indicate the desired precision in most cases.

There is a *Student Solutions Manual for Elementary Mathematical Modeling* that gives complete, worked-out solutions to all of the odd-numbered problems. My students have found this manual to be valuable. This manual is best used in an active manner, that is, work problems on your own and use the manual when you are unsure of the next step or are stuck. To become an independent problem-solver, you should avoid too much dependence on the manual.

To the Teacher

My overarching goals are that students (1) develop and use models of simple situations, (2) understand what can and cannot be implied by mathematical models, and (3) develop an appreciation for mathematics and how it can be helpful in understanding the world. These goals are strongly supported by the last model discussed in Chapter 7. This model analyzes the disastrous negative eugenics movement of the early 20th Century. This movement resulted in the forced sterilization of hundreds of thousands of people, worldwide, including tens of thousands in the United States. In addition to the moral issues involved, careful analysis of the model developed in this chapter shows that eugenics is essentially ineffective in the way it was applied. This brings home the point that careful mathematical modeling is an important tool in most fields of study and that understanding modeling helps us make informed decisions.

The *Instructor's Resource Manual* is an invaluable resource when teaching this course. This manual has worked-out solutions for all problems and projects, and gives additional problems that can be used for homework or tests. This manual gives teaching suggestions and possible classroom demonstrations that I have found effective. There is also a student version of this manual with worked-out solutions to all of the odd-numbered problems.

The material in this text requires access to a graphing calculator or spreadsheet. Effective teaching requires that students have access to the calculator or spreadsheet during class as well as when they are doing homework. While my preference would be to teach this course in a computer lab using spreadsheets, logistics at Georgetown University make using the graphing calculators easier. The TI-83+ calculator, which is what I use, has a "sequence" mode, an essential feature that many graphing calculators, including the TI-85 and TI-86 do not have. To aid in using the TI-83+ calculator, the *Instructor's Resource Manual* includes detailed calculator instructions that can be copied and distributed to students. These instructions are organized by the chapter in which they are needed.

This material can be used to teach a one-semester course to non-science majors. Because of the varied ability of these students, I pace the class depending on the particular class. I usually cover some of each chapter, omitting a few sections as time requires. Sections 3.6, 4.5–4.8, and 6.5 can be omitted without losing continuity. Chapters 6 and 7 are the capstone for the course, using most of the previous material on extended applications, so at least one of these chapters should be covered. Section 7.4 is time consuming, but important. I pace the course so that I have time to discuss this section, although I often just discuss its highlights on the last 2 days of class.

About the Author

James Sandefur received his B.A. degree from Vanderbilt University, his M.A. from University of Denver, and his Ph.D. from Tulane University. Currently, he is Professor of Mathematics at Georgetown University. He has been Visiting Professor at the Center for Applied Math at Cornell University, the University of Iowa, and the Freudenthal Institute at the University of Utrecht. He is the author of *Discrete Dynamical Modeling* and *Discrete Dynamical Systems: Theory and Applications* as well as numerous research and expository papers on differential equations and dynamical systems. He was on the writing team for National Council of Teachers of Mathematics's *Principles and Standards for School Mathematics* and is on the Editorial Board for the *Mathematics Teacher.* He was twice the recipient of a Georgetown University Sony Award for Excellence in Science Education and is currently Faculty Chair of the Georgetown Honor Council.

Acknowledgments

While I appreciate the help of the many people who contributed to the development of this text, there are a few I would like to single out for special thanks.

I want to thank Professor Mary Erb for her support in the development of this material. Her careful reading of the solutions and feedback from teaching this course herself has been invaluable to the development of this text. I want to thank my friend and colleague, Professor Rosalie Dance, for all her help and support. Thanks also go to Professors Ray Bobo, Ron Rosier, and Haitao Fan, who have been supportive from the early days of my math modeling course. This text would not have been possible without the many students who have taken this class from me over the past several years. I have learned a great deal from these students and their feedback on the material has helped me in writing material that most students can read and enjoy.

There were many people who helped me understand a number of the models in this text. Among these are: Dr. Carl Peck, Professor of Pharmacology and Medicine, Georgetown University, and Dr. Wendall Weber, Emeritus Professor of Pharmacology at Northwestern University, who were invaluable in helping me understand models related to chemical buildup and elimination in the body; Dr. Michael Kaiser, a clinical psychologist and certified addictions counselor, who provided valuable information about alcohol and problems related to alcoholism; Dr. Barry Anderson, M.D., and Professor Marcela Parra, who provided valuable information concerning the relationship between Sickle Cell Anemia and malaria.

I want to thank all of the people at Brooks/Cole who have believed in and supported this text, particularly, Robert Pirtle, Jennifer Huber, and, most recently, John-Paul Ramin. I give special thanks to Anne Seitz at Hearthside Publishing Services, copyeditor Andrew Potter, and Senior Project Manager Janet Hill who are largely responsible for turning my manuscript into a quality text, despite having to put up with me through the whole process.

My heartfelt thanks also goes to the many reviewers of my text, including

Daniel S. Alexander, Drake University
Jeff Bonn, Michigan Technological University
Henry P. Decell, Jr., University of Houston
Charles Doering, University of Michigan
William Fox, Francis Marion University
Joel Haack, University of Northern Iowa
Joseph Harkin, SUNY College at Brockport
John Haverhals, Bradley University
Gloria Hitchcock, Georgia Perimeter College
John G. Koelzer, Rockhurst College
Robert H. Lewis, Fordham University
William Martin, North Dakota State University

James Sandefur

1 Introduction to Modeling

1.1 Introduction to Dynamical Systems

In working through this text, you are going to learn to use mathematics to answer important questions about real situations, such as determining how much you can afford to borrow given your monthly salary or how to control your diet to achieve and maintain a desired weight. In other cases, you will use mathematics to get a better understanding of certain situations, such as how medicines build up and are eliminated from a body.

The situations studied in this text are those in which some quantity is changing over time. Such situations are **dynamic.** In Section 1.2, you will learn to model the situations using what is called a **discrete dynamical system.** A discrete dynamical system is an equation that relates a quantity at one point in time to the same quantity at an earlier point in time. For example, suppose an object is at a position of 10 feet from a starting point and that it is traveling away from the start at 2 feet per second. In one second it will be 12 feet from the start, in 2 seconds it will be 14 feet from the start, and so on.

The situations studied in Section 1.3 involve one or more quantities that are changing over time. The change may involve the addition or subtraction of a constant amount in addition to the addition or subtraction of a constant proportion. One simple example would be a bank account to which a constant deposit is being added in addition to interest, which is a proportion of what was already in the account.

In Section 1.4, you will learn to generate tables and graphs using a calculator or computer to answer difficult but important questions about dynamic situations. The use of computers or calculators is overkill when studying simple situations, but for some of the more difficult situations in Section 1.4, and in the rest of the book, the use of computational technology will be indispensable.

Section 1.5 is devoted to the study of finance.

Dynamic refers to situations changing over time, such as, a frog going through its stages of life as in Figure 1.1.

FIGURE 1.1 A dynamic situation.

1.2 Examples of Modeling

In this section, you will learn the notation that will be used throughout the book and will use this notation to begin describing some simple dynamic situations, that is, situations in which something is changing. You will then study these situations using 2 different approaches. When studying the situations, I will point out some advantages and disadvantages of each approach.

Consider the following simple situation. Suppose you earn $15 per hour tutoring mathematics. You wish to know how much money you will earn for tutoring different amounts of time. Table 1.1 gives some possible results.

TABLE 1.1 **Amount earned for tutoring different numbers of hours.**

hours worked	0	1	2	3	4
money earned	0	15	30	45	60

The first step in modeling is to assign variables. Let n represent the number of hours worked. The variable n is called the **independent variable** and usually represents time.

Let u represent the amount of money earned tutoring. The variable u depends on n, the number of hours worked, and will usually be written as $u(n)$, meaning the amount of money earned for tutoring n hours. Therefore u is called the **dependent variable.** The dependent variable is a **function** of the independent variable. Table 1.2 gives the value of the dependent variable for several values of the independent variable. For example, $u(3) = 45$ is read as "the amount of money earned for working 3 hours is 45 dollars."

TABLE 1.2 **Values of $u(n)$ for several values of n.**

hours worked	$n = 0$	$n = 1$	$n = 2$	$n = 3$	$n = 4$
money earned	$u(0) = 0$	$u(1) = 15$	$u(2) = 30$	$u(3) = 45$	$u(4) = 60$

The dependent variable represents the size or amount of what is changing as the independent variable changes, such as the distance our car is from home after several hours, the amount of money in an account after several years, or the prevalence of a particular genetic trait after several generations.

When modeling, you should always write a specific description of what each variable represents. In the case of tutoring, you might give the following descriptions of the variables.

Independent variable: n represents the integer number of hours worked.

Dependent variable: $u(n)$ represents the amount of money earned working n hours.

You should get into the habit of writing precise descriptions of what the independent and dependent variables represent. Do not make vague descriptions, such as n is time or $u(n)$ is money, because these descriptions are incomplete and will lead to problems later. Also note that the description of the dependent variable should refer back to the independent variable.

Different tutors could have different hourly rates. For example, 1 tutor might charge \$10 per hour and another tutor \$20 per hour. We could let r represent the hourly rate for a tutor. Since you charge \$15 per hour, $r = 15$ when modeling your earnings. The variable r is called a **parameter** that varies depending on the particular situation.

The **domain** of the function is the set of n-values for which $u(n)$ is defined. In this book, the domain will consist of the set of n-values for which the context makes sense, usually the positive integers $n = 1, 2, \ldots$ or the nonnegative integers $n = 0, 1, \ldots$. In the tutoring example, the domain is $n = 0, 1, \ldots$.

Plotting the points $(0, u(0))$, $(1, u(1))$, ... on a graph helps in understanding the relationship between the independent and dependent variables. The n-values are in the horizontal direction and the u-values in the vertical direction. For the tutoring example, the points $(0,0)$, $(1,15)$, $(2,30)$, ..., $(20,300)$ are plotted on Figure 1.2. Connecting consecutive points with a line helps us see a pattern. But remember that the only points that have meaning are those for which the independent variable is an integer. Such a graph is called a **time graph** because the dependent variable is plotted over time.

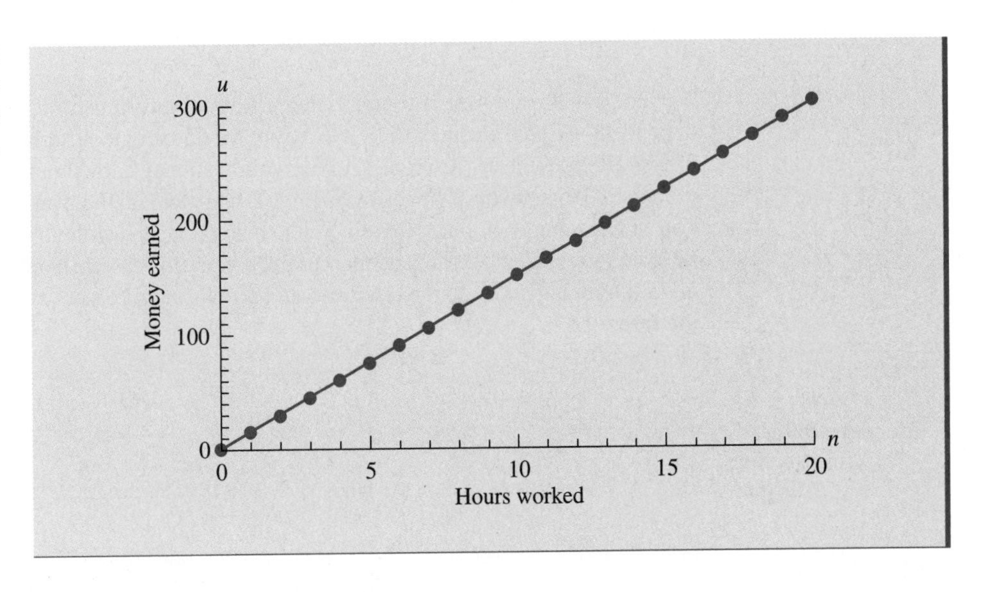

FIGURE 1.2 Graph of hours worked versus money earned.

We will use 2 approaches to model situations such as earning money by tutoring. The most direct approach is to try to develop an explicit expression for $u(n)$ in terms of n. This will be called the **explicit** approach, since $u(n)$ is explicitly given as an expression of n. In the tutoring situation, it is relatively easy to develop the expression for $u(n)$. If you get \$15 per hour and work n hours, then you earn

$$u(n) = 15n$$

dollars. This expression shows that $u(n)$ is a linear function of n. This is not surprising, given the graph in Figure 1.2. Since the equation for a line in slope/intercept form is $y = mx + b$, it is clear that the slope of this line is 15 and the vertical intercept is 0. Note that the slope of this line, 15, is the amount of money earned each hour. The vertical intercept is 0, since 0 dollars are earned for working 0 hours.

The second approach to this situation will be called the **dynamic** approach. This approach may at first seem more difficult than the explicit approach, but you will find that when the situations are more complex, the dynamic approach will be much easier. This is because in many cases, it is difficult or even impossible to write an explicit expression for the function $u(n)$.

The dynamic approach considers how the dependent variable **changes** as the independent variable changes. We usually study the change in the dependent variable as the independent variable changes by 1. In this tutoring example, that means we study how the amount of money earned changes as you work 1 more hour. For example, we could consider how $u(5)$ differs from $u(4)$ or how $u(15)$ differs from $u(14)$. To understand this change, it often helps to make a written description of the change.

Written Description

> If you work one more hour, you earn 15 more dollars, that is, the amount earned increases by 15 dollars.

For many examples, a figure called a **flow diagram** can help you visualize how the dependent variable changes. A flow diagram for this tutoring example is seen in Figure 1.3. The oval represents a "reservoir," a reservoir of money earned in this case. The amount of money in the reservoir changes over time. The arrow into the oval describes how the reservoir is changing each hour; that is, \$15 is being added each hour. Ovals or reservoirs will be used to represent the dependent variables, u in this case. Arrows into an oval represent quantities being added to the reservoir, and arrows out represent amounts being taken from the reservoir.

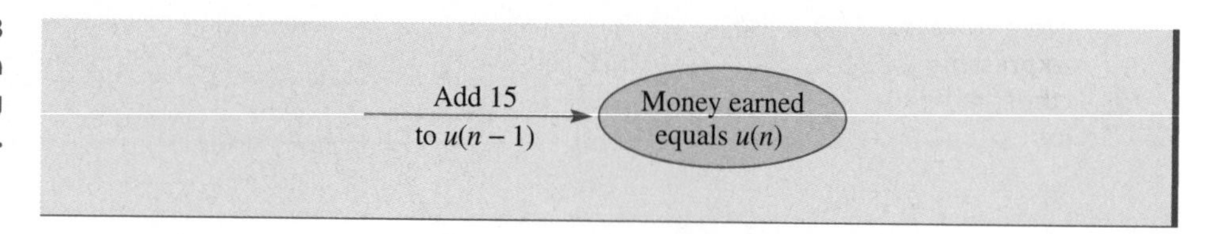

FIGURE 1.3 Flow diagram for tutoring example.

Finally, we can describe the change in u with an equation, called a **discrete dynamical system.**

Discrete Dynamical System: $u(n) = u(n-1) + 15.$

This equation is a direct translation of the written statement and the flow diagram. In particular, n is 1 more hour than $n-1$, and the amount earned for working n hours, $u(n)$, is \$15 more than the amount earned for working 1 fewer hours, $u(n-1)$.

A **discrete dynamical system** is an equation in which the dependent variable at one point in time, $u(n)$, is given in terms of the dependent variable at previous points in time, $u(n-1)$ in this case. In many situations, we will develop the discrete dynamical system by considering how things were added to and subtracted from the previous total, that is,

$$u(n) = u(n-1) + \text{amounts added} - \text{amounts taken away.}$$

The "amounts added" are represented by arrows going into the oval in the flow diagram, and the "amounts taken away" are represented by arrows going out of the oval.

A discrete dynamical system can be written many different ways. For example, you could have considered the 2 times $n+1$ and n. The important point is that these 2 numbers of hours are only 1 hour apart (see Figure 1.4, 1.5). Then the 2 amounts earned would be $u(n+1)$ and $u(n)$. In this case, the discrete dynamical system would be written as

$$u(n+1) = u(n) + 15$$

The main point to notice is that both equations, $u(n) = u(n-1) + 15$ and $u(n+1) = u(n) + 15$, say the same thing, that you get \$15 for working 1 more hour. When $u(n)$ is given in terms of $u(n-1)$, we say the dynamical system is written in **standard form.** The dynamical system $u(n) = u(n-1) + 15$ is written in standard form; the dynamical system $u(n+1) = u(n) + 15$ is not. We could rewrite $u(n+1) = u(n) + 15$ as $u(n) = u(n+1) - 15$. This also is not in standard form, since $u(n)$ is given in terms of $u(n+1)$, not $u(n-1)$.

The discrete dynamical system $u(n) = u(n-1) + 15$ can be used to determine the amount of money earned after working any number of hours. To use it, you need to know the first amount earned, which is $u(0) = 0$, meaning that if you work 0 hours, you earn 0 dollars. Now substitute 1 for n into the discrete dynamical system. This gives

$$u(1) = u(1-1) + 15 \quad \text{or} \quad u(1) = u(0) + 15$$

Now substitute 0 for $u(0)$, giving $u(1) = 0 + 15 = 15$. To compute $u(2)$, substitute 2 for n into the dynamical system. This gives

$$u(2) = u(2-1) + 15 \quad \text{or} \quad u(2) = u(1) + 15 = 15 + 15 = 30$$

FIGURE 1.4 Different ways of expressing 2 times that are one hour apart.

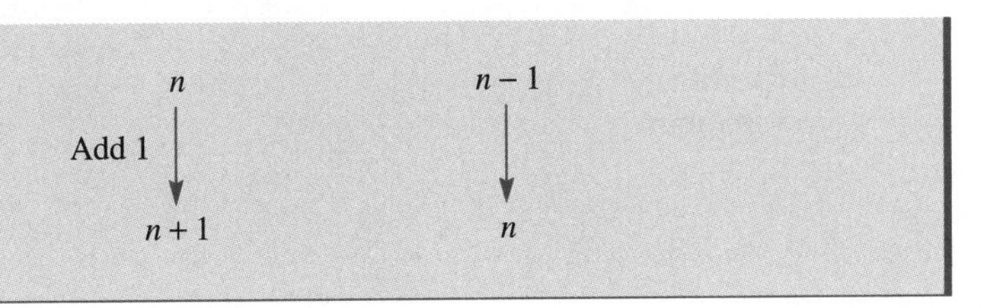

FIGURE 1.5 Different ways of expressing two times that are one day apart.

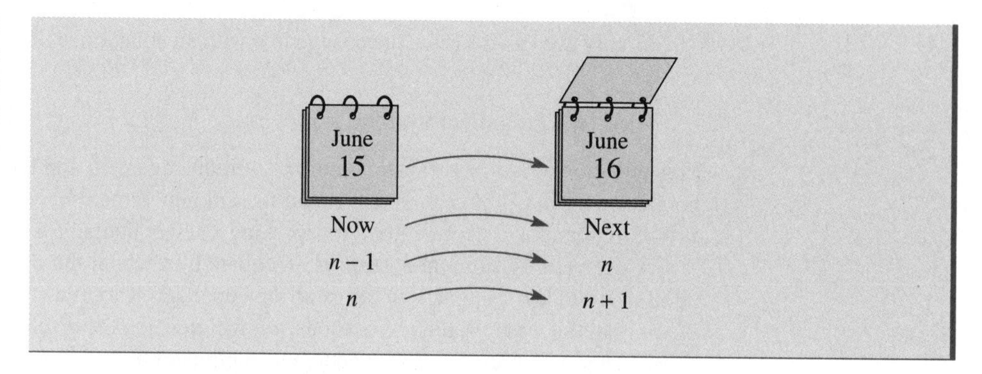

By repeating this process, you can determine the amount earned for any given integer number of hours. It seems that this process is very time-consuming. But with the use of computer spreadsheets or calculators, you will see that it is relatively easy to investigate $u(n)$ in many complex situations.

Let's review. We are considering situations in which 1 variable (u in this tutoring example) is a function of another variable (n in this tutoring example). The function is the actual collection of values, $u(n)$, for each value of n in the domain. Sometimes we can use a dynamic approach to actually compute the values for $u(1)$, $u(2)$, $u(3)$, ... Other times we can use an expression for $u(n)$ to compute the values. Using either approach, we can make a table of values for $u(n)$ and a time graph of the function by plotting the points $(n, u(n))$, as was seen in Figure 1.1. (When graphing the function, we often connect the points so that it is easier to see patterns formed by the function.)

Consider another situation.

Example 1.1

In this example, we will investigate the height of different sized stacks of Styrofoam cups, as seen in Figure 1.6. To understand this problem better, get some cups and actually measure the heights of different-sized stacks.

FIGURE 1.6 Investigating the height of a stack of cups.

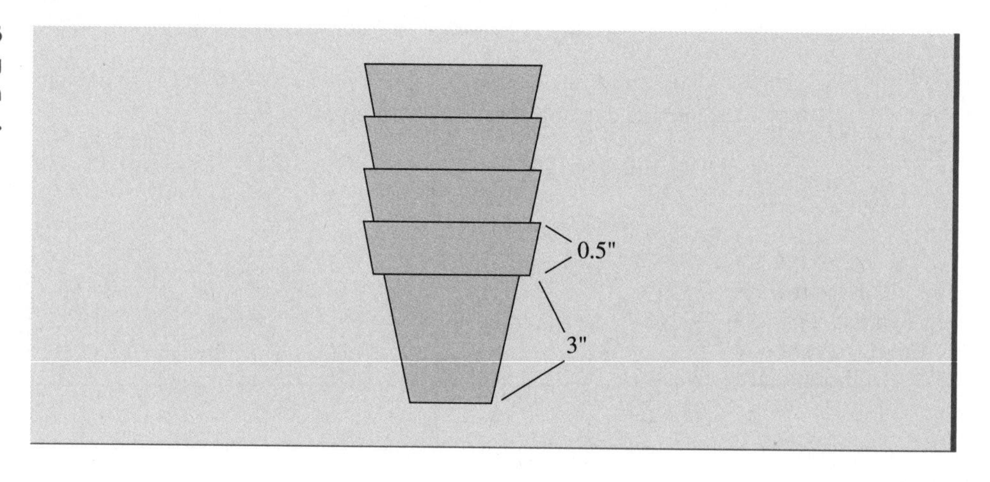

The first step is to define the variables. The independent variable n is the number of cups. The dependent variable u is the height of the stack of cups. In particular, $u(n)$ represents the height of a stack of n cups.

Let's approach this problem using the explicit approach. By looking at Figure 1.6 or your own stack of cups, it should be clear that a stack of n cups consists of one base and n "lips." For the cups in Figure 1.5, the base is 3 inches high and the lips are $\frac{1}{2}$ inch high. This leads to the expression

$$u(n) = 3 + 0.5n$$

This means that u is a linear function of n. A graph of this function is seen in Figure 1.7.

FIGURE 1.7 Heights of a stack of cups.

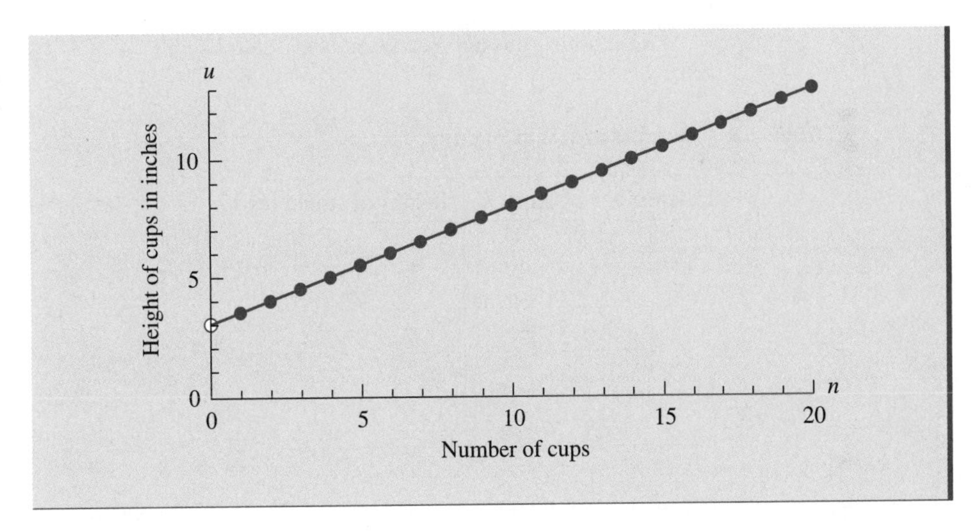

This is the graph of a line with slope $\frac{1}{2}$, which is also the height of the lip of a cup. Note that the vertical intercept is $u(0) = 3$, the height of the base. The expression is actually incorrect when $n = 0$, since zero cups have zero height. This is why the point $(0, 3)$ is intercept open. In fact the graph only makes sense at points where n is a positive integer, so the domain of the function is $n = 1, 2, \ldots$. This means that only the points noted on the graph have any physical meaning.

Now we will study this problem using the dynamic approach. To do this, observe how the height of the stack changes when one more cup is added. A written description of the change is "when 1 cup is added to a stack of cups, the height of the stack increases by $\frac{1}{2}$ inch, which is the height of the lip of a cup." A flow diagram is seen in Figure 1.8. The discrete dynamical system is developed by considering

$$u(n) = u(n-1) + \text{things added} - \text{things taken away} = u(n-1) + 0.5 - 0$$

This simplifies to

$$u(n) = u(n-1) + 0.5 \tag{1.1}$$

To use this dynamical system, we need to know that $u(1) = 3.5$. Substitution of 2 for n into (1.1) gives

$$u(2) = u(1) + 0.5 = 3.5 + 0.5 = 4$$

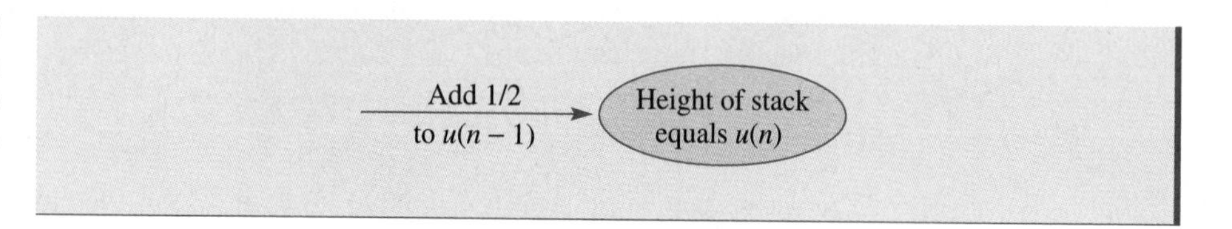

FIGURE 1.8 Flow diagram for stacking cup problem.

Continuing this approach results in Table 1.3.

TABLE 1.3 Heights of stacks of cups.

number of cups, n	height of stack, $u(n)$
1	3.5
2	4
3	4.5
4	5
⋮	⋮

Note that we did not begin with $u(0) = 0$ when using equation (1.1). If we had, we would have computed $u(1) = u(0) + 0.5 = 0 + 0.5 = 0.5$, which is incorrect. It often requires some thought about the context to know which n-value to begin with.

Since $u(0) = 0$ does not fit the pattern whether we are using the dynamic approach or the explicit approach, we could define the domain of the function to be $n = 1, 2, \ldots$. Alternatively, we could let the domain be the nonnegative integers, define $u(0) = 0$ and $u(1) = 3.5$, and then use the dynamic or explicit approach to find $u(n)$ for $n = 2, 3, \ldots$. The point is that we sometimes have latitude in determining the domain. But we must be careful to be consistent with the context. ■

In considering the money earned for tutoring, the domain could be considered all values of n that are greater than or equal to 0, since you can tutor for any amount of time. For the cup problem, the domain is the set of positive integers, since we don't consider parts of a cup. An advantage of the explicit approach is that the expression can be evaluated using any n-value. In the tutoring problem, you can use the expression $u(n) = 15n$ to compute the money earned for working any amount of time. A disadvantage of the dynamic approach is that it can be used to find $u(n)$ only when n is an integer. For the cup problem, this was not a disadvantage, since $u(n)$ only exists when n is an integer.

Consider 1 more example.

Example 1.2

The cost for having a party at a local club is \$300 plus \$30 per person. What is the cost of the party if 200 people come? How many people can you invite if you can afford to pay \$2000?

Again, the first step to studying this problem is to define the variables. Let n represent the number of people coming to the party, and let $c(n)$ represent the cost of the party if n people come.

First approach this problem using the explicit approach. The cost for having n people at the party is the flat fee of \$300 plus \$30 for each of the n people. This translates into the expression

$$c(n) = 300 + 30n$$

To find the cost of having $n = 200$ friends just substitute 200 for n, giving

$$c(200) = 300 + 30 \times 200 = 6300 \text{ dollars.}$$

To find the number of friends you can invite for \$2000, substitute 2000 for the cost, $c(n)$, and solve for n, giving

$$2000 = 300 + 30n \quad \text{or} \quad 1700 = 30n$$

which becomes $56.66 = n$. Since you must invite an integer number of friends and limit the cost to at most \$2000, you will invite at most 56 friends.

Now consider the dynamic approach. The written description is that, in order to invite 1 more friend to the party, you must pay \$30 more. This translates into the flow diagram in Figure 1.9.

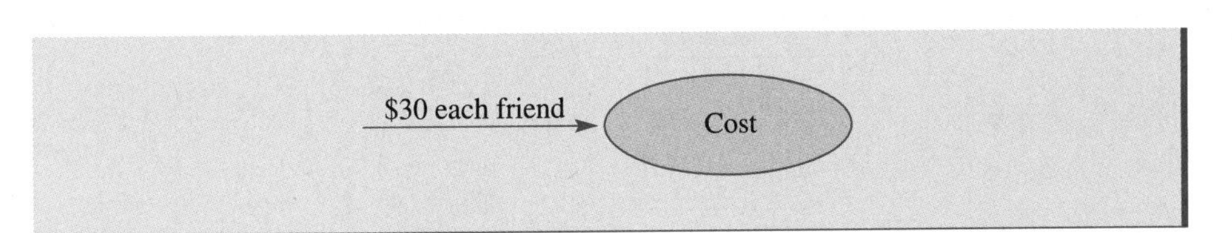

FIGURE 1.9 Flow diagram for party problem.

This results in the discrete dynamical system,

$$c(n) = c(n-1) + 30$$

If you had no friends coming to your party, the cost would be $c(0) = 300$. To find the cost for having 1 friend, you use the preceding expression with $n = 1$; that is

$$c(1) = c(0) + 30 = 330$$

To compute the cost for 2 friends, you compute

$$c(2) = c(1) + 30 = 330 + 30 = 360$$

By doing lots of computations, you could then find the cost for having any number of friends, although it might take you a while to compute the cost of having 100 friends using this method. ■

As you may have noticed from the examples in this section, it seems to be much easier to answer relevant questions using the explicit approach than using the dynamic approach. In fact, it seems that it would be quite difficult, using the dynamic approach, to answer questions such as, How much would a party cost if 200 people came? How many cups can be in a stack of cups that must fit into a box that is 4 feet long? How many hours must you tutor in order to earn the $3000 down payment for a car? But you will soon see how to answer these and other questions using the dynamic approach and a computer spreadsheet or an appropriate calculator. You may still be wondering why you should go to all this trouble when the explicit approach is so easy. The reason is that it is actually much easier to develop a discrete dynamical system than it is to find an expression for the function in situations that are only slightly more complex than the ones presented in this section. Beginning in the next section, the situations will be much easier to study using the dynamic approach. In fact, later in this book you will analyze situations that would be impossible to study using the explicit approach.

1.2 Problems

1. In each part, write a discrete dynamical system associated with the flow diagram. The independent variable is n, and the dependent variable is $u(n)$. Simplify the expression if possible.

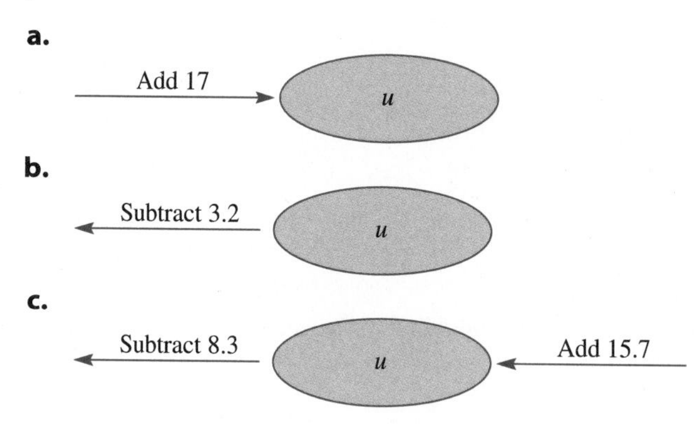

2. In each of the following parts, use the dynamical system and the given value to compute $u(2)$ through $u(5)$. In each part, plot the points $(2, u(2))$, $(3, u(3))$, $(4, u(4))$, $(5, u(5))$. Describe the shape you see.

a. $u(n) = u(n-1) + 4$ with $u(1) = -3$.
b. $u(n) = 2u(n-1) + 1$ with $u(0) = 2$.
c. $u(n) = u(n-1) - 1$ with $u(-1) = 10$.
d. $u(n+1) = u(n) + 3$ with $u(1) = 5$.
e. $u(n) = -u(n-1) + 6$ with $u(0) = 15$.
f. $u(n) = 5u(n-1) - 8$ with $u(0) = 2$.

3. Suppose a basketball player gets a contract for $5000 per game and a $30,000 signing bonus. You are going to study this player's earnings over the season.

a. Define the independent and dependent variables. What is the domain?

b. Develop an expression for the dependent variable in terms of the independent variable. Use that expression to determine how much this player will earn if he plays 120 games and how many games he needs to play to earn at least $1 million.

c. Use the expression to complete the following table.

Number of games played	0	1	2	3	4
Money made					

d. Graph the function you developed in part (b). It may help to plot points from the table developed in part (c). While your graph may be a smooth line, note that the only points that have meaning in this context are ones in which the independent variable is a nonnegative integer.

e. What are the slope and y-intercept of the line graphed in part (d), and what is their physical significance in the context of this problem?

f. Draw a flow diagram to model this situation.

g. Develop a discrete dynamical system to model this situation.

4. Suppose your car holds 15 gallons of gas and the tank is full. Your car also gets 20 miles to a gallon of gas. Your are going to study the amount of gas left in your car's tank after driving for some number of miles.

a. Define the independent and dependent variables. What is the domain?

b. Develop an expression for the dependent variable in terms of the independent variable. (Think about how much gas it takes to drive 1 mile if the car gets 20 miles per gallon.) Use that expression to determine how much gas will be left in your tank after you have driven 127 miles and how many miles you can drive before running out of gas.

c. Use the expression to complete the following table.

Number of miles driven	0	1	2	3	4
Amount of gas remaining in tank					

d. Graph the function you developed in part (b). It may help to plot points from the table developed in part (c). Note that all points on this graph have meaning in this context.

e. What are the slope and y-intercept of the line graphed in part (d), and what is their physical significance in the context of this problem?

f. Describe how the amount of gasoline in the car changes when you drive one more mile. Draw a flow diagram to model this situation.

g. Develop a discrete dynamical system to model this situation.

5. Air cools by about 0.01°C for each meter rise above ground level, up to about 12 km. You are going to study the temperature in relation to height above the ground.

a. Define the independent and dependent variables. What is the domain?

b. Suppose the current temperature at ground level is 28°C. Develop an expression that gives the temperature in terms of the height above the ground. Use this expression to find the temperature 1 km above the ground. Also, find the height above the ground at which the temperature is about 0°C.

c. Use the expression to complete the following table.

Number of kilometers above ground	0	1	2	3	4
Temperature in degrees Centigrade					

d. Graph the function you developed in part (b).

e. What are the slope and y-intercept of the line graphed in part (d), and what is their physical significance in the context of this situation?

f. Draw a flow diagram to model this situation.

g. Develop a discrete dynamical system to model this situation.

6. There is evidence that the number of chirps some crickets make depends on the temperature. At 68°F, one species of crickets makes about 124 chirps per minute. For each degree the temperature rises, the number of chirps increases by 4 per minute.

a. Define the independent and dependent variables. What is the domain?

b. Develop an expression that relates the number of chirps per minute in terms of the temperature. Use this expression to approximate the temperature if you count about 24 chirps in 10 seconds.

c. Graph the function you developed in part (b).

d. What is the slope and what is its physical significance? Note that the y-intercept is negative, which would mean that the crickets make a negative number of chirps. This is because 0 is out of the range of the domain of the function. The crickets will be dead by the time the temperature reaches 0°F.

e. Draw a flow diagram to model how the number of chirps per minute changes.

f. Develop a discrete dynamical system to model this situation.

7. When you travel toward a sound, the frequency of the sound seems higher. This is called the Doppler effect. The frequency of middle C on a piano keyboard is 256 cycles per second. For each mile per hour you increase your speed, the apparent frequency of the sound increases by $256/760$ cycles per second. Note that the speed of sound is 760 miles per hour.

a. Define the independent and dependent variables. What is the domain?

b. Develop an expression that relates the frequency of the sound to your speed toward the object. How fast would you need to travel for the middle C of a keyboard to sound like C#, which is 271 cycles per second? How fast would you need to travel for the middle C of a keyboard to sound like the C that is 1 octave higher? This C is 512 cycles per second.

c. Graph the function you developed in part (b).

d. What are the slope and y-intercept of the line graphed in part (c), and what is their physical significance?
e. Draw a flow diagram to model how the frequency changes as your speed increases by 1 mile per hour.
f. Develop a discrete dynamical system to model this situation.

8. At 78°F, the pavement on a bridge is 1000 ft long. For each degree rise in temperature, the length of the pavement grows by 0.012 ft.
a. Define the independent and dependent variables.
b. Develop an expression that gives the length of the bridge in terms of the temperature. Use this expression to determine how long the bridge is if the temperature reaches 105°F. The highway designers will need to leave a series of expansion joints that totals the extra length so that the bridge will not buckle.
c. Use the expression to complete the following table.

Temperature in degrees Fahrenheit	0	20	40	60	80
Length of bridge in feet					

d. Graph the function you developed in part (b).
e. What are the slope and y-intercept of the line graphed in part (d), and what is their physical significance?
f. Draw a flow diagram to model how the length of the bridge changes as the temperature changes.
g. Develop a discrete dynamical system to model this situation.

9. The air pressure at sea level is 1030 g/cm^2. For each centimeter of depth in salt water, the pressure increases by 1.026 g/cm^2.
a. Define the independent and dependent variables.
b. Develop an expression that gives the pressure, in g/cm^2, in terms of the depth in salt water. Use this to find the pressure at a depth of 500 cm.
c. Draw a flow diagram.
d. Develop a discrete dynamical system to model this situation.

1.2 Projects

Project 1. A new set of stairs is going to be built from the first floor to the second floor of a building. The height of a step should be between 6 and 8 in., and each step should be the same height. Let the height of a step be denoted by h. The symbol h will be considered a constant (See Figure 1.10). Let the independent variable n be the number of steps. The dependent variable $s(n)$ is the height of a set of n stairs.

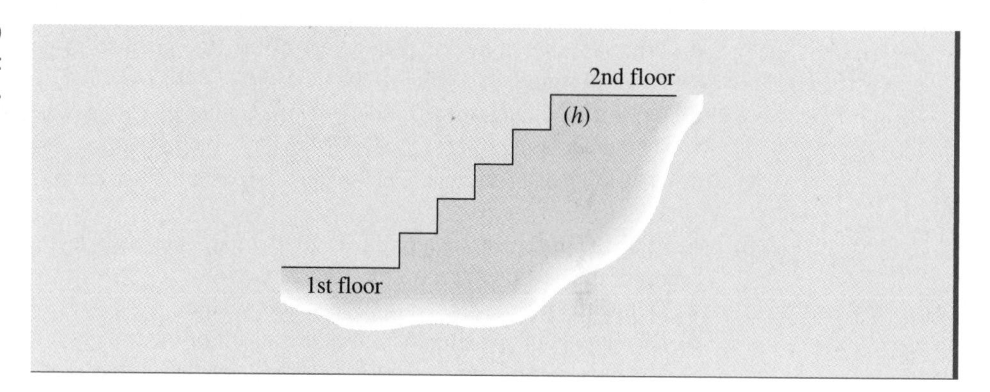

FIGURE 1.10 h is a constant in Project 1.

a. Develop an expression for $s(n)$ in terms of the variable n and the constant h. Graph the function using step heights of size $h = 6$ inches. On the same axes, graph the function using steps of height $h = 8$ inches. What other graphs are possible using different values for h?

b. The set of stairs is going to be 110 inches high. Determine the lowest number of steps needed, and what the height h of each step, to the nearest hundredth of an inch, should be in this case. Graph the function using that value for h on the same axes as the first 2 functions.

c. What is the greatest number of steps needed and what should the height of each step be, to the nearest hundredth of an inch?

d. What are the different possible heights that can be chosen for the steps, to the nearest hundredth of an inch?

e. Draw a flow diagram indicating the change in the height of a set of steps if 1 more step is needed. Use the constant h, not a particular value for h.

f. Develop a discrete dynamical system for this situation. Note that it would be difficult to answer parts (b) and (c) using this system.

g. Find the height of steps in several different places.

Project 2. In this problem, we are going to study how water is wasted in the brushing of teeth. Let n be the number of people that let the water run while brushing their teeth. Let $w(n)$ be the amount of water that is wasted per day by these people while brushing their teeth.

a. Estimate how much water is wasted when you leave the faucet running while brushing your teeth. Assuming you brush your teeth 3 times per day, estimate how much water would be wasted per day if you leave the water running while brushing your teeth.

b. Assuming that everyone wastes about the same amount of water as you, develop an expression for $w(n)$ in terms of n. Use this expression to estimate the amount of water wasted in your city if everyone left the water running.

c. In part (a), you estimated the amount of water wasted brushing your teeth. This value may vary from person to person. Make what you think is an underestimate for the amount of water wasted each day, and rework part (b) using this underestimate.

d. Make what you think is an overestimate for the amount of water wasted brushing your teeth, and rework part (b) using this overestimate.

e. Draw a flow diagram to indicate how the amount of wasted water increases as the number of people who leave the water running increases by 1, using your estimate from part (a).

f. Develop a discrete dynamical system to model this situation, using your estimate from part (a).

1.3 Affine Dynamical Systems

In the previous section we learned to develop discrete dynamical systems of the form

$$u(n) = u(n-1) + b$$

where b was some constant number that arose from the context of the situation. In this section we are going to consider more complex situations that lead to discrete dynamical systems of the form

$$u(n) = au(n-1) + b$$

where a and b are both constants that depend on the context of the situation. An example is

$$u(n) = 1.2u(n-1) + 3$$

A dynamical system that has this form is called an **affine dynamical system.**

Before doing any computations, we need to know the value for $u(n)$ for some value of n. Usually we know $u(0)$ or $u(1)$. This beginning value for $u(n)$ is called the **initial value.** Suppose we are given the initial value $u(0) = 10$ for $u(n) = 1.2u(n-1) + 3$. We can now use it by substituting the value $n = 1$ into the dynamical system, giving the equation

$$u(1) = 1.2u(0) + 3 = 1.2(10) + 3 = 15$$

We can then use $u(1)$ to find $u(2)$ by substituting $n = 2$, giving

$$u(2) = 1.2u(1) + 3 = 1.2(15) + 3 = 21$$

You should compute $u(3)$ through $u(5)$ and compare your results with those in Table 1.4.

TABLE 1.4 Computations for $u(n) = 1.2u(n-1) + 3$ with $u(0) = 10$.

$n =$	0	1	2	3	4	5
$u(n) =$	10	15	21	28.2	36.84	47.208

A time graph of the values up to $n = 10$ is seen in Figure 1.11.

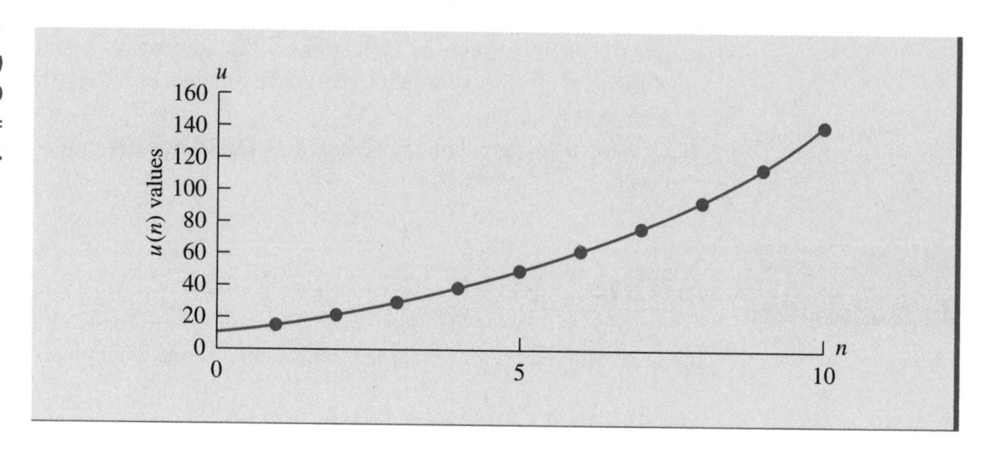

FIGURE 1.11 Points $(n, u(n))$ for $n = 0, \ldots, 10$ for $u(n) = 1.2u(n-1) + 3$.

In this case, the points do not lie on a line. This nonlinear behavior of the values means that it is much more difficult to find an expression for $u(n)$ in terms of n. This is why we are using the dynamic approach. For fun, you might try to find an expression for $u(n)$. Check your expression to see if it gives the same values for $u(1)$ through $u(5)$ as are in Table 1.4. Don't be discouraged if you can't find the correct expression. In Chapter 3, you will learn to find the correct expression for $u(n)$.

In some situations, two related things vary over time, such as the size of a deposit into a bank account and the total value of the account. To model such situations requires using 2 or more dependent variables that are changing simultaneously, say $u(n)$ and $v(n)$. In these cases, we need to develop a dynamical system consisting of 2 or more equations, such as

$$u(n) = 1.2u(n-1) + 0.9v(n-1) + 2$$
$$v(n) = 0.4u(n-1) + 1.5v(n-1) - 9$$

To use equations such as these, you need to be given an initial value for u and for v, such as $u(0) = 9$ and $v(0) = 7$. Then you can use these equations to find $u(1)$ and $v(1)$, then use these values to find $u(2)$ and $v(2)$, and so on. For example,

$$u(1) = 1.2u(0) + 0.9v(0) + 2 = 1.2(9) + 0.9(7) + 2 = 19.1$$
$$v(1) = 0.4u(0) + 1.5v(0) - 9 = 0.4(9) + 1.5(7) - 9 = 5.1$$

Making a flow diagram of the situation will often help in developing the equations. A dynamical system of the form

$$u(n) = a_1u(n-1) + b_1v(n-1) + c_1$$
$$v(n) = a_2u(n-1) + b_2v(n-1) + c_2$$

is called an **affine dynamical system of 2 equations.**

Let's consider an example of affine dynamical systems applied to the area of medicine.

Example 1.3

In this example we consider how to prescribe correct doses of medicines in order to safely and effectively treat an ailment. In particular, we will consider problems related to prescribing the medicine digoxin.

For your general information, digoxin is a drug that is often used to increase the contractile power of the heartbeat in patients who have congestive heart failure. In layman's terms, it makes the heartbeat stronger. This is needed when a patient's heart does not pump enough blood. Digoxin is also used to control the ventricular rate in patients who have atrial fibrillation, that is, it slows down the heart rate when the heart is beating dangerously fast. The medical conditions that are treated by digoxin are usually found in older patients, but a number of young people have heart conditions that require the use of digoxin or similar medicines. Some of these young people got damaged heart valves from rheumatic fever. Others were born with various congenital defects.

Digoxin is filtered from the blood by the kidneys. Each 24-hour period, the kidneys filter out about one-third of the digoxin that was in the blood at the beginning of that 24-hour period. Typically, at the beginning of each 24-hour period a patient might be given 1 mg of digoxin, which is rapidly absorbed into the blood. Let's assume our patient, Oliver, had no digoxin in his blood prior to the first dose. We wish to determine the amount of digoxin in his blood after several days of treatment.

First, we define our variables. Let n represent the number of days after taking the first dose. Let $d(n)$ be the number of milligrams of digoxin in Oliver's blood n days after taking the first dose and immediately after taking that day's dose. Note how carefully $d(n)$ was defined.

We now use words to describe how the amount of digoxin in Oliver's blood is changing. Each day the kidneys remove one-third of the digoxin from his blood. Then the patient receives a dose of digoxin, which results in 1 mg of digoxin being added to his blood. This description can be seen in the flow diagram in Figure 1.12.

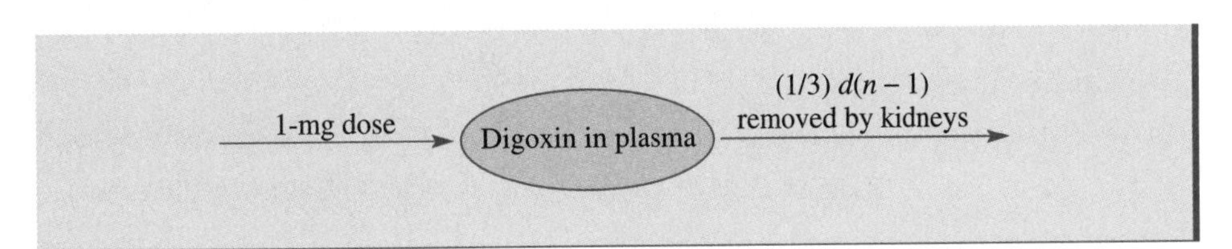

FIGURE 1.12 Flow diagram of medication with digoxin.

We can now write the dynamical system describing the amount of digoxin in Oliver's blood as $d(n) = d(n-1)$ – arrow out + arrow in, or

$$d(n) = d(n-1) - \tfrac{1}{3}d(n-1) + 1$$

Since $1d(n-1) - (\frac{1}{3})d(n-1) = (1-\frac{1}{3})d(n-1) = (\frac{2}{3})d(n-1)$, this equation simplifies to

$$d(n) = \tfrac{2}{3}d(n-1) + 1$$

Notice that there was one term subtracted from $d(n-1)$ and one term added, the 1 mg.

Now we can use this affine dynamical system to approximate the amount of digoxin in Oliver's blood for the first week. To use this dynamical system, we need to know $d(0)$, which is the amount of digoxin in Oliver's blood at the beginning of treatment. This is just the amount of digoxin in the first dose, so $d(0) = 1$. This gives the values in Table 1.5, with values rounded off to 2 decimal places. A graph of this function is seen in Figure 1.13.

TABLE 1.5 Number of milligrams of digoxin in blood after *n* days.

$n =$	0	1	2	3	4	5	6	7
$d(n) =$	1	1.67	2.11	2.41	2.60	2.74	2.82	2.88

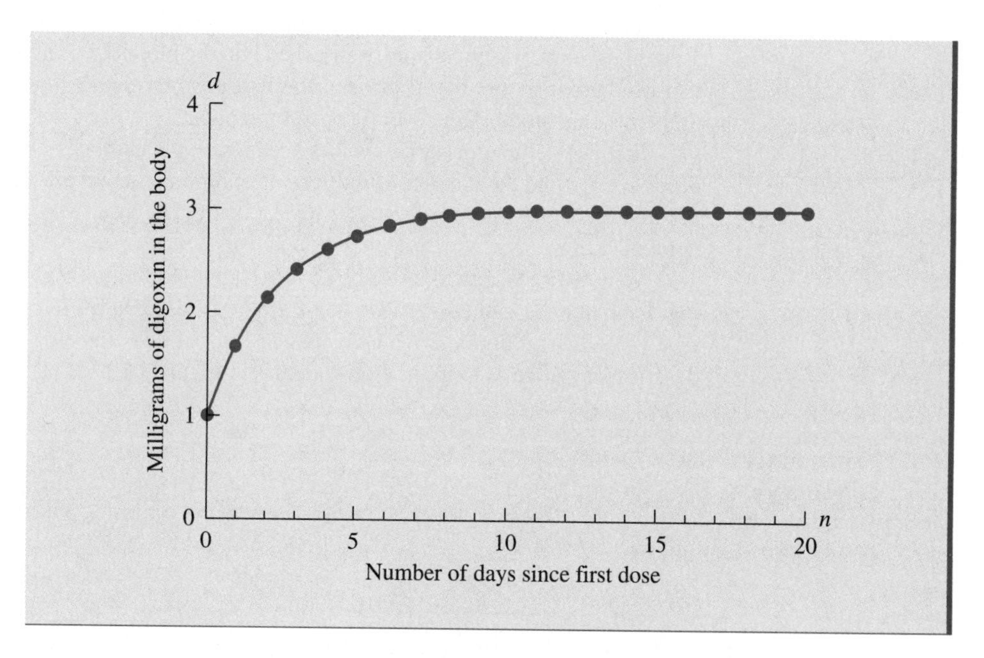

FIGURE 1.13 Digoxin in Oliver's plasma.

Note that this graph indicates that the amount of digoxin in Oliver's blood seems to level off at around 3 mg. In fact, the physician had previously determined that 3 mg was likely to be the proper amount of digoxin in Oliver's blood and that daily doses of 1 mg would achieve this goal. The desired amount of the medicine in the blood, 3 mg in this case, is called the **target goal.**

It can happen that someone has side effects from the medicine. In this case, it might be decided to lower the dosage to achieve a lower target goal. Or it might be that the target goal does not achieve the desired benefits. In this case, it might be decided to raise the target goal.

We have been finding the amount of digoxin in the patient's blood just after taking his medicine each day. A somewhat more difficult problem would be to predict the amount of medicine in the patient's blood at any point in time. During the period after taking a dose, the amount of medicine will decrease continuously. At the time the dose is taken, the amount of medicine will increase by the amount of the dose. A graph of what this might look like is seen in Figure 1.14. An expression for this function can be derived using a bit more mathematics than we will do in this text. ■

FIGURE 1.14 **Amount of digoxin in plasma, continuous model.**

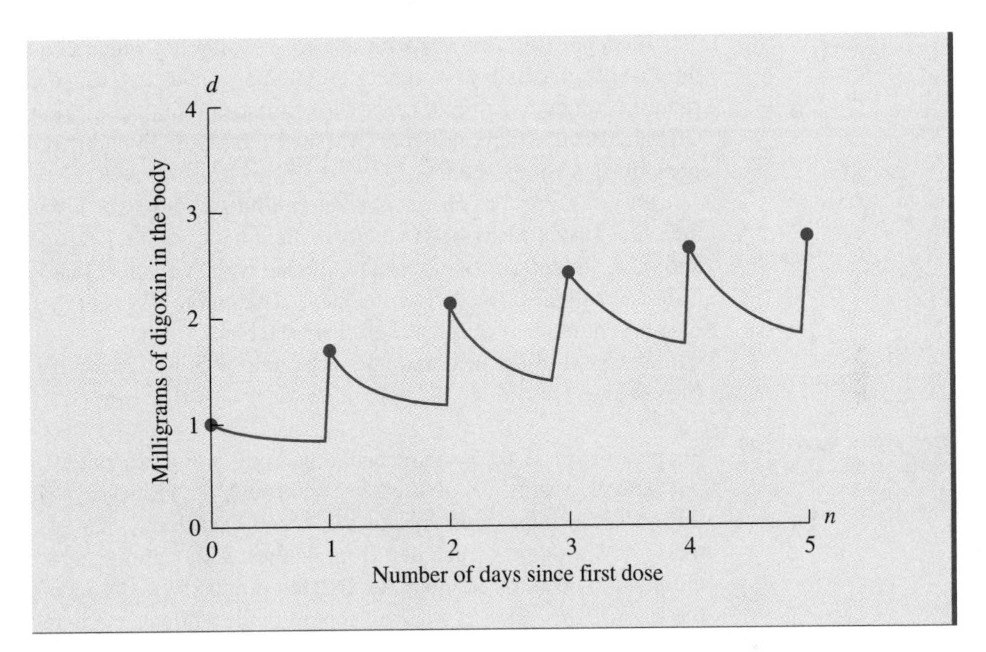

Let's continue our discussion of the buildup and elimination of chemicals in the body. Primary methods for elimination of drugs from the body are through filtration by the kidneys and liver metabolism. When the liver metabolizes a drug, it turns it into new chemical compounds. Sometimes these new chemicals are inactive in the body and are eliminated over time. Other times the new chemical also interacts with the body. For example, some of the asthma medication theophylline is metabolized into caffeine.

One interesting phenomenon is when the body converts one chemical into a second chemical and converts the second chemical back into the first chemical. This process is called **interconversion.** In this case, a person could be given 1 drug and end up being treated with 2 drugs.

Consider vitamin K, which is essential for the blood to coagulate. Without vitamin K in our system, we would continue bleeding from just a small injury and would be prone to hemorrhaging. The body metabolizes vitamin K into **vitamin K–epoxide,** which is inactive in the body, and converts vitamin K–epoxide into vitamin K. People who have had blood clots, pulmonary embolism, and heart attacks are often treated with anticoagulants. One popular anticoagulant, Warfarin, works by preventing vitamin K–epoxide from converting back to vitamin K, resulting in a reduction in the amount of vitamin K in the blood, which then decreases the blood's ability to clot.

Another pair of chemicals for which interconversion occurs is Prednisone and Prednisolone. Prednisone is an adrenal corticosteroid used to treat an almost endless list of disorders, such as skin rashes, asthma, arthritis, blood disorders, and ulcerative colitis. When chemicals convert to each other, over time the amounts of each of these chemicals in the body can stabilize at a constant ratio. Because of this, physicians can prescribe either of the 2 medications, Prednisone or Prednisolone, and reach the same therapeutic level of Prednisone.

Interconversion occurs with the drug Clofibrate, which is used to help lower cholesterol and triglycerides. For this drug, interconversion can cause a problem. In people with reduced kidney function, there is often reduced elimination from the blood of the chemical paired with Clofibrate. This can cause Clofibrate to build up in the body if the amount prescribed is not reduced.

One more pair of chemicals that exhibit this behavior is Sulindac and Sulindac sulfide. Sulindac is prescribed to reduce inflammation and relieve pain from arthritis, bursitis, and other inflammatory diseases. Sulindac is the chemical that is prescribed, but, in fact, Sulindac sulfide is the active chemical. This interconversion helps moderate and sustain the concentration of Sulindac sulfide in the blood.

One goal of the next example is to help you understand the dynamics of drug interconversion.

Example 1.4

Suppose (1) there are 2 chemicals in the body, which we call U and V; (2) the body filters out 10% of U and 15% of V each day through the kidneys; (3) liver enzymes metabolize 40% of the U into V and 30% of the V into U each day; and (4) the body absorbs 50 mg of U and 23 mg of V each day from its diet. Although the consumption, conversion, and elimination occur throughout the day, for simplicity we assume it all occurs at the end of the day. We then let $u(n)$ and $v(n)$ represent the amounts of chemicals U and V in the body at the beginning of day n, just after the elimination, conversion, and consumption from the previous day. These assumptions are summarized in the flow diagram of Figure 1.15.

FIGURE 1.15 Flow diagram for interconversion of 2 drugs, U and V.

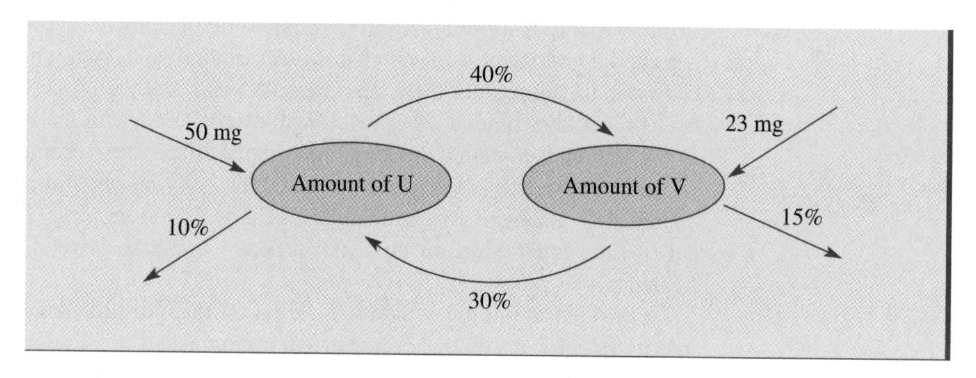

Note that there is an arrow going from reservoir U to reservoir V and vice versa. These indicate that something is being subtracted from U and also added to V and vice versa.

In real situations, it is difficult to compute the percentage of each chemical that is converted to the other. It is easier to compute the amount of each chemical that is eliminated in the urine. By doing some analysis, such as we will be doing later, scientists can make inferences about the conversion rates.

The question we wish to answer is, over time, how much U and V will be in the body at the beginning of each day? To answer this question, we let $u(n)$ and $v(n)$ represent the amount of U and V, respectively, in the body at the beginning of the nth day. We wish to compute $u(n)$ and $v(n)$, the amount of U and V, respectively, in the body at the beginning of that day in terms of the amounts of U and V that were in the body the previous day. To do this, we use the same approach as in Example 1.3, that is,

$$u(n) = u(n-1) + \text{arrows in} - \text{arrows out}$$
$$v(n) = v(n-1) + \text{arrows in} - \text{arrows out}$$

There are 2 arrows into each reservoir and 2 arrows out of each reservoir. This gives the following equations, which result from the flow diagram.

$$u(n) = u(n-1) - \text{eliminated} - \text{converted to V} + \text{converted from V} + \text{consumed}$$
$$v(n) = v(n-1) - \text{eliminated} - \text{converted to U} + \text{converted from U} + \text{consumed}$$

or

$$u(n) = u(n-1) - 0.1u(n-1) - 0.4u(n-1) + 0.3v(n-1) + 50$$
$$v(n) = v(n-1) - 0.15v(n-1) - 0.3v(n-1) + 0.4u(n-1) + 23$$

which simplify to

$$u(n) = 0.5u(n-1) + 0.3v(n-1) + 50$$
$$v(n) = 0.4u(n-1) + 0.55v(n-1) + 23$$

Suppose that $u(0) = 150$ and $v(0) = 100$. Then

$$u(1) = 0.5u(0) + 0.3v(0) + 50 = 0.5(150) + 0.3(100) + 50 = 155$$
$$v(1) = 0.4u(0) + 0.55v(0) + 23 = 0.4(150) + 0.55(100) + 23 = 138$$

We can plot the points $(n, u(n))$ and the points $(n, v(n))$ to see how the amounts of these chemicals in the body change over time, as can be seen in Figure 1.16. It seems that, over time, the amounts of U and V in the body level out to about 260 and 300 mg, respectively. In the next section, you will learn how to easily make time graphs such as this. ■

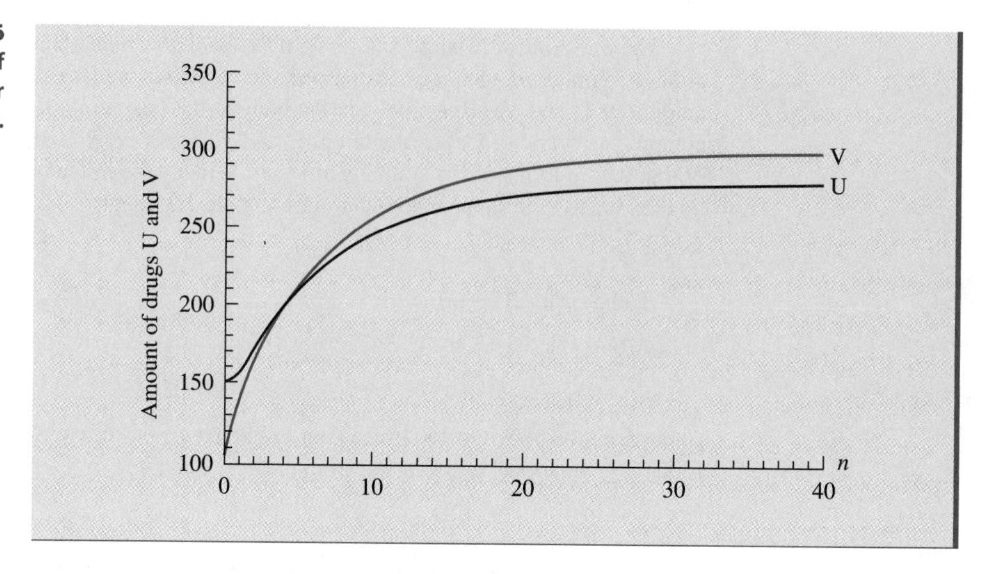

FIGURE 1.16 Amount of drugs U and V over time.

Presently, we cannot do much with the dynamical systems we are developing. In Section 1.4, we will learn how to easily answer important questions using a combination of dynamical systems and spreadsheets or appropriate calculators.

1.3 Problems

1. Consider the affine dynamical system

$$u(n) = 1.1u(n-1) + 3$$

a. Given that $u(1) = 10$, find $u(2)$, $u(3)$, and $u(4)$.
b. Given that $u(1) = -30$, find $u(2)$, $u(3)$, and $u(4)$.

2. Consider the affine dynamical system

$$v(n) = 1.5v(n-1) - 3$$

a. Given that $v(0) = 7$, find $v(4)$.
b. Given that $v(0) = 6$, find $v(4)$.

3. Consider the affine dynamical system

$$w(n) = 0.7w(n-1) + 3$$

a. Given that $w(0) = 1$, find $w(3)$.
b. Suppose you know that $w(0) = w(1)$. What is the value for $w(0)$?

4. Consider the affine dynamical system

$$u(n) = -1.4u(n-1) + 3$$

a. Given that $u(0) = 4$, find $u(3)$.
b. Suppose you know that $u(0) = u(1)$. What is the value for $u(0)$?

5. Consider the affine dynamical system of two equations

$$u(n) = 3u(n-1) - 2v(n-1) + 2$$
$$v(n) = u(n-1) + v(n-1) - 2$$

a. Suppose that $u(0) = 3$ and $v(0) = -1$. Find $u(1), v(1), u(2)$, and $v(2)$.
b. Suppose that $u(0) = 2$ and $v(0) = 3$. Find $u(1), v(1), u(2)$, and $v(2)$.

6. Consider the affine dynamical system of two equations

$$u(n) = u(n-1) + 2v(n-1) + 4$$
$$v(n) = -u(n-1) + v(n-1) + 3$$

a. Suppose that $u(0) = 0$ and $v(0) = 0$. Find $u(1), v(1), u(2)$, and $v(2)$.
b. Suppose that $u(0) = 3$ and $v(0) = -2$. Find $u(1), v(1), u(2)$, and $v(2)$.

7. A savings account earns 5% annual interest, compounded quarterly, meaning that it earns 1.25% interest each quarter of a year. You have \$50,000 in that account on January 1. At the beginning of every quarter of a year you withdraw \$1000 from that account, with the first withdrawal occurring on April 1. Your account is therefore earning interest each quarter on what was in the account at the beginning of the quarter but is losing money from the amount you withdraw. Let n represent the number of quarters of a year after January 1. Let $a(n)$ represent the amount of money in this account at the beginning of the nth **quarter** after January 1, just after making the withdrawal for that quarter. This means that $a(0) = 50,000$ and $a(1) = 50,000 + 0.0125 \times 50,000 - 1000 = 49,625$.
a. Draw a flow diagram of the situation.
b. Use the flow diagram to generate a dynamical system that models this account.
c. Use the dynamical system to determine the amount in that account after 1 year.

8. Cadmium is an extremely dangerous heavy-metal pollutant and is most toxic when inhaled. Initial symptoms include chest pains and nausea. Symptoms can progress to emphysema and fatal pulmonary edema. One source of inhalation of cadmium is through cigarettes. A person who smokes one pack of cigarettes a day will absorb about 2.7 mg of cadmium each year from smoking. While the cadmium is being absorbed continuously throughout the year, assume for simplicity that it is all absorbed at the end of the year. The numerical results using either a continuous model or this simplifying assumption are similar. Some people eliminate about 7% of the cadmium from their body each year. Develop a dynamical system for $c(n)$, the amount of cadmium in such a person's body after n years as a result of smoking. Drawing a flow diagram may help. Use that system to find the amount of cadmium in such a person's body as a result of smoking for 3 years.

9. Presently the U.S. growth rate is about 1% per year. In addition, there are approximately 700,000 immigrants into the United States each year and approximately 160,000 emigrants from the United States each year (see flow diagram in Figure 1.17). For simplicity, assume all of the immigration and emigration occurs at the end of the year. Draw a flow diagram to model the changing U.S. population. Develop a discrete dynamical system to model the population size. Given that the population in January 1999 was about 271 million, use this dynamical system to predict the U.S. population in January 2003. We must note that the numbers given in this problem are very volatile and do not remain fixed over time. Thus, other factors would need to be considered to get a better model for predicting U.S. population size.

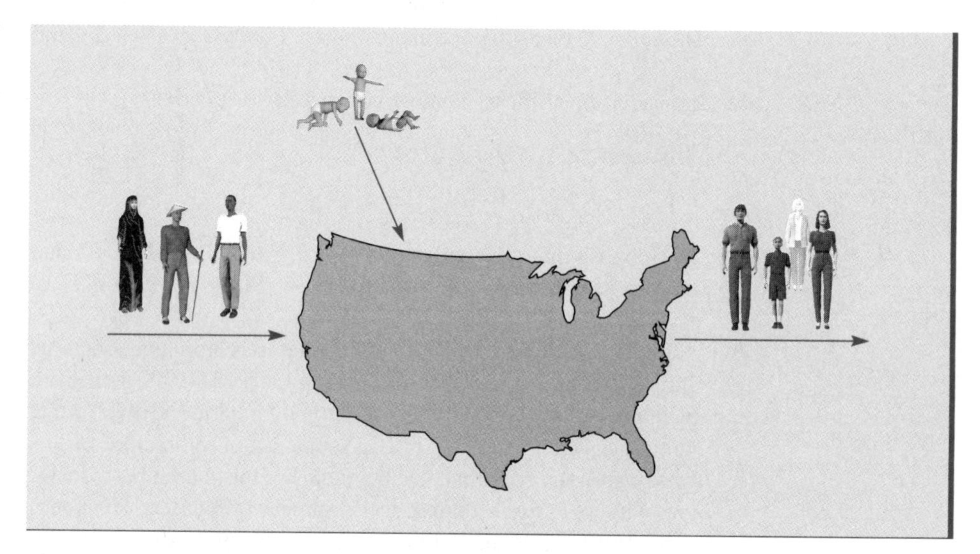

FIGURE 1.17 Flow diagram modeling U.S. population growth.

10. Presently you weigh 150 pounds. You consume 4 pounds worth of calories each week. Make the unrealistic assumption that all of the consumption of calories occurred at the end of the week. Assume your body burns off the equivalent of 3% of its weight each week through normal metabolism. In addition, you burn off $\frac{1}{4}$ pound of weight through daily exercise each week. Again, assume for simplicity that all of the exercise occurred at the end of the week. Draw a flow diagram to model this situation. Develop a discrete dynamical system to model this weight loss. Use this model to predict your weight in 4 weeks. (The simplifying assumptions that calorie consumption and exercise occurred at the end of the week have little effect on the numerical results obtained in this exercise. The numbers given—4 pounds worth of calories, 3% loss, and $\frac{1}{4}$ pound burned off through exercise—are realistic.)

11. Vitamin A is stored primarily in our plasma and our liver. Suppose that 40% of the vitamin A in the plasma is filtered out by the kidneys each day and that 30% of the vitamin A in the plasma is absorbed into the liver each day. Also assume that 1% of the vitamin A in the liver is absorbed back into the plasma each day. Suppose you have a daily intake of 1 mg of vitamin A each day, which goes directly into the plasma.

a. Draw a flow diagram of the situation.

b. Determine equations for $p(n)$ and $l(n)$, the number of milligrams of vitamin A in the plasma and liver, respectively, on day n, in terms of $p(n-1)$ and $l(n-1)$, the amount of vitamin A in the plasma and liver, respectively, on day $n-1$.

c. Suppose that $p(0) = 5$ and $l(0) = 80$ mg. Find $p(1)$, $l(1)$, $p(2)$, and $l(2)$.

12. In nature, one radioactive material often decays into another radioactive material. A sequence of radioactive materials in which each material decays into the next material in the sequence is called radioactive chains. Suppose that each day, 2% of material A decays into material B and 10% of material B decays into lead. Suppose that initially, there are 40 grams of A and 5 grams of B. Draw a flow diagram for this situation. Develop a dynamical system of two equations to model this situation. How much of each material will be left after 3 days?

13. Suppose, starting today, a young child absorbs 0.3 mg of lead into her plasma each day. It happens that about 34% of the lead in her plasma each day is absorbed into her bones and 8% of the lead in her plasma is eliminated in the urine each day. Also, about 0.173% (or 0.00173 as a decimal) of the lead in her bones is absorbed back into her plasma each day. Let $p(n)$ represent the amount of lead in this girl's plasma at the beginning of day n and $b(n)$ represent the amount of lead in this girl's bones at the beginning of day n. [Assume $p(0) = 0$ and $b(0) = 0$.]

a. Draw a diagram of this situation.

b. Set up a dynamical system to model this situation. Find $p(3)$ and $b(3)$.

14. Suppose you have a roll of paper, such as paper towels. Let the radius of the cardboard core be 0.5 inches. Suppose the paper is 0.002 inches thick. Let $r(n)$ represent the radius, in inches, of the roll when the paper has been wrapped around the core n times. Let $l(n)$ be the total length of the paper when it is wrapped about the core n times. Note that $r(0) = 0.5$ and $l(0) = 0$. Remember, the circumference of a circle is given by $c = 2\pi r$.

a. Develop a dynamical system for $r(n)$ in terms of $r(n-1)$.

b. Develop a dynamical system for $l(n)$ in terms of $r(n-1)$ and $l(n-1)$.

15. Suppose a person's body burns 130 kcal per week for each pound it weighs. Suppose this person presently weighs 165 pounds and consumes about 21,000 kcal per week. Suppose this person decides to eat 200 kcal less each week than the week before. Let $c(n)$ represent the number of kilocalories this person consumes during the nth week of this diet. Then $c(1) = 20,800$. Let $w(n)$ be the weight of this person after n weeks of this diet. Then $w(0) = 165$. Develop a dynamical system of two equations, one for c and one for w. To work this problem, you need to know that burning 3600 kcal would reduce a person's weight by 1 pound and consuming 3600 kcal would result in the person gaining 1 pound.

1.4 Parameters

In this section, you will learn to investigate a dynamical system using different values of a parameter, such as the different possible dosages of a drug that you could take. You will also learn how an appropriate calculator or a computer spreadsheet can help in the investigation.

In the last section, you studied the situation in which a patient takes 1 mg of digoxin and eliminates one-third of the digoxin each day. This situation was described by the discrete dynamical system

$$d(n) = \tfrac{2}{3}d(n-1) + 1$$

It appeared that the amount of digoxin in this patient's plasma was leveling off at 3 mg. Are you sure? A doctor might also be interested in how to adjust the patient's daily dosage so that the amount of digoxin will level off at a higher or lower level. For example, what dosage will cause the amount to level off at 4 mg? You will now see how to answer these and other questions using an appropriate calculator or a spreadsheet.

Let's first determine how to estimate the long-term amount of digoxin in this patient's plasma. Recall that in this case, $d(0) = 1$.

Spreadsheet Users If you are using a spreadsheet, you will need 1 column consisting of the n-values. Put these under the A column. Put 0 in cell A1. Put the formula

+A1 + 1

into cell A2. Then copy this formula down as far as you want, say to A201. Now you need to put the dynamical system into the spreadsheet. Put the first value, 1, into cell B1. Then put the formula

+(2/3)*B1 + 1

into cell B2 and copy this formula down. You will now have a table giving the amount of digoxin in this patient's plasma at the beginning of the next 200 days. You can now look at the B cell, cell B11, that is next to the A cell containing a 10 to see the amount of digoxin after 10 days, which is 2.965 mg, to three decimal places. ■

Calculator Users If you are using a calculator, it needs to have a **sequence mode.** Once in sequence mode, you can input that the starting or minimum n-value is 0 and the starting u-value is 1. You put in the equation $d(n) = (2/3)d(n-1) + 1$. You then generate a table and scroll down until you find $n = 10$. Next to it, you will find the amount of digoxin, 2.965 mg. See the Instruction Manual for more details. ■

Spreadsheet and Calculator Users You should learn how to generate graphs of the functions. When generating a graph, you should first determine a reasonable range for the n-values, considering the questions you want to answer. In Figure 1.13, I decided to use the range $0 \leq n \leq 20$. I could have just as easily used $0 \leq n \leq 50$ or $0 \leq n \leq 100$. Spreadsheets and calculators will automatically make reasonable first choices for the range of the vertical axis. See manual for details. You will often have to make minor adjustments to the axes to obtain a graph that looks good. Picking an appropriate viewing window requires some art.

Spreadsheets will automatically make reasonable first choices for the horizontal and vertical axis. You will sometimes have to adjust those choices, though. ■

Example 1.5

Suppose the patient does not get the desired benefits from taking 1 mg of digoxin per day. The doctor decides that the proper target goal is 5 mg of digoxin in the blood. The doctor also wants the digoxin in the blood to reach this level within 14 days; that is, $d(14) \approx 5$. We can't expect $d(14)$ to be exactly equal to 5, so the doctor needs to decide what is acceptable, say $4.9 < d(14) < 5.1$ What is an acceptable dose (see Figure 1.18)?

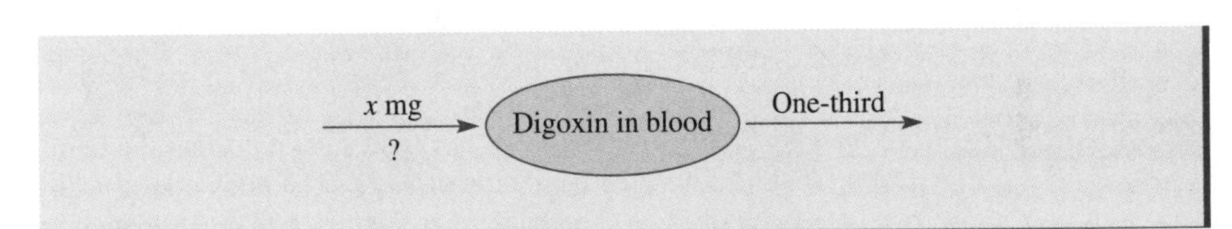

FIGURE 1.18 Goal is $d(14) = 5$.

From Figure 1.18 we see that the dynamical system is

$$d(n) = \tfrac{2}{3}d(n-1) + x$$

where x represents the correct dose and $d(0) = x$. To determine the proper dose, we make a guess at the appropriate dose, say $x = 1.5$ mg. We use the dynamical system $d(n) = (2/3)d(n-1) + 1.5$, along with the initial value of $d(0) = 1.5$. In this case, a computer or calculator gives that $d(14) = 4.49$ mg, which is not enough. We now try $x = 2.0$ mg. To do this, we need to change 1.5 to 2.0 in the dynamical system and also for the starting value. In this case, we get $d(14) = 5.99$, which is too much. In Table 1.6 are my results using different values for x. The results are given in the order in which I tried them. From this table, we can see that a dosage of 1.7 mg gives $d(14) = 5.09$, which is within the acceptable range.

TABLE 1.6 **Size of dose and resulting amount of digoxin in blood after 2 weeks.**

Size of dose, x	1	1.5	2.0	1.8	1.6	1.7
$d(14) =$	2.99	4.49	5.99	5.39	4.79	5.09

Note that we could get an even more accurate answer, but we are limited by the available sizes of pills for digoxin. For digoxin, pills are available in 0.1-mg increments, so 1.7 mg per day is the only practical dosage that achieves the desired results. ■

An important aspect of this section is learning to use parameters to answer questions about a situation. A **parameter** is a variable that is part of the dynamical system. The variable x is a parameter in $u(n) = (\frac{2}{3})u(n-1) + x$. By trying different values for the parameter, we can observe different outcomes from the dynamical system. We tried different values for x, which resulted in different amounts of digoxin in the patient's blood. We then picked the value for the parameter that gave the most acceptable result: $x = 1.7$ mg results in approximately 5 mg of digoxin in the patient's blood after 2 weeks.

One difficulty with using parameters is that we often do not get an exact answer. For example, our value of $x = 1.7$ gave $d(14) = 5.09$ mg, not 5 mg. In practice, there will be times in which an exact answer for the parameter cannot be found and so we need to decide how close to be to the correct answer, such as to two decimal places.

When using parameters, it is important to vary their values systematically until you find a value that gives the desired result. Setting up a table of values for the parameter and the corresponding results, as we did in Table 1.6, is usually helpful. If you observe the results and think about them as you vary your parameter, you can save a great deal of time.

Example 1.6

There are 2 drugs, C and F. Let $c(n)$ and $f(n)$ represent the number of milligrams of each drug in the body at the beginning of day n. The body converts 20% of C to F each day and converts 10% of F to C each day. Assume that 800 mg of F are consumed each day. Assume the body eliminates 100 mg of C and 200 mg of F each day. (There are some chemicals of which the body eliminates a constant amount instead of a constant percentage, as long as there is a relatively large amount of the chemical in the body.) We are going to use the parameter x to represent the number of milligrams of C consumed each day. We are then going to investigate what happens for different values of x—in particular, how we can vary x to obtain a desired result.

A flow diagram for this situation is seen in Figure 1.19. There is an arrow going into C, indicating the x mg consumption. There is an arrow going into F, representing the 800 mg consumed. There is an arrow going out of C and into F, meaning some amount of C is subtracted and the same amount is added to F. Similarly, the arrow going from F to C indicates that F is decreasing and C is gaining. Finally, there are arrows going out of C and F to indicate the amounts eliminated by the body. This means that the equations

$$c(n) = c(n-1) + x - 100 - 0.2c(n-1) + 0.1f(n-1)$$
$$f(n) = f(n-1) + 800 - 200 + 0.2c(n-1) - 0.1f(n-1)$$

describe this situation. These equations can be simplified to

$$c(n) = 0.8c(n-1) + 0.1f(n-1) + x - 100$$
$$f(n) = 0.2c(n-1) + 0.9f(n-1) + 600$$

Let's see what happens when $x = 500$. Then the equations become

$$c(n) = 0.8c(n-1) + 0.1f(n-1) + 400$$
$$f(n) = 0.2c(n-1) + 0.9f(n-1) + 600$$

Let's assume that at the beginning, $n = 0$, $c(0) = 200$, and $f(0) = 500$. Then $c(10) = 3781.67$ and $f(10) = 6918.33$. You should use these equations to determine the amounts after 30 days.

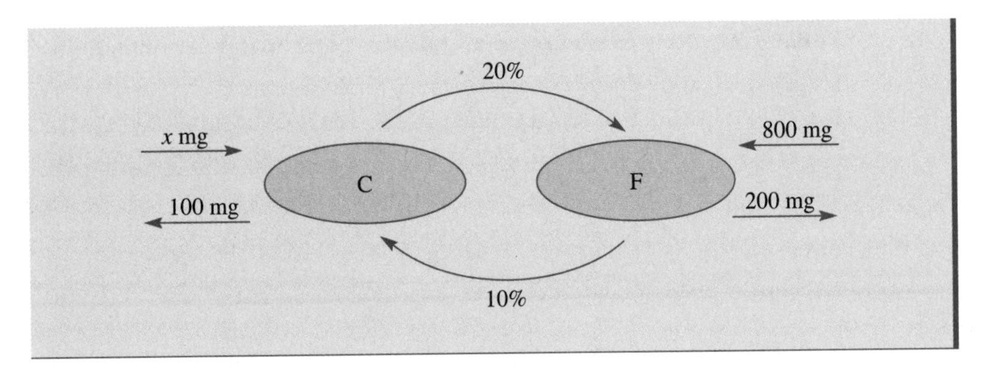

FIGURE 1.19 Flow diagram for changing amounts of drugs C and F in the body.

In Figure 1.20 are time graphs of the daily amounts of each drug. The graphs look linear. If you look at a table of values, you find out that they are not linear; that is, each drug does not increase by the same amount each day. But after about 15 days, they are close to linear in that C increases by about 333.33 mg each day and F increases by about 666.67 mg each day.

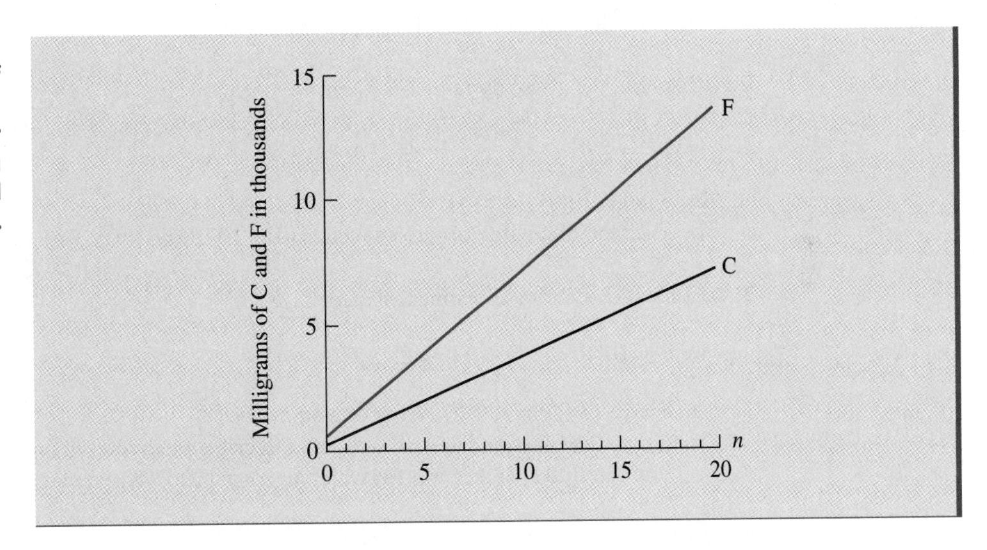

FIGURE 1.20 Number of milligrams of C and F in the body over time when 500 mg of C are consumed each day.

Suppose we want the amount of drug C in the body to be 15,000 mg after 20 days. We would experiment by using the equations

$$c(n) = 0.8c(n-1) + 0.1f(n-1) + x - 100$$
$$f(n) = 0.2c(n-1) + 0.9f(n-1) + 600$$

with $c(0) = 200$ and $f(0) = 500$. We would try different values for x until we achieved $c(20) \approx 15,000$. Suppose we want x to the nearest 10 mg. Table 1.7 gives my tries for x and the resulting value for $c(20)$.

TABLE 1.7 **Daily consumption of chemical C and 20 day accumulation.**

$x =$	500	1100	1600	1400	1380	1390
$c(20) =$	7122	12,454	16,898	15,120	14,943	15,032

Thus, this person needs to consume about 1390 mg of chemical C each day. ■

1.4 Problems

1. Consider the affine dynamical system

$$u(n) = 1.1u(n-1) + 3 \quad \text{with} \quad u(1) = 10$$

a. Find $u(30)$.
b. Find the first value of n for which $u(n) > 2000$.

2. Consider the affine dynamical system

$$v(n) = 1.5v(n-1) - 3 \quad \text{with} \quad v(1) = 7$$

a. Find $v(20)$.
b. Find the first value of n for which $v(n) > 100,000$.

3. Consider the affine dynamical system

$$w(n) = 0.7w(n-1) + 3 \quad \text{with} \quad w(0) = 1$$

a. Find $w(30)$ and $w(50)$. What do you notice?
b. Graph the function for $n = 0, \ldots, 30$. Describe what you see on the graph. Do you think you can find a value for n for which $w(n) > 1000$? Why or why not?

4. Consider the affine dynamical system

$$u(n) = -1.1u(n-1) + 3 \quad \text{with} \quad u(0) = 4$$

a. Find $u(30)$ and $u(31)$.
b. Graph the function for $n = 0, \ldots, 20$. Describe what you see on the graph.
c. Find the first value of n for which $|u(n)| > 1000$. Completing a table of values, such as the one that follows, will help in quickly finding the correct value for n.

$n =$	40	80	60		
$u(n) =$	118	5269	?		

5. Consider the affine dynamical system

$$v(n) = 1.1v(n-1) + x \quad \text{with} \quad v(1) = 10$$

Find a value for x so that $932 < v(30) < 934$.

$x =$	50	25			
$v(30) =$	7590				

6. Consider the affine dynamical system

$$w(n) = 0.8w(n-1) + x \quad \text{with} \quad w(1) = 1.$$

Find a value for x so that $31 < w(20) < 32$.

7. Consider the affine dynamical system

$$u(n) = 1.1u(n-1) + 1 \quad \text{with} \quad u(1) = x.$$

Find a value for x so that $474 < u(30) < 476$.

8. Consider the affine dynamical system

$$v(n) = 0.8v(n-1) + 3 \quad \text{with} \quad v(1) = x.$$

a. Find a value for x so that $170 < v(10) < 180$.
b. Find a value for x so that $14.9 < v(30) < 15.1$.

9. Suppose, starting today, a young child absorbs 0.3 mg of lead into her plasma each day. It happens that about 34% of the lead in her plasma each day is absorbed into her bones and that 8% of the lead in her plasma is eliminated in the urine each day. Also, about 0.173% (or 0.00173 as a decimal) of the lead in her bones is absorbed back into her plasma each day. Let $p(n)$ represent the amount of lead in this girl's plasma at the beginning of day n and $b(n)$ represent the amount of lead in this girl's bones at the beginning of day n.
a. Find $p(365)$ and $b(365)$.
b. Note that 0.9 mg of lead in the plasma would require treatment. After how many days would this girl require treatment?

10. Recall that smoking a pack of cigarettes per day results in absorbing about 2.7 mg of cadmium each year from smoking. Assume that about 7% of the cadmium is eliminated from the body each year. Find the amount of cadmium in this person's body as a result of smoking for 30 years.

11. Presently you weigh 160 pounds. You consume x pounds worth of calories each week. Assume your body burns off the equivalent of 3% of its weight each week through normal metabolism. In addition, you burn off $\frac{1}{4}$ pound of weight through daily exercise each week. Find x to one-decimal place if you want to weigh between 144 and 146 pounds in 1 year

(52 weeks). How many calories per day can you average if 1 pound is the equivalent of 3600 calories?

12. Suppose that each day, 2% of material A decays into material B and 10% of material B decays into lead.

a. Suppose $a(0) = 40$ and $b(0) = 5$. Make a graph of $a(n)$ and $b(n)$ for n going from 0 to 50, and observe how they behave.

b. Suppose that after 30 days, there are 20 grams of material B left, but there were only 10 grams of B to start with. How many grams of material A were there to begin with, to the nearest gram?

13. Suppose you have a roll of paper, such as paper towels. Let the radius of the cardboard core be 0.5 inches. Suppose the paper is 0.002 inches thick. Let $r(n)$ represent the radius, in inches, of the roll when the paper has been wrapped around the core n times. Let $l(n)$ be the total length of the paper when it is wrapped about the core n times. Note that $r(0) = 0.5$ and $l(0) = 0$. What is the length of paper on the roll when it has a radius of 2 inches?

1.5 Financial Models

In this section, we will use what we have learned to answer questions related to finance, such as, What will the payment be on a loan? How much can you afford to borrow? Which of 2 investments is better? You will be faced with these and similar questions throughout life.

We begin with an example in which we determine the monthly payment on a car loan. When we make monthly payments to pay back a loan, it is said that we are **amortizing** our loan.

Example 1.7

Suppose you decide to borrow \$15,000 to buy a car. The bank will loan you the money at 9% annual interest, compounded monthly. The loan is to be amortized or paid back in 48 equal monthly payments over the next 4 years. The question is, what should you pay each month so that the amount you owe the bank after 4 years is zero?

Let n be the number of monthly payments made and $b(n)$ be the amount you owe the bank just after making the nth monthly payment. Each month, the bank adds interest onto what you owe, but the amount you owe is reduced by your monthly payment. The model is visualized in the flow diagram of Figure 1.21. The arrow going out, "Payment," represents the constant payment. The point of this example is to determine what this payment is. Until then, we will refer to the payment with the parameter p.

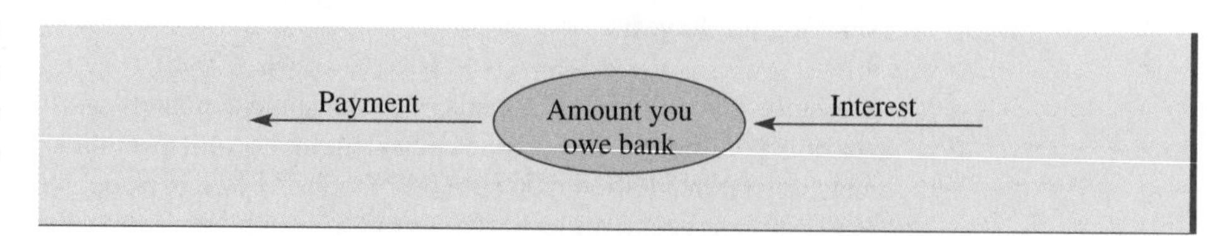

FIGURE 1.21 Flow diagram for amortizing a loan.

The arrow that goes into the oval represents interest being added onto what you owe. The yearly interest rate is 9%. Since you make monthly payments, the amount you owe changes each month. Thus, the bank adds 1 month's interest onto what you owe, each month. One month's interest rate is $\frac{1}{12}$ of the yearly interest rate, that is,

$$\frac{9\%}{12} = 0.75\%$$

The dynamical system is then $b(n) = b(n-1) + \text{interest} - \text{payment}$, or

$$b(n) = b(n-1) + 0.0075b(n-1) - p$$

Combining terms gives

$$b(n) = 1.0075b(n-1) - p$$

The amount you owe before making any payments is \$15,000, which means that $b(0) = 15,000$. After making 48 payments, you will have amortized, or paid off, your entire debt. This means that $b(48) = 0$. The 1 unknown is the monthly payment, p.

You must first make a reasonable estimate for p and then use the dynamical system to see if this estimate gives the required result of $b(48) = 0$. If it doesn't, make a new estimate based on the result of the last guess.

You need to pay back \$15,000 in 48 payments. If there were no interest, this would be \$312.50. Since you also have to pay back interest, it is clear the payments must be more than this. One reasonable first guess is $p = 400$. There are other reasonable first guesses that could be made for p.

Using the dynamical system $b(n) = 1.0075b(n-1) - 400$, the initial value $b(0) = 15,000$, and a computer, I found that $b(48) = -1573.20$. This means that \$400 payments will result in overpaying the loan by \$1573.20. Thus the payments must be less than \$400, but more than \$312.50. In Table 1.8 are my results. You should try your own guesses for p and see if you get the same result.

TABLE 1.8 Amount owed after 4 years on \$15,000 loan with monthly payments of \$*p*.

$p =$	400	375	373	373.30	373.27	373.28
$b(48) =$	−1573.20	−99.19	15.85	−1.40	0.32	−0.25

We now interpret these results. Monthly payments of \$373.28 for 48 months will overpay the loan by 25 cents. Monthly payments of \$373.27 for 48 months will underpay the loan by 32 cents. When there are 2 choices, the bank usually picks the choice that overpays the loan. In order for you to avoid being overcharged, the bank will reduce your last payment by the amount that would be overpaid, that is by 25 cents. Thus, you make 47 monthly payments of \$373.28. Your last payment will be \$373.28 − \$0.25 = \$373.03.

In Figure 1.22 is a graph that depicts how the amount of money owed decreases as payments are made. The lower graph shows the decrease with \$400 payments; the higher graph shows the decrease with \$373 payments. The goal is for the graph to go through the point $(48, 0)$. ■

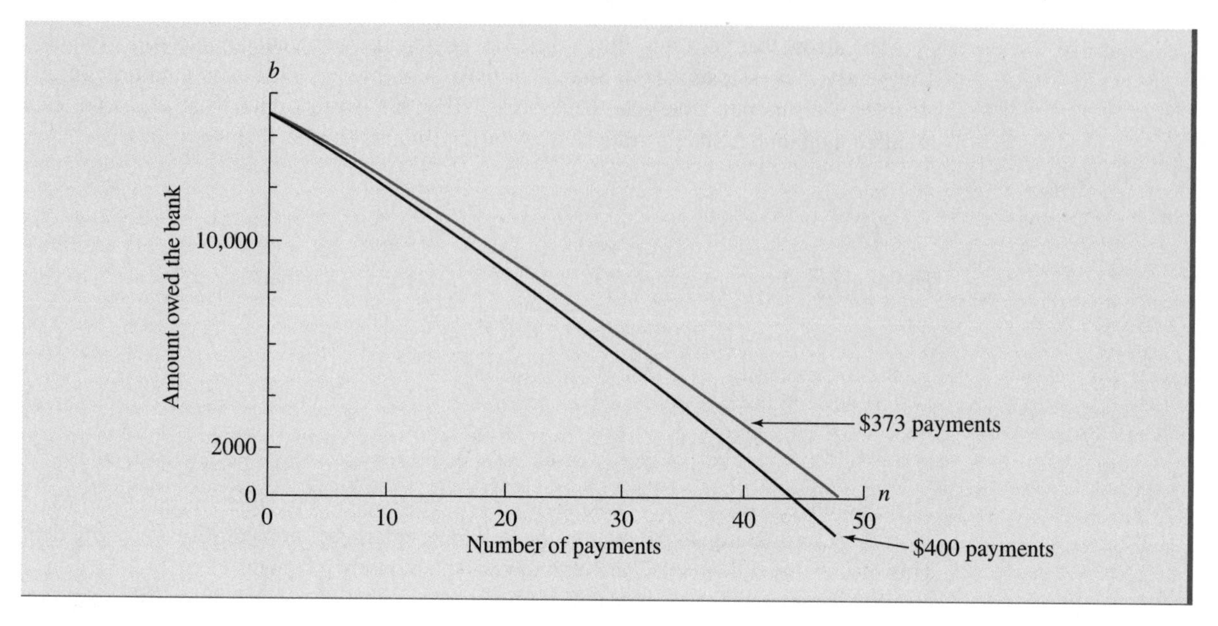

FIGURE 1.22 Amount of debt remaining after n payments.

As you may notice, this process to determine the monthly payment on a loan is the same as the process used in the last section for finding the proper dose of medicine. One of the powers of mathematics is that many seemingly different questions can be answered using the same mathematics.

Example 1.8

Suppose you decide to borrow money to buy a car. The bank will loan you money at 9% annual interest, compounded monthly. The loan is to be amortized in 48 equal monthly payments over the next 4 years. The problem is that you can only afford to pay about $250 per month. The question is, How much can you afford to borrow?

As in Example 1.7, let n be the number of monthly payments made and let $b(n)$ be the amount you owe the bank just after making the nth monthly payment. As before, each month, the bank adds interest onto what is owed and the amount owed is reduced by the monthly payment. In Example 1.7 we determined that the interest rate is 0.75% per month. The monthly payment is $250. The model is visualized in the flow diagram in Figure 1.23.

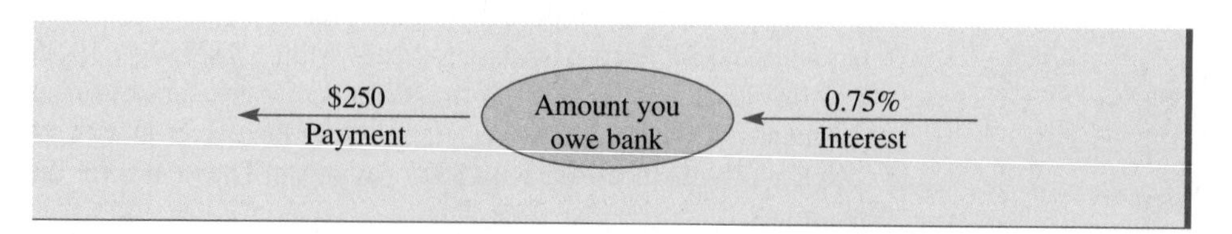

FIGURE 1.23 Flow diagram for loan with $250 monthly payments.

The arrow that goes into the oval represents interest being added onto what is owed. We can now write our dynamical system as

$$b(n) = b(n-1) + \text{interest} - \text{payment}$$

or as

$$b(n) = 1.0075b(n-1) - 250$$

To use a dynamical system, you need to know an initial value. In this case, the amount owed before making any payments is unknown. Let's use the parameter l for the amount of the loan. This means that $b(0) = l$. After making 48 payments, you will have amortized, or paid off, the entire debt. This means that $b(48) = 0$. The one unknown is the initial loan, l.

You need to pay back l dollars in 48 payments of $250 each. If there were no interest, this would be 48 × $250 = $12,000. Since you also have to pay back interest, it is clear you can't borrow this much.

I made a first estimate of $l = 8000$ for the amount that can be borrowed. There is no one best value for the first estimate of the parameter. Using the dynamical system $b(n) = 1.0075b(n-1) - 250$, the initial value $b(0) = 8000$, and a calculator, I found that $b(48) = -2929.94$. This means that $250 payments will result in overpaying the $8000 loan by $2929.94. Thus you can afford to borrow more than $8000 but less than $12,000. In Table 1.9 are my results, in the order that I found them. You should try your own guesses for l and see if you get the same results.

TABLE 1.9 **Amount owed after 4 years on loan of *l* dollars with $250 monthly payments.**

$l =$	8000	11,000	10,000	10,100	10,050	10,046
$b(48) =$	−2929.94	1365.28	−66.12	77.02	5.45	−0.28

Let's interpret these results. If you make monthly payments of $250 per month for 48 months, you will overpay a loan of $10,046 by 28 cents. So you can borrow about $10,046.

In Figure 1.24 is a graph that depicts how the amount of money you owe the bank decreases as you make your monthly payments of $250. The lower graph shows the decrease from an initial loan of $8000; the higher graph shows the decrease from an initial loan of $10,000. The goal is for the graph to go through the point $(48, 0)$. This is the mathematical version of trying to aim at a target. ■

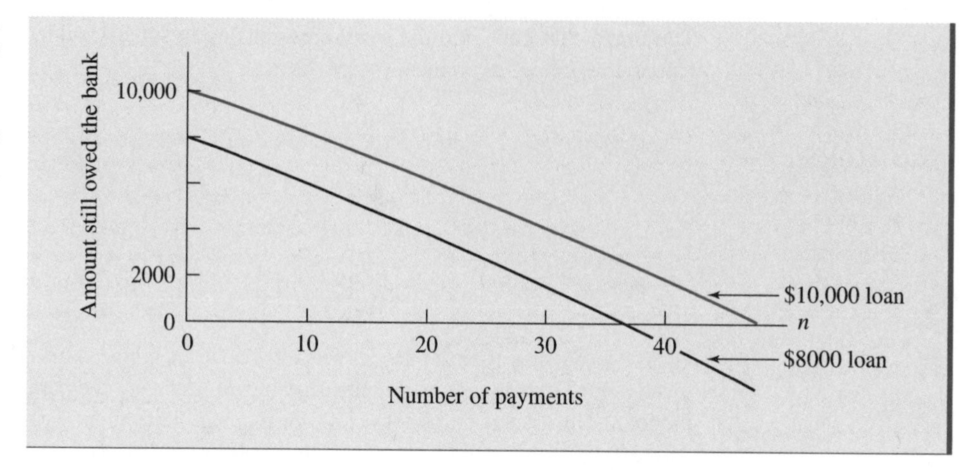

FIGURE 1.24 Amount of debt remaining after making *n* payments of $250 each.

In Example 1.7, we were determining the monthly payment for a loan. If we work for the bank, we must get the correct payment to the exact penny. On the other hand, in Example 1.8, we were determining about how much we could afford to borrow. We might be willing to pay a few dollars over or under $250. So we only need to find an approximate value for the initial amount of the loan, say to the nearest $1000 or nearest $500. Once we find a car for about that price, we can then use the method of Example 1.7 to find the exact monthly payment.

One additional concept we deal with in this section is the concept of **present value.** There are a number of different situations in which we could receive a number of payments over several years. Because of inflation, money we receive in the future will have less buying power than money we receive today. Often we will be interested in what all these future payments are worth today. For example, you may own a bond that pays you a fixed income each year for the next 10 years. You decide to sell this bond to someone. What is a fair price for this bond? Or a company projects that a new franchise will earn a profit of about $800,000 a year for the next 10 years. Is it worth spending $6,000,000 to open the franchise?

There is a relatively easy approach to determining the present value of money to be received in the future. The idea is to determine how much money you would need to put into your favorite investment to earn the same amount of money in the future. Let's consider a particular example.

Example 1.9

If you win a lottery, you will often be given the choice of a one-time cash payment or annual payments paid over some specified time period. Suppose you win a lottery in the sense that you can receive $500,000 as a cash payment today or you can receive an annuity that pays $50,000 today, and another $50,000 a year for the next 19 years. To know which is the better deal, you need to determine if the present value of the annuity exceeds $500,000.

To know which is the better deal, you need to know what interest you can earn on the money if you invest it yourself. Let's assume you can earn 9% interest, compounded annually.

Present value problems can be rephrased as bank account problems, which we do as follows. Suppose you receive w dollars. You decide to keep \$50,000 of it and put the remainder in a bank account, which pays 9% interest, compounded annually. Each year for the next 19 years, you withdraw \$50,000 from the bank, at which point the money will all be spent. How much money did you put into the bank? How much money did you receive?

Let n represent the number of years since initially depositing the money in the bank. Let $b(n)$ represent the amount in this bank account after n years, just after making that year's withdrawal of \$50,000 and receiving that year's interest.

Each year the bank adds interest and you make a withdrawal. This is described in the flow diagram in Figure 1.25.

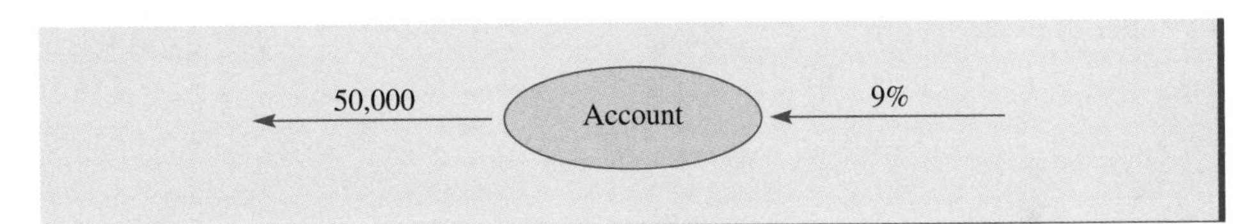

FIGURE 1.25 Flow diagram for present value problem.

It is now easy to write the dynamical system that models this bank account as

$$b(n) = b(n-1) + 0.09b(n-1) - 50,000 \quad \text{or} \quad b(n) = 1.09b(n-1) - 50,000$$

The parameter w will represent the amount of money you won. You keep \$50,000 and deposit the rest into the account, so $b(0) = w - 50,000$. After 19 years, the money is gone, meaning that $b(19) = 0$.

As before, keep estimating values for w until $b(19)$ is close to 0. I computed $w =$ \$498,000, to the nearest \$1000. You should check my answer. This means that \$448,000 was deposited into the bank account. We say that the present value of \$50,000 for 20 years (starting this year), assuming a 9% interest rate, is \$498,000.

Thus, for you, the present value of the annuity for this lottery is slightly less than \$500,000, so the cash option is a little better.

The present value of the 20 payments of \$50,000 depends on what interest you can earn on your money. You should rework this problem to see that the present value of the 20 \$50,000 payments is about \$530,000 if you can invest at only 8% interest. Thus, if you can earn only 8% on your money, the annuity of 20 annual payments is better. It is interesting to note that the smaller the interest rate, the larger the present value of the money earned in the future. ■

Realistic models of savings accounts should consider that you are able to increase what you deposit into your savings account each year. If you are living on an annuity, then because of inflation you may need to withdraw more from that annuity each year in order to keep the same standard of living. Let's consider how we can make realistic models of these situations.

Example 1.10 Suppose you make an initial deposit of \$1000 into an account that pays 5% annual interest, compounded annually. You estimate that you will get yearly raises of about 4%, so you will be able to deposit 4% more into your savings account each year than you deposited in the previous year. Let $d(n)$ be the amount you deposit at the end of the nth year, so $d(0) = 1000$ and $d(1) = 1040$. Let $t(n)$ be the total amount in your account at the end of the nth year, just after that year's interest and your new deposit have been added. So $t(0) = 1000$ and

$$t(1) = \text{previous total} + \text{interest on previous total} + \text{new deposit} = 2090$$

This situation is represented in the flow diagram in Figure 1.26.

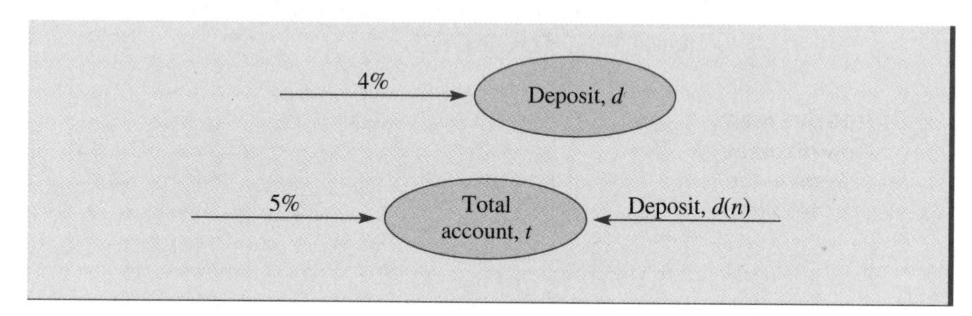

FIGURE 1.26 Flow diagram for increasing deposit model.

The arrow into the deposit reservoir indicates that each year's deposit is larger by an amount that depends on the previous year's deposit. The "Deposit" arrow into "Total account" indicates that the amount in the account is being increased by the deposit. The other arrow that goes into "Total account" represents the interest being added.

The 1 arrow into "Deposit" indicates that the dynamical system for "Deposit" is

$$d(n) = d(n-1) + 0.04d(n-1)$$

which can be simplified to

$$d(n) = 1.04d(n-1)$$

The 2 arrows into "Total account" indicate that the dynamical system for "Total account" is

$$t(n) = t(n-1) + d(n) + 0.05t(n-1)$$

which can be simplified to

$$t(n) = 1.05t(n-1) + d(n)$$

While an equation such as this can be used with spreadsheets, you cannot use $d(n)$ or $t(n)$ on the right-hand side of an equation on most calculators. Thus, when using a calculator, you must use your first equation to substitute $1.04d(n-1)$ into the second equation for $d(n)$. Then the second equation simplifies to

$$t(n) = 1.05t(n-1) + 1.04d(n-1)$$

For conformity, we will always make substitutions such as this, so that we never have the dependent variable in terms of just n on the right side of our equations.

Using these equations and $t(0) = d(0) = 1000$ gives $d(10) = 1480.24$ and $t(10) = 17{,}088.53$. This means that we deposit $1480.24 into our account at the end of the tenth year, at which point we will have $17,088.53 in our account. Use your computer or calculator to determine the total amount of your deposit and the total amount in your account at the end of the 20th year.

It is often valuable to sketch graphs involving both dependent variables. Depending on the values of the 2 variables, you can sometimes put them on the same graph. But if their values are too disparate, as in this example, you may need to put them on different axes, such as is seen in Figure 1.27.

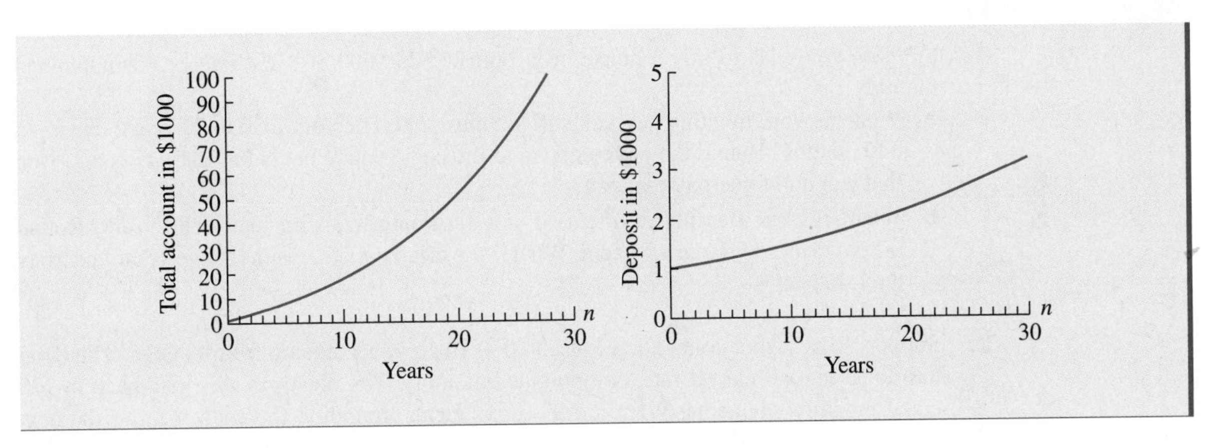

FIGURE 1.27 Left graph gives total account and right graph gives deposit.

Let's change this situation slightly to see how we could estimate an unknown parameter. Instead of making an initial deposit of $1000, let's determine what initial deposit would result in our account having $500,000 after 30 years. We want to estimate the initial deposit to the nearest $100.

For this problem, we use the same equations

$$d(n) = 1.04d(n-1)$$
$$t(n) = 1.05t(n-1) + 1.04d(n-1)$$

but we use the parameter x for the initial deposit. This means that $d(0) = x$ and $t(0) = x$. I tried $x = 3000$ and got $t(30) = 349{,}472$. The initial deposit needs to be more than $3000. I tried $x = 6000$, resulting in $t(30) = 698{,}944$. This amount is too much. I continued making guesses, the results of which are recorded in Table 1.10. This implies that the initial deposit should be about $4300.

TABLE 1.10 Initial deposit and amount in account after 30 years.

$d(0) =$	3000	6000	4000	4500	4300	4200
$t(30) =$	349,472	698,944	465,962	524,208	500,910	489,261

Suppose you initially deposit \$3000 and increase the deposit by *I*% each year, with the goal of having \$500,000 in the account after 30 years. Experiment to find the percentage that works, to the nearest 0.1%. ■

1.5 Problems

1. Suppose you wish to buy a house. You borrow \$250,000 at 8.1% interest, compounded monthly.
   - **a.** What are your monthly payments if you must amortize your loan in 20 years, that is, in 240 months? Round the payments up to the next cent. What is the last payment, given that you don't overpay the loan?
   - **b.** What are your monthly payments if you must amortize your loan in 30 years? Round the payments up to the next cent. What is the last payment, given that you don't overpay the loan?

2. Suppose you have a college loan of \$30,000 when you graduate from college. This loan has a 6% annual interest rate, compounded monthly. You must pay this loan back in 120 equal monthly payments. What are those payments, rounding the answer up to the next cent? What is the last payment, given that you don't overpay the loan?

3. Suppose you wish to buy a house but can only afford \$1100 a month.
   - **a.** If you can get a loan at 8.1% annual interest, compounded monthly, for 20 years, how large can your loan be, to the nearest thousand dollars?
   - **b.** How large can your loan be if you amortize your loan in 30 years, to the nearest thousand dollars? Note that by paying off your home loan over a longer period of time, you can actually afford to borrow more money and thus buy a more expensive home.

4. We have owned a bond for several years. This bond will pay us \$350 for each of the next 10 years, with the first payment coming one year from today. At the end of that time, it will return our original investment, which was \$5000.
   - **a.** What is the present value of this bond, to the nearest \$10, if we can get 9% interest, compounded annually, from our bank? *Hint:* This problem can be rephrased as follows. We deposit some money into a bank. Each year for the next 10 years, we withdraw \$350 from this bank account. At the end of 10 years, there is \$5000 in this account. How much did we deposit originally?
   - **b.** What is the present value of this bond if our bank only pays 7% interest? Give your answer to the nearest dollar.

5. We purchase our house for \$150,000. Ten years later, we sell our house for \$250,000. What is the average yearly rate of return on this investment; that is, what is the average percentage by which the value of our house has increased? Give your answer to the nearest

0.1%. *Hint:* Let $u(n)$ be the value of the house after n years. Then $u(0)$ =\$150,000 and $u(10)$ =\$250,000. The following flow diagram might help.

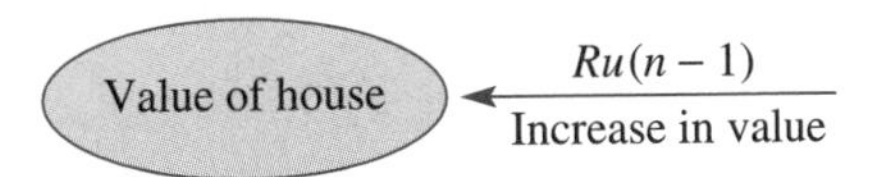

6. Suppose you initially deposit \$5000 into a savings account. Each year, for the next 20 years, you deposit an additional \$2000 into this account. Your goal is to have \$100,000 in this account at the end of the 20 years. To the nearest 0.1%, what annual interest (compounded annually) must you earn to reach your goal?

7. You purchase 100 shares of stock for \$2000. Each year, for the next 5 years, you earn a dividend of \$50. At the end of 5 years, you sell the stock for \$3500. What was the average annual interest earned on this stock investment, to the nearest 0.1%? *Hint:* Consider a bank account in which interest is added, but from which you withdraw \$50 per year. The present value of the account is $a(0) = 2000$. Its value in 5 years, just after receiving that year's \$50, is $a(5) = 3500$. What is the interest rate the bank is paying?

8. You win the lottery. The sponsors give you a choice. They will give you \$1 million today. Or they will give you \$100,000 for each of the next 15 years with the first \$100,000 being given to you today. If you can invest your money at 7.5% annual interest, compounded annually, which choice will you take and why? Ignore problems with taxes.

9. When you were 5 years old, your parents bought you \$200 worth of comics. For each of the next 13 years you got an additional \$100 worth of comics. At the end of 13 years, you decide to sell your entire comic book collection and use the money to pay for college.

a. A dealer pays you \$2000 for all of your comics. What was the average rate of return earned on your comics, to the nearest 0.1%? *Hint:* Let $c(n)$ be the value of your comics after n years and let I represent the percentage increase, as a decimal, in the value of your comics each year.

b. You didn't keep very good care of your comics, and the dealer only gave you \$1000 for all of your comics. What is the rate of return, to the nearest 0.1%? What does that mean?

10. It costs you \$50,000 per year to live today. Assume that inflation is 3.5%, meaning that each year it will cost you 3.5% more to live than the previous year.

a. How much will it cost per year for you to live 20 years from now?

b. How much will it cost per year to live in 20 years if inflation is 5% per year?

11. Suppose you retire today. You take \$40,000 out of your savings to make up for the lack of a salary. Each of the next 30 years, you take an additional \$40,000 out of your savings account. How much, to the nearest \$1000, is the minimum you would need to have in your savings account today to be able to do this, given that your account pays 8% annual interest, compounded annually.

12. You would like to be a millionaire within 25 years. You will make a deposit into a savings account today and will make the same deposit each of the next 25 years. Your savings account pays 8.3% annual interest, compounded annually. What must this deposit be, to the nearest $100, to reach your goal?

13. What annual interest rate, to the nearest 0.1%, must you get to double your money in 10 years?

14. You deposit $3000 (in year 0) into an account that pays 6.5% annual interest, compounded annually. Each year you deposit $1000 more than you deposited in the previous year. Let $d(n)$ be the amount you deposit at the end of the nth year, so $d(0) = 3000$, and let $t(n)$ be the total amount in your account at the end of the nth year, just after that year's interest and your new deposit have been added, so $t(0) = 3000$.

a. Draw a flow diagram to represent this situation.

b. Develop a dynamical system for $d(n)$ in terms of $d(n-1)$ and a dynamical system for $t(n)$ in terms of $t(n-1)$ and $d(n-1)$. Use these equations to determine your deposit and the total amount in your account after 3 years.

15. A savings account earns 5% annual interest compounded quarterly, meaning that it earns 1.25% interest each quarter of a year. You had $50,000 in that account on January 1. Every quarter of a year you withdraw $1000 from that account, with the first withdrawal occurring on April 1. Your account is therefore earning interest each quarter on what was in the account at the beginning of the quarter but is losing money from the amount you withdraw. Let n represent the number of quarters of a year after January 1. Let $a(n)$ represent the amount of money in this account n **quarters** after January 1, just after making the withdrawal for that quarter. This means that $a(0) = 50,000$ and $a(1) = 50,000 + 0.0125 \times 50,000 - 1000 = 49,625$.

a. Determine the amount in that account after 10 years, which is 40 quarters.

b. How long will it take to deplete the entire account? How much can you withdraw the last quarter?

c. Instead of withdrawing $1000 each quarter, you will withdraw an amount x that results in your account being depleted in 5 years. Approximate x to the nearest $10.

16. You deposit $3000 (in year 0) into an account that pays 6.5% annual interest, compounded annually. Let $d(n)$ be the amount you deposit at the end of the nth year, so $d(0) = 3000$, and let $t(n)$ be the total amount in your account at the end of the nth year, just after that year's interest and your new deposit have been added, so $t(0) = 3000$.

a. Each year you deposit x dollars more than you deposited in the previous year. Develop a dynamical system of 2 equations for $t(n)$ and $d(n)$. The equations will involve the parameter x. To the nearest $100, what should x be so that you have $1 million in your account at the end of the 20th year?

b. Instead of making an initial deposit of $3000, suppose you make an initial deposit of $$x$. You increase your deposit by $1000 per year. What should x be, to the nearest $100, so that $t(20) = 800,000$?

17. On the day you retire you will have x dollars in a bank account that pays 7% interest, compounded annually. You will withdraw \$40,000 from your account on that day and use it to live for the next year. Each year you will need to withdraw 3% more than the year before, because of inflation. Let $w(n)$ be the amount you withdraw from your account n years after you retire, so $w(0) = 40,000$ and $w(1) = 41,200$. Let $a(n)$ represent the amount in your account n years after you retire, just after making that year's withdrawal. This means

$$a(0) = x - 40,000 \quad \text{and} \quad a(1) = 1.07(x - 40,000) - 41,200$$

You want to have enough money in your account to be able to make a total of 30 withdrawals, including the one you make on the day you retire.

a. Develop a dynamical system of 2 equations for $a(n)$ and $w(n)$. The parameter x occurs only in the initial value $a(0)$.

b. How much money (correct to the nearest \$1,000) must be in your account immediately before you make your first withdraw?

1.5 Projects

Project 1. Find the cost of a car you would like to buy. Find out the interest rate and length of time for car loans. Use those values to determine the monthly payment for your car.

Project 2. Find the value of some state lottery. The lottery should give you 2 values, a cash value and an annuity value. For example, a recent Powerball paid \$5.4 million cash or an annuity of \$10 million, which is paid in 25 yearly payments of \$400,000 with the first payment being made immediately. To get the annuity value of \$10 million, the \$5,400,000 is invested in an account that pays a fixed annual interest rate (compounded annually). This account then pays out \$400,000 per year to the winner for a total of 25 payments, with the first payment being made today. For the example given here, the account is paying 6.04%. The question you need to answer is, What is the interest rate at which the money is invested, to the nearest 0.01%. Use the values that you find for your lottery, not the values given here. Describe how you solved the problem, including giving a table of the guesses you made at the interest rates. After 25 payments, the account should have 0 dollars. In your table, give how close the end amount was to 0 for each interest rate you tried. Give your dynamical system, and draw a diagram of the situation. The reason the interest rate is important is that if you can get a higher interest rate, you might consider taking the lump sum.

Project 3. When you lease a car, you normally make some initial payment. You then make monthly payments for a fixed period of time. At the end of the leasing period, you can purchase the car at a reduced cost. In reality, leasing is a method for car dealers to charge an interest rate for a loan without revealing it. For example, a recent ad for a \$16,635 car said that you could make an initial payment of \$949 and then make 39 monthly payments of \$230 starting in 1 month. At the end of the 39 months, you could purchase the car for \$9481.95 or return the car. This is in reality the same as purchasing the car with a \$949 down payment and monthly payments of \$230, with the amount still owed being \$9481.95 after 39 months.

From this information, you can determine what interest rate you are being charged for this car. In this example, you would have found that the interest rate was 6.7%. For some car you would like to have, find its current sales amount, the initial lease fee, the monthly payments, and the amount it would cost to purchase the car at the end of the lease. Use these values to find the interest rate being charged if you purchased the car with these conditions. You can probably get these numbers out of a newspaper ad.

2 Analysis of Dynamical Systems

2.1 Introduction to Analysis

In Chapter 1, you learned to develop dynamical systems to model a number of different real-world situations. From those systems, you were able to answer many numerical questions, such as, How much is the monthly payment on a loan? What dose of medicine will result in the optimal level in the blood? These are **quantitative** aspects of a problem.

In Chapter 2, we will study the general behavior of a dynamical system. In Section 2.2, we will look for situations in which the quantity we are studying does not change over time but instead remains constant. For example, if we have a bank account in which we withdraw the interest each month, then this account is in equilibrium and will not grow or decrease. If we take a dose of medicine that exactly replaces what was eliminated by our body since taking the last dose of medicine, then our medicine is in equilibrium.

In Section 2.3, we will look at the long-term behavior of the quantity being studied. We may find that the quantity eventually levels off at some particular value, no matter where it started. Other times, we may find that the quantity continues to grow or decrease by ever increasing amounts.

In Section 2.4, we will study how quickly the dependent variable levels off at some particular value. In some situations, where the quantity is growing by ever increasing amounts, we will find that the quantity is growing by a fixed percentage each time period. We will learn to determine what that percentage is.

In Section 2.5, we will consider cases in which there are several dependent variables and they are all growing. In these cases, the amounts often grow at the same rate, with the amount always being in about the same proportion to one another.

In Section 2.6, we will study dynamical systems in which the quantity oscillates in a wave-like manner.

All of these aspects of a model—how it grows, where it levels off, if it oscillates—are **qualitative** aspects of the situation being modeled.

2.2 Equilibrium

In this section we are going to learn what equilibrium means and how to find it. Consider the dynamical system

$$u(n) = 1.2u(n-1) - 3$$

Table 2.1 displays the values for $u(1)$ through $u(5)$ for 3 different $u(0)$-values, $u(0) = 5$, $u(0) = 15$, and $u(0) = 20$. The values are rounded to 2 decimal places. Figure 2.1 shows the time graphs of the 3 functions corresponding to the 3 initial values. Although $u(n)$ only exists for integer values of n, it is easier to see the pattern if the points are connected by line segments, as is done in this figure.

TABLE 2.1 ***u*(*n*) values for different initial values for $u(n) = 1.2u(n-1) - 3$.**

$n =$	0	1	2	3	4	5
$u(n) =$	5	3	0.6	−2.28	−5.74	−9.88
$u(n) =$	15	15	15	15	15	15
$u(n) =$	20	21	22.2	23.64	25.37	27.44

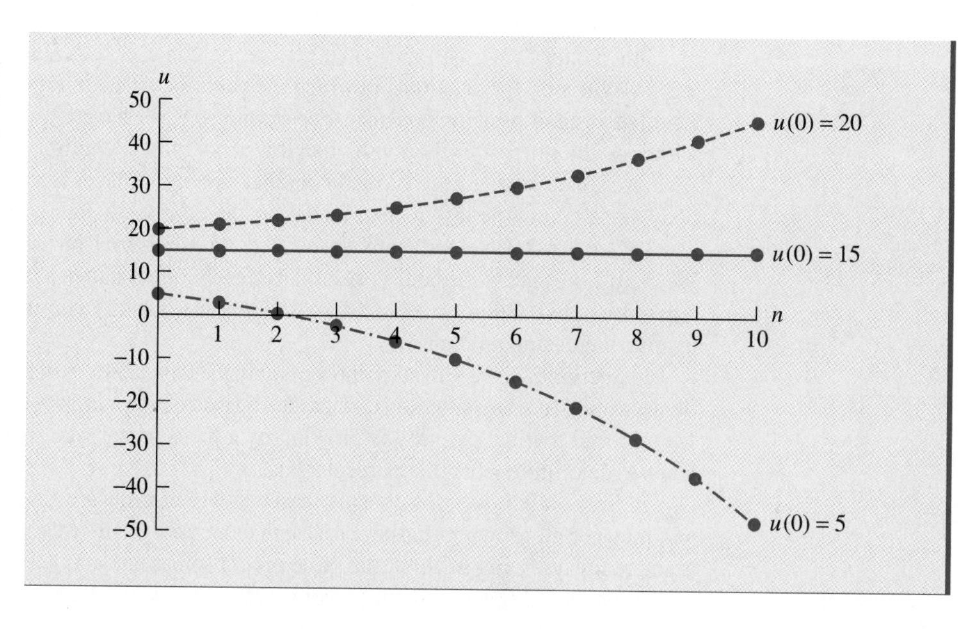

FIGURE 2.1 Three functions that satisfy $u(n) = 1.2u(n-1) - 3$.

An interesting thing happened when $u(0) = 15$. All of the other $u(n)$-values also equaled 15. If you kept computing $u(n)$ for $n = 6, 7, \ldots$, you would find they were all equal to 15. (Think about why this must be true.) Thus, as is seen in Figure 2.1, when $u(0) = 15$, the points $(n, u(n))$ lie on a horizontal line. You should compute $u(1)$ by hand when $u(0) = 15$ to make sure you understand what is happening. If you try other values for $u(0)$, you will observe that if $u(0) > 15$, the $u(n)$-values get larger and larger, and that if $u(0) < 15$, then the $u(n)$-values become more and more negative.

Consider the dynamical system

$$u(n) = 0.7u(n-1) + 9$$

In Table 2.2 are the values for $u(1)$ through $u(5)$ when $u(0) = 0$, $u(0) = 30$, and $u(0) = 50$. The values are again rounded to two decimal places.

TABLE 2.2 ***u(n)* values for different initial values for *u(n)* = 0.7*u(n* – 1) + 9.**

$n =$	0	1	2	3	4	5
$u(n) =$	0	9	15.3	19.71	22.80	24.96
$u(n) =$	30	30	30	30	30	30
$u(n) =$	50	44	39.8	36.86	34.80	33.61

For this dynamical system, when $u(0) = 30$, all of the other $u(n)$-values also equal 30. In Figure 2.2, you can see the $u(n)$-values for these 3 choices of $u(0)$. Note that when $u(0) = 30$, the points lie on a horizontal line. You should compute $u(1)$ by hand when $u(0) = 30$.

When $u(0) \neq 30$, as in the other two cases in Table 2.2, then the $u(n)$-values get closer and closer to 30. This can be seen in the time graphs of the functions in Figure 2.2. In this case, we say that the line $u = 30$ is a horizontal asymptote for each of the functions. We say that the limit of $u(n)$, as n goes to infinity, is 30, which is written as

$$\lim_{n \to \infty} u(n) = 30.$$

You should pick your own values for $u(0)$ and see if the $u(n)$-values approach 30.

DEFINITION

> If the constant function $u(n) = E$ satisfies the dynamical system, then the value E is called an **equilibrium value** for the dynamical system.

The value $E = 15$ is an equilibrium value for the dynamical system $u(n) = 1.2u(n-1) - 3$. The value $E = 30$ is an equilibrium value for the dynamical system $u(n) = 0.7u(n-1) + 9$.

Let's investigate equilibrium values within a context. Suppose you borrow $u(0)$ dollars from a friend at 1% per month interest and you pay your friend \$20 at the end of each month. The dynamical system that describes what you owe your friend each month is

$$u(n) = 1.01u(n-1) - 20$$

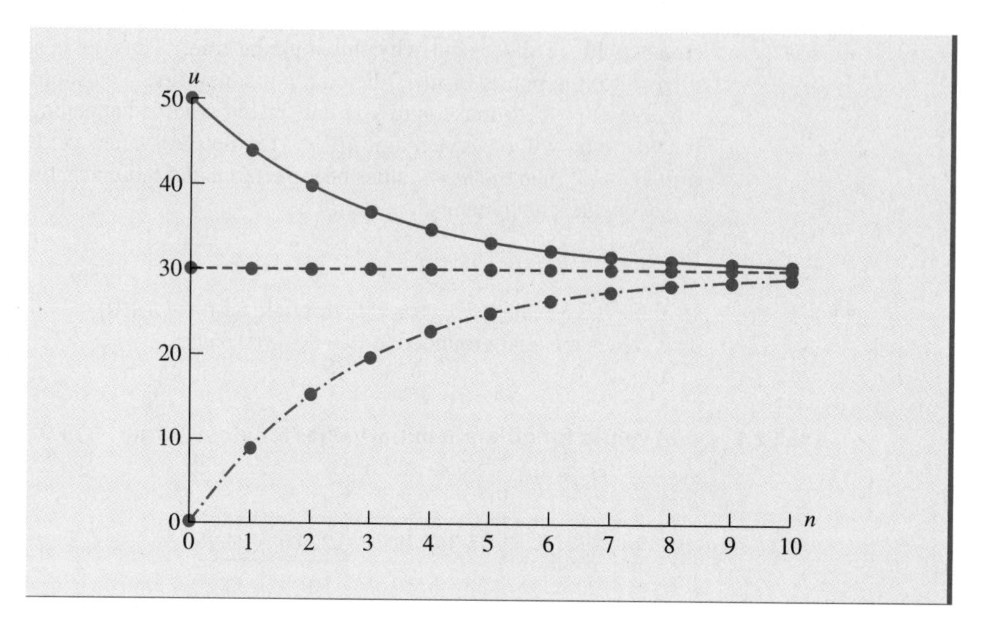

FIGURE 2.2 Solutions to $u(n) = 0.7u(n-1) + 9$ for three different $u(0)$ values.

Here, $u(n)$ represents the amount of money you owe your friend just after you make your nth payment and the interest for that month has been added on. Notice that if the original debt was $u(0) = 1000$, then $u(1) = 990$, $u(2) = 979.90$, and so forth. When you reach a point where $u(n)$ is negative, you will have paid off, or **amortized,** your loan. On the other hand, suppose that $u(0) = 3000$. Then $u(1) = 3010$, $u(2) = 3020.10$, and so forth. You now observe that you will never pay off your loan and in fact will owe your friend more and more money as time goes on.

The problem is that if you restrict your payments to \$20 a month, then there is an upper limit to what you can borrow if you ever want to pay back the loan. This limit is \$2000, the point at which your monthly payments equal your 1% monthly interest. Notice that if $u(0) = 2000$, then $u(1) = 2000$, $u(2) = 2000$, and in fact $u(n) = 2000$ for every value of n (see Figure 2.3).

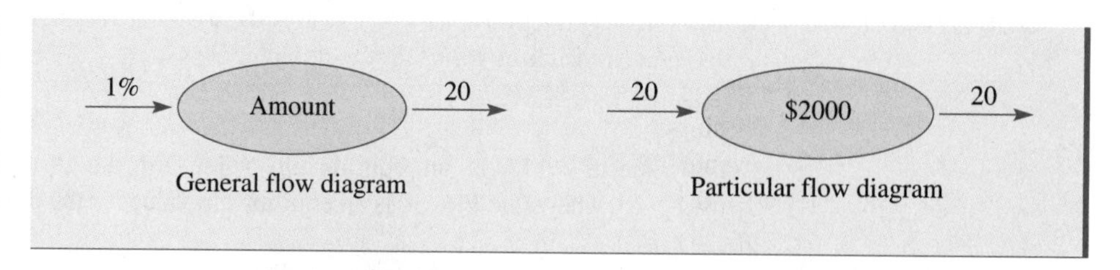

FIGURE 2.3 In arrow balances out arrow so amount in account is constant or in equilibrium.

The constant function $u(n) = 2000$ satisfies $u(n) = 1.01u(n-1) - 20$, so the number $E = 2000$ is an **equilibrium value** for this dynamical system. Notice that if you substitute

2000 for $u(n)$ and for $u(n-1)$ into the dynamical system $u(n) = 1.01u(n-1) - 20$, you get

$$2000 = 1.01(2000) - 20$$

which is balanced.

If you didn't know that 2000 was the equilibrium value, you could find it by substituting E for $u(n)$ and for $u(n-1)$ in the dynamical system to get

$$E = 1.01E - 20.$$

When you solve this equation, you get $E = 2000$, meaning that 2000 is the equilibrium value for this dynamical system.

Equilibrium values are of extreme importance in that they often tell us what will eventually happen to solutions to a dynamical system. Recall the digoxin model of Section 1.3 in which one-third of the digoxin is eliminated each day. Suppose that the patient takes a dose of 0.2 mg each day. This gives the dynamical system

$$u(n) = \tfrac{2}{3}u(n-1) + 0.2.$$

Pick any starting value; for example, $u(0) = 0$. Use your calculator or computer to find what $u(n)$ equals for relatively large values of n, say $n = 10$ or $n = 20$. You should observe that $u(n) \approx 0.6$. What this means is that no matter what the initial amount of digoxin in the plasma, over time the amount of digoxin in the body stabilizes at 0.6 mg. Notice that if $u(0) = 0.6$, then $u(1) = u(2) = \cdots = 0.6$. In this case, 0.6 is an **equilibrium value.** Notice that substitution of 0.6 for $u(n)$ and for $u(n-1)$ in the dynamical system gives equality:

$$0.6 = \tfrac{2}{3}(0.6) + 0.2$$

If we didn't know that 0.6 was the equilibrium value, we could find it by substituting E for $u(n)$ and for $u(n-1)$ in the dynamical system, getting

$$E = \tfrac{2}{3}E + 0.2$$

Solving gives the equilibrium value $E = 0.6$.

Using any starting value in the digoxin model , such as $u(0) = 3$, the values $u(1)$, $u(2)$, ... eventually get close to 0.6. In this case, we say that as n goes to infinity, the limit of $u(n)$ is 0.6, which is written as

$$\lim_{n \to \infty} u(n) = 0.6$$

This means that the line $u = 0.6$ is a horizontal asymptote for any function that satisfies the dynamical system, as is seen in Figure 2.4.

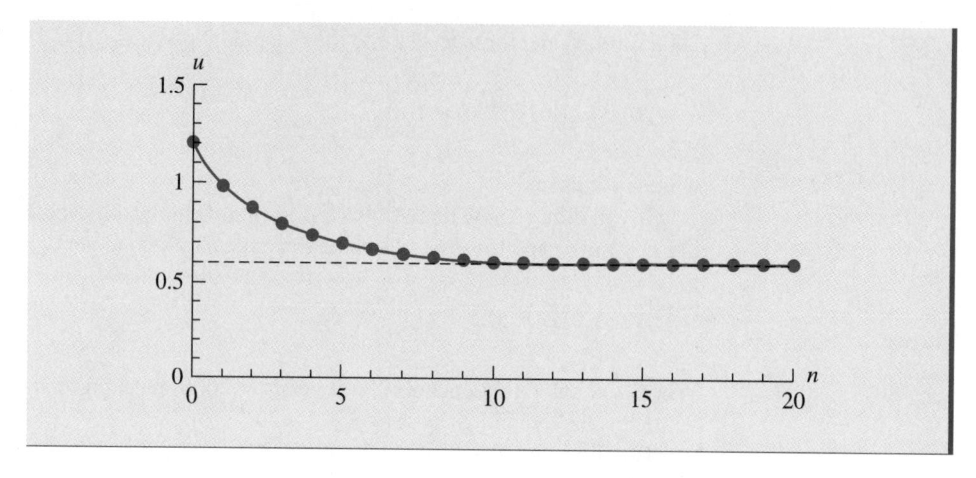

FIGURE 2.4 Function that satisfies the digoxin model with $u(0) = 1.2$. The line $u = 0.6$ is a horizontal asymptote.

In financial examples, constant solutions often tell us what you don't want to happen. In the earlier example in which you pay off a loan to your friend, the constant solution $u(n) = 2000$ means you can never pay off this loan if you owe \$2000. In the digoxin model, the equilibrium value is the actual desired level of digoxin in the body.

Example 2.1 Consider the equation

$$u(n) = 2u(n-1) - 3$$

The solution to

$$E = 2E - 3$$

is $E = 3$. Note that if $u(0) = 3$, then $u(1) = 2u(0) - 3 = 2 \times 3 - 3 = 3$. By repeating this argument, you see that $u(2) = 3$, $u(3) = 3$, and so on. Thus, $E = 3$ is an equilibrium value for this dynamical system. ■

A dynamical system may have many equilibrium values. In general, the more nonlinear the dynamical system is, the more equilibrium values that system may have. For example, if there is a $u(n)u(n) = u^2(n)$ term, then there may be up to 2 equilibrium values, if there is a $u^3(n)$ term then there may be up to 3 equilibrium values, and so on. The simple dynamical system $u(n) = u(n-1)$ is an exception in that it has an infinite number of equilibrium values, because every number is an equilibrium value. Later in this text, you will explore more complicated dynamical systems that have more than 1 equilibrium.

Now let's consider the meaning of equilibrium for a dynamical system of two equations, for example

$$u(n) = 0.4u(n-1) + 0.6v(n-1) + 6 \quad \textbf{(2.1)}$$

$$v(n) = 0.5u(n-1) + 0.2v(n-1) + 4$$

DEFINITION

> We call the pair of values (E, F) the **equilibrium point** for a dynamical system consisting of 2 equations if, when $u(0) = E$ and $v(0) = F$, then $u(n) = E$ and $v(n) = F$ for every value of n.

This means that for a pair of equations with 2 dependent variables, we need to find a pair of equilibrium values, one for u and one for v. We will denote the equilibrium value for u by E and the equilibrium value for v by F. We substitute E for $u(n)$ and $u(n-1)$ and F for $v(n)$ and $v(n-1)$ in both equations. This gives the equations

$$E = 0.4E + 0.6F + 6 \tag{2.2}$$

$$F = 0.5E + 0.2F + 4$$

There are several methods for solving a system of equations. One method is to solve for E in the first equation. To do this we subtract $0.4E$ from both sides of the first equation to get $0.6E = 0.6F + 6$. Next, we divide both sides by 0.6, giving

$$E = F + 10$$

We then substitute $F + 10$ for E in the second equation, giving

$$F = 0.5(F + 10) + 0.2F + 4 \text{ or } F = 0.7F + 9$$

Subtracting $0.7F$ from both sides, then dividing by 0.3 gives

$$F = 30$$

Substitution of 30 for F in the equation $E = F + 10$ gives

$$E = 40$$

Some people prefer to solve systems of equations by bringing all of the unknowns to the left side of the equations. In this case, we would rewrite equations (2.2) as

$$0.6E - 0.6F = 6$$

$$-0.5E + 0.8F = 4$$

We could then multiply the first equation by 5 and the second equation by 6, getting

$$3E - 3F = 30$$

$$-3E + 4.8F = 24$$

Adding the equations gives $1.8F = 54$. Dividing by 1.8 gives $F = 30$, as before. Substitution of 30 for F into either of the original equations and solving for E again gives $E = 40$. You can use any valid method to solve for equilibrium.

You should now let $u(0) = 40$ and $v(0) = 30$, and compute several u- and v-values by hand. When you do, you will observe that $u(1) = 40$ and $v(1) = 30$, $u(2) = 40$ and $v(2) = 30$, and so on.

Recall that in Section 1.3 we developed the dynamical system of two equations

$$u(n) = 0.5u(n-1) + 0.3v(n-1) + 50$$

$$v(n) = 0.4u(n-1) + 0.55v(n-1) + 23$$

to model interconversion between two chemicals, U and V, in a body. You should solve for the equilibrium point. If you do, you will get that it is $(E, F) = (280, 300)$, meaning that if there is 280 mg of chemical U and 300 mg of chemical V in the body one day, then there will be 280 mg of U and 300 mg of V every day.

In general, interconversion combined with constant dosage of one or the other or both chemicals leads to equilibrium amounts in our bodies. The equilibrium amounts correspond to physicians' target goal for one or both of the medicines.

There are other factors that complicate the elimination of chemicals from our bodies. One of these factors, which you will explore in the exercises, is that often a chemical is deposited in several places in the body. One chemical for which this is of concern is lead. Vitamin A is another chemical that is difficult to model, because it is stored in the blood and in the liver, and interconversion also occurs.

Vitamin A deficiency and protein deficiency are the human race's 2 most serious nutritional problems. Vitamin A deficiency is common in Southeast Asia, the Middle East, Africa, Central America, and South America. It is particularly common among children. Vitamin A deficiency can result in night blindness (inability to see well in medium-dim light), increased respiratory infections, skin lesions, and diarrhea.

While the numbers we use for the transfer of vitamin A between the liver and plasma in Exercises 26 and 27 are not quite right, they give a good sense of what actually happens; that is, the equilibrium amount of vitamin A in the liver is much greater than the equilibrium amount in the plasma. This is true for many animals. In fact, the vitamin A concentration in polar bears' livers is so high that people have had toxic reactions to vitamin A from eating them (see Figure 2.5).

FIGURE 2.5 Vitamin A concentration in polar bears' livers is extremely high.

When we consider the fact that interconversion may occur simultaneously with the two chemicals being absorbed into several different parts of the body, as is the case with vitamin A, it is clear that much work must be done to totally understand the dynamics of any chemical being studied. Another complicating factor is that it is often impossible to directly measure how much of a chemical is changing or moving from one part of the body to another. Much of the evidence gathered about chemicals is circumstantial, and the conclusions made are through inference. But the proper use of mathematics can help researchers develop a better understanding of interactions taking place in our bodies.

You may remember that a system of 2 linear equations with 2 unknowns may have exactly 1 solution, no solution, or infinitely many solutions. For the dynamical system (2.1), the only equilibrium point is (40, 30). Example 2.2 gives a dynamical system that has no equilibrium points. Example 2.3 gives a dynamical system that has infinitely many equilibrium points.

Example 2.2 Consider the dynamical system consisting of the two equations

$$u(n) = u(n-1) + v(n-1) + 2$$
$$v(n) = 2v(n-1) + 3$$

Simplifying $E = E + F + 2$ gives $F = -2$. Solving $F = 2F + 3$ gives $F = -3$. These are inconsistent results. This dynamical system does not have an equilibrium point. ■

Example 2.3 Consider the dynamical system consisting of the two equations

$$u(n) = u(n-1) + v(n-1) - 2$$
$$v(n) = 0.5v(n-1) + 1$$

The first equation is

$$E = E + F - 2$$

which simplifies to $F = 2$. The second equation is

$$F = 0.5F + 1$$

which also simplifies to $F = 2$. Thus, F must equal 2, but the equations are balanced for any value of E. In this case, $(E, 2)$ is an equilibrium point for any number E. Thus, this system has infinitely many equilibrium points. ■

While the definition of an equilibrium point applies to a dynamical system of 2 equations, it can easily be extended to a dynamical system of 3 equations (or even more), one for $u(n)$, one for $v(n)$, and one for $w(n)$. In this case, an equilibrium point would consist of 3 values, (E, F, G). To find the equilibrium point, we would substitute into the 3 equations E for $u(n)$ and $u(n-1)$, F for $v(n)$ and $v(n-1)$, and G for $w(n)$ and $w(n-1)$. We would then have to solve 3 equations for the 3 unknowns.

2.2 Problems

1. Find the equilibrium values, if any exist, for each of the following dynamical systems, if any

a. $u(n) = 0.5u(n-1) + 3$
b. $u(n) = 3u(n-1) - 8$
c. $u(n) = -3u(n-1) + 8$
d. $u(n) = u(n-1) + 3$
e. $u(n) = -0.8u(n-1) + 9$
f. $u(n) = 0.7u(n-1) - 15$
g. $u(n) = u(n-1)$
h. $u(n) = -u(n-1) + 8$

2. Suppose that the equilibrium value is $E = 5$ for the dynamical system

$$u(n) = 1.3u(n-1) + b$$

What is b?

3. Suppose that the equilibrium value is $E = -3$ for the dynamical system

$$u(n) = 0.8u(n-1) + b$$

What is b?

4. Suppose that the equilibrium value is $E = 110.3$ for the dynamical system

$$u(n) = -0.8u(n-1) + b$$

What is b?

5. Suppose that the equilibrium value is $E = 25$ for the dynamical system

$$u(n) = au(n-1) + 40$$

What is a?

6. Suppose that the equilibrium value is $E = 4$ for the dynamical system

$$u(n) = au(n-1) - 6$$

What is a?

7. Suppose Sara burns 130 kcal per pound per week. She consumes 16,000 kcal per week. (3600 kcal equals about 1 pound in body weight). Develop a dynamical system to model Sara's weight. What is the equilibrium value for this dynamical system? Suppose Sara presently weighs 110 pounds. How many weeks will it take for Sara to be within 1 pound of her equilibrium weight?

8. A bank account is earning 6% annual interest, compounded quarterly. In addition, $300 is being added to the account at the end of every quarter. Develop a dynamical system to model this situation. What is the equilibrium value for this dynamical system?

9. Suppose a patient takes 40 mg of a medicine every day. Through a blood test, it is determined that the amount of this drug in the patient's blood has stabilized at about 140 mg. About what percentage of the drug does this patient's kidneys eliminate each day?

10. A person who smokes one pack of cigarettes a day will absorb about 2.7 mg of cadmium each year from smoking. Assume that 7% of this cadmium is removed from the body each year. Find the equilibrium value for the dynamical system that models this situation. What does this value mean for the person who is smoking?

11. Suppose a patient's body eliminates 40% of a medicine every 8 hours. Suppose the patient takes a pill containing the medicine every 8 hours. How many milligrams of medicine should the pill contain if the doctor wants the equilibrium amount of drug in this patient's body to equal 45 mg?

12. Suppose the body eliminates 20% of a medicine every 6 hours. A drug company determines that the optimal level of drug in the bloodstream is 40 ml, which should be the equilibrium value. How many ml of drug should be taken every 6 hours so that the body reaches an equilibrium of 40 ml?

13. Consider the dynamical system of 2 equations

$$u(n) = 0.9u(n-1) + 0.3v(n-1) + 5$$

$$v(n) = -0.3u(n-1) + 0.4v(n-1) + 6$$

Find the equilibrium point or points, if any exist.

14. Consider the dynamical system

$$u(n) = 1.1u(n-1) + 3v(n-1) + 1$$

$$v(n) = 2v(n-1)$$

Find the equilibrium point or points, if any exist.

15. Consider the dynamical system

$$r(n) = 0.5r(n-1) + 0.6s(n-1) + 31$$

$$s(n) = 0.5r(n-1) - 0.2s(n-1) - 37$$

Find the equilibrium point or points, if any exist.

16. Consider the dynamical system

$$u(n) = 1.2u(n-1) + w(n-1) - 11$$

$$w(n) = -u(n-1) + 0.8w(n-1) + 7$$

Find the equilibrium point or points, if any exist.

17. Consider the dynamical system

$$u(n) = 0.6u(n-1) + 1.2v(n-1) + 18$$
$$v(n) = 0.2u(n-1) + 0.4v(n-1) + 1$$

Find the equilibrium point or points, if any exist.

18. Consider the dynamical system

$$p(n) = 0.6p(n-1) + 1.2q(n-1) + 18$$
$$q(n) = 0.2p(n-1) + 0.4q(n-1) - 9$$

Find the equilibrium point or points, if any exist.

19. Consider the dynamical system

$$u(n) = u(n-1) + 0.4v(n-1) + 2.4$$
$$v(n) = -u(n-1) + 0.6v(n-1) + 2.6$$

Find the equilibrium point or points, if any exist.

20. Consider the dynamical system

$$u(n) = 1.2u(n-1) - 0.8w(n-1) + a$$
$$w(n) = 0.3u(n-1) + 1.5w(n-1) + b$$

What values for a and b would result in an equilibrium point of $(3.1, 4.3)$?

21. Consider the dynamical system

$$u(n) = 0.4u(n-1) + 0.3w(n-1) + a$$
$$w(n) = 0.5u(n-1) + 0.7w(n-1) + b$$

What values for a and b would result in an equilibrium point of $(-1.2, 3.3)$?

22. Consider the dynamical system

$$u(n) = -1.3u(n-1) + rw(n-1) + 6.1$$
$$w(n) = su(n-1) + w(n-1) + 2.5$$

What values for r and s would result in an equilibrium point of $(5, 27)$?

23. Consider the dynamical system

$$u(n) = ru(n-1) + 0.6w(n-1) + 6.4$$
$$w(n) = su(n-1) + 1.8w(n-1) - 8.4$$

What values for r and s would result in an equilibrium point of $(11, -7)$?

24. Assume that (1) there are 2 chemicals in the body, U and V, (2) the body filters out 10% of U and 15% of V each day through the kidneys, and (3) liver enzymes metabolize 40% of the U into V and 30% of the V into U each day. In addition, we assume that no U is ingested but each day x mg of V are taken (see Flow diagram in Figure 2.6).

a. What should x be so that the equilibrium value for V is 200 mg?

b. What will be the resulting equilibrium value for U given the prescribed dosage of V found in part (a)?

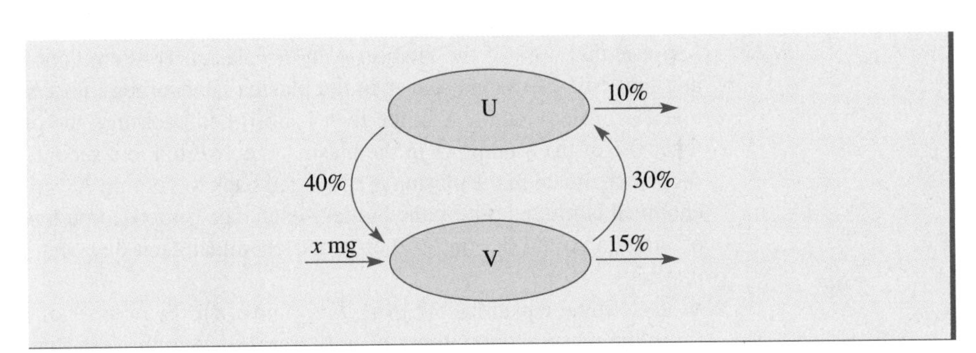

FIGURE 2.6 Flow diagram for Problem 24.

25. Vitamin A is stored primarily in our plasma and our liver. Suppose that 40% of the vitamin A in the plasma is filtered out by the kidneys each day and that 30% of the vitamin A in the plasma is absorbed into the liver each day. Also assume that 1% of the vitamin A in the liver is absorbed back into the plasma each day. Suppose you have a daily intake of 1 mg of vitamin A each day, which goes directly into the plasma (see Flow diagram in Figure 2.7).

a. Determine equations for $p(n)$ and $l(n)$, the number of milligrams of vitamin A in the plasma and liver, respectively, on day n in terms of $p(n-1)$ and $l(n-1)$.

b. Find the equilibrium amounts of vitamin A in the plasma and liver.

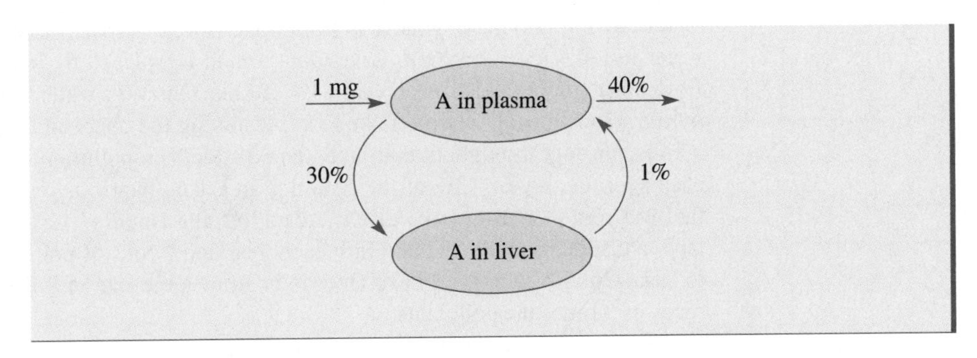

FIGURE 2.7 Flow diagram to use with Problem 25.

26. Vitamin A is stored primarily in our plasma and our liver. Suppose that 40% of the vitamin A in the plasma is filtered out by the kidneys each day and that 30% of the vitamin A in the plasma is absorbed into the liver each day. Also assume that 1% of the vitamin A in the liver is absorbed back into the plasma each day.

a. How much vitamin A should you ingest each day so that the equilibrium amount of vitamin A in the plasma is 4 mg? Assume all of the vitamin A ingested goes into the plasma.

b. What is the equilibrium amount of vitamin A in your liver given the answer to part (a)?

27. Suppose that 40% of the vitamin A in the plasma is filtered out by the kidneys each day and that 30% of the vitamin A in the plasma is absorbed into the liver each day. Assume that 1% of the vitamin A in the liver is absorbed back into the plasma each day. Assume that 10% of the vitamin A in the plasma is converted to a second chemical and 5% of the second chemical in the plasma is converted back to vitamin A. Suppose 20% of this second chemical is filtered out by the kidneys each day. Suppose you have a daily intake of 1 mg of vitamin A and 0.5 mg of the second chemical each day, and both go directly into the plasma.

a. Determine equations for $p(n)$, $l(n)$, and $s(n)$, the number of milligrams of vitamin A in the plasma, the number of milligrams of vitamin A in the liver, and the number of milligrams of the second chemical in the plasma, respectively, on day n in terms of $p(n-1)$, $l(n-1)$, and $s(n-1)$. A flow diagram will help.

b. Find the equilibrium amounts of vitamin A in the plasma and liver, and the equilibrium amount of the second chemical in the plasma.

c. Suppose that you ingest 1.4 mg of vitamin A each day. You want the equilibrium for vitamin A in your plasma to be 3 mg. How much of the second chemical should you ingest each day to accomplish this? What will be the equilibrium amount of the second chemical in the plasma and what will be the equilibrium amount of vitamin A in your liver with this consumption?

28. Let $E(n)$ and $O(n)$ be the total amount of pollution in Lake Erie and Lake Ontario, respectively, in year n. It has been determined that each year the percentage of water replaced in Lakes Erie and Ontario is approximately 38% and 13%, respectively. Since most of the water flowing into Lake Ontario is from Lake Erie, this means that each year, 38% of the water in Lake Erie flows into Lake Ontario and is replaced by rain and water flowing in from other sources. Also, each year 13% of Lake Ontario's water flows out and is replaced by the water flowing in from Lake Erie. Assuming the concentration of pollution in each lake is constant throughout that lake, then 38% of the pollution in Lake Erie is removed each year. Each year 13% of the pollution in Lake Ontario is removed, but the pollution that was removed from Lake Erie is added to Lake Ontario. Assume that 3 tons of pollutants are added directly to Lake Erie each year and 9 tons of pollutants are added directly to Lake Ontario each year. Lake Ontario is 3 times the size of Lake Erie, so we assume it receives 3 times the pollutants.

a. Make a flow diagram that summarizes the preceding information.

b. Develop a dynamical system for $E(n)$ and $O(n)$.

c. Find the equilibrium point for Lake Erie and Lake Ontario.

d. Suppose it is determined that an equilibrium level of a total of 5 tons of pollutants in Lake Erie and a total of 15 tons in Lake Ontario would be acceptable. In order to achieve these equilibrium levels, what restrictions should be placed upon the total amounts of additional pollutants that are added directly to Lake Erie and Lake Ontario?

2.3 Stability

You have learned to develop dynamical systems to model simple situations. Once you have developed a dynamical system, you use it to analyze the situation being modeled. In the previous sections, you used calculators and computers to get a number that answered questions about the situation being modeled (e.g., your monthly payment or the present value of a lottery).

In Section 2.2, you learned to find equilibrium values and equilibrium points for dynamical systems. Equilibrium gives important information about a dynamical system. In this section, you are going to learn a new concept that can be used to help understand the situation being modeled. Specifically, you are going to learn to determine if $u(n)$-values go toward or away from equilibrium values.

For a dynamical system with a given $u(0)$-value, you can compute $u(1)$, $u(2)$, and so forth. Often the particular values are not as important as is knowing the general long-term behavior, that is, what will eventually happen to $u(n)$. There are many types of long-term behaviors that can occur. We will only discuss a few of them in this section.

DEFINITION

If the $u(n)$-values for a discrete dynamical system eventually get arbitrarily close to the equilibrium value E, no matter what the initial value, then the equilibrium value is **globally stable** (see Figure 2.8). Graphically this means that whenever horizontal lines are drawn above and below the line $u = E$, then the points $(n, u(n))$ are eventually between those two lines. This must be true no matter what the initial value or how close the two horizontal lines are to equilibrium.

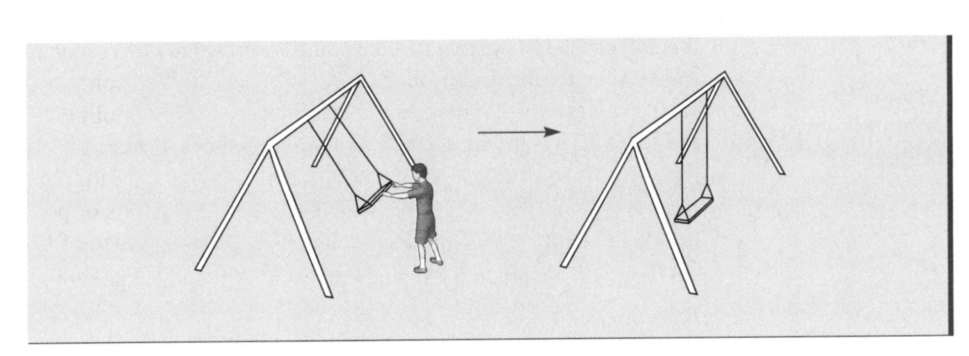

FIGURE 2.8 A swing at rest is at its globally stable equilibrium.

Example 2.4 In Figure 2.9 are the graphs of two functions that satisfy

$$u(n) = 0.7u(n-1) + 15$$

In one case, $u(0) = -100$, and in the other, $u(0) = 100$. Notice that in both cases, the line $u = 50$ is a horizontal asymptote for the function; that is,

$$\lim_{n\to\infty} u(n) = 50$$

The equilibrium value, $E = 50$, is globally stable.

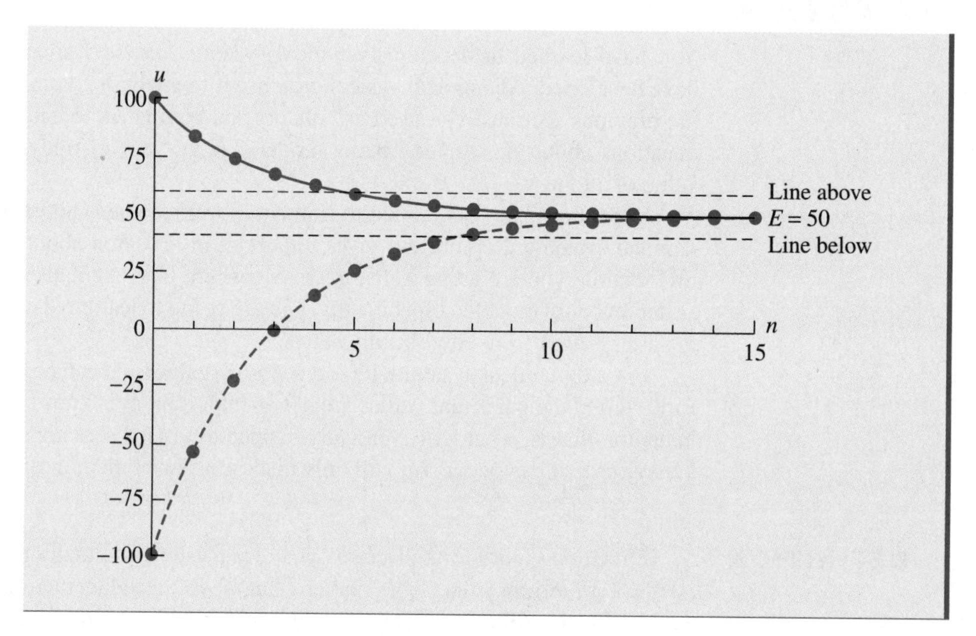

FIGURE 2.9 Solutions to $u(n) = 0.7u(n-1) + 15$ move between the 2 lines and remain between them.

Also note in Figure 2.9 that there are 2 lines, one above equilibrium and one below. Both functions moved between those lines and remained between them. We have observed $u(n)$ approaching equilibrium for only 2 initial values. For an equilibrium value to be globally stable, $u(n)$ must approach equilibrium for all choices of $u(0)$. While we cannot yet "prove" that $E = 50$ is globally stable, we can try enough values for $u(0)$ to be fairly certain it is. ■

It is important to note that it is the equilibrium value that is globally stable, not the dynamical system nor the initial value.

Example 2.5 In Figure 2.10 is the time graph of a function that satisfies

$$u(n) = -0.8u(n-1) + 9$$

with $u(0) = 100$. These $u(n)$-values oscillate above and below the equilibrium value $E = 5$, but they also approach 5. Again, $u = 5$ is a horizontal asymptote for the function; that is,

$$\lim_{n\to\infty} u(n) = 5$$

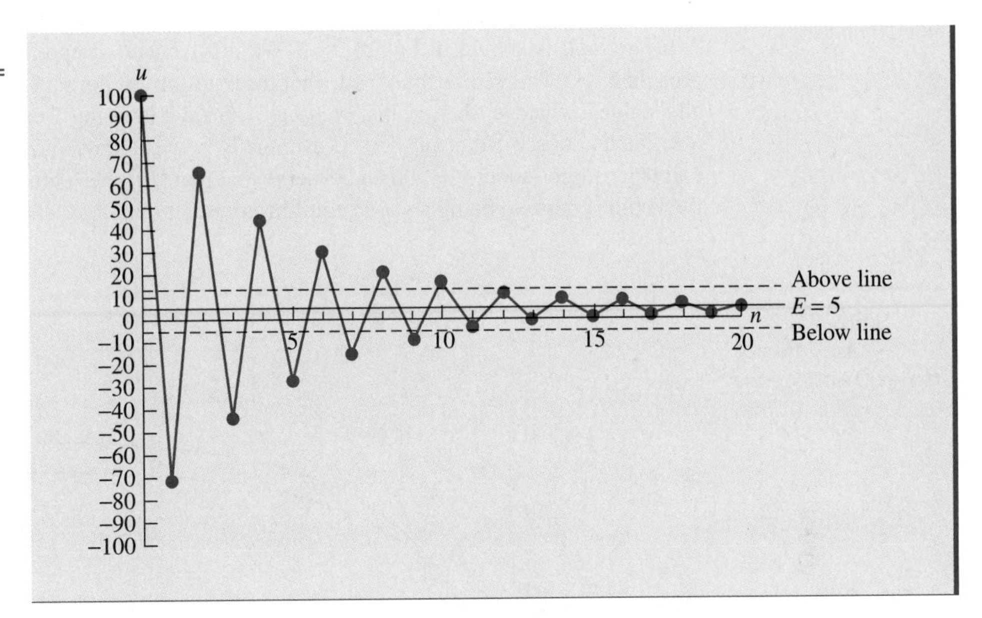

FIGURE 2.10 Solution to $u(n) = -0.8u(n-1) + 9$.

and the equilibrium value is globally stable. The graph of $u(n)$ is eventually between the 2 arbitrary lines drawn in that figure. You should try other values for $u(0)$ until you are convinced this is always true. ■

For an equilibrium value to be globally stable, we are not interested in how $u(n)$ approaches the equilibrium value but if $u(n)$ eventually gets "infinitely" close to the equilibrium value.

Example 2.6 Suppose we have a dynamical system of two equations with the 2 dependent variables u and v, say

$$u(n) = 0.5u(n-1) + 0.3v(n-1) + 50$$
$$v(n) = 0.4u(n-1) + 0.55v(n-1) + 23$$

which model interconversion of the two chemicals U and V. Recall that the equilibrium point is $(E, F) = (280, 300)$. Table 2.3 gives some resulting values when $u(0) = 150$ and $v(0) = 100$.

TABLE 2.3 **Values for interconversion model.**

$n =$	0	1	2	3	4
$u(n) =$	150	155	169	183	195
$v(n) =$	100	138	161	179	195

A time graph is shown in Figure 2.11. We note that $u(n)$ approaches 280 and $v(n)$ approaches 300. This seems to indicate that the equilibrium point $(280, 300)$ might be "globally stable," whatever that means. We might consider drawing 1 pair of lines about the line $u = 280$ and observe that $(n, u(n))$ is eventually between those lines. We could draw a second pair of lines about $v = 300$ and observe that $(n, v(n))$ is eventually between those lines. But there is more to being a stable equilibrium point than that.

FIGURE 2.11 Amounts of drugs U and V over time.

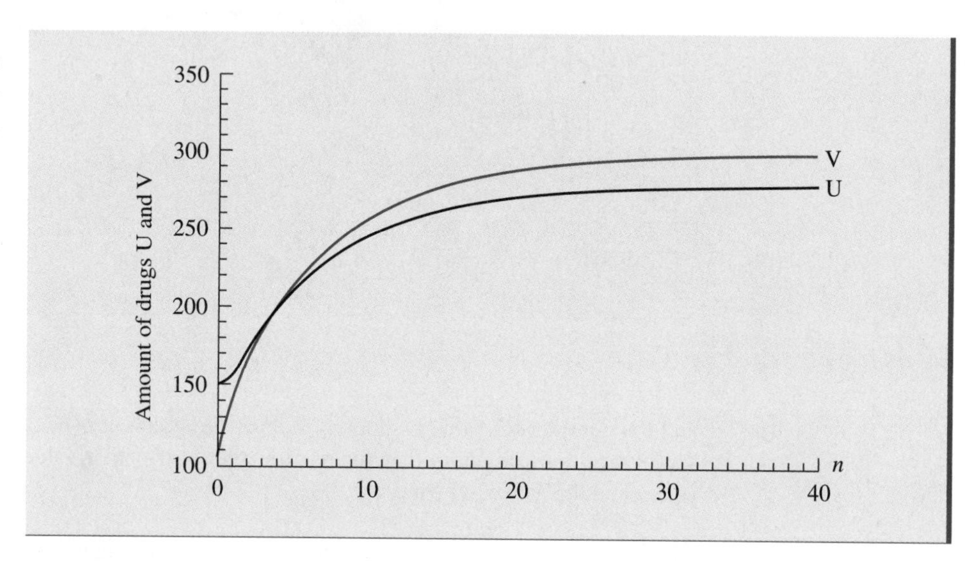

The fact that equilibrium is a point consisting of a u-value and a v-value indicates that it might be better to plot the points $(u(0), v(0))$, $(u(1), v(1))$, ..., that is, the points $(150, 100)$, $(155, 138)$, ..., in the uv-plane, as in Figure 2.12. In this graph, time is not plotted. Instead, time is implicit in the movement from $(u(0), v(0))$ to $(u(1), v(1))$ to $(u(2), v(2))$, and so on. This movement is indicated by the arrows in Figure 2.12. Note that the points approach the point $(280, 300)$. This will be true no matter what the initial values. Like equilibrium values, an equilibrium point is globally stable if for any initial values, the points $(u(n), v(n))$ approach the equilibrium point. This means that the points $(u(n), v(n))$ will eventually be inside any circle centered at (E, F), as seen in Figure 2.12. ■

DEFINITION

Suppose that (E, F) is an equilibrium point for a dynamical system of 2 equations. If the points $(u(n), v(n))$ eventually get arbitrarily close to the equilibrium point (E, F), no matter what the initial point, then the equilibrium point is **globally stable.** Graphically, this means that for any circle drawn about the point (E, F), eventually the points $(u(n), v(n))$ will be inside that circle.

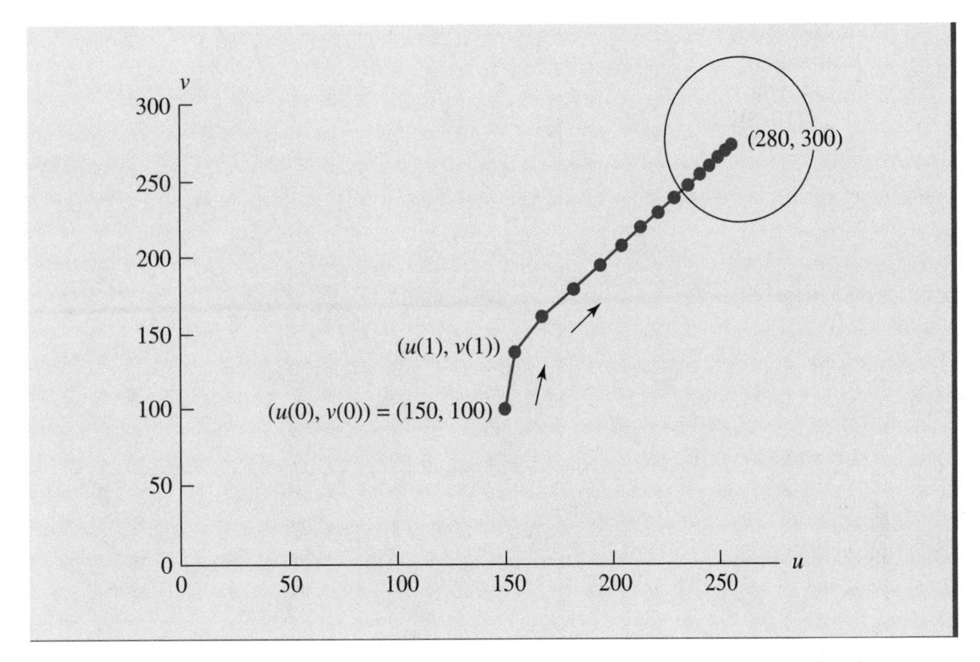

FIGURE 2.12 Points $(u(n), v(n))$ approach the equilibrium point (280, 300).

When an equilibrium point is globally stable, then $u(n)$ must approach E and $v(n)$ must approach F.

Example 2.7 Consider the dynamical system

$$u(n) = 0.6u(n-1) - 0.5v(n-1) + 5.5$$
$$v(n) = 0.68u(n-1) + 0.5v(n-1) - 5.3$$

The equilibrium point is $(E, F) = (10, 3)$. Pick your own values for $u(0)$ and $v(0)$, and see if you get similar results to mine. I used $u(0) = 3$ and $v(0) = 4$. If you make a table of values, you will see that the values for both $u(n)$ and $v(n)$ rise and fall. In Figure 2.13 are the points $(u(0), v(0)) = (3, 4)$, $(u(1), v(1)) = (5.3, -1.26), \ldots$. Note that these points spiral in closer to the equilibrium point $(10, 3)$ and are all eventually inside of the circle centered at $(10, 3)$. Thus, the equilibrium point appears to be globally stable. To be sure, you should try other values for $u(0)$ and $v(0)$.

Time graphs for the points $(n, u(n))$ and $(n, v(n))$ are seen in Figure 2.14. Note that the points $(n, u(n))$ are oscillating above and below the line $u = 10$. Similarly, $v(n)$ becomes arbitrarily close to 3. Another way to say this is that $u = 10$ is a horizontal asymptote for the function $u(n)$ and $v = 3$ is a horizontal asymptote for the function $v(n)$. You should try other values for $u(0)$ and $v(0)$ until you become convinced that this always happens. ■

When the points $(u(n), v(n))$ form a spiral as in Figure 2.13, then the time graphs for u and v will oscillate, as in Figure 2.14.

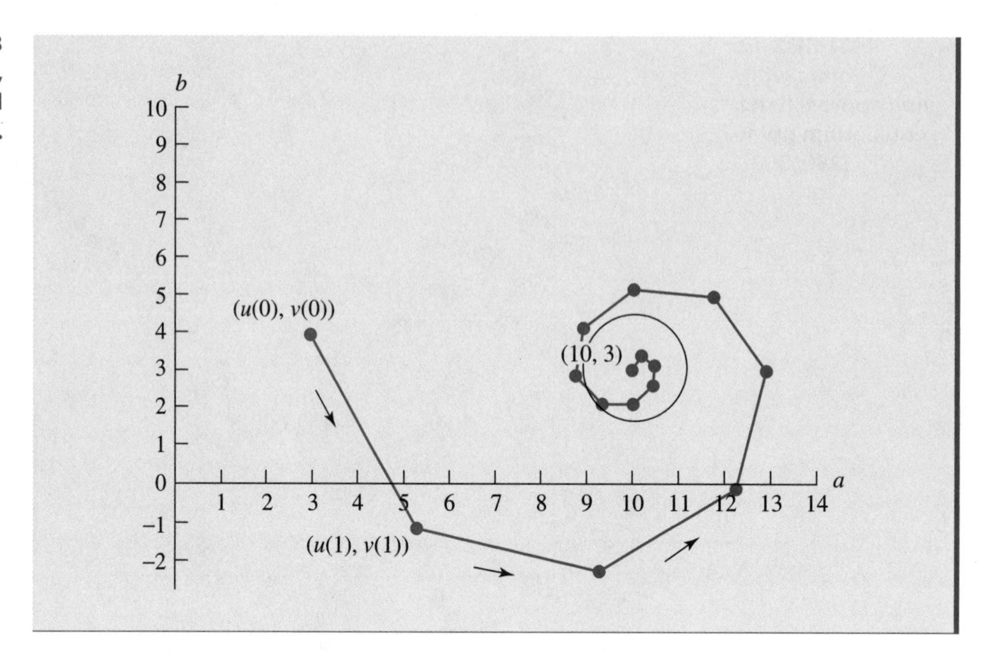

FIGURE 2.13 Points ($u(n)$, $v(n)$) go toward (10, 3).

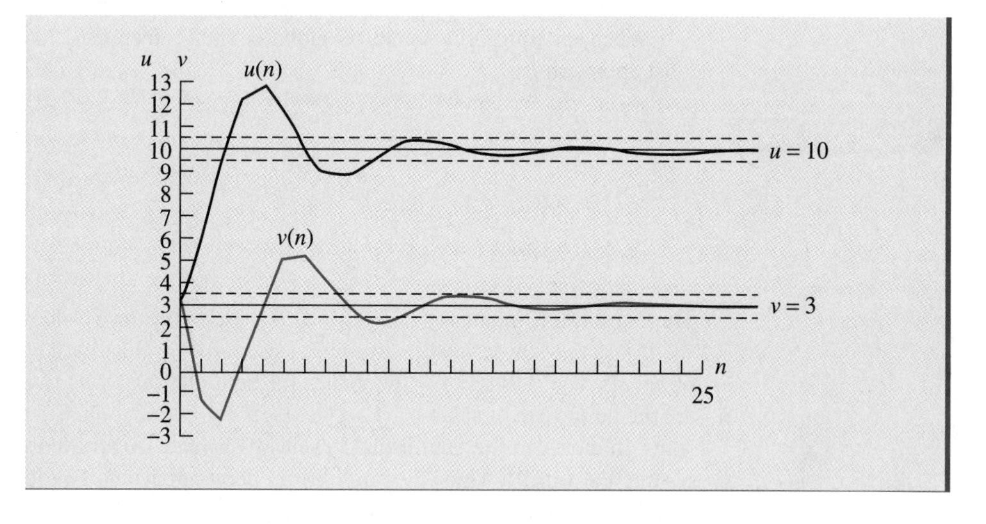

FIGURE 2.14 Oscillations toward equilibrium.

DEFINITION

Suppose that whenever the initial value is not equal to the equilibrium value E, then the $u(n)$-values eventually get "far" away from equilibrium. The equilibrium value is then said to be **globally unstable.** Graphically this means that for any pair of horizontal lines, with one above and the other below $u = E$, eventually the points $(n, u(n))$ will not be between those two lines.

Example 2.8 Consider the dynamical system

$$u(n) = 3u(n-1) - 8$$

The equilibrium value is $E = 4$. In Figure 2.15 are time graphs for $u(0) = 3$ and $u(0) = 5$. Note that not only are the graphs of both functions getting far from $E = 4$, $|u(n)|$ is going to infinity in both cases. The equilibrium value $E = 4$ is globally unstable. Note that the graphs of both functions are eventually not between the 2 lines. ■

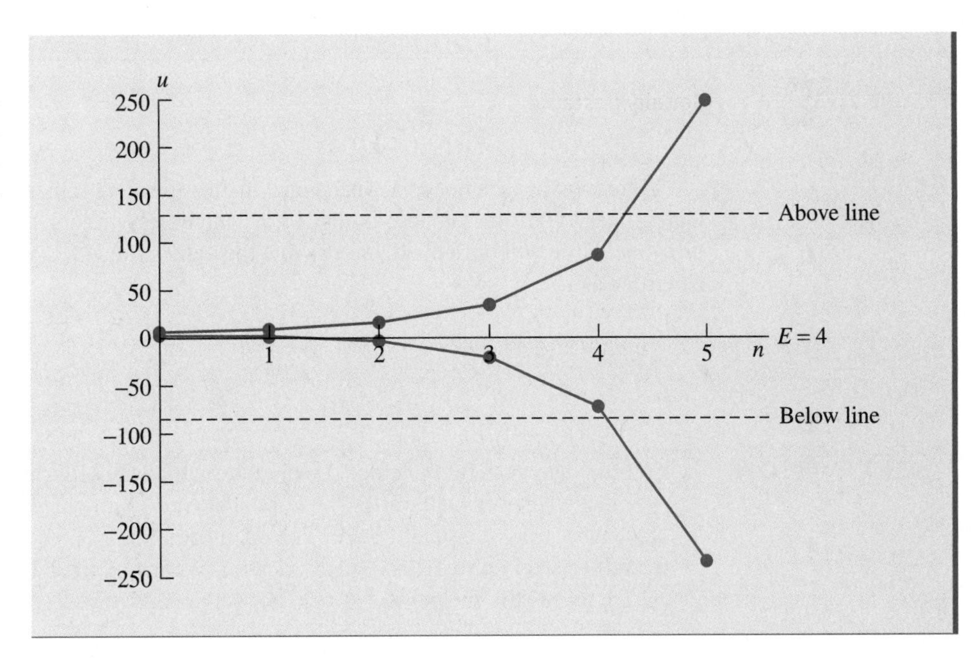

FIGURE 2.15 $E = 4$ is a globally unstable equilibrium value for $u(n) = 3u(n-1) - 8$.

Example 2.9 In Figure 2.16 is the time graph of a function that satisfies

$$u(n) = -2u(n-1) + 3$$

with $u(0) = 5$. The equilibrium value is $E = 1$. Note that the points $(n, u(n))$ eventually get outside of the lines. In fact, the $u(n)$-values eventually alternate between large positive and large negative values. ■

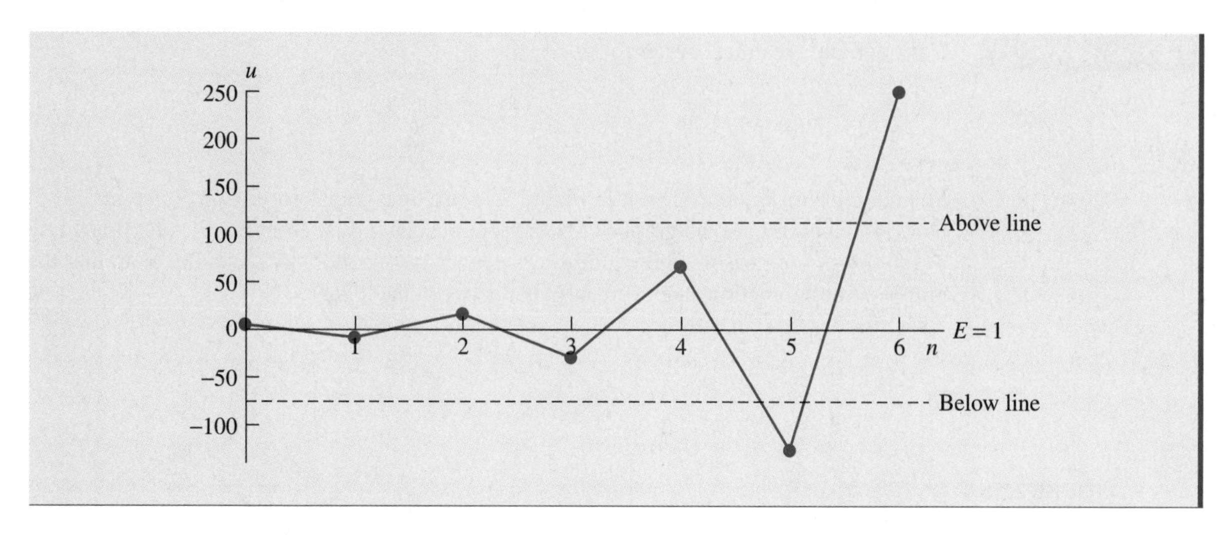

FIGURE 2.16 $E = 1$ is globally unstable.

At present, we are not concerned with whether the $u(n)$-values are positive or negative. Our primary concern for now is that $|u(n)|$ increases without bound. When the values $|u(n)|$ get large without bound, we say that the values for $|u(n)|$ go toward infinity, which is written as

$$\lim_{n \to \infty} |u(n)| = \infty$$

DEFINITION

Consider a dynamical system of 2 equations with the equilibrium point (E, F). Suppose that whenever $(u(0), v(0)) \neq (E, F)$, then the points $(u(n), v(n))$ eventually get "far" away from the point (E, F). The equilibrium point is then said to be **globally unstable** (see Figure 2.17). Graphically this means that for any circle, centered at (E, F), eventually the points $(u(n), v(n))$ get outside of that circle.

FIGURE 2.17 A coin in unstable equilibrium.

Example 2.10 Consider the dynamical system consisting of the 2 equations

$$u(n) = 2u(n-1) - 2.6v(n-1) + 47$$
$$v(n) = 2u(n-1) - 2v(n-1) + 50$$

You should be able to find that the equilibrium point is $(5,20)$, meaning that if $u(0) = 5$ and $v(0) = 20$, then $u(n) = 5$ and $v(n) = 20$ for every n-value. To study the stability, pick your own values for $u(0)$ and $v(0)$. If you make a table of values, you will notice that $u(n)$ and $v(n)$ begin oscillating, with larger and larger oscillations. The points $(u(n), v(n))$ are plotted in Figure 2.18, with $u(0) = 4$ and $v(0) = 19$. These points get further from the equilibrium point. For any circle centered at $(5,20)$, the points will eventually get outside of the circle. Thus, $(5,20)$ is apparently a globally unstable equilibrium point.

FIGURE 2.18 Equilibrium point (5, 20) is globally unstable.

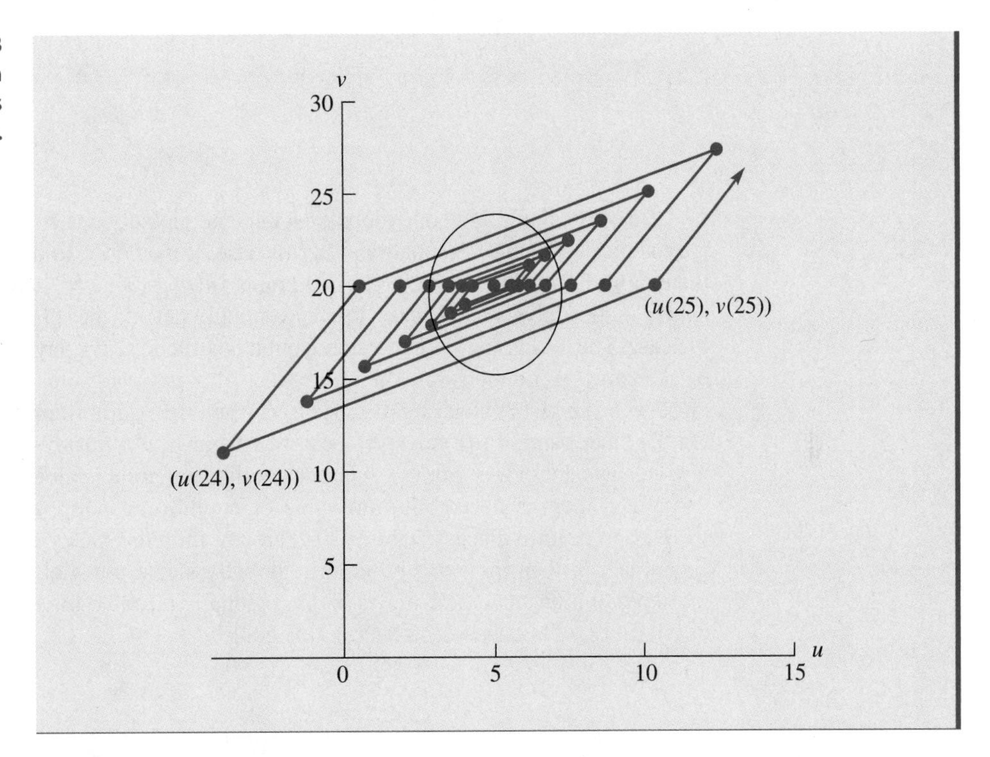

Even though an equilibrium point (E,F) is globally unstable, $u(n)$-values can still be close to E and $v(n)$-values can be close to F for large values of n. Notice the points $(n, u(n))$ and $(n, v(n))$ in Figure 2.19. Lines are drawn on both sides of the line $u = 5$, which corresponds to the equilibrium value for $u(n)$. Similarly, lines are drawn on each side of the line $v = 20$, which corresponds to the equilibrium value for $v(n)$. Note that some of the points $(n, v(n))$ are between the "$v = 20$" lines, even as n gets larger. Similarly, there will be points $(n, u(n))$ between the "$u = 5$" lines for some large values of n. Thus, it is often difficult to determine the stability of an equilibrium point from time graphs. ■

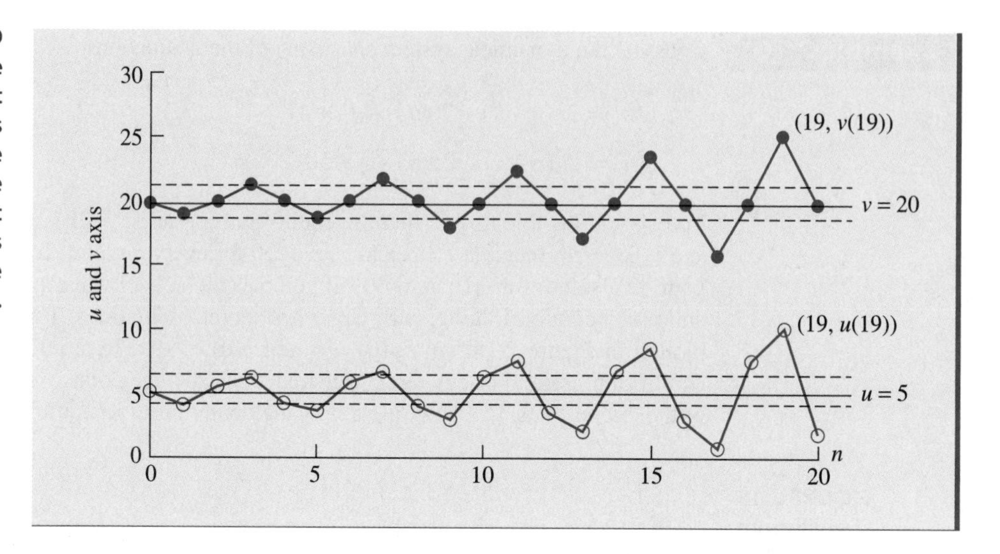

FIGURE 2.19 Some of the points $(n, u(n))$ get outside of lines bounding the line $u = 5$. Some of the points $(n, v(n))$ get outside of lines bounding the line $v = 20$.

We could similarly define globally stable and globally unstable equilibrium points for a dynamical system of 3 equations. In this case, it would mean that for any sphere about the equilibrium point (E, F, G), then all points $(u(n), v(n), w(n))$ are eventually inside the sphere in the globally stable case, or are eventually outside the sphere in the globally unstable case. The idea is easy to understand, but it is difficult to display because of the difficulty of sketching in 3 dimensions. In reality, if $u(n)$ always gets close to E, $v(n)$ always gets close to F, and $w(n)$ always gets close to G, then the equilibrium point is globally stable. On the other hand, if $u(n)$ always goes away from E, $v(n)$ always goes away from F, and $w(n)$ always goes away from G, then the equilibrium point is globally unstable.

When there is no equilibrium value or equilibrium point, as in Example 2.2, then there is no need to discuss stability. If there are infinitely many equilibrium points, as in Example 2.3, then they will be neither globally stable nor globally unstable. There are more possibilities than globally stable and globally unstable for equilibria. Some of these possibilities will be discussed later in this book.

2.3 Problems

1. You found the equilibrium values for the following dynamical systems in Problem 1 of Section 2.2. Determine the stability of the equilibrium value.

a. $u(n) = 0.5u(n-1) + 3$

b. $u(n) = 3u(n-1) - 8$

c. $u(n) = -3u(n-1) + 8$

d. $u(n) = -0.8u(n-1) + 9$

e. $u(n) = 0.7u(n-1) - 15$

2. Suppose that the equilibrium value is $E = 5$ for the dynamical system

$$u(n) = 1.3u(n-1) + b$$

What is the value for b? Is $E = 5$ stable or unstable?

3. Suppose that the equilibrium value is $E = -3$ for the dynamical system

$$u(n) = 0.8u(n-1) + b$$

What is the value for b? Is $E = -3$ stable or unstable?

4. Suppose that the equilibrium value is $E = 110.3$ for the dynamical system

$$u(n) = -0.8u(n-1) + b$$

What is the value for b? Is $E = 110.3$ stable or unstable?

5. Suppose that the equilibrium value is $E = 25$ for the dynamical system

$$u(n) = au(n-1) + 40$$

What is the value for a? Is $E = 25$ stable or unstable?

6. Suppose that the equilibrium value is $E = 4$ for the dynamical system

$$u(n) = au(n-1) - 6$$

What is the value for a? Is $E = 4$ stable or unstable?

7. You deposit some money into a bank account that pays 6% annual interest, compounded monthly. Each month, you withdraw $200 from that account. Develop a dynamical system that models this account. Find the equilibrium value for this dynamical system. Determine the stability of the equilibrium value. Predict what will happen to this account for different initial deposits.

8. Suppose Sue presently weighs 110 pounds and burns 130 kcal per pound per week. She consumes 16,000 kcal per week. What is her equilibrium weight? What is the stability of this equilibrium value? What does that mean for Sue?

9. Consider the dynamical system of 2 equations

$$u(n) = 0.9u(n-1) + 0.3v(n-1) + 5$$

$$v(n) = -0.3u(n-1) + 0.4v(n-1) + 6$$

Find the equilibrium point and determine its stability.

10. Consider the dynamical system

$$u(n) = 1.1u(n-1) + 3v(n-1) + 1$$

$$v(n) = 2v(n-1)$$

Find the equilibrium point and determine its stability.

11. Consider the dynamical system

$$u(n) = 0.5u(n-1) + 0.6v(n-1) + 31$$
$$v(n) = 0.5u(n-1) - 0.2v(n-1) - 37$$

Find the equilibrium point and determine its stability.

12. Consider the dynamical system

$$u(n) = 1.2u(n-1) + w(n-1) - 11$$
$$w(n) = -u(n-1) + 0.8w(n-1) + 7$$

Find the equilibrium point and determine its stability.

13. Consider the dynamical system

$$u(n) = 0.4u(n-1) + 0.3w(n-1) + a$$
$$w(n) = 0.5u(n-1) + 0.7w(n-1) + b$$

What values for a and b would result in an equilibrium point of $(-1.2, 3.3)$? What is the stability of $(-1.2, 3.3)$?

14. Consider the dynamical system

$$u(n) = 1.2u(n-1) - 0.8w(n-1) + a$$
$$w(n) = 0.3u(n-1) + 1.5w(n-1) + b$$

What values for a and b would result in an equilibrium point of $(3.1, 4.3)$? What is the stability of $(3.1, 4.3)$?

15. Consider the dynamical system

$$u(n) = ru(n-1) + 0.6w(n-1) + 6.4$$
$$w(n) = su(n-1) + 1.8w(n-1) - 8.4$$

What values for r and s would result in an equilibrium point of $(11, -7)$? What is the stability of $(11, -7)$?

16. Consider the dynamical system

$$u(n) = -1.3u(n-1) + rw(n-1) + 6.1$$
$$w(n) = su(n-1) + w(n-1) - 2.5$$

What values for r and s would result in an equilibrium point of $(5, 27)$? What is the stability of $(5, 27)$?

2.4 Ratios and Proportional Change

In the previous section, you learned to determine whether the dependent variables were going toward or away from equilibrium. In this section, we will investigate how the dependent variables are **changing.**

Suppose you are earning \$15 per hour tutoring. Recall that this situation could be modeled by the dynamical system

$$u(n) = u(n-1) + 15$$

or by the linear function

$$u(n) = 15n$$

In this case, there is no equilibrium, so there is no stability to be determined. Table 2.4 gives the amounts earned for working different numbers of hours.

TABLE 2.4 Amount earned for tutoring *n* hours.

$n =$	0	1	2	3	4
$u(n) =$	0	15	30	45	60

It is easy to see that the dependent variable is changing by a constant amount each hour. But we also know that the dependent variable is a linear function of n. In fact, a linear relationship is one in which there is constant change per unit change in the independent variable.

LINEAR PATTERN

If the differences of consecutive values of the dependent variable are constant then the dependent variable is increasing or decreasing linearly. The slope of the line equals the constant difference.

In some cases, the change in the dependent variable is almost constant, but not quite. Consider the dynamical system

$$c(n) = 0.8c(n-1) + 0.1f(n-1) + 400$$

$$f(n) = 0.2c(n-1) + 0.9f(n-1) + 600$$

of Example 1.6. There are no equilibrium points for this dynamical system. Suppose we let $c(0) = 200$ and $f(0) = 500$. Then we get the results in Table 2.5, rounded to the nearest integer.

TABLE 2.5 Amounts of *c*(*n*) and *f*(*n*) over time.

$n =$	0	1	2	3	...	15	16	17	18
$c(n) =$	200	610	997	1368	...	5454	5788	6122	6455
$f(n) =$	500	1090	1703	2332	...	10,246	10,912	11,578	12,245

For the first few values of $c(n)$ and $f(n)$, no pattern emerges. But after some time, we observe that

$$c(16)-c(15) \approx c(17)-c(16) \approx c(18)-c(17) \approx 334$$

and

$$f(16)-f(15) \approx f(17)-f(16) \approx f(18)-f(17) \approx 666$$

If you do the computations more accurately, you will find that

$$c(n)-c(n-1) \approx 333.33$$

and

$$f(n)-f(n-1) \approx 666.67$$

when n is relatively large. Thus, the dependent variables are almost increasing by a constant amount. In this case, the expressions for $c(n)$ and for $f(n)$ are almost linear, with the approximate slopes being 333.33 and 666.67, respectively. This means that the dependent variables can be approximated by functions of the form

$$c(n) \approx 333.33n + a \quad \text{and} \quad f(n) \approx 666.67n + b$$

This is why the time graphs of these functions appear linear, as seen in Figure 2.20 (which was also Figure 1.20 in Chapter 1).

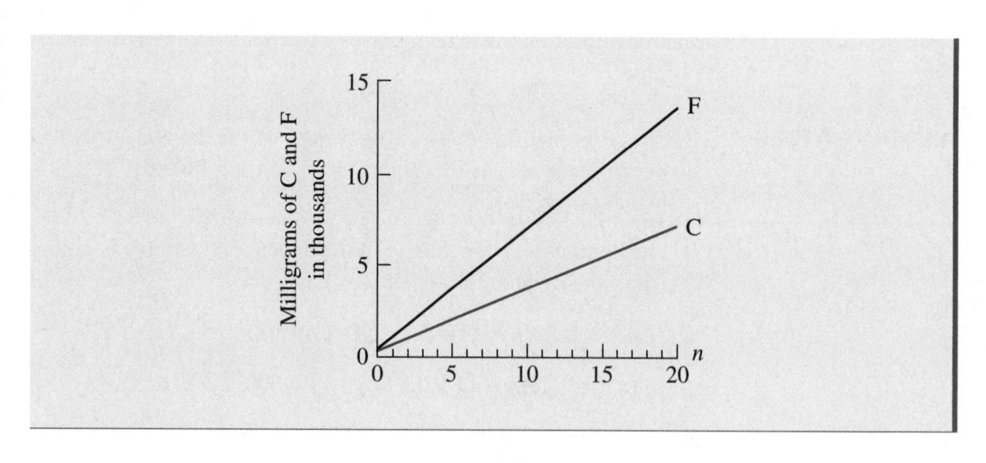

FIGURE 2.20 $c(n)$ and $f(n)$ are almost linear functions with slopes 333.33 and 666.67, respectively.

We could approximate the vertical intercepts if we wanted, but it is the linear behavior and slope that we are most interested in.

APPROXIMATELY LINEAR PATTERN

If differences of consecutive values of the dependent variable approach a constant m as n increases, then the dependent variable is approximately linear. The constant m approximates the slope of the line.

A second common type of change is proportional change. In particular, we look for a proportional change in the distance the dependent variable is from equilibrium, that is,

$$\frac{u(n)-E}{u(n-1)-E}$$

for different values of n.

Example 2.10 Consider the dynamical system

$$u(n) = 1.4u(n-1) - 2$$

The equilibrium value is $E = 5$. Let's start with $u(0) = 10$. The first several values of $u(n)$ are given in the second row of Table 2.6. You should check that the consecutive differences of the $u(n)$-values are not constant. The third row of Table 2.6 gives the distances of each value from equilibrium.

TABLE 2.6 Distances from equilibrium for $u(n) = 1.4u(n-1) - 2$.

$n =$	0	1	2	3
$u(n) =$	10	12	14.8	18.72
$u(n) - 5 =$	5	7	9.8	13.72

Let's look at the quotient of consecutive differences from equilibrium. We see that

$$\frac{u(1)-5}{u(0)-5} = \frac{7}{5} = 1.4 \qquad \frac{u(2)-5}{u(1)-5} = \frac{9.8}{7} = 1.4 \quad \text{and} \quad \frac{u(3)-5}{u(2)-5} = \frac{13.72}{9.8} = 1.4$$

This says that each value is 1.4 times as far from equilibrium as the previous value, or that each value is 140% as far from equilibrium as the previous value, or that each value is 40% further from equilibrium as the previous value. ■

Do not use the equilibrium value E for $u(0)$ when looking for ratios, since that would result in a division by 0, which is not allowed. We are looking for how $u(n)$ is going toward or away from equilibrium. If $u(0) = E$, then $u(n)$ does not go toward or away from equilibrium, it stays at equilibrium.

Example 2.11 Consider the dynamical system

$$u(n) = 3u(n-1) - 8$$

The equilibrium value is $E = 4$. I chose $u(0) = 20$ and found that $u(1) = 52$, $u(2) = 148$, $u(3) = 436$, and $u(4) = 1300$. This gives

$$\frac{u(1)-4}{u(0)-4} = \frac{52-4}{20-4} = 3 \qquad \frac{u(2)-4}{u(1)-4} = \frac{148-4}{52-4} = 3$$

$$\frac{u(3)-4}{u(2)-4} = \frac{436-4}{148-4} = 3 \quad \frac{u(4)-4}{u(3)-4} = \frac{1300-4}{436-4} = 3$$

In fact

$$\frac{u(n)-4}{u(n-1)-4} = 3$$

for every value of n. This means that each value is 3 times as far from equilibrium as the previous value. This can also be stated as "each value is 300% as far from equilibrium as the previous value" or as "each value is 200% further from equilibrium than the previous value."

You should repeat this example using different values for $u(0)$ and see if you get the same result. Remember when you are checking ratios that $u(0)$ should not equal the equilibrium value. ■

Example 2.12 In this example, we are going to consider what happens when the ratio is negative. Consider the dynamical system

$$u(n) = -0.8u(n-1) + 9.$$

The equilibrium value is $E = 5$. I used $u(0) = -4$, although the result will be the same no matter what value you use for $u(0)$, other than 5. I found that

$$\frac{u(1)-5}{u(0)-5} = \frac{12.2-5}{-4-5} = -0.8 \quad \frac{u(5)-5}{u(4)-5} = \frac{7.94912-5}{1.3136-5} = -0.8$$

$$\text{and} \quad \frac{u(9)-5}{u(8)-5} = -0.8$$

We can use values for n other than 1, 2, 3, and 4.

Notice that

$$\left|\frac{u(1)-5}{u(0)-5}\right| = 0.8 \quad \left|\frac{u(5)-5}{u(4)-5}\right| = 0.8 \quad \text{and} \quad \left|\frac{u(9)-5}{u(8)-5}\right| = 0.8$$

This means that each value is only 80% as far from equilibrium as the previous value or that each value is 20% closer to equilibrium. The negative sign in the previous ratios means that each value is on the other side of equilibrium from the previous value, as can be seen in the up-and-down pattern of points in Figure 2.21.

Note that $u(0) < 5$, $u(1) > 5$, $u(2) < 5$, and so on. This is analogous to a swing that on each swing only goes 80% as far as the previous swing but goes from side to side. ■

We will see in Chapter 3 that when the ratios of differences from equilibrium are constant, then the dependent variable can be written explicitly using an exponential function of the form

$$u(n) = ca^n + b$$

where a, b, and c are all constants that are unique to the dynamical system and the initial value.

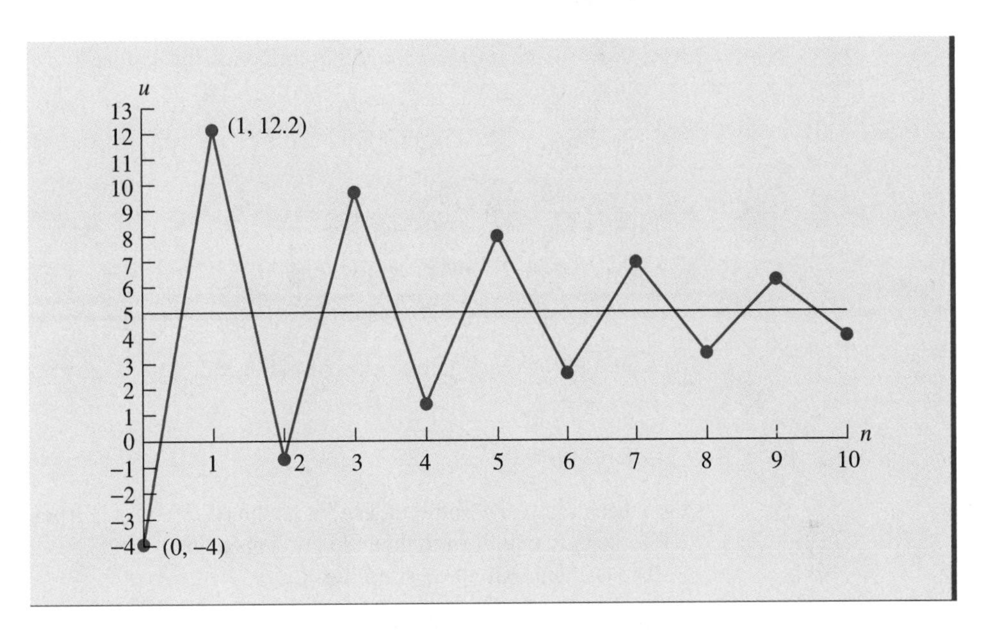

FIGURE 2.21 Points $(n, u(n))$ for system $u(n) = -0.8u(n-1) + 9$, along with line $u = 5$.

EXPONENTIAL PATTERN If ratios of the differences of consecutive values of the dependent variable from the equilibrium value E are constant, then the dependent variable grows or decays exponentially.

You may have observed that the ratios in Examples 2.10, 2.11, and 2.12 equaled the coefficient of the $u(n-1)$-term in the dynamical system. This is not usually true when considering a dynamical system of 2 equations. In fact, for dynamical systems of 2 equations, the ratios are not constant, but they often become approximately constant as n increases.

Example 2.13

Recall the dynamical system

$$u(n) = 0.5u(n-1) + 0.3v(n-1) + 50$$
$$v(n) = 0.4u(n-1) + 0.55v(n-1) + 23$$

which modeled the buildup of 2 chemicals in the body. The globally stable equilibrium point is $(E, F) = (280, 300)$. We now want to determine the rate at which $u(n)$ approaches 280. We do this the same way we did in the previous examples, by computing

$$\frac{u(n) - 280}{u(n-1) - 280}$$

for several different values of n. We are interested in finding how far each value is from 280.

I let $u(0) = 50$ and $v(0) = 23$, and got the following results.

$$\frac{u(1) - 280}{u(0) - 280} = \frac{81.9 - 280}{50 - 280} = 0.861$$

$$\frac{u(9) - 280}{u(8) - 280} = \frac{213.8 - 280}{204.1 - 280} = 0.872$$

$$\frac{u(24) - 280}{u(23) - 280} = \frac{271.4709 - 280}{270.2224 - 280} = 0.873$$

The values for $u(n)$ are getting closer to 280, and

$$\frac{u(n) - 280}{u(n-1) - 280} \approx 0.873$$

This means that each value of $u(n)$ is about 87.3% as far from equilibrium or is about 12.7% closer to equilibrium than the previous value.

The last approximation simplifies to

$$u(n) \approx 0.873u(n-1) + 35.56$$

This means that although $u(n)$ satisfies a dynamical system of 2 equations, it behaves like functions that satisfy the much simpler dynamical system

$$u(n) = 0.873u(n-1) + 35.56$$

as seen in Figure 2.22.

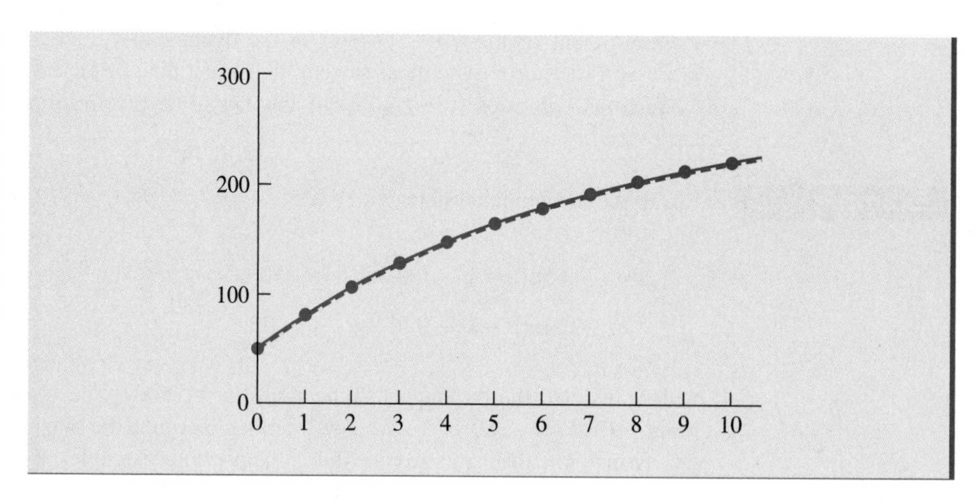

FIGURE 2.22 The solid curve gives the actual values for $u(n)$ while the dotted curve gives the values resulting from $u(n) = 0.873u(n-1) + 35.56$, both with $u(0) = 50$.

You should repeat this process for the $v(n)$-values. In this case, you will find that

$$\frac{v(n) - 300}{v(n-1) - 300} \approx 0.873$$

when the n-values are moderately large. So the $v(n)$-values also get about 12.7% closer to equilibrium each day. This means that $v(n)$ behaves like functions that satisfy

$$v(n) = 0.873v(n-1) + 38.1$$

■

Approximately Exponential Patterns

If ratios of the differences of consecutive values of the dependent variable from the equilibrium value E approach a constant as n increases, then the dependent variable can be approximated with an exponential function.

If such patterns exist, then the functions that satisfy the dynamical system of two equations behave like the simpler functions that satisfy one simple dynamical system.

REMARK Sometimes, the ratio you obtain differs depending on the initial values used. This means that you should compute the ratios using several different choices of initial values. When this happens, we are interested only in the largest ratio.

Suppose that ratios approach a constant. If the equilibrium value is unstable, then the constant ratio (or if the ratio is negative, the absolute value of the rate) will be greater than 1; that is,

$$\left|\frac{u(n)-E}{u(n-1)-E}\right| = 1+r$$

If the equilibrium value is stable, the absolute value of the constant ratio will be less than 1; that is,

$$\left|\frac{u(n)-E}{u(n-1)-E}\right| = 1-r$$

This can be summarized any of the following ways: Each value is $100(1 \pm r)\%$ as far from equilibrium as the previous value, or each value is $100r\%$ further from (when unstable) or closer to (when stable) equilibrium than the previous value. The number r is the **rate** of growth or decay. When the ratio is negative, we can also say the values are oscillating.

A special case of the Exponential Pattern is when the equilibrium value is 0, $E = 0$. In this case

$$\left|\frac{u(n)-E}{u(n-1)-E}\right| = \left|\frac{u(n)}{u(n-1)}\right|$$

so ratios of consecutive values are constant. If the constant ratio is greater than 1,

$$\left|\frac{u(n)}{u(n-1)}\right| \approx 1+r$$

then each value is $1+r$ times the previous value, or each value is $100r\%$ larger than the previous value. If the constant ratio is less than 1,

$$\left|\frac{u(n)}{u(n-1)}\right| = 1-r$$

then each value is $1 - r$ times the previous value, or each value is $100r\%$ less than the previous value.

SUMMARY Suppose you have an affine dynamical system. To analyze the dynamical system, apply the following five steps.

1. Find equilibrium values or points, if any.
2. Determine the stability of any equilibria.
3. Determine how the dependent variables are changing by looking for approximately constant differences $u(n) - u(n-1)$ or constant ratios

$$\frac{u(n) - E}{u(n-1) - E}$$

for increasingly large values of n until a pattern emerges. You should use more than 1 set of initial values.

4. Interpret your result. For example, if you find that

$$\left| \frac{u(n) - E}{u(n-1) - E} \right| \approx 1 \pm r$$

then $u(n)$ is getting $100r\%$ closer to or further away from equilibrium each time period. (If you get more than 1 ratio, choose the largest one.)

5. If the constant ratio is negative, note that the values are oscillating.

I leave you with a final word of warning. Sometimes when computing

$$\frac{u(n) - E}{u(n-1) - E}$$

for increasingly large values of n, no pattern emerges. This is another reason why it is important to compute these ratios for a variety of n-values. When no 1 ratio can be found, no conclusion can be made. We will deal with these situations in more detail in Section 2.6.

Let's apply these techniques to study the growth in the population size of American Bison. When studying the population growth for some species, it is often helpful to break the species down into age groups that have similar survival and fertility rates. American Bison can be broken into 3 distinct age groups. Those under 1 year old are called calves, those between 1 and 2 years old are called yearlings, and those over 2 years old are called adults. Only adults can reproduce. Studies indicate that about 60% of the calves survive to become yearlings. About 75% of yearlings survive to become adults, and 5% of adults die each year. On the average, about 42 calves are born each year for every 100 adults at the beginning of the year. We then say that the adults have a fertility rate of 0.42 or 42%. Let $c(n)$, $y(n)$, and $a(n)$ represent the number of calves, yearlings, and adults, respectively, at the beginning of the nth year, just after calves have been born. The flow diagram in Figure 2.23 will help us develop a dynamical system of 3 equations.

The first equation is

$$\begin{aligned} c(n) &= c(n-1) - \text{“die”} - \text{“year older”} + \text{“born”} \\ &= c(n-1) - 0.4c(n-1) - 0.6c(n-1) + 0.42a(n-1) \end{aligned}$$

This becomes

$$c(n) = 0.42a(n-1)$$

Another way to think about developing this equation is that to be between 0 and 1 year old at the beginning of year n, you had to be born during year $n-1$, and $0.42a(n-1)$ gives the number of calves born during that year.

FIGURE 2.23 Flow diagram for bison population.

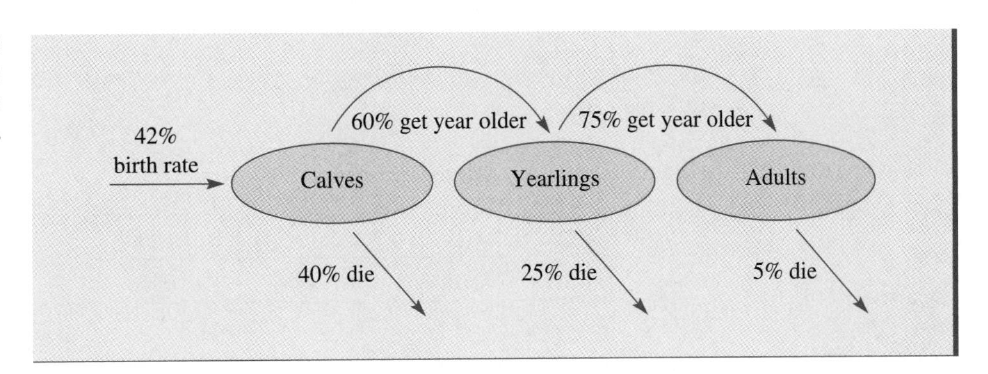

Similarly, we get the equations

$$\begin{aligned} y(n) &= 0.6c(n-1) \\ a(n) &= 0.75y(n-1) + 0.95a(n-1) \end{aligned}$$

The equilibrium point for these 3 equations is $(c, y, a) = (0, 0, 0)$. This makes sense. If there are no bison in year 0, there will be no bison every year. They are extinct.

You should experiment with different positive values for $c(0)$, $y(0)$, and $a(0)$ until you are convinced that all three values go to infinity. The equilibrium point is unstable. I tried $c(0) = 100$, $y(0) = 150$, and $a(0) = 200$, and got the results in Figure 2.24.

Knowing that the bison population is growing is helpful, but it is not very much information. So we will determine the percentage by which the bison population is growing each year. In particular, we want to determine what percentage larger each age group is 1 year compared to the previous year. To estimate the percentage increase in the calves each year, we compute the ratio

$$\frac{c(n)-0}{c(n-1)-0} = \frac{c(n)}{c(n-1)}$$

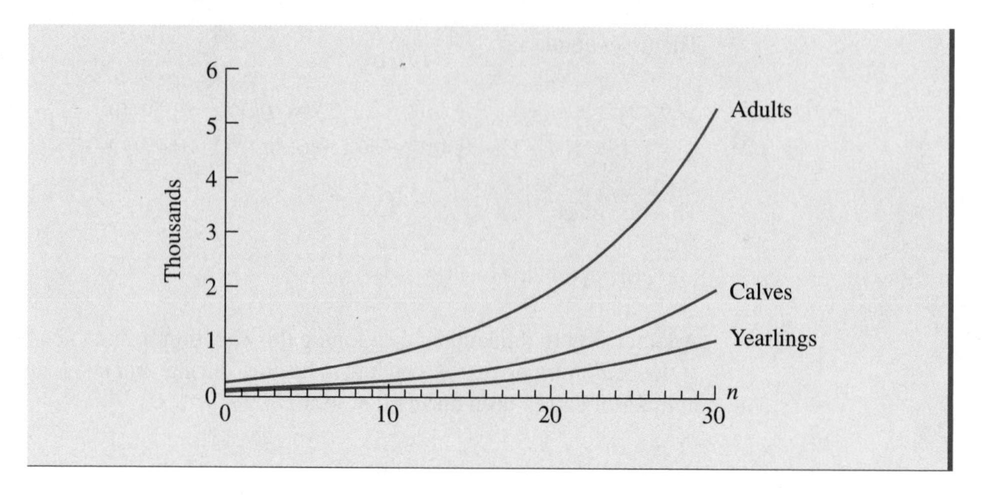

FIGURE 2.24 Population of bison.

for several n values. Letting $c(0) = 100$, $y(0) = 150$, and $a(0) = 200$, I obtained the results in Table 2.7.

TABLE 2.7 Ratios of calf populations in consecutive years.

c(1)/c(0)	c(2)/c(1)	c(10)/c(9)	c(25)/c(24)	c(81)/c(80)
0.84	1.5125	1.104858	1.104834	1.104834

They seem to indicate that

$$\frac{c(n)}{c(n-1)} \approx 1.104834 \quad \text{or} \quad c(n) \approx 1.1c(n-1)$$

for n reasonably large. This means that the size of the population of calves is increasing by about 10% each year. You should try different initial populations and see if it makes a difference in the percentage increase per year for calves.

The fact that $c(n) \approx 1.1c(n-1)$ means that the graph of the function $c(n)$ will be approximately exponential, which appears reasonable from its graph in Figure 2.24. This will be discussed in more detail in Chapter 3.

Observe that the ratios converge to a fixed number, 1.1, which was not one of the numbers in the dynamical system.

You should develop your own table for

$$\frac{y(n)}{y(n-1)} \quad \text{and} \quad \frac{a(n)}{a(n-1)}$$

You may be surprised to discover that the yearlings and the adults are also increasing by about 10% per year. We have now discovered that the size of the population of bison that satisfies the assumptions seen in Figure 2.24 is increasing by about 10% per year, since each age group is increasing by 10% per year.

2.4 Problems

1. Consider the dynamical system

$$u(n) = 0.7u(n-1) + 15$$

The equilibrium value is $E = 50$, which is stable. Pick your own value for $u(0)$, other than 50, and compute the ratios

$$\frac{u(n) - 50}{u(n-1) - 50}$$

for several values of n. What do you observe? Describe how the values are approaching the equilibrium value, using percents.

2. Consider the dynamical system

$$u(n) = 1.07u(n-1) - 7$$

The equilibrium value is $E = 100$, which is unstable. Pick your own value for $u(0)$, other than 100, and compute the ratios

$$\frac{u(n) - 100}{u(n-1) - 100}$$

for several values of n. What do you observe? Using percentages, describe how the values are approaching the equilibrium value.

3. Consider the dynamical system

$$u(n) = 0.85u(n-1) + 60$$

a. Find the equilibrium value.
b. Find the stability of the equilibrium value.
c. Compute ratios of the differences of consecutive values from equilibrium and use them to determine what percentage **closer** to equilibrium each value is than the previous value.

4. Consider the dynamical system

$$u(n) = -0.5u(n-1) + 90$$

a. Find the equilibrium value.
b. Find the stability of the equilibrium value.
c. Compute ratios of the differences of consecutive values from equilibrium and use them to determine what percentage **closer** to equilibrium each value is than the previous value.
d. Why are the ratios negative?

5. Consider the dynamical system

$$u(n) = -2.5u(n-1) + 7$$

a. Find the equilibrium value.
b. Find the stability of the equilibrium value.
c. Compute ratios of the differences of consecutive values from equilibrium and use them to determine what percentage each value is **further** from equilibrium than the previous value.
d. Why are the ratios negative?

6. Consider the dynamical system

$$u(n) = u(n-1) + 0.03$$

Pick your own value for $u(0)$, and use it to determine how the $u(n)$-values for this dynamical system are changing.

7. Consider the dynamical system

$$u(n) = u(n-1) - 1.07$$

Pick your own value for $u(0)$, and use it to determine how the $u(n)$-values for this dynamical system are changing.

8. Consider the dynamical system

$$u(n) = 1.15u(n-1)$$

Pick your own value for $u(0)$, and use it to determine how the $u(n)$-values for this dynamical system are changing.

9. Consider the dynamical system

$$u(n) = 0.3u(n-1)$$

Pick your own value for $u(0)$, and use it to determine how the $u(n)$-values for this dynamical system are changing.

10. Consider the dynamical system

$$u(n) = 1.8u(n-1) + 1.6$$

Pick your own value for $u(0)$, and use it to determine how the $u(n)$-values for this dynamical system are changing.

11. Consider the dynamical system

$$u(n) = -0.97u(n-1) + 29.55$$

Pick your own value for $u(0)$, and use it to determine how the $u(n)$-values for this dynamical system are changing.

12. Suppose Sue presently weighs 110 pounds and burns 130 kcal per pound per week. She consumes 16,000 kcal per week. What percentage closer to her equilibrium weight does she get each week?

13. You deposit some money into a bank account that pays 6% annual interest, compounded monthly. Each month, you withdraw $200 from that account. Develop a dynamical system that models this account.

a. Suppose you initially deposited $30,000 into this account. What percentage further from equilibrium is the amount in the account each month compared to the previous month?

b. Suppose you initially deposited $50,000 into this account. What percentage further from equilibrium is the amount in the account each month compared to the previous month?

c. In parts (a) and (b), you found that each month the account was the same percentage further from equilibrium than the previous month. What is different about what is happening to the amounts in the account in parts (a) and (b)?

14. Recall that a person who smokes one pack of cigarettes a day will absorb about 2.7 mg of cadmium each year from smoking, and that about 7% of this cadmium is removed from the body each year. Develop a dynamical system for $c(n)$, and find the equilibrium value for that dynamical system. Describe how the amount of cadmium approaches the equilibrium amount.

15. Consider the dynamical system

$$u(n) = 0.3u(n-1) + 0.5v(n-1) + 6$$
$$v(n) = 0.7u(n-1) + 0.5v(n-1) - 3$$

a. There is no equilibrium point. The values for $u(n)$ (eventually) change by approximately the same amount. How much is that change?

b. There is no equilibrium point. The values for $v(n)$ (eventually) change by approximately the same amount. How much is that change?

16. Consider the dynamical system

$$u(n) = 1.2u(n-1) + 0.6v(n-1) + 10$$
$$v(n) = -0.2u(n-1) + 0.4v(n-1) + 20$$

a. There is no equilibrium point. The values for $u(n)$ (eventually) change by approximately the same amount. How much is that change?

b. There is no equilibrium point. The values for $v(n)$ (eventually) change by approximately the same amount. How much is that change?

17. Consider the dynamical system

$$u(n) = 1.2u(n-1) + 3v(n-1) + 2$$

$$v(n) = 1.5v(n-1) - 1$$

a. Find the equilibrium point (E, F), and determine its stability.
b. Find the approximate ratio $(u(n) - E)/(u(n-1) - E)$ for a variety of n-values. Does this ratio seem to become constant? What does this say about the $u(n)$-values?
c. Find the approximate ratio $(v(n) - F)/(v(n-1) - F)$ for a variety of n-values. Does this ratio seem to become constant? What does this say about the $v(n)$-values?
d. Find an affine dynamical system (one equation) that is satisfied, approximately, by the function $u(n)$.

18. Consider the dynamical system

$$u(n) = 1.5u(n-1) - 100v(n-1) - 2$$

$$v(n) = 1.2v(n-1) - 3$$

a. Find the equilibrium point (E, F), and determine its stability.
b. The equilibrium point is unstable. Find the approximate ratio $(u(n) - E)/(u(n-1) - E)$ for a variety of n-values. Does this ratio seem to become constant? What does this say about the $u(n)$-values?
c. Find the approximate ratio $(v(n) - F)/(v(n-1) - F)$ for a variety of n values. Does this ratio seem to become constant? What does this say about the $v(n)$-values?
d. Find an affine dynamical system (1 equation) that is satisfied, approximately, by the function $u(n)$.

19. Consider the dynamical system

$$u(n) = 1.1u(n-1) - v(n-1) + 4$$

$$v(n) = 0.8v(n-1) + 1.4$$

a. Find the equilibrium point (E, F), and determine its stability.
b. The equilibrium point is unstable. Find the approximate ratio $(u(n) - E)/(u(n-1) - E)$ for a variety of n-values. Does this ratio seem to become constant? What does this say about the $u(n)$-values?
c. Find the approximate ratio $(v(n) - F)/(v(n-1) - F)$ for a variety of n-values. Does this ratio seem to become constant? What does this say about the $v(n)$-values?
d. Find an affine dynamical system (1 equation) that is satisfied, approximately, by the function $u(n)$.

20. Consider the dynamical system

$$u(n) = 1.2u(n-1) + w(n-1) - 18$$

$$w(n) = -1.1u(n-1) + 0.7w(n-1) + 21$$

a. Find the equilibrium point (E, F), and determine its stability.
b. Find the approximate ratio $(u(n)-E)/(u(n-1)-E)$ for a variety of n-values. Does this ratio seem to become constant? What does this say about the $u(n)$-values?
c. Find the approximate ratio $(w(n)-F)/(w(n-1)-F)$ for a variety of n-values. Does this ratio seem to become constant? What does this say about the $w(n)$-values?

21. Consider the dynamical system

$$u(n) = 0.6u(n-1)+10v(n-1)-2$$

$$v(n) = 0.9v(n-1)+3$$

a. Find the equilibrium point (E, F), and determine its stability.
b. Find the approximate ratio $(u(n)-E)/(u(n-1)-E)$ for a variety of n-values. Does this ratio become constant? What does this say about the $u(n)$-values?
c. Find the approximate ratio $(v(n)-F)/(v(n-1)-F)$ for a variety of n-values. Does this ratio become constant? What does this say about the $v(n)$-values?
d. Find an affine dynamical system (1 equation) that is satisfied, approximately, by the function $u(n)$.

22. Consider the dynamical system

$$u(n) = 0.8u(n-1)+v(n-1)+9$$

$$v(n) = 0.7v(n-1)-6$$

a. Find the equilibrium point (E,F), and determine its stability.
b. Find the approximate ratio $(u(n)-E)/(u(n-1)-E)$ for a variety of n-values. Does this ratio become constant? What does this say about the $u(n)$-values?
c. Find the approximate ratio $(v(n)-F)/(v(n-1)-F)$ for a variety of n-values. Does this ratio become constant? What does this say about the $v(n)$-values?
d. Find an affine dynamical system (1 equation) that is satisfied, approximately, by the function $u(n)$.

23. Consider the dynamical system

$$u(n) = 0.5u(n-1)-1.2v(n-1)+2.9$$

$$v(n) = 0.5u(n-1)+0.6v(n-1)-0.3$$

a. Find the equilibrium point (E,F), and determine its stability.
b. Find the approximate ratio $(u(n)-E)/(u(n-1)-E)$ for a variety of n-values. Does this ratio become constant? What does this say about the $u(n)$-values?
c. Find the approximate ratio $(v(n)-F)/(v(n-1)-F)$ for a variety of n-values. Does this ratio become constant? What does this say about the $v(n)$-values?

24. Consider the dynamical system

$$u(n) = 0.4u(n-1)+0.3v(n-1)+0.9$$

$$w(n) = 0.5u(n-1)+0.7v(n-1)-0.5$$

a. Find the equilibrium point (E,F), and determine its stability.

b. Find the approximate ratio $(u(n)-E)/(u(n-1)-E)$ for a variety of n-values. Does this ratio become constant? What does this say about the $u(n)$-values?

c. Find the approximate ratio $(v(n)-F)/(v(n-1)-F)$ for a variety of n-values. Does this ratio become constant? What does this say about the $v(n)$-values?

d. Find an affine dynamical system (1 equation) that is satisfied, approximately, by the function $u(n)$.

25. Consider the dynamical system

$$u(n) = 0.3u(n-1)+0.4v(n-1)$$

$$v(n) = 0.7u(n-1)+0.5v(n-1)$$

The equilibrium point is (0, 0), which is stable.

a. Find the approximate ratio $u(n)/u(n-1)$ for a variety of n-values. Does this ratio seem to become constant? What does this say about the $u(n)$-values?

b. Find the approximate ratio $v(n)/v(n-1)$ for a variety of n-values. Does this ratio seem to become constant? What does this say about the $v(n)$-values?

c. Find an affine dynamical system (1 equation) that is satisfied, approximately, by the function $u(n)$.

26. Consider the dynamical system

$$u(n) = 2u(n-1)-0.8v(n-1)+a$$

$$v(n) = 0.5u(n-1)+0.5v(n-1)+b$$

a. Let $a=1$ and $b=3$. The equilibrium point is unstable. Approximate the rate $(u(n)-E)/(u(n-1)-E)$.

b. Let $a=7$ and $b=22$. The equilibrium value is unstable. Approximate the rate $(u(n)-E)/(u(n-1)-E)$.

c. Let $a=-15$ and $b=5$. The equilibrium value is unstable. Approximate the rate $(u(n)-E)/(u(n-1)-E)$.

d. What conclusions do you think you can make? Try other values for a and b and see if your conclusions hold.

2.5 Stable Distributions

In the previous sections, we have learned to find the equilibrium for a dynamical system, to determine its stability, and to approximate the rate at which the dependent variables are growing or approaching equilibrium. In this section, we are going to consider cases in which the dependent variables, say u and v, go toward infinity or toward zero as n increases. In particular, we are interested in the fraction each dependent variable is of the total for different values of n, that is,

$$\frac{u(n)}{u(n)+v(n)} \quad \text{and} \quad \frac{v(n)}{u(n)+v(n)}$$

Let's recall the dynamical system of 3 equations that model the growth of the American Bison,

$$
\begin{aligned}
c(n) &= 0.42a(n-1) \\
y(n) &= 0.6c(n-1) \\
a(n) &= 0.75y(n-1) + 0.95a(n-1)
\end{aligned}
$$

The equilibrium point for these 3 equations is $(c, y, a) = (0, 0, 0)$, which is unstable. By computing ratios for different values of n, we found that

$$\frac{c(n)}{c(n-1)} \approx 1.1$$

for "large" values of n. Similarly,

$$\frac{y(n)}{y(n-1)} \approx 1.1 \quad \text{and} \quad \frac{a(n)}{a(n-1)} \approx 1.1$$

This meant that each age group is increasing by 10% per year, and so the size of the bison population is increasing by 10% per year.

We want to find out even more about our bison population. Using the initial values $c(0) = 100$, $y(0) = 150$, and $a(0) = 200$, I found that after 10 years, there will be about $c(10) \approx 273$ calves, $y(10) \approx 148$ yearlings, and $a(10) \approx 717$ adults, for a total population of $273 + 148 + 717 = 1138$ bison.

In this section, we want to compute what fraction of the total is in each age group. The fraction of the total population that are calves is

$$\frac{c(10)}{c(10)+y(10)+a(10)} = \frac{273}{1138} \approx 0.24$$

This means that after 10 years, 24% of the population are calves. Similarly,

$$\frac{y(10)}{c(10)+y(10)+a(10)} = \frac{148}{1138} \approx 0.13$$

and

$$\frac{a(10)}{c(10)+y(10)+a(10)} = \frac{717}{1138} \approx 0.63$$

so about 13% of the population are yearlings and about 63% of the population are adults.

In year 25, I found that $c(25) = 1216$, $y(25) = 660$, and $a(25) = 3198$, so the total population is predicted to be $c(25) + y(25) + a(25) = 5074$. The fractions of the population in each age group are now

$$\frac{c(25)}{c(25)+y(25)+a(25)} = \frac{1216}{5074} \approx 0.24 \quad \frac{y(25)}{5074} \approx 0.13 \quad \text{and} \quad \frac{a(25)}{5074} = 0.63$$

This means that in year 25, 24% of the population will be calves, 13% will be yearlings, and 63% will be adults.

If you try this for other moderately large values of n, you will find the same results; that is,

$$\frac{c(n)}{c(n)+y(n)+a(n)} \approx 0.24 \quad \frac{y(n)}{c(n)+y(n)+a(n)} \approx 0.13 \quad \text{and}$$
$$\frac{a(n)}{c(n)+y(n)+a(n)} \approx 0.63$$

In Figure 2.25 you can see how each age group converges to a fixed fraction of the total population. You should compare Figure 2.25 to Figure 2.24, in which the total size of each age group is exponentially increasing. Try finding the fractions using other initial values for $c(0)$, $y(0)$, and $a(0)$.

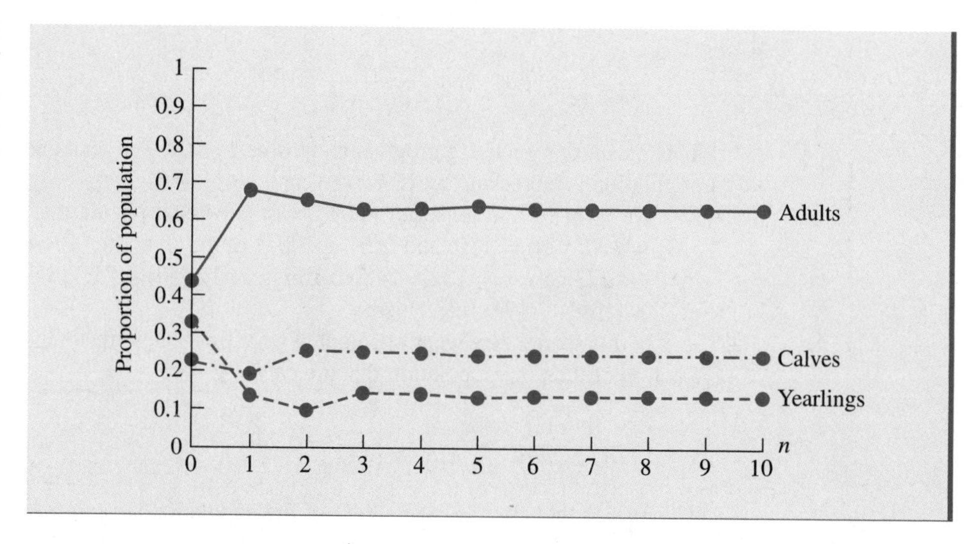

FIGURE 2.25 Proportions of population that are calves, yearlings, and adults in year n.

Even though the size of the population is exponentially increasing, the proportions of the population in each age group are about the same over time. We say there is a **stable distribution.**

Let me summarize. You can often estimate the ratio at which the dependent variables are growing or decaying by computing the ratios $(u(n)-E)/(u(n-1)-E)$ and $(v(n)-F)/(v(n-1)-F)$ for several reasonably large values of n. This is what we did for the bison to determine that the population was growing by 10% per year. When the $u(n)$-values and the $v(n)$-values are both going to infinity or 0, you should also determine if there is a stable distribution, and if so, estimate what it is. This can be done by computing

$$\frac{u(n)}{u(n)+v(n)} \quad \text{and} \quad \frac{v(n)}{u(n)+v(n)}$$

for several increasingly large values for n. We did this for bison, using 3 variables instead of 2, to determine that there was a stable distribution over time of about 24% calves, 13% yearlings, and 64% adults regardless of the initial distribution of the population.

DEFINITION

For a dynamical system of 2 equations with 2 dependent variables, say u and v, there is a **stable distribution** if

$$\lim_{n\to\infty} \frac{u(n)}{u(n)+v(n)} = p \quad \text{and} \quad \lim_{n\to\infty} \frac{v(n)}{u(n)+v(n)} = q$$

for some numbers p and q, where $p+q=1$. In this case, p and q give the relative amounts of $u(n)$ and $v(n)$, respectively, for large n (See Figure 2.26).

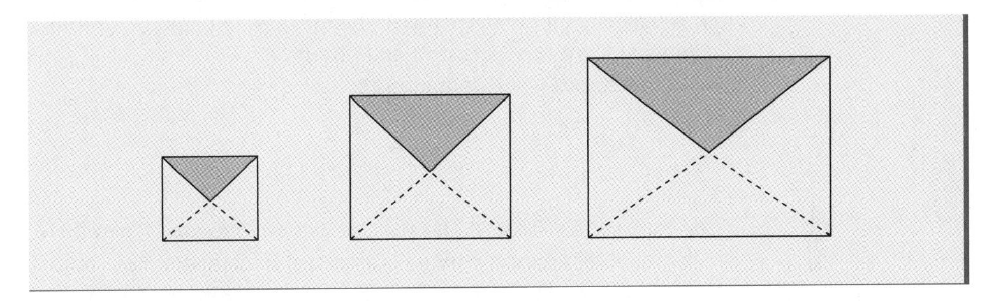

FIGURE 2.26 Stable distribution: Triangle is always one-fourth of area as rectangle grows.

Example 2.14

Consider the affine dynamical system of 2 equations

$$u(n) = u(n-1) + 0.5v(n-1) - 3$$
$$v(n) = u(n-1) + 1.5v(n-1) - 5$$

You should be able to find that the equilibrium point is $(2, 6)$. You should also be able to find that the rate of growth for $u(n)$ and for $v(n)$ is approximately 2 by computing

$$\frac{u(n)-2}{u(n-1)-2} \quad \text{and} \quad \frac{v(n)-6}{v(n-1)-6}$$

for a variety of n-values, using your own choices for the initial values. (For some exceptional choices of initial values, you may get a ratio of 0.5 instead of 2. Don't worry about this.) This means each value is approximately twice the previous value or that each value is about 100% bigger than the previous value.

Now that we know both $u(n)$ and $v(n)$ are growing at the same rate, we try to find if there is a stable distribution. My results are seen in Table 2.8, with $u(0) = 15$ and $v(0) = 20$.

TABLE 2.8 **Proportion of total that is $u(n)$ and $v(n)$ for different n-values.**

$n =$	0	5	20	50
$u(n) =$	15	290.125	9,437,186	1.01331×10^{16}
$v(n) =$	20	581.875	18,874,374	2.02662×10^{16}
$u(n) + v(n)$	35	872	28,311,560	3.03993×10^{16}
$u(n)/[u(n) + v(n)] =$	0.4286	0.3327	0.3333	0.3333
$v(n)/[u(n) + v(n)] =$	0.5724	0.6673	0.6667	0.6667

It appears there is a stable distribution, with $u(n)$ being about 1/3 of the total and $v(n)$ being about 2/3 of the total once n becomes sufficiently large. Try other initial values and see if you get the same results. ■

To get the correct distribution, you should only use initial values that you used to get the correct (largest) rate. When looking at $u(n)/(u(n)+v(n))$, you should try a variety of progressively larger n-values until a pattern emerges. For "small" n-values, you may not get the eventual distribution. If the n-values are too large, you may get an overload error on your computer or calculator. There is no rule of thumb for what is too small or too large for n. It depends on the particular dynamical system and the initial values you choose. Thus, you must learn to experiment and observe.

Sometimes when computing

$$\frac{u(n)}{u(n)+v(n)}$$

for increasingly large values of n, no pattern emerges. Then there is no stable distribution. This is another reason why it is important to compute these ratios for a variety of n-values. When no one ratio can be found, no conclusion can be made. We will deal with these situations in more detail in the next section.

When $u(n)$ and $v(n)$ are going toward a stable equilibrium point $(E, F) \neq (0, 0)$, you do not need to look for a stable distribution because the stable distribution will always be

$$\frac{u(n)}{u(n)+v(n)} \approx \frac{E}{E+F} \quad \text{and} \quad \frac{v(n)}{u(n)+v(n)} \approx \frac{F}{E+F}$$

If the equilibrium point is $(0, 0)$ and it is stable, then we try to find a stable distribution just as we did when there was an unstable equilibrium point.

Example 2.15 In nature, 1 radioactive material often decays into another radioactive material, a process called a radioactive chain. Suppose that each day, 8% of material U decays into material V and 15% of material V decays into lead. This can be modeled with the dynamical system of 2 equations

$$u(n) = 0.92u(n-1)$$
$$v(n) = 0.08u(n-1) + 0.85v(n-1)$$

where $u(n)$ and $v(n)$ are the number of grams of materials U and V left after n days. The equilibrium value is $(E, F) = (0, 0)$, which is stable. This means that over a long enough time period, the amounts of U and V are decreasing to 0. I found that the rates at which they are decaying are

$$u(n)/u(n-1) \approx 0.92$$
$$v(n)/v(n-1) \approx 0.92$$

(If the initial amount of U is 0, then $v(n)/v(n-1) \approx 0.85$, but in this case, there is no radioactive chain.)

Using $u(0) = 10$ and $v(0) = 10$, I found that

$$\frac{u(1)}{u(1)+v(1)} = 0.4973 \quad \frac{u(10)}{u(10)+v(10)} = 0.4812$$

$$\frac{u(50)}{u(50)+v(50)} = 0.4673 \quad \frac{u(1)}{u(1)+v(1)} = 0.4667$$

It appears that $u(n)/(u(n)+v(n)) \approx 0.47$ for "large" n-values. This means that $v(n)/(u(n)+v(n)) \approx 0.53$. So material U is about 47% of the total and B is about 53% of the total.

In real life, some radioactive materials behave in this manner. This means that over time the 2 have become close to the stable distribution. Therefore, if we know the amount present of one material, we can estimate the amount of the other. Radon gas and uranium form a chain similar to the materials in this problem. Even though radon has a short half-life, it is still present in significant quantities because it exists as a proportion of the uranium in the ground. This is why radon gas can be a problem in many areas, seeping into homes from underground. ■

2.5 Problems

1. Consider the dynamical system

$$u(n) = 1.2u(n-1) + 3v(n-1) + 2$$

$$v(n) = 1.5v(n-1) - 1$$

The equilibrium point is unstable. Is there a stable distribution; that is, do $u(n)$ and $v(n)$ each become a fixed proportion of the total? If so, what is the stable distribution?

2. Consider the dynamical system

$$u(n) = 1.5u(n-1) - 100v(n-1) - 2$$

$$v(n) = 1.2v(n-1) - 3$$

The equilibrium point is unstable. Is there a stable distribution; that is, do $u(n)$ and $v(n)$ each become a fixed proportion of the total? If so, what is the stable distribution?

3. Consider the dynamical system

$$u(n) = 1.1u(n-1) - v(n-1) + 4$$

$$v(n) = 0.8v(n-1) + 1.4$$

The equilibrium point is unstable. Is there a stable distribution; that is, do $u(n)$ and $v(n)$ each become a fixed proportion of the total? If so, what is the stable distribution?

4. Consider the dynamical system

$$u(n)=2u(n-1)-v(n-1)+1$$

$$v(n)=0.5u(n-1)+0.5v(n-1)+3$$

There is no equilibrium point. Is there a stable distribution; that is, do $u(n)$ and $v(n)$ each become a fixed proportion of the total? If so, what is the stable distribution?

5. Consider the dynamical system

$$u(n)=1.2u(n-1)+v(n-1)-18$$

$$v(n)=-1.1u(n-1)+0.7v(n-1)+21$$

The equilibrium point is unstable. Is there a stable distribution; that is, do $u(n)$ and $v(n)$ each become a fixed proportion of the total? If so, what is the stable distribution?

6. Consider the dynamical system

$$u(n)=0.3u(n-1)+0.4v(n-1)$$

$$v(n)=0.7u(n-1)+0.5v(n-1)$$

The equilibrium point is $(0, 0)$, which is stable. Although the total is getting small, is there a stable distribution; that is, do $u(n)$ and $v(n)$ each become a fixed proportion of the total? If so, what is the stable distribution?

7. Recall that Vitamin A is stored primarily in our plasma and our liver. Suppose that 40% of the vitamin A in the plasma is filtered out by the kidneys each day and that 30% of the vitamin A in the plasma is absorbed into the liver each day. Also assume that 1% of the vitamin A in the liver is absorbed back into the plasma each day. Suppose you stop ingesting vitamin A. The flow diagram is similar to the flow diagram in Problem 25 Section 2.2.

a. Determine equations for $p(n)$ and $l(n)$, the number of milligrams of vitamin A in the plasma and liver, respectively, on day n in terms of $p(n-1)$ and $l(n-1)$, the amount of vitamin A in the plasma and liver, respectively, on day $n-1$.

b. The equilibrium amounts of vitamin A in the plasma and liver are $(0, 0)$. Find the rate at which $p(n)$ and $l(n)$ decrease to zero.

c. Find the stable distribution for $p(n)$ and $l(n)$. When choosing the initial amounts, recall that the amount of vitamin A in the liver is normally much greater than the amount of vitamin A in the plasma. The point behind this is that when a person has a lack of vitamin A in his or her diet, it may not be apparent from a blood test, since the amount of vitamin A in the plasma will be a certain percentage of the vitamin A in the liver. Until the vitamin A in the liver decreases to a dangerously low level, the vitamin A in the plasma may appear nearly normal.

8. Consider the dynamical system

$$u(n)=2u(n-1)-0.8v(n-1)+a$$

$$v(n)=0.5u(n-1)+0.5v(n-1)+b$$

The equilibrium is unstable for every choice of a and b.

a. Let $a = 1$ and $b = 3$. Approximate the stable distribution.

b. Let $a = 7$ and $b = 22$. Approximate the stable distribution.

c. Let $a = -15$ and $b = 5$. Approximate the stable distribution.

d. What conclusions do you think you can make? Try other values for a and b and see if your conclusions hold.

2.6 Cycles

Sometimes the dependent variables for a dynamical system do not approach constant differences or constant ratios, but oscillate instead. In this section we will examine oscillations and discuss what they mean.

Example 2.16 Consider the dynamical system

$$u(n) = -0.8u(n-1) + 3.6$$

The equilibrium value is $E = 2$, which is stable. A time graph for $u(n)$ is given in Figure 2.27 with $u(0) = 10$.

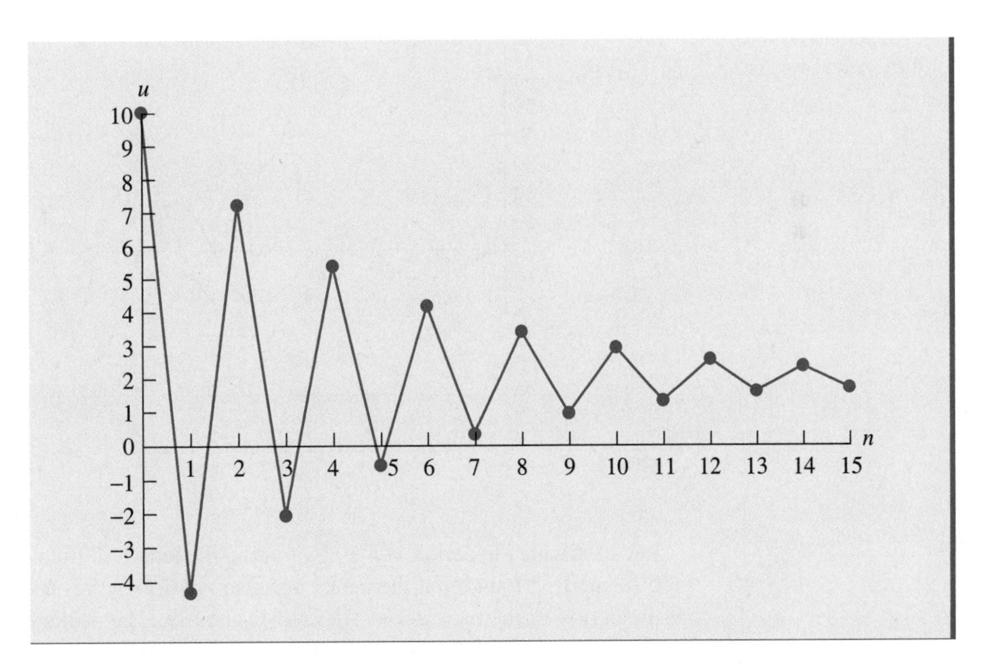

FIGURE 2.27 Points $(n, u(n))$ oscillate.

When the dependent variable oscillates, as in Figure 2.27, we should try to determine the period of the oscillation, which is the time between peaks of the oscillations. In this case that is relatively easy. The peaks occur at times $n = 2$, $n = 4$, $n = 6$, and so on. The distance between these values is always 2, so we say that $u(n)$ oscillates with a period of 2. ■

DEFINITION

The **period** of an oscillation is the length of time it takes to make 1 complete oscillation.

The period is not always an integer number, so you should approximate it as closely as you can. This can be done by looking at peaks on a time graph or, in the case of a dynamical system of 2 equations, looking at spirals formed by plotting one dependent variable versus another.

Example 2.17 Consider the dynamical system

$$u(n) = 0.8u(n-1) + 0.7v(n-1) + 3$$

$$v(n) = -0.7u(n-1) + 0.9v(n-1) + 0.3$$

The equilibrium point is $(1, -4)$, which is unstable. Pick your own values for $u(0)$ and $v(0)$ and see if you get similar results to what I obtained using $u(0) = 2$ and $v(0) = 0$. My results can be seen in the oscillations in the time graph in Figure 2.28.

FIGURE 2.28 Oscillations for dynamical system

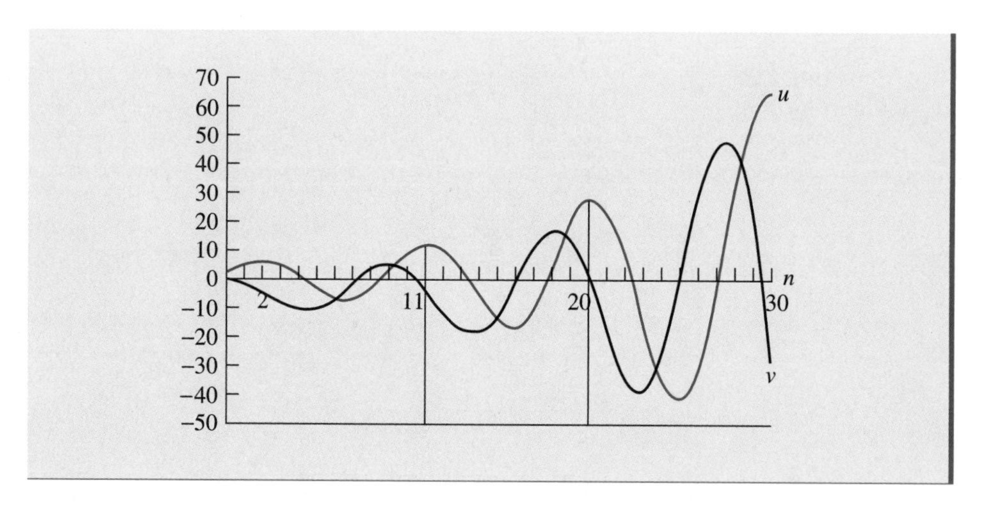

Let's estimate the period of $u(n)$ by finding the length of time between the peaks of the graph $(n, u(n))$. I found that the peaks occur at $n = 2$, $n = 11$, and $n = 20$. This indicates that the period of the oscillations for $u(n)$ is about 9. The peaks for $v(n)$ are also about 9 units apart, so $v(n)$ also has a period of about 9. The period is actually a little larger than 9, so sometimes the peaks will be 10 units apart. To get a better estimate for the period, we should determine the time between several peaks and then average those times.

When the dependent variable is oscillating, then the points $(u(n), v(n))$ spiral, as seen in Figure 2.29. The points $(u(n), v(n))$ are spiraling away from equilibrium, verifying that the equilibrium point $(1, -4)$ is unstable.

FIGURE 2.29 Cycles formed by points $(u(n), v(n))$.

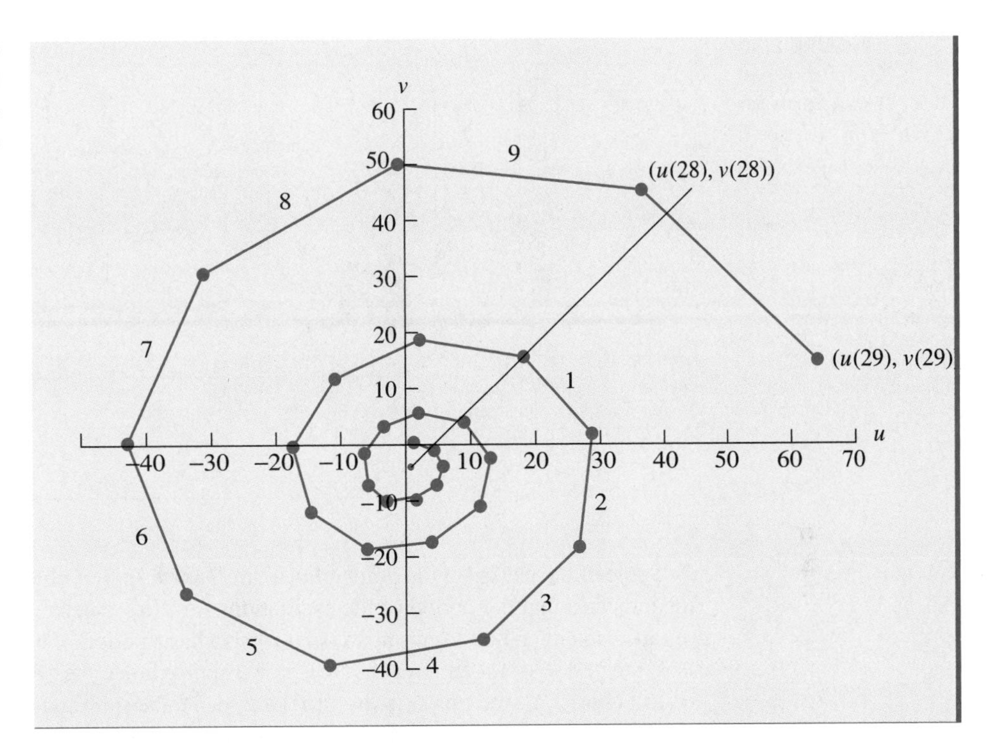

When the dependent variables are oscillating, we can estimate the period by counting the number of line segments it takes for the dependent variables to make one complete loop in the spiral.

In Figure 2.29, the points $(u(n), v(n))$ spiral outward from the point $(1, -4)$. It appears that 1 "loop" of the spiral consists of a little more than 9 line segments, which can be seen from the numbered line segments. You should check that the number of line segments is the same no matter which loop you use. We can estimate the period using the number of line segments, about 9.1 in this case. The number of line segments that form a loop equals the period of the dependent variables.

When considering a dynamical system of two equations, it is often easier to estimate the period using the uv-graph, as we did using Figure 2.29. ■

We could also estimate the period by observing when the largest values of the dependent variable occur in a table of values. This is often more difficult than using the graphs.

Example 2.18 Consider the dynamical system

$$u(n) = 0.6u(n-1) - 0.5v(n-1) + 5.5$$

$$v(n) = 0.68u(n-1) + 0.5v(n-1) - 5.3$$

The equilibrium point is $(10, 3)$. Pick your own values for $u(0)$ and $v(0)$ and see if you get similar results to mine. I used $u(0) = 3$ and $v(0) = 4$ and obtained the time-graph in Figure 2.30, which shows that the dependent variables are oscillating.

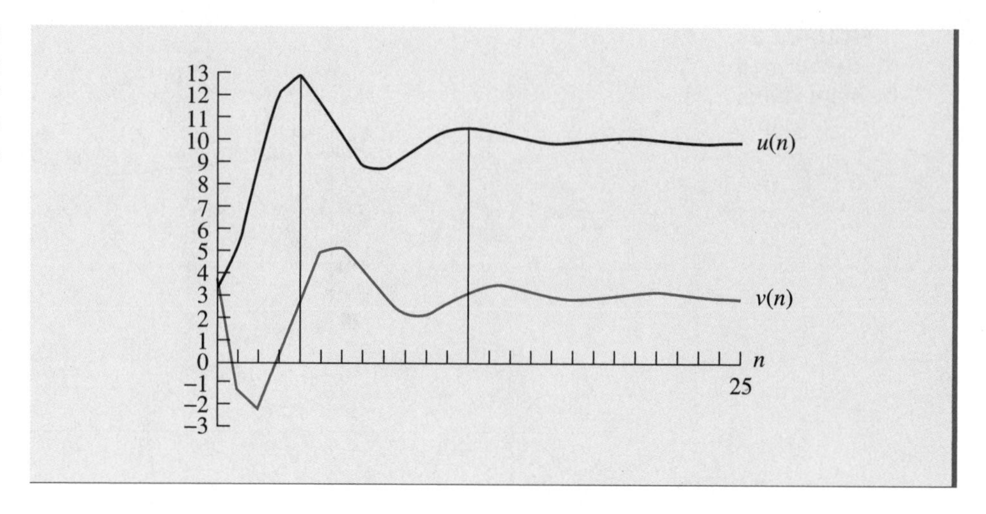

FIGURE 2.30 Oscillations in time-graph for dynamical system in Example 2.18.

Some of the peaks for the graph of $u(n)$ are 7 units apart. Other peaks are 8 units apart. This indicates that the period of the oscillations for $u(n)$ is between 7 or 8. The peaks for $v(n)$ are also either 7 or 8 units apart, so $v(n)$ also has a period of between 7 and 8. In many cases, the period is not an integer value, so we approximate the period as well as we can.

In Figure 2.31, the points $(u(n), v(n))$ are seen to be spiraling closer to the equilibrium point $(10, 3)$, indicating that it is a stable equilibrium point.

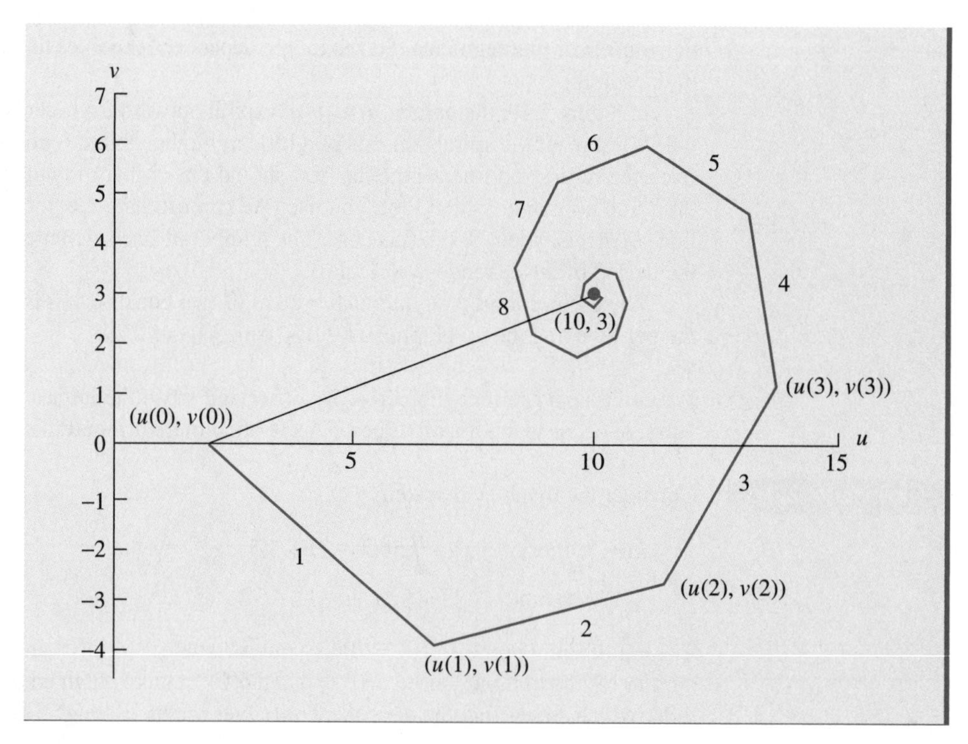

FIGURE 2.31 Spirals of points (u(n), v(n)) toward equilibrium point (10, 3).

Note in Figure 2.31 that it takes about $7\frac{3}{4}$ line segments to complete a cycle, as seen by the line drawn from the equilibrium point to the point $(u(0), v(0))$. This indicates that the period is about 7.75. ∎

Example 2.19 Consider the dynamical system

$$u(n) = 3u(n-1) - 2v(n-1) - 10$$
$$v(n) = 5u(n-1) - 3v(n-1) - 27$$

The equilibrium point is $(7, 2)$. Using $u(0) = 4$ and $v(0) = -2$, I obtained the time graph in Figure 2.32. In this case, the oscillations don't appear to be getting larger or smaller. The approximate size of the oscillations remains constant. In this case, the same 4 numbers keep repeating for $u(n)$,

$$u(0) = u(4) = u(8) = \cdots = 4, \quad u(1) = u(5) = \cdots = 6$$
$$u(2) = u(6) = \cdots = 10, \text{ and } u(3) = u(7) = \cdots = 8.$$

This means that $u(n)$ has a period of 4. Similarly, $v(n)$ keeps repeating the same 4 numbers

$$v(0) = v(4) = v(8) = \cdots = -2, \quad v(1) = v(5) = \cdots = -1$$
$$v(2) = v(6) = \cdots = 6, \quad \text{and} \quad v(3) = v(7) = \cdots = 5.$$

so $v(n)$ also has a period of 4. You should check that no matter what the initial values, other than the equilibrium point, both $u(n)$ and $v(n)$ repeat the same 4 values.

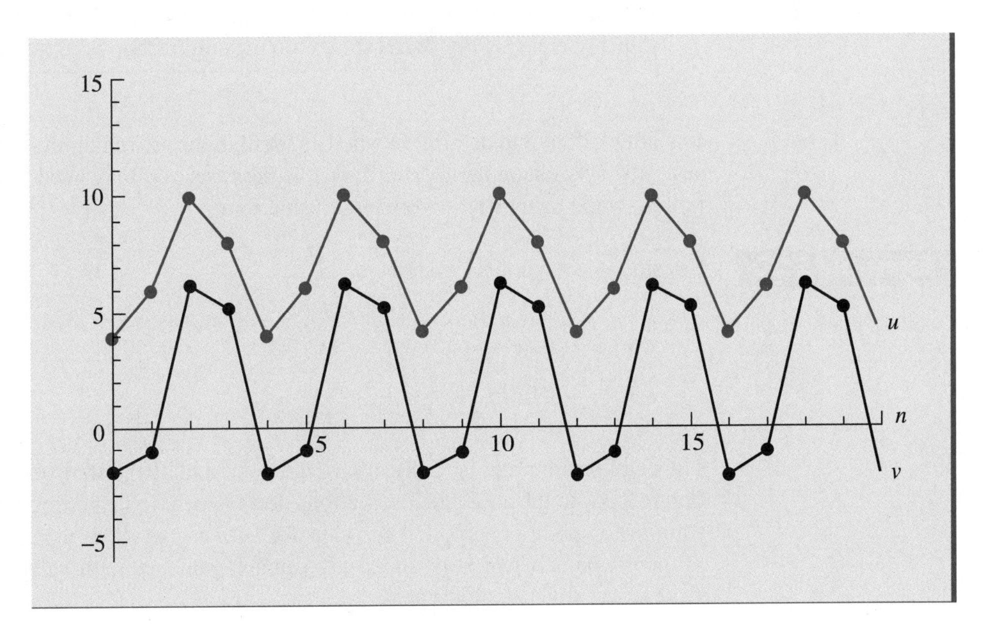

FIGURE 2.32 $u(n)$ and $v(n)$ have a period of 4. The size of the oscillations does not change.

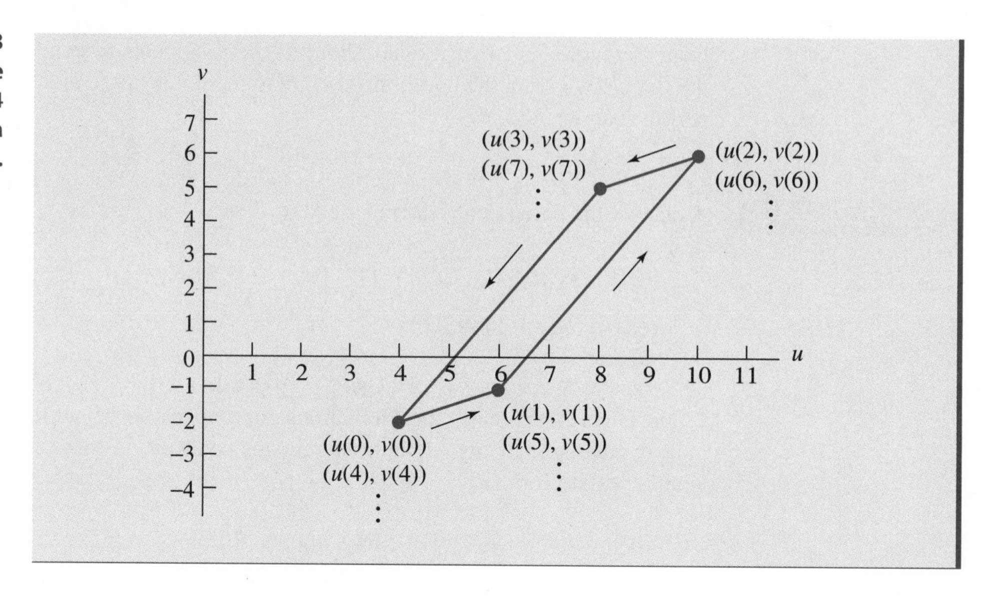

FIGURE 2.33 $(u(n), v(n))$ have a period of 4 and form a parallelogram.

We can also see that the period is 4 from Figure 2.33. In this case, since the points $(u(n), v(n))$ repeat the same values, the points form a parallelogram. ■

DEFINITION

Suppose a dynamical system has an equilibrium value or an equilibrium point. If the values of the dependent variable oscillate, with the amplitude of the oscillations neither decreasing nor increasing, then the equilibrium is **neutral.**

It is often difficult to determine whether oscillations are remaining approximately the same size or slowly changing in size. If such is the case, you may need to generate a time graph or a *uv*-graph using a relatively large value for *n*.

Example 2.20 Consider the dynamical system

$$u(n) = 1.2u(n-1) + 0.4v(n-1) - 4.2$$
$$v(n) = -0.4u(n-1) + 0.7v(n-1) + 4.4$$

The equilibrium point is $(5, 8)$. Using $u(0) = 0$ and $v(0) = 6$, I obtained the time graph in Figure 2.34. In this case, the oscillations don't appear to be getting larger or smaller, so the equilibrium point is neutral. The peaks for $(n, u(n))$ occur at $n = 13$, $n = 33$, and $n = 52$, so the period is between 19 and 20. If you generate a table of values, you will see that the same exact values do not repeat.

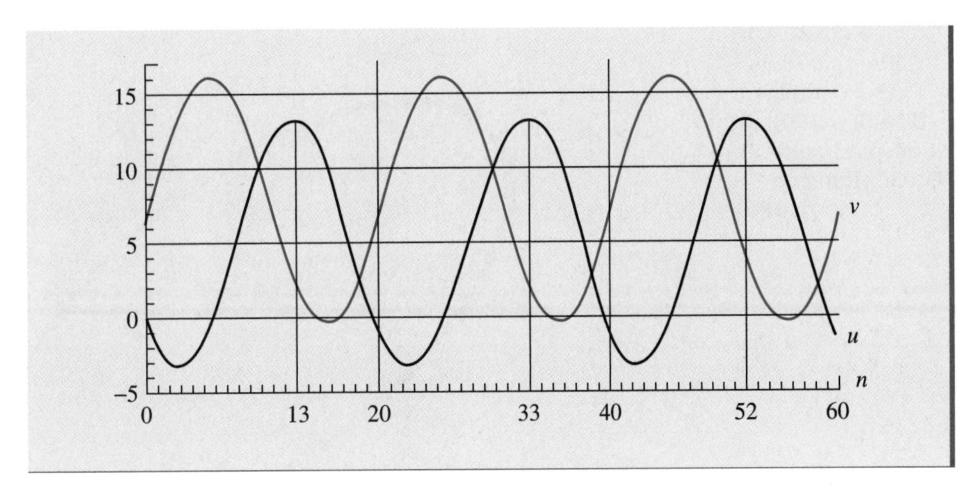

FIGURE 2.34 $u(n)$ and $v(n)$ have a period of between 19 and 20. Equilibrium is neutral.

The points $(u(0), v(0))$ through $(u(20), v(20))$ are plotted in Figure 2.35. From this, we can see that it takes nearly 20 line segments to complete 1 cycle, indicating the period is almost 20.

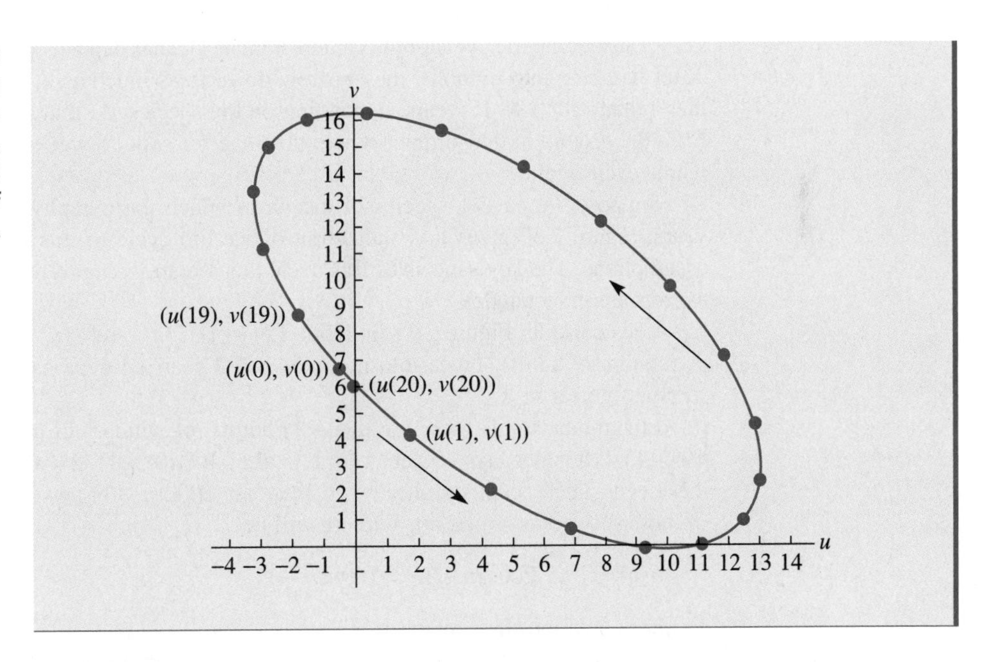

FIGURE 2.35 It takes almost 20 line segments to complete one cycle, so $u(n)$ and $v(n)$ have a period of almost 20.

The points $(u(0), v(0))$ through $(u(100), v(100))$ are plotted in Figure 2.36. The fact that the points are neither spiraling out nor spiraling in indicates that the equilibrium point $(5, 8)$ is neutral. ■

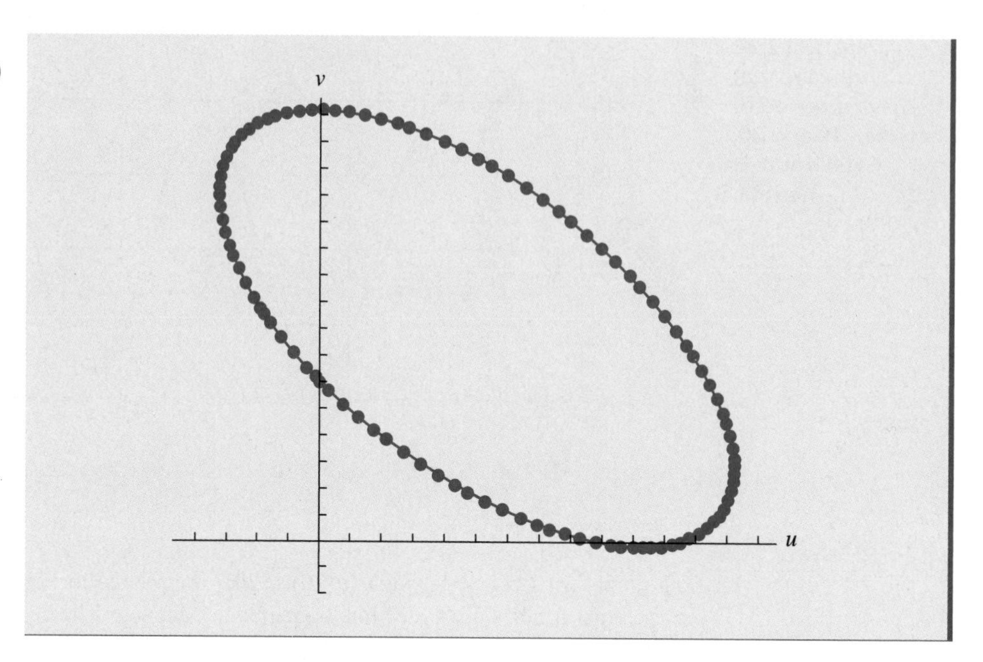

FIGURE 2.36 Points $(u(0), v(0))$ through $(u(100), v(100))$ do not spiral in or out. Equilibrium point is neutral.

Let's see how cycling relates to population models.

Consider the life cycle of the cicada. Female cicadas deposit eggs in twigs or branches. After hatching into nymphs, they burrow down to as much as 6 feet underground where they remain for 1 to 17 years, depending on the species. At that point, the adults emerge from the ground in the summer where they live for about 5 weeks, lay eggs, and start the whole cycle again.

Suppose we have 1 species of cicada in which the nymphs live underground for 3 years. Figure 2.37 gives a visualization of the life cycle of this species, along with our assumptions. The key aspects of this cycle that we are going to model are in ovals. Other aspects are in rectangles.

Let's consider Figure 2.37 in detail. Let $a(n)$, $b(n)$, and $c(n)$ represent the size of the population of adults, 1-year-old nymphs, and 2-year-old nymphs, respectively, all in the summer of year n.

In summer $n-1$, there are $a(n-1)$ adults, of which half are females. Each of the $a(n-1)/2$ females lays 400 eggs for a total of $400a(n-1)/2 = 200a(n-1)$ eggs. All of these eggs hatch, so in summer $n-1$, there are $200a(n-1)$ newly hatched nymphs. Half of these survive to summer n, so there will be

$$b(n) = (0.5)200a(n-1) = 100a(n-1)$$

1-year-old nymphs in summer n.

In summer $n-1$, there are $b(n-1)$ 1-year-old nymphs. Since only 10% survive to the next summer, there will be

$$c(n) = 0.1b(n-1)$$

2-year-old nymphs in summer n.

FIGURE 2.37 Life cycle of cicadas.

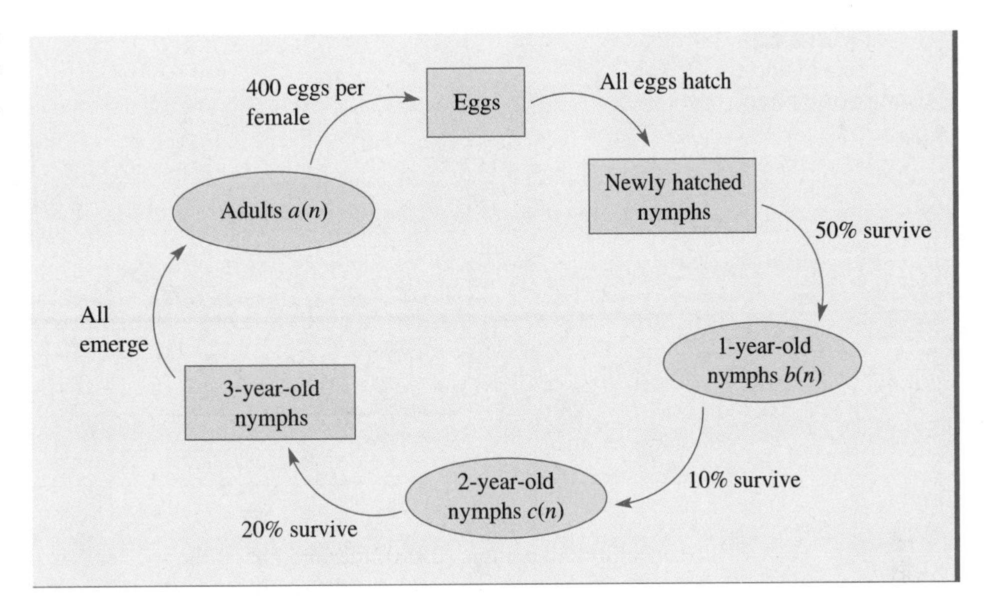

In summer $n-1$, there are $c(n-1)$ 2-year-old nymphs. Since 20% of these survive to the next summer, there will be $0.2c(n-1)$ 3-year-old nymphs in summer n. All of these will emerge from the ground, so there will be

$$a(n) = 0.2c(n-1)$$

adults in summer n.

We summarize the life cycle of this cicada species with the dynamical system consisting of 3 equations

$$a(n) = 0.2c(n-1)$$
$$b(n) = 100a(n-1)$$
$$c(n) = 0.1b(n-1)$$

Let's investigate how the number of each group evolves over time in the particular case where $a(0) = 300, b(0) = 200$, and $c(0) = 100$. A time graph for the adults is seen by plotting the points $(n, a(n))$ in Figure 2.38. From this graph, we can see that $a(n)$ oscillates with a period of 3.

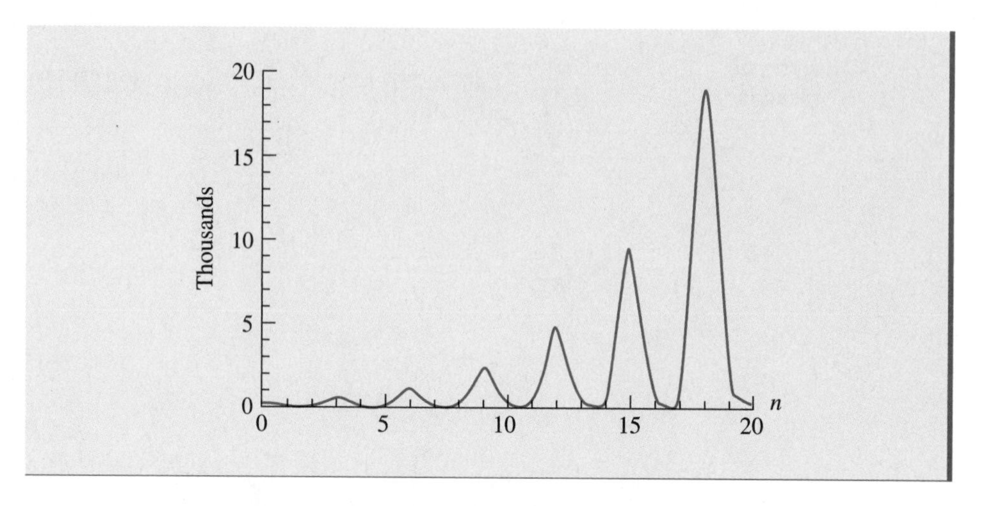

FIGURE 2.38 Size of adult cicada population.

The equilibrium point is $(0,0,0)$ meaning that if the species is extinct, it remains that way. Figure 2.38 indicates that $a(n)$ oscillates with increasing amplitude or size. Similar results hold for $b(n)$ and $c(n)$. If we could plot the points $(a(n), b(n), c(n))$ on a 3-dimensional graph, we would see the points spiraling outward, indicating that $(0,0,0)$ is an unstable equilibrium point. It is easy to believe this since one of the values, $a(n)$, $b(n)$, or $c(n)$ is large for every value on n.

Table 2.9 gives the sizes of each age group of cicada for several years. Note that the 1-year nymphs have a population peak in year 1. Following this, the 2-year nymphs have a population peak the following year, year 2, and the adults have a population peak in year 3. Then in year 4, the process repeats itself, with the 1-year nymphs having another population peak. These population peaks moving through the age groups of the cicada population are called **population waves** and are believed to explain large variations in population sizes for a number of different species. One example of a population wave is the 17-year cicadas. There are few adults of these cicadas most years, but every 17 years, there is a large influx of the adults. Human populations also appear to exhibit population waves for similar reasons, one example being the Baby Boomer population peak moving through the U.S. population age groups over the past 50 years.

TABLE 2.9 Size of adult cicada population over several years.

$n =$	0	1	2	3	4	5	6	7	8
$a(n) =$	300	20	4	600	40	8	1200	80	16
$b(n) =$	200	30,000	2000	400	60,000	4000	800	120,000	8000
$c(n) =$	100	20	3000	200	40	6000	400	80	12,000

To summarize, when a dependent variable oscillates, you should (1) determine the stability of the equilibrium and (2) try to find the period of the oscillation.

2.6 Problems

1. Consider the dynamical system

$$u(n) = -u(n-1) + 8$$

a. Find the equilibrium value for this dynamical system.
b. Find the stability of the equilibrium value.
c. Find the period of the oscillations.

2. Consider the dynamical system

$$u(n) = -1.5u(n-1) + 10$$

a. Find the equilibrium value for this dynamical system.
b. Find the stability of the equilibrium value.
c. Find the period of the oscillations.

3. Consider the dynamical system

$$u(n) = 2u(n-1) - 5v(n-1) + 20$$
$$v(n) = u(n-1) - 2v(n-1) + 10$$

a. Find the equilibrium point for this dynamical system and find its stability.
b. What is the length of a period for $u(n)$? Obtain this result using both a time graph and a uv-graph.

4. Consider the dynamical system

$$u(n) = u(n-1) + 0.5858v(n-1)$$
$$v(n) = -u(n-1) + 0.4142v(n-1)$$

a. Find the equilibrium point for this dynamical system.
b. Determine the stability of the equilibrium point.
c. If $u(n)$ is oscillating, approximate its period using both a time graph and a uv-graph.

5. Consider the dynamical system

$$u(n) = 2u(n-1) - 3v(n-1) + 7$$
$$v(n) = u(n-1) - v(n-1) + 5$$

a. Find the equilibrium point for this dynamical system. Find the stability.
b. If $u(n)$ oscillates, find its period using both a time graph and a uv-graph.

6. Consider the dynamical system

$$u(n) = 0.2u(n-1) + 3.7v(n-1) + 16.3$$
$$v(n) = -0.3u(n-1) + 0.5v(n-1) + 8$$

a. Find the equilibrium point for this dynamical system.
b. Determine the stability of the equilibrium point.
c. Approximate the period of $u(n)$.

7. Consider the dynamical system

$$u(n) = 0.2u(n-1) - 0.7v(n-1) + 7$$
$$v(n) = 0.8u(n-1) + 0.4v(n-1) + 6$$

a. Find the equilibrium point for this dynamical system.
b. Determine the stability of the equilibrium point.
c. Approximate the period of $u(n)$.

8. Consider the dynamical system

$$u(n) = 0.8u(n-1) - 0.5v(n-1) + 2$$
$$v(n) = 0.4u(n-1) + v(n-1) - 2$$

a. Find the equilibrium point for this dynamical system.
b. Determine the stability of the equilibrium point.
c. Approximate the period of $u(n)$.

9. Consider the dynamical system

$$u(n) = 0.5u(n-1) + 0.8v(n-1) + 0.7$$
$$v(n) = -u(n-1) + 0.4v(n-1) + 3.6$$

a. Find the equilibrium point for this dynamical system.
b. Determine the stability of the equilibrium point.
c. Approximate the period of $u(n)$.

10. Consider the dynamical system

$$u(n) = 0.8u(n-1) - 0.5v(n-1) + 5.6$$
$$v(n) = 0.5u(n-1) + 0.7v(n-1) - 1.6$$

a. Find the equilibrium point for this dynamical system.
b. Determine the stability of the equilibrium point.
c. Approximate the period of $u(n)$.

11. Consider the dynamical system

$$u(n) = 1.1u(n-1) + 1.5v(n-1) - 17.2$$
$$v(n) = -0.3u(n-1) + 0.9v(n-1) + 3.2$$

a. Find the equilibrium point for this dynamical system.
b. Determine the stability of the equilibrium point.
c. Approximate the period of $u(n)$.

12. Consider the dynamical system

$$u(n) = 0.8u(n-1) - 0.3v(n-1) + 4.6$$
$$v(n) = 0.3u(n-1) + 0.9v(n-1) + 0.8$$

a. Find the equilibrium point for this dynamical system.
b. Determine the stability of the equilibrium point.
c. Approximate the period of $u(n)$.

13. Consider the dynamical system

$$u(n) = 0.5u(n-1) + v(n-1) + 5$$
$$v(n) = -0.9u(n-1) + 0.2v(n-1) + 9$$

a. Find the equilibrium point for this dynamical system.
b. Determine the stability of the equilibrium point.
c. Approximate the period of $u(n)$.

14. Consider the dynamical system

$$u(n) = 0.1u(n-1) - 0.7v(n-1) + 1.6$$
$$v(n) = 1.4u(n-1) + 0.2v(n-1) - 0.6$$

a. Find the equilibrium point for this dynamical system.
b. Determine the stability of the equilibrium point.
c. Approximate the period of $u(n)$.

15. Consider the dynamical system

$$u(n) = u(n-1) - 0.8v(n-1) - 4$$
$$v(n) = 0.3u(n-1) + 1.2v(n-1) + 0.1$$

a. Find the equilibrium point for this dynamical system.
b. Determine the stability of the equilibrium point.
c. Approximate the period of $u(n)$.

16. Consider the dynamical system

$$u(n) = 0.9u(n-1) - 0.9v(n-1) + 1$$

$$v(n) = 0.5u(n-1) + 1.1v(n-1) - 0.6$$

a. Find the equilibrium point for this dynamical system.
b. Determine the stability of the equilibrium point.
c. Approximate the period of $u(n)$.

17. Consider the dynamical system

$$u(n) = -0.5u(n-1) - 0.8v(n-1) + 3$$

$$v(n) = u(n-1) - 0.4v(n-1) - 2$$

a. Find the equilibrium point for this dynamical system.
b. Determine the stability of the equilibrium point.
c. Approximate the period of $u(n)$.

18. Consider the dynamical system

$$u(n) = 0.5u(n-1) - 0.3v(n-1) + 7$$

$$v(n) = 0.2u(n-1) + 0.8v(n-1)$$

a. Find the equilibrium point for this dynamical system.
b. Determine the stability of the equilibrium point.
c. Approximate the period of $u(n)$.

3 Function Approach

3.1 Introduction to Function Approach

Modeling usually involves making numerous trips through each of the 3 major forms of a situation: 1) the actual real world situation being analyzed, 2) our idealized or theoretical understanding of the situation, and 3) our data driven understanding of the situation. This is visualized in Figure 3.1.

The situations we have been modeling in Chapters 1 and 2 are **theory based.** This means that we made some assumptions about how things change in a particular situation and translated our assumptions into a dynamical system. We then made predictions about the future, using either the dynamical system or the expression we had found. At that point, we should look at real data and compare our predictions to the actual data to see how well they match. We could then refine our model (see Figure 3.1).

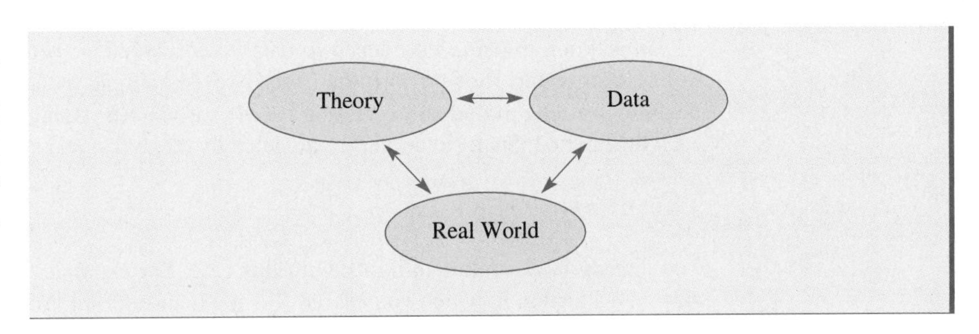

FIGURE 3.1 Visualization of interplay between a situation and our different understandings of the situation.

In this chapter, we will also consider **data-based** modeling. For example, we have a situation in which something is changing, such as a cup of coffee that is cooling over time. In data-based modeling, we will collect data about the situation, the temperature of the cup of coffee after 1, 2, 3, ... minutes, but we won't know what assumptions to make about how the situation is changing (see Figure 3.2). We will try to fit an expression to our data. If we are successful, we will then find a dynamical system that gives the same values as the expression. Finally, we will determine what assumptions would lead to that dynamical system. From this, we can hypothesize that these assumptions are important in determining how this situation is changing. In real work, we would need to consider if these assumptions actually make sense and would design additional experiments to test our hypotheses. We will often skip this last step in this book, since we would need to know a considerable amount about the situation being studied.

FIGURE 3.2 Cooling coffee is an example of data-based modeling.

In addition to being introduced to data-based modeling, you are going to learn how to find expressions for functions that satisfy several simple dynamical systems. You will begin by learning to find an expression for $u(n)$ when $u(n)$ changes by a constant amount each time period. Such situations are modeled by the dynamical system

$$u(n) = u(n-1) + b$$

You already have some familiarity with this case. For example, in Section 1.2 you considered a situation in which you earned \$15 per hour tutoring. You modeled this situation using the dynamical system

$$u(n) = u(n-1) + 15$$

and expressed $u(n)$ explicitly as the function

$$u(n) = 15n$$

Next, you will learn to find an expression for $u(n)$ in the case where each value is a multiple of the previous value. These situations are modeled by the dynamical system

$$u(n) = au(n-1)$$

An example of this situation would be a bank account on which you earned 6% interest, compounded annually. Let n represent the number of years after the initial deposit and $u(n)$ be the amount in this account after n years. Then this situation is modeled by the dynamical system

$$u(n) = 1.06u(n-1)$$

Finally, you will learn to find an expression for $u(n)$ in the general case in which $u(n)$ satisfies the affine dynamical system

$$u(n) = au(n-1) + b$$

An example of this situation would be the bank account that is earning 6% interest, compounded annually, but with the added condition that you deposit an additional \$1000 at the end of each year. This situation is modeled by the dynamical system

$$u(n) = 1.06u(n-1) + 1000$$

Once you know how to find expressions that satisfy certain dynamical systems, you will learn how to compare and contrast theory-based modeling and data-based modeling. In theory-based modeling, you develop a dynamical system and find an expression for $u(n)$ from assumptions about the situation. In data-based modeling, you find an expression that fits actual data relatively well and then determine a dynamical system and corresponding assumptions that would lead to that same expression.

3.2 Linear Functions

Recall that when $u(n)$ was the height of a stack of n cups, you found 2 ways of representing the situation. One method, which was called the dynamic approach, was to express the height of a stack of cups in terms of the height of a stack that had 1 less cup. This led to the equation

$$u(n) = u(n-1) + 0.5$$

with $u(1) = 3.5$ inches. This was called a dynamical system. A second method for representing this situation was finding that

$$u(n) = 3 + 0.5n$$

You knew that this formula worked because a stack of n cups consists of 1 base that is 3 inches tall and n lips that are each 0.5 inches tall. This is an expression for $u(n)$ in terms of n. An advantage of using the expression is that you can compute $u(100)$ directly, instead of computing $u(1), \ldots, u(99)$ first.

In this section, we are going to consider dynamical systems of the form

$$u(n) = u(n-1) + b \qquad (3.1)$$

These dynamical systems can be visualized with the flow diagram in Figure 3.3.

FIGURE 3.3 Flow diagram for $u(n) = u(n-1) + b$.

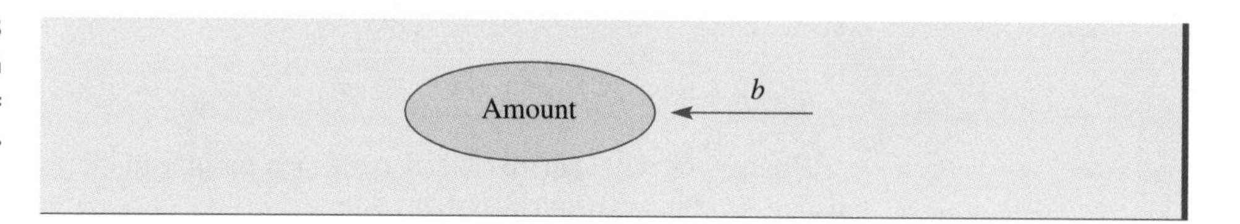

We can also give a written description of this situation.

Written Description

Each time period, add the constant amount b.

We now want to write an expression for $u(n)$ that satisfies the dynamical system (3.1). From the written description, we know that we add b to u each time n increases by 1. $u(n)$ is the amount after n increases. Thus, we must have added b a total of n times. This means we have added a total of bn. So the expression for $u(n)$ must include the term bn. The other part of the expression for $u(n)$ is the amount to which nb was added. Let's denote this amount by c. Thus, the expression for $u(n)$ should be of the form

$$u(n) = bn + c \qquad (3.2)$$

All that is needed at this point is to find the value for c. That can often be accomplished through substitution.

DEFINITION

When an expression gives the same values as the dynamical system for some given initial value, then the expression is said to be the **solution** to the dynamical system with that initial value.

Example 3.1

Suppose that it costs 99 cents to talk long distance for the first 20 minutes. It then costs 8 cents per minute for each minute over 20 minutes. Let n represent the number of minutes talked and $u(n)$ be the total cost of the call, in cents. We know the cost for $n \leq 20$ is 99 cents. So we will only consider values for $n \geq 20$. The dynamical system is

$$u(n) = u(n-1) + 8$$

since it costs 8 cents more to talk 1 minute more. This dynamical system is just (3.1) with $b = 8$. By (3.2), the expression for $u(n)$ is of the form

$$u(n) = 8n + c$$

To determine c, we use the fact that $u(20) = 99$. Substitution of 20 for n in the expression gives

$$u(20) = 8 \times 20 + c = 160 + c$$

Since $u(20) = 99$ and $u(20) = 160 + c$, we set these two expressions equal to each other, giving

$$99 = 160 + c$$

or $c = -61$. Substitution of this value for c gives the expression

$$u(n) = 8n - 61.$$

You should put the dynamical system into your calculator or spreadsheet, along with the initial value of $u(20) = 99$, and observe the values for $u(21)$, $u(22)$, and so forth. You should also insert the expression $u(n) = 8n - 61$ into your calculator or spreadsheet and see that it gives the same values for $u(n)$ when $n = 20, 21, \ldots$. This means that $u(n) = 8n - 61$ is the solution to $u(n) = u(n-1) + 8$ with $u(20) = 99$ and $n \geq 20$.

The expression $u(n) = 8n - 61$ gives $u(0) = -61$, $u(1) = -53$, $u(2) = -45, \ldots, u(19) = 91$, $u(20) = 99$. Note that this linear expression does not give the correct costs for $n < 20$. This is because the assumption of adding 8 cents for each additional minute does not apply when $n < 20$. So we must qualify our function with the statement that the cost is $u(n) = 99$ when $n < 20$ and is $u(n) = 8n - 61$ when $n \geq 20$ (see Figure 3.4). ■

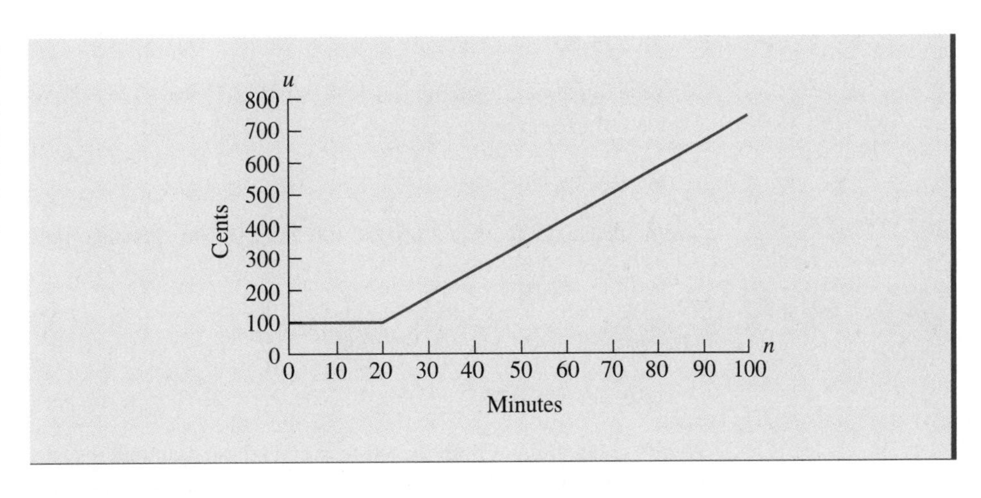

FIGURE 3.4 Graph of function giving cost for talking *n* minutes. Equation $u(n) = 8n - 61$ is only valid for $n \geq 20$.

We summarize our previous discussion.

THEOREM

If $u(n)$ satisfies the dynamical system (3.1) with some given initial value, then $u(n)$ can be expressed as an expression of the form (3.2); that is, $u(n)$ is a linear function.

Figure 3.5 is a visualization of this theorem. Let's consider another example of this result.

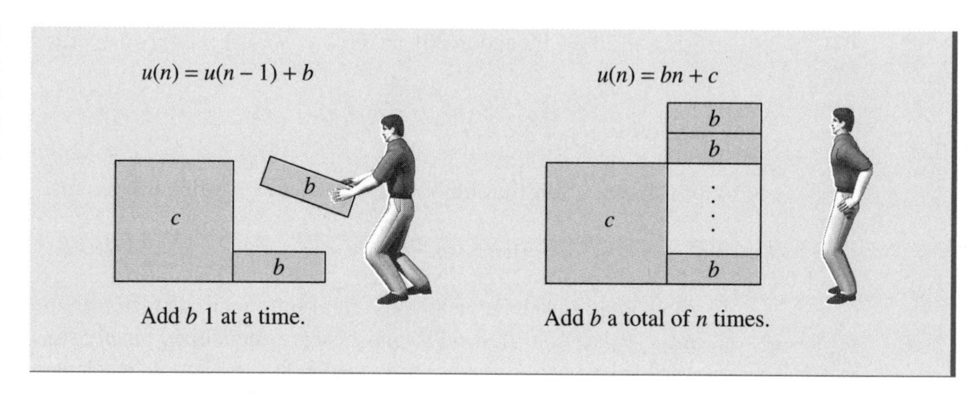

FIGURE 3.5 The dynamic process of adding b 1 at a time and the end result.

Example 3.2

Suppose you have \$400 and are going to pay \$2 each for your favorite trading cards. Let n be the number of cards you purchase and let $u(n)$ represent the amount of money you have left after buying n cards. For each additional card you purchase, you have \$2 less. This translates into the dynamical system

$$u(n) = u(n-1) - 2$$

But this is the same as

$$u(n) = u(n-1) + (-2)$$

Now the dynamical system is in the form of (3.1) with $b = -2$. You can express $u(n)$ using expression (3.2), giving

$$u(n) = (-2)n + c = -2n + c$$

To find c, use the facts that $u(0) = 400$ and that the expression gives

$$u(0) = -2(0) + c = c$$

This gives $c = 400$, and the solution is

$$u(n) = -2n + 400$$

Again, you could put both the expression and the dynamical system into your calculator or spreadsheet to see that they give the same values for the function $u(n)$. ■

From the previous discussion, we know that the following 4 statements are equivalent.

1. $u(n)$ satisfies the dynamical system $u(n) = u(n-1) + b$.
2. $u(n)$ changes by the constant amount b for each unit change in n.
3. $u(n)$ is modeled by the flow diagram in Figure 3.3.
4. $u(n)$ is given by the linear equation $u(n) = bn + c$.

This means that if we have a situation in which we know $u(n)$ changes by a constant amount, then we can model $u(n)$ with the dynamical system $u(n) = u(n-1) + b$ or with the linear function $u(n) = bn + c$. Equally important, if we discover that $u(n)$ can be approximated by a linear function $u(n) = bn + c$, then we can conclude that $u(n)$ changes by a constant amount for each unit change in n and satisfies the dynamical system $u(n) = u(n-1) + b$. This also means that the constant change is the slope of the line.

It should not be surprising that $u(n) = u(n-1) + b$ and $u(n) = bn + c$ are equivalent equations. This is what slope means; that is, the slope of a line b gives the change in the dependent variable for one unit change in the independent variable.

Example 3.3

Suppose you measure the diameter and circumference of several circles. You could use cans, buckets, tops, rolls of paper towels, and so on. Let n be the diameter of the circle and $u(n)$ be the circumference of the circle. Suppose you collected data similar to that in Table 3.1.

TABLE 3.1 **Comparisons of radius and circumference of various circles.**

diameter n in cm	5	7	12.5	17
circumference $u(n)$ in cm	15.7	22.0	39.3	53.4

You might plot your points on a graph, similar to Figure 3.6 and see that these points apparently lie on a line.

You then find the equation of this line using two of the points. If you are accurate in your computations and measurements, you will get that the equation is about

$$u(n) = 3.14n$$

From this equation, you see that the circumference is about 3.14 times the diameter, or circumference equals pi times diameter, which you already knew.

We now know that $u(n)$ satisfies the dynamical system

$$u(n) = u(n-1) + 3.14$$

To summarize, from knowing that $u(n)$ was a linear function of n, we could conclude that $u(n)$ satisfies the dynamical system $u(n) = u(n-1) + 3.14$ and therefore the circumference increases by 3.14 cm for each 1 cm increase in diameter. ■

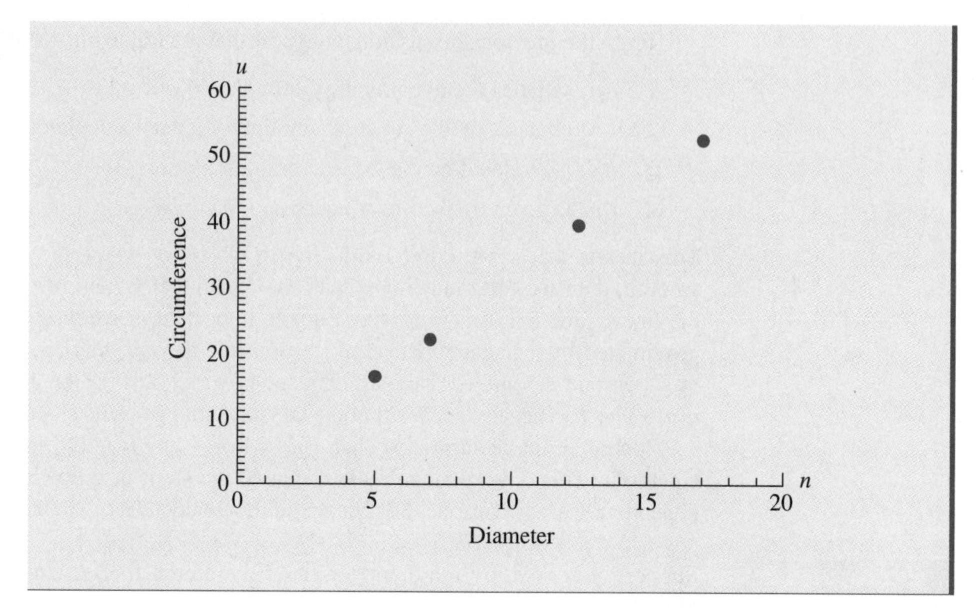

FIGURE 3.6 Diameter versus circumference.

In previous work, we have assumed knowledge about a situation, such as that when we add 1 cup to a stack, we increase the height by 1 lip. From this we were able to develop a linear function to give the height of any number of cups. Suppose instead we are investigating some situation and have little understanding of the process of how something is changing. We can collect data and plot points. If it appears from the data that the amount is growing linearly, then we know something about how this process changes; that is, it is growing by a constant amount. Let's consider an example.

Example 3.4

Dendrochronology is the study of tree rings. The variation in the size of tree rings can be used to date archaeological remains, to study prehistoric climatic changes, and to study forest ecosystems. In this example we want to develop a basic understanding of how a tree grows. We might make any number of assumptions. We could assume that the volume of the tree increases by the same amount each year, or we could assume that as the tree gets larger, it grows faster. Any assumption we would make is just a guess. So let's look at some data to get a better understanding of the process. In particular, let's measure the length of tree rings for some tree.

I had a piece of pine similar to that pictured in Figure 3.7.

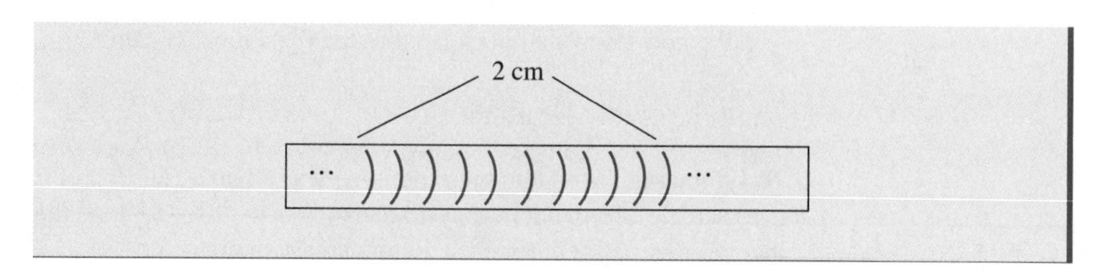

FIGURE 3.7 Tree rings in section of wood.

I found that the thickness of the first 10 rings was 2 cm. The thickness of the first 20 rings was 4.5 cm. Table 3.2 gives all of my results.

TABLE 3.2 **thickness of tree rings.**

number of rings, n	0	10	20	30	40	50	60
thickness in cm, $u(n)$	0	2	4.5	6.3	8.7	11.3	13.8

The points in Table 3.2 are plotted in Figure 3.8.

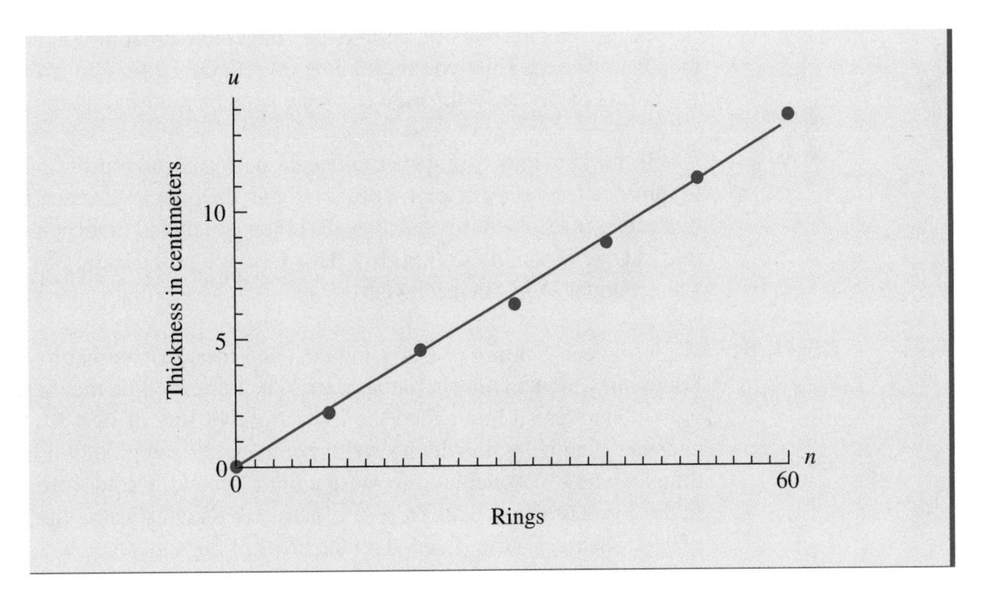

FIGURE 3.8 Line approximating tree rings and thickness.

You can see that these points are nearly in a line. When dealing with real data, we don't expect to get points exactly on a line.

I concluded that $u(n)$ could be approximated reasonably well with a linear function. I drew in a line that I thought approximated the points fairly well. To approximate the slope of this line, I estimated that 2 points on this line are $(0,0)$ and $(60,14)$. I then computed that the slope is

$$\frac{14-0}{60-0} = \frac{7}{30} \approx 0.23$$

cm per ring. This means the line would look like $u(n) = 0.23n + c$. I drew the line so it went through the origin because I figured that if there were $n = 0$ rings, then the length would be $u(0) = 0$. Thus, I approximated $u(n)$ with the line

$$u(n) = 0.23n$$

Because of Theorem 3.1, we can now conclude that each year, as the tree adds a new ring, the thickness of the new ring is about 0.23 cm, that is

$$u(n) = u(n-1) + 0.23$$

This does not say that each ring will be exactly 0.23 cm. In years of drought, a ring may be less than 0.23 cm, while in other years, rings may be more than 0.23 cm thick.

We now have some knowledge about how a tree might grow, that is, increasing its radius linearly by about 0.23 cm per year. We should now do more experiments. Do the radii of all pine trees grow linearly? Is that growth always around 0.23 cm per year? Do other trees grow linearly? I measured rings that were between the center and the outer edge. Do rings grow faster near the center of the tree? Do they grow differently after the tree has reached maturity?

In the lumber industry, people are concerned about how much total volume of wood they can harvest. Thus you might now investigate by how much the volume of the tree is increasing each year if the radius is increasing by a constant amount. ■

In the previous example, I collected data and plotted the points on a graph. Such a graph is called a **scatter plot.** I observed that the points were nearly linear and found a line that represented the points fairly well. I then concluded that each unit change in n resulted in a change of about 0.23 in $u(n)$. This means that the radius of this tree changed by the same amount, 0.23 cm, each year.

REMARK We have been plotting points using data and then approximating those points with a line. Sometimes, we can draw what appears to us to be the line that best fits the points. Usually practitioners use a line called the **least squares line of best fit.** To understand this line, consider Figure 3.9, in which several points have been plotted. First, I drew in a line that I thought was a reasonable fit to these points. Second, I drew vertical lines from each point to the approximating line. Third, I constructed squares, using the vertical lines as one side of each square. Fourth, I added up the areas of the squares.

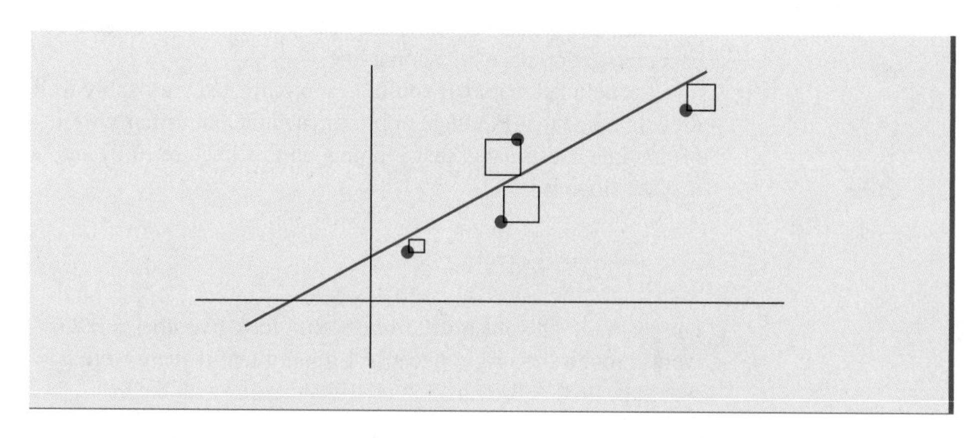

FIGURE 3.9 Data points, approximating line, and squares going from points to line.

In Figure 3.10, I repeated these steps, using a different approximating line.

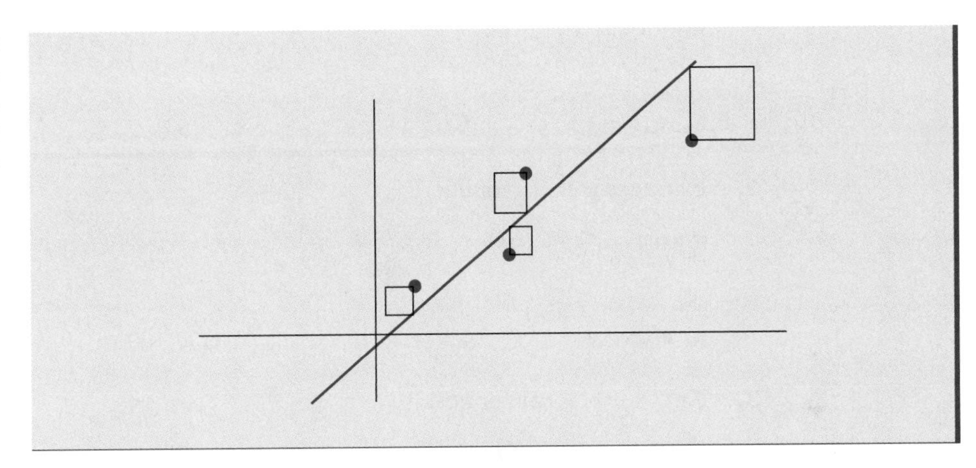

FIGURE 3.10 Data points, approximating line, and squares going from points to line.

The figure in which the sum of the areas of the squares is smallest or "least" is considered to be the best "least squares" fit. The least squares line of best fit is the line, of all possible approximating lines, that results is the least total area for the squares. Calculators and computers will usually find the equation of this line for you. See your instruction manual for details.

Once a line of best fit has been found, you need to decide if it really describes the behavior. Calculators or computers will usually give you a number called the **coefficient of determination.** This number, often denoted by r^2, gives some assessment on how well the linear equation fits the data. The closer this number is to 1, the better the fit. The closer to 0, the worse the fit. But you can't go just by this number. You must actually **look** at the points and line and decide if a linear function truly describes the pattern of the points. Almost any book on statistics will give you more details about this.

For the tree ring data points, the least squares line of best fit is

$$u(n) = 0.23n - 0.22$$

where the numbers are given to 2-decimal-place accuracy. The coefficient of determination is $r^2 = 0.998$ to 3 decimal places, which means it is a pretty good fit. There is a problem with this line in that this formula gives $u(0) = -0.22$, meaning that no tree rings have a length of -0.22 cm instead of 0, which we know it should be. We will ignore such problems in this material. The slope of the least squares line is 0.23, meaning that an increase of 1 cm for n will result in the length of all n rings increasing by 0.23 cm. This agrees with the slope I got by drawing an approximate line. Normally, drawing a line by hand will get you reasonably close to the least squares line.

3.2 Problems

1. Find an expression for $u(n)$ if

$$u(n) = u(n-1) + 0.3$$

and

a. $u(0) = 15$ **b.** $u(1) = 9$ **c.** $u(53) = 17$

2. Find an expression for $u(n)$ if

$$u(n) = u(n-1) - 12$$

and

a. $u(0) = 2$ **b.** $u(1) = 1032$ **c.** $u(100) = 10,348$

3. Find an expression for $u(n)$ if

$$u(n) = u(n-1) + 3/2$$

and

a. $u(0) = -12$ **b.** $u(1) = 0$ **c.** $u(-10) = 123$

4. Find an expression for $u(n)$ if it satisfies the following flow diagram

and

a. $u(0) = 0$ **b.** $u(1) = 35$ **c.** $u(1000) = 0$

5. Find an expression that satisfies the dynamical system

$$u(n) = u(n-1) + b \quad \text{with} \quad u(0) = 7 \quad \text{and} \quad u(100) = 1043$$

What is the value for the parameter b?

6. Find an expression that satisfies the dynamical system

$$u(n) = u(n-1) + b \quad \text{with} \quad u(10) = 15 \quad \text{and} \quad u(16) = 39$$

What is the value for the parameter b?

7. Suppose you presently have $u(0) = \$120$ and a job in which you earn \$10 an hour. Let $u(n)$ represent the amount of money you will have after you have worked n hours.

a. Model this situation with a dynamical system.

b. Find an expression that models this situation.

c. Compute $u(1)$, $u(2)$, and $u(3)$ using the dynamical system and again using the expression. Do the computations by hand, not with a calculator or spreadsheet. Do you get the same results both ways?

8. Suppose a friend has been earning \$15 an hour. After working 80 hours she has \$2000.

a. Model this situation by developing a dynamical system, that is, a relationship involving $u(n)$ and $u(n-1)$.

b. Write down an expression that models this situation.

c. How much money did your friend have before starting to work?

9. You have signed a contract to write a book. You have already written the first 75 pages. Your publisher wants the book to be about 400 pages long. Each day, starting with today, you will write b pages of your book. Let $u(n)$ represent the total number of pages you will have written after n days, with $u(0) = 75$.

a. Write a dynamical system for $u(n)$, using the unknown parameter b in that system.

b. Find an expression that models this situation, using the unknown b. Your publisher wants the book to be finished in 6 weeks, or 42 days (no weekends off for you). Use your expression to determine the approximate number of pages b you need to write each day.

10. The following table gives the average number of miles each car in America was driven in the corresponding years. (This data was obtained from the U.S. Department of Transportation web site.)

year	1987	1988	1989	1990	1991
miles per car	9,878	10,119	10,330	10,562	10,726

a. Let n represent years since 1987, so data is given for $n = 0, 1, 2, 3$, and 4. Let $u(n)$ represent the number of miles driven per car in year n. Make a scatter plot from this data. If the data seems linear, draw a line that approximates the data. Find an equation that approximates this line.

b. Write a dynamical system that $u(n)$ might satisfy. Make an assumption about how driving is changing from year to year for Americans.

c. Obtain a least squares line and graph it. Compare your line from part (a) to the least squares line.

11. The following table gives the approximate amount of petroleum used for transportation in the United States in the corresponding years. The amount of petroleum is given in quadrillion Btu. (This data was adapted from information obtained on the Department of Transportation web site.)

year	1970	1975	1980	1985	1990	1995
quadrillion Btus of petroleum	15.3	17.5	19.4	20.0	21.6	23.8

a. Let n represent years since 1970 so that data is given for $n = 0, 5, 10, 15, 20$, and 25. Make a scatter plot from this data. Sketch a line that approximates the points. Approximate an equation for that line.

b. Write a dynamical system that corresponds to the line from part (a). What assumptions might you make about the increase in the use of petroleum for transportation use?

c. Obtain a least squares line and graph it. Compare your line from part (a) to the least squares line.

12. The following table gives the average number of miles each person in the United States traveled by plane in the corresponding years. (This data was obtained from the U.S. Department of Transportation web site.)

year	1987	1988	1989	1990
air miles per person	1710	1773	1813	1895

a. Let n represent years since 1987 so that data is given for $n = 0, 1, 2$, and 3. Let $a(n)$ represent the average number of miles traveled by air per capita in year n. Make a scatter plot from this data. Draw a line that approximates the data. Find an equation that approximates this line.

b. Write a dynamical system that $a(n)$ might satisfy. Make an assumption about how the Americans flying habits were changing from year to year.

c. Obtain a least squares line and graph it. Compare your line from part (a) to the least squares line.

13. The following table gives waste generation in the United States for several years. (This data was obtained from the U.S. Environmental Protection Agency web site.)

year	1960	1970	1980	1990
millions of tons of waste	88.1	121.0	151.6	196.9

Let n represent years since 1960 (so that data is given for $n = 0, 10, 20$, and 30). Let $w(n)$ represent the number of millions of tons of waste generated in year n.

a. Make a scatter plot from this data. Draw a line that approximates the data. Find an equation that approximates this line.

b. Write a dynamical system that $w(n)$ might satisfy. Make an assumption about how the amount of waste generated in the United States is changing from year to year.

c. Obtain a least squares line and graph it. Compare your line from part (a) to the least squares line.

14. The following table gives the number of orbits the space shuttle made and the corresponding time of the mission to the nearest tenth of an hour. The data comes from 20 missions dating from 4/12/81 to 8/27/85. They are listed in terms of the number of orbits, not in chronological order. (This data comes from **Mathematics Explorations I,** which was a joint NASA-AMATYC-NSF project.)

orbits	36	36	48	80	81	96	97	97	107	109
time	54.4	54.2	73.6	122.2	122.2	144.9	146.4	145.2	167.7	167.9

orbits	110	111	111	112	126	126	127	129	132	166
time	168.2	169.7	170.3	169.2	191.8	190.8	191.3	192.1	197.4	247.8

Fit this data with a line of best fit. Does this line fit the data well? What does it say about the relationships between number of orbits and time of mission? About how long does the shuttle take to complete one orbit? Why should a linear function fit the data well? Why doesn't a linear relation fit the data exactly?

3.3 Exponential Functions

The first part of this section is structured to help you understand why functions that satisfy the dynamical system

$$u(n) = au(n-1) \tag{3.3}$$

can be written as an expression of the form

$$u(n) = ca^n \tag{3.4}$$

Example 3.5 Suppose you have a sheet of paper. You fold it in half. You then fold it in half again and again and again. Let $u(n)$ represent the number of layers after making n folds. This situation can be modeled by the dynamical system

$$u(n) = 2u(n-1)$$

with $u(0) = 1$. Simple computations show that $u(1) = 2$, $u(2) = 4 = 2^2$, $u(3) = 8 = 2^3$, and so on. You should see that each time you fold the paper, you double the previous number of layers, which means multiplying the last number by 2. So if you fold the paper in half n times, then you have multiplied by 2 n times, or

$$u(n) = 2^n$$

Suppose you do this same folding procedure but begin with three sheets of paper (see Figure 3.11). In this case, $u(0) = 3$ but the dynamical system is still $u(n) = 2u(n-1)$. When you fold the three sheets of paper in half, you get $3\times 2 = 6$ layers. When you fold in half again, you get $3\times 2\times 2 = 12$ layers. Each fold results in multiplying by another 2. $u(n)$ is the number of layers after making n folds, which means you have multiplied by 2 n times. This gives the expression

$$u(n) = 3\times 2^n$$

FIGURE 3.11 Repeated folding of 3 sheets of paper.

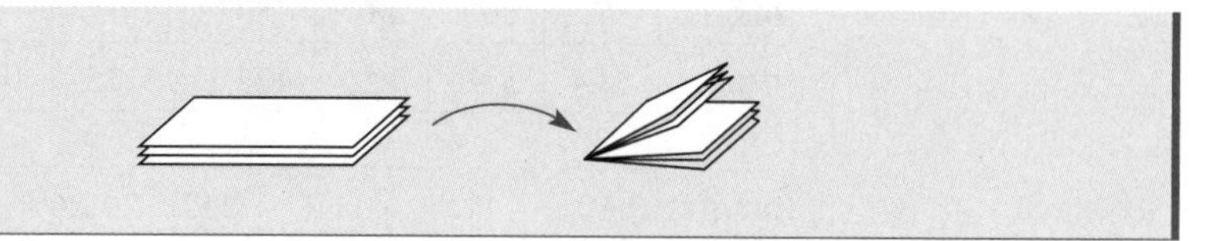

You should check that this expression gives the correct number of layers after $n = 1, 2, 3,$ and 4 folds. ■

Recall that you can always check if your expression makes sense by using the dynamical system to compute $u(1)$, $u(2)$, ... and then using your expression to compute $u(1)$, $u(2)$, ... If both get the same results, it is likely you have the correct expression. If any values differ, you know you have the wrong expression.

From Example 3.5, you should be developing an understanding of why a function that satisfies the dynamical system

$$u(n) = au(n-1)$$

has the form

$$u(n) = ca^n$$

Let me summarize. The dynamical system $u(n) = au(n-1)$ means that each u-value is a times the previous u-value. If $u(n) = ca^n$, then $u(n-1) = ca^{n-1}$, and we again see that $u(n)$ is a times $u(n-1)$, since $a^n = aa^{n-1}$.

We have actually proven the following.

THEOREM

The exponential function

$$u(n) = ca^n$$

satisfies the dynamical system

$$u(n) = au(n-1)$$

The particular value of c depends on the given initial value.

Example 3.6

Suppose we want to find an expression for $u(n)$, knowing that $u(n) = 5u(n-1)$ and $u(0) = 7$. We can then write

$$u(n) = c5^n$$

Substitution of 0 for n gives $u(0) = c5^0 = c(1) = c$. But we know that $u(0)$ equals 7, so $c = 7$ and

$$u(n) = 7 \times 5^n = 7(5)^n$$

Suppose we don't know $u(0)$ but know that $u(2) = 100$. In this case we substitute 2 for n into $u(n) = c5^n$, giving

$$u(2) = c5^2 = 25c$$

Now we equate the two forms of $u(2)$ to get

$$25c = 100 \quad \text{or} \quad c = 4$$

In this case,

$$u(n) = 4(5)^n$$

In Figure 3.12 is a time graph of $u(n) = 4(5)^n$. We should only be plotting the points where n is an integer, but are actually plotting points for $u(n) = 4(5)^n$ for all values of n. One advantage of the expression is that we can estimate $u(n)$ for values of n that are not integers, if these values make sense in the context of the problem. ■

FIGURE 3.12 The function $u(n) = 4(5)^n$.

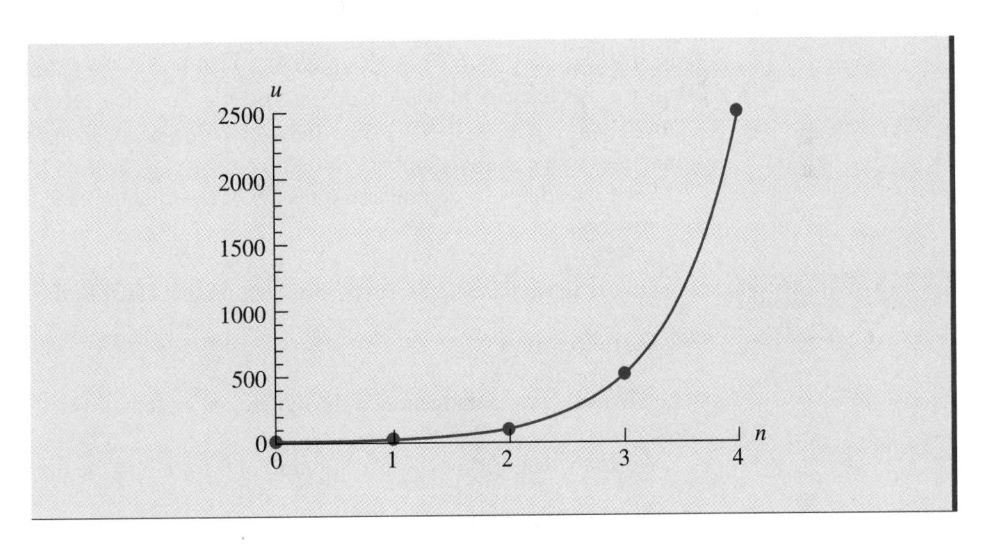

Example 3.7 Let's find an expression that satisfies the dynamical system $u(n) = -0.8u(n-1)$, given that $u(0) = 5$.

We know that

$$u(n) = c(-0.8)^n$$

Using $u(0) = 5$ gives $u(0) = c(-0.8)^0 = c(1) = c$, so $c = 5$. This means that

$$u(n) = 5(-0.8)^n$$

This function is graphed in Figure 3.13. ■

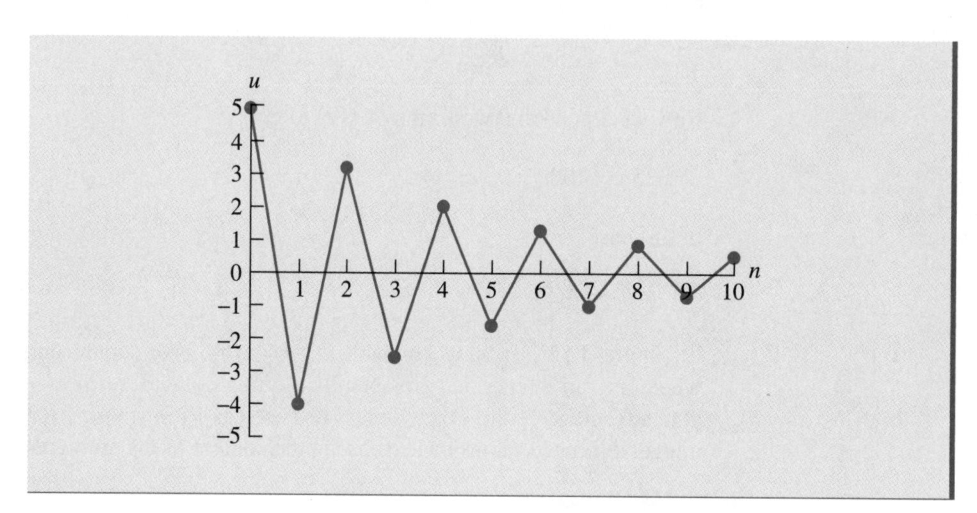

FIGURE 3.13 The function $u(n) = 5(-0.8)^n$. This function only exists when n is an integer. The lines connecting the points are there only for visual purposes.

In contrast to $u(n) = 4(5^n)$ in Example 3.6, $u(n) = 5(-0.8)^n$ makes physical sense only when n is an integer. In particular, suppose $n = \frac{1}{2}$. Then the function $u(n) = 5(-0.8)^n$ results in $u(\frac{1}{2}) = 5(-0.8)^{1/2} = 5\sqrt{-0.8}$ which doesn't even exist since we can't take the square root of a negative number.

Understanding how to find an expression for $u(n)$ helps us to find unknown parameters in a problem.

Example 3.8

We will use the exponential form of the function that satisfies the dynamical system

$$u(n) = au(n-1)$$

to find the value for parameter a in the dynamical system, given that $u(0) = 2$ and $u(30) = 2000$.

We know that $u(n) = ca^n$. Substitution of 0 for n into the function gives

$$u(0) = ca^0 = c \times 1 = c$$

Since we know that $u(0)$ equals 2, we have that $c = 2$, so

$$u(n) = 2a^n$$

We now substitute 30 for n into the function to get

$$u(30) = 2a^{30} = 2000$$

Dividing both sides by 2 gives

$$a^{30} = 1000$$

Finally, we take the 30th root of both sides to get that

$$a = (1000)^{1/30} = \sqrt[30]{1000} \approx 1.26$$

We now know the parameter $a = (1000)^{1/30} \approx 1.26$ and so

$$u(n) = 2a^n = 2\left((1000)^{1/30}\right)^n = 2(1000)^{n/30} \approx 2(1.26)^n$$

If we were computing $u(3)$, we would probably use $u(3) \approx 2(1.26)^3 \approx 4.00$ to 2 decimal places. On the other hand, if $n = 90$, we would probably compute

$$u(90) = 2(1000)^{90/30} = 2(1000)^3 = 2{,}000{,}000{,}000 = 2 \times 10^9$$

The form in which you write an expression often depends on what you are going to do with it.

Suppose we were given that $u(10) = 500$ and $u(30) = 2000$. How would we find the value of the parameter a for the dynamical system

$$u(n) = au(n-1)$$

We know that $u(n) = ca^n$. Using the initial values, we get the 2 equations

$$500 = ca^{10} \quad \text{and} \quad 2000 = ca^{30}$$

In this case, we divide the second equation by the first, giving

$$\frac{2000}{500} = \frac{ca^{30}}{ca^{10}}$$

Reducing the left side of this equation to 4 and canceling the c's on the right gives

$$4 = \frac{a^{30}}{a^{10}}$$

We now subtract the exponents on the right side, giving

$$4 = a^{20}$$

Finally, we take the 20th root of both sides, getting

$$a = 4^{1/20} = 4^{0.05} \approx 1.07$$

We can now use the equation $500 = ca^{10}$ to find c. In particular, substitution for a gives

$$500 = c\left(4^{1/20}\right)^{10} = c4^{1/2} = 2c$$

Dividing by 2 gives $c = 250$ and

$$u(n) = 250\left(4^{n/20}\right) \approx 250(1.07)^n$$

■

To appreciate being able to convert from the dynamical system (3.3) to expression (3.4), we need to know where dynamical systems of the form $u(n) = au(n-1)$ arise in practice.

Often dynamical systems of the form $u(n) = au(n-1)$ arise from situations in which a certain percentage of the amount is being added or subtracted. For example, suppose you have a bank account that earns 6% interest, compounded annually. This is seen visually in the flow diagram of Figure 3.14.

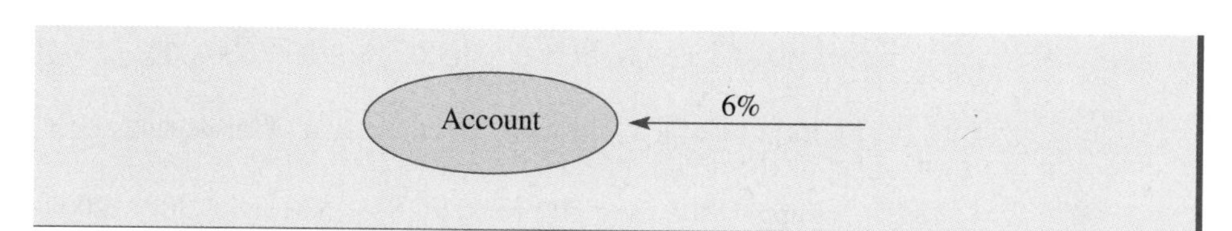

FIGURE 3.14 Flow diagram for bank account.

This can be written as "add 6% of the amount each year." The dynamical system that models this account is $u(n) = 1.06u(n-1)$, where $u(n)$ is the amount in the account at the end of n years. We now know that the amount in the account is given by

$$u(n) = c(1.06)^n$$

Once we know the amount in the account at some point in time, we can find the value for c and determine the amount in the account at any point in time. A similar result occurs when a fixed percentage is removed each time period. Let's consider the situation of caffeine being removed from the body.

Various foods and drinks popular around the world contain caffeine. Caffeine is an alkaloid compound that comes from plants, including coffee, tea, kola nuts, maté, cacao and guarana. Many people drink caffeinated drinks because they like the taste of them, others for the physical effect of the caffeine. Most people are aware of differences in the way they feel as a result of drinking caffeine, which stimulates the central nervous system, the heart muscles, and the respiratory system. The way individuals interpret the effects of caffeine as a stimulant varies widely. For many, the effect is pleasant and energizing, a "wake up" or a "pick-me-up," and it can delay fatigue. For others, the effects are unpleasant. Laboratory tests indicate that 1 to 3 cups of coffee can produce an increased capacity for sustained intellectual effort and decrease reaction time but may adversely affect tasks involving delicate muscular coordination and accurate timing.

Table 3.3 lists an approximate amount of caffeine in some drinks. The amount of caffeine in coffee and tea varies widely from cup to cup, but these numbers give some idea. The caffeine levels in commercial sodas tend to be consistent. Table 3.4 gives some of these values for a 12-oz drink.

TABLE 3.3 Approximate amounts of caffeine in coffee and tea.

	8-oz cup of coffee			2 oz of Espresso	8-oz cup of tea		12 oz of iced tea
	brewed	instant	drip		brewed	instant	
mg of caffeine	80–135	65–100	115–175	100	40–80	30	70

TABLE 3.4 Caffeine in popular soft drinks.

Brand	Coca-Cola	Dr. Pepper	Pepsi	Surge	Jolt	Mountain Dew
mg of caffeine	45.6	39.6	37.2	51.0	71.2	55.0

Information in these tables came from the National Soft Drink Association and the U.S. Food and Drug Administration.

The effects of caffeine can only be felt when the caffeine is present in sufficient amounts. For most people, 32–200 mg of caffeine acts as a minor stimulant; these amounts have been shown to speed up reactions in simple routine tasks in laboratory experiments. Steadiness of the hand has been shown to be worse after 200 mg of caffeine. More than 300 mg is enough to produce temporary insomnia. 480 mg has been known to cause panic attacks in panic disorder patients. Amounts of 5 to 10 g (5000–10,000 mg) of caffeine cause death.

Our bodies eliminate caffeine primarily by the functioning of the kidneys. The kidneys tend to filter out a constant proportion of a chemical, that proportion depending on the particular chemical and individual. In the "average" person, about 13% of the caffeine in the body is eliminated each hour.

Example 3.9 You start the day by quickly drinking 3 cups of coffee containing 130 mg of caffeine each. The flow diagram of Figure 3.15 shows how caffeine is eliminated.

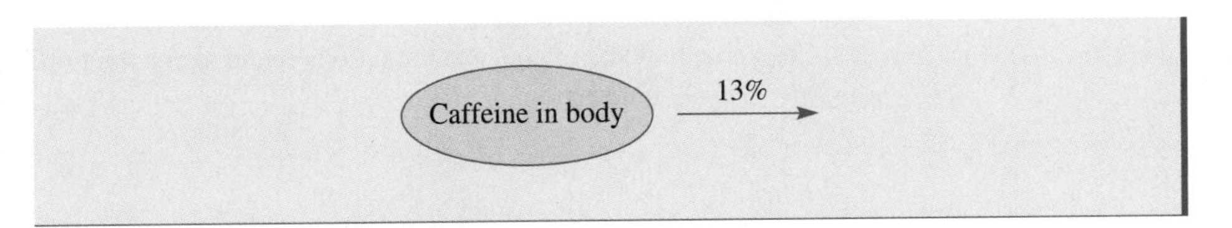

FIGURE 3.15 Flow diagram for elimination of caffeine.

This is written as "the body eliminates 13% of the caffeine each hour."

Let n be the number of hours since you consumed your 3 cups of coffee. Let $C(n)$ be the amount of caffeine left in your body after n hours. This means that $C(0) = 390$. The dynamical system that gives the amount of caffeine in your body is

$$C(n) = C(n-1) - 0.13C(n-1) \text{ or } C(n) = 0.87C(n-1)$$

The expression that gives the amount of caffeine is then

$$C(n) = c(0.87)^n$$

From the fact that $C(0) = 390$ and $C(0) = c(0.87)^n = c$, we get that

$$C(n) = 390(0.87)^n$$

See the graph of this function in Figure 3.16.

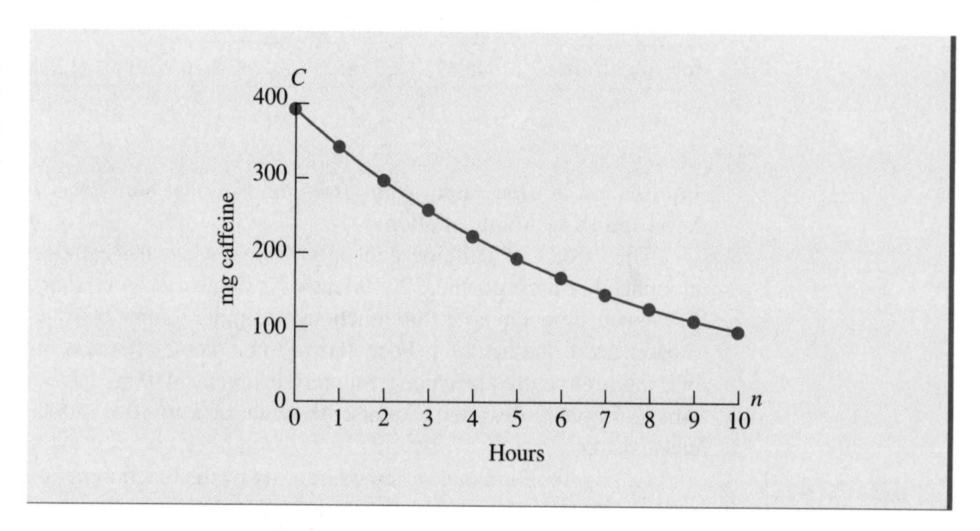

FIGURE 3.16 Caffeine in body n hours after consuming 3 cups of coffee.

From this, we can not only estimate the amount of caffeine remaining in the body after an integer number of hours, but we can also estimate the amount remaining at any point in time. For example, after 3.5 hours, there is about

$$C(3.5) = 390(0.87)^{3.5} \approx 239.5$$

mg of caffeine left.

Let's note how the exponential functions relate to constant ratios of consecutive values of $C(n)$. You should see that

$$\frac{C(n)}{C(n-1)} = 0.87$$

for any value of n that you choose. One reason this is true is that the ratio is just the original dynamical system, written a different way. Another reason the ratio is constant can be seen by use of the exponential form of the function. In particular,

$$\frac{C(5)}{C(4)} = \frac{390(0.87)^5}{390(0.87)^4} = \frac{0.87^5}{0.87^4}$$

Subtracting exponents gives

$$\frac{C(5)}{C(4)} = 0.87$$

Recall that this means that the amount of caffeine in the body at any hour is 87% of the amount of caffeine in the body the previous hour. It also means that the amount of caffeine in the body at any hour is 13% less than the amount of caffeine in the body the previous hour. ■

Let's summarize what we have learned in this section. Let n represent a unit of time and $u(n)$ represent the amount of something at time n. Consider the following statements.

1. $u(n)$ satisfies the dynamical system $u(n) = au(n-1)$.
2. Each time n increases by one, I percent of the current amount is added to the amount

These statements are equivalent, with $a = 1 + 0.01I$.

This means that the dynamical system $u(n) = 1.06u(n-1)$ is equivalent to adding 6%. These statements are also equivalent if I is negative. For caffeine elimination, the dynamical system $C(n) = 0.87C(n-1)$ is equivalent to subtracting 13% or adding a negative 13%. In the case of folding paper, the number of layers satisfied the dynamical system $u(n) = 2u(n-1) = (1+1)u(n-1)$, which is the same as adding 100% each time.

When either of the 2 statements is true, we know that $u(n)$ can be written as the exponential function

$$u(n) = ca^n$$

3.3 Problems

1. Find the expression that satisfies the dynamical system

$$u(n) = 1.3u(n-1)$$

given that

a. $u(0) = 12$. Graph the function for n going from 0 to 10.
b. $u(1) = -26$. Graph the function for n going from 0 to 10.
c. $u(15) = 3$. Find c to 3 decimal places.
d. What percentage is being added to or subtracted from the amount each time?

2. Find the expression that satisfies the dynamical system

$$u(n) = 0.8u(n-1)$$

given that

a. $u(0) = -15$. Graph the function for n going from 0 to 10.
b. $u(1) = 2.4$. Graph the function for n going from 0 to 10.
c. $u(20) = 5$. Find c to 3 decimal places.
What percentage is being added to or subtracted from the amount each time?

3. Find the expression that satisfies the dynamical system

$$v(n) = -0.3v(n-1)$$

given that
a. $v(0) = 2$. Graph the function for n going from 0 to 10.
b. $v(1) = -9$. Graph the function for n going from 0 to 10.
What percentage is being added to or subtracted from the amount each time?

4. Find the expression that satisfies the dynamical system

$$w(n) = -w(n-1)$$

given that
a. $w(0) = 5$. Graph the function for n going from 0 to 10.
b. $w(1) = 8$. Graph the function for n going from 0 to 10.

5. Find the expression that satisfies the dynamical system

$$u(n) = -4u(n-1)$$

given that
a. $u(0) = 0.1$. Graph the function for n going from 0 to 5.
b. $u(1) = 0.16$. Graph the function for n going from 0 to 5.

6. Find the value for a in the dynamical system

$$u(n) = au(n-1)$$

given that
a. $u(0) = 5$ and $u(10) = 35$.
b. $u(5) = 15$ and $u(10) = 75$.

7. Find the value for a in the dynamical system

$$u(n) = au(n-1)$$

given that
a. $u(0) = 2$ and $u(8) = 6$.
b. $u(6) = 30$ and $u(18) = 270$.

8. Lead poisoning is one of the more serious environmental dangers for young children. One problem with lead poisoning is that once a person has absorbed a dangerous amount of lead, only about 1.5% of the lead is removed from the blood each day.

a. Develop a dynamical system to model the elimination of lead from a child's blood.
b. Assume that a 60-pound girl has 2 mg of lead in her blood today. This is a significant amount. Write an expression for the amount of lead in this girl's blood n days from today.
c. Graph the function you found in part (b) over the next month (assuming it is a 31-day month).
d. Graph the function from part (b) over the next year.

9. Suppose you have a certain amount of a radioactive material. You begin with 5 grams of this material and discover that every minute, 8% of the material decays into some other material. Let n be the number of minutes and $u(n)$ the amount of this material remaining after n minutes.
a. Develop a dynamical system to model this situation.
b. Write an exponential function that gives the amount of this material after n minutes.
c. Find the amount of this material left after 2000 **seconds.**

10. Suppose a distant relative deposited money into a bank account that earns 7% interest, compounded annually.
a. Develop a dynamical system that models the amount of money in this account.
b. Assume that 20 years after depositing this money, your relative dies and leaves you the money. Your inheritance is \$4837.11. Find an expression for the amount in this account n years after your relative deposited the money. How much money did your relative originally deposit?

11. You bought stock for \$2000 and sold it 20 years later for \$7500. Assume the stock gained the same percentage in value each year. Let n be the number of years since you bought your stock and $s(n)$ be its value after n years. Then $s(0) = 2000$ and $s(20) = 7500$.
a. Develop a dynamical system for the value of your stock, using a fixed parameter for the yearly interest rate.
b. Find an exponential function for the value of the stock, and find the value of the parameter using the known information. What is the average yearly growth rate, as a percentage, for this stock? (While the stock probably did not increase in value by the same percentage each year, this problem shows how you can find the average percentage gain over the time you have owned the stock.)

12. A boy has a blood test, which determines that he has 80 μg of lead per deciliter of blood. Assume he ingests no additional lead. This boy takes a treatment for lead poisoning, which is a drug that increases the rate at which he eliminates lead from his body. After 1 week, his blood test indicates he has 40 μg of lead per deciliter of blood. Assume that he eliminates the same percentage of the lead each day. Develop a dynamical system for $l(n)$, the number of micrograms of lead per deciliter of blood in his blood after n days. Use a parameter for the elimination rate. Find an explicit function for $l(n)$ and use that function to estimate the percentage of lead eliminated from his blood each day.

3.4 Exponential Growth and Decay

In this section we will learn how to use the exponential form of the function that satisfies the dynamical system $u(n) = au(n-1)$ to analyze real-world situations. One concept we will explore is **half-life.** Half-life refers to the period of time it takes for the amount of something to be cut in half. For example, the period of time it takes for half the amount of the caffeine to be eliminated from our plasma or for half of an amount of a radioactive material to decay. A second use of the function $u(n) = ca^n$ will be to understand when a situation is modeled by a dynamical system of the form $u(n) = au(n-1)$ and what that means in context.

Let's begin our discussion of half-life with an example.

Example 3.10

You already know that about 13% of the caffeine is eliminated from the plasma each hour. What is the half-life of caffeine in the body; that is, how long does it take for the amount of caffeine to be cut in half? To answer this question, let's recall that if $C(n)$ is the amount of caffeine in the body n hours after consuming some caffeine, then the function $C(n)$ satisfies the dynamical system

$$C(n) = 0.87C(n-1)$$

and can also be written as

$$C(n) = c(0.87)^n$$

If $C(0) = 390$, then the half-life of caffeine is the value of n for which $C(n) = 195$. If you use your calculator or computer, you will find that $C(5) = 194.38$, so the half-life of caffeine is just under 5 hours. If you start with any other value for $C(0)$, you will again find that $C(5)$ is slightly less than half of $C(0)$. You will also find that $C(10)$ is slightly less than half of $C(5)$. In fact, you will find that if you pick any starting time, say time k, then $C(k+5)$ will be slightly less than half of $C(k)$. This is what half-life means (see Figure 3.17). Note: If you pick any point on the graph in Figure 3.17, then the point 5 hours to the right is only half as high. ■

In Example 3.10, you were introduced to the idea of half-life. Finding the half-life of caffeine in the body was not difficult, but some chemicals and radioactive materials have **very** long half-lives, thousands of days for some chemicals and thousands of years for some radioactive materials. We want to develop a simple algebraic means for finding half-life.

In finding half-life, we often need to solve an equation for an unknown exponent. One important use of logarithms is to help solve an equation when the exponent is the unknown. You should review some of the properties of logarithms. The main property we will use is the exponent logarithm rule

$$\log(a^b) = b\log(a)$$

Example 3.11

Solve for x in the equation

$$6 = 2(4)^x$$

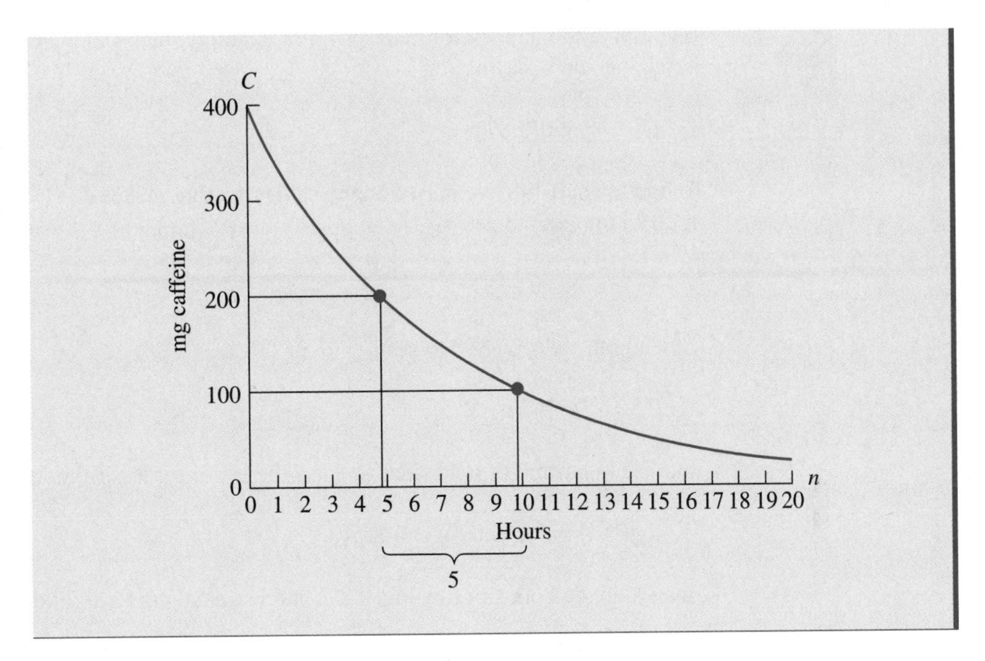

FIGURE 3.17 Take any point on the graph. Half-life being 5 hours means the point that is 5 hours to the right will be half as high.

To do this, we divide both sides by 2, getting

$$3 = 4^x$$

We then take the logarithm (to the base 10) of both sides, getting

$$\log(3) = \log(4^x)$$

We then apply the exponent logarithm rule, giving

$$\log(3) = x \log(4)$$

Dividing both sides of the equation by $\log(4)$ gives

$$x = \frac{\log(3)}{\log(4)} \approx 0.79$$

to 2 decimal places. We could have taken the logarithm to the base e of both sides and found that

$$x = \frac{\ln(3)}{\ln(4)} \approx 0.79$$

You should check that $4^{0.79} \approx 3$. It is not exactly 3, since we rounded x to 2-decimal places.

■

Once you know this method for solving for an unknown exponent, you will find it easy to compute the half-life of materials. For example, recall that the amount of caffeine in a certain person's plasma was

$$C(n) = 390(0.87)^n$$

To find the half-life, we must find the value of n that makes $C(n) = 195$. To do this, substitute 195 for $C(n)$ to get

$$195 = 390(0.87)^n$$

Divide both sides by 390 to get

$$0.5 = (0.87)^n$$

Take the logarithm of both sides and use the exponent logarithm rule to get

$$\log(0.5) = \log(0.87^n) = n \log(0.87)$$

Finally, divide both sides by $\log(0.87)$ and use a calculator to find that

$$n = \frac{\log(0.5)}{\log(0.87)} = 4.97$$

to 2 decimal places. So you see that the actual half-life of caffeine in the body is 4.97 hours.

In the case of caffeine, the elimination rate of 13% is only an approximation and varies from person to person, so it is probably best to give the half-life of caffeine as about 5 hours. For other problems, such as radioactive decay, it can be important to find the half-life as accurately as possible.

The advantage of this algebraic method is we can determine the half-life as precisely as we want. We were only able to approximate the half-life to the nearest integer value when using a calculator and the dynamical system.

Many researchers use the half-life as a convenient time interval. For example, they might describe the elimination of caffeine from the body as "50% of the caffeine is removed every 5 hours." In this case, they would let $u(n)$ be the amount of caffeine in the body after n 5-hour periods. They would then use the dynamical system

$$u(n) = 0.5u(n-1)$$

as the one that describes the elimination of caffeine from the body, and they would write the function as

$$u(n) = 390(0.5)^n$$

Refer to Figure 3.18.

An advantage of working with the half-life as the unit of time is that it is easy to compute integer powers of one-half, while you would probably want a calculator to compute

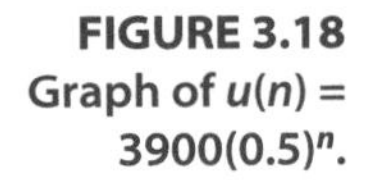

FIGURE 3.18 Graph of $u(n) = 3900(0.5)^n$.

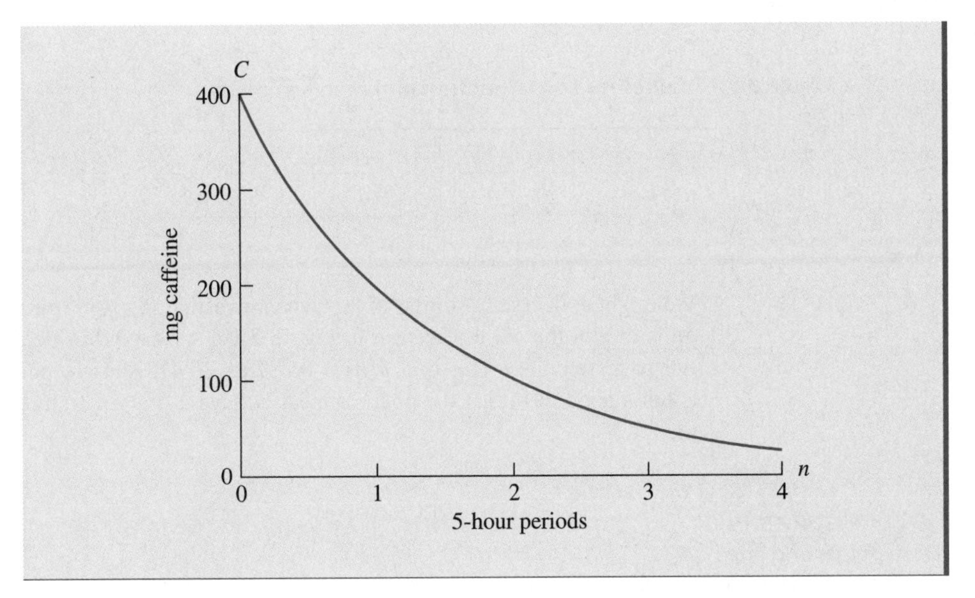

powers of 0.87. It also implies that after 10 hours, one-fourth is left, after 15 hours, or three 5-hour periods, only one-eighth is left, and so on.

We now come to the second part of this section. Previously, we have started with the dynamical system $u(n) = au(n-1)$ and have used it to find the exponential form of the function $u(n) = ca^n$. We are now going to consider situations in which we know the function is of the form $u(n) = ca^n$. We then know that the function satisfies the dynamical system $u(n) = au(n-1)$. Once we know it satisfies this dynamical system, then we know that $u(n)$ is increasing (or decreasing) at a fixed rate each time period, since the ratios of consecutive u-values are constant,

$$\frac{u(n)}{u(n-1)} = a$$

In practice, we are often given a set of data-points and want to fit these points with a function. In Section 3.2 the data points were relatively linear, so we approximated them with a linear function and then concluded that in this situation, the amount seems to be changing by a constant each time period. We are now going to consider data that appears to be growing or decaying exponentially. We are then going to find an exponential function that fits the points reasonably well. From that, we write a dynamical system that is consistent with the data. From that dynamical system, we can conclude that the amount is increasing or decreasing by a fixed percentage each time period.

The problem with this plan is being able to determine if a set of data is exponentially increasing or decreasing; that is, if it fits an exponential function. Often it is not clear if data fits an exponential function, a quadratic function, or some other function. Example 3.12 illustrates the main idea.

Example 3.12 Consider the data in Table 3.5.

TABLE 3.5 Values for some function $u(n)$.

$n =$	0	1	2	3	4	5
$u(n) =$	3	3.6	4.32	5.184	6.2208	7.46496

We might collect data similar to this when investigating some situation. The first thing we do is to plot the points, as seen in Figure 3.19. As we did in Section 3.2, we try fitting a line to these points. This line, $u(n) = 0.887n + 2.747$, is also seen in Figure 3.19. This line is not a good fit in that the points appear to have a "curve" to them that is not reflected in the line.

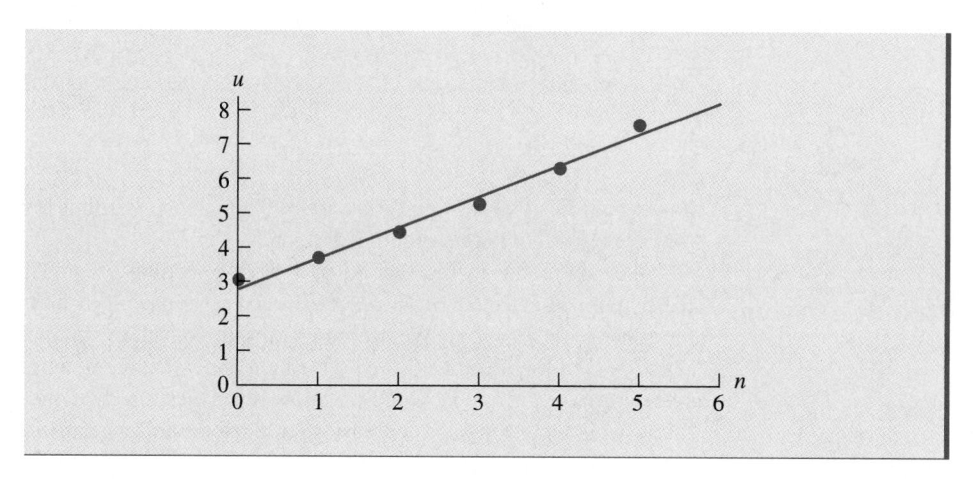

FIGURE 3.19 Data points and line of best fit.

It is difficult to look at a set of points and determine if the data is exponential, and if it is, what the exponential function is. One approach is to use a computer or calculator to generate a best fit exponential regression curve, which is similar to the best fit linear function. We then graph that curve on the same axis as the data points and observe if the graph seems to be a good fit. My calculator gave me the exponential regression function

$$u(n) = 3(1.2^n)$$

By comparing values using this function to the values in Table 3.5 and comparing the graph of this function to the data points, seen in Figure 3.20, we see that we not only have a good fit, we have a perfect fit.

We now know that $u(n)$ satisfies the dynamical system

$$u(n) = 1.2u(n-1)$$

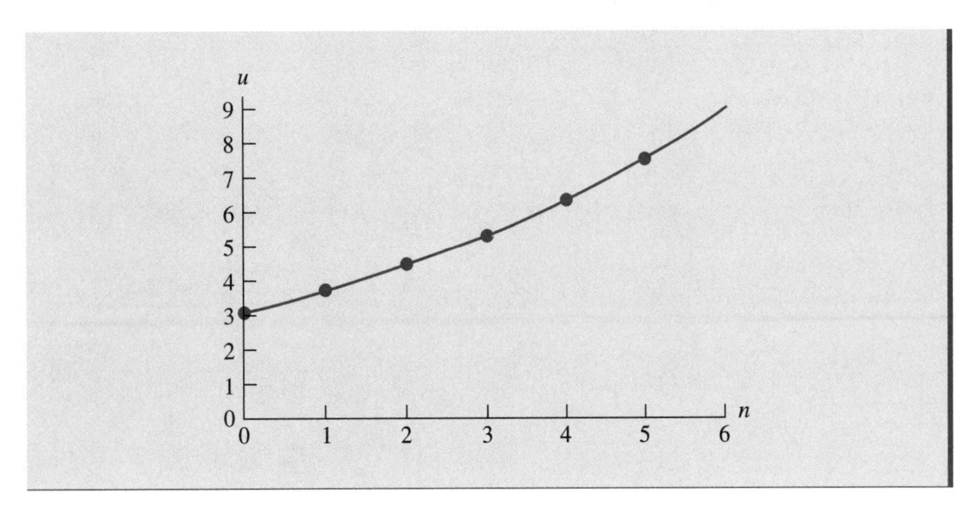

FIGURE 3.20 Data from Table 3.5 and $u(n) = 3(1.2)^n$.

This means that the ratios of consecutive u-values is constant,

$$\frac{u(n)}{u(n-1)} = 1.2$$

This constant ratio means that each u-value is 120% of the previous value, or that each value is 20% larger than the previous value. ■

One question is, How do we know if an exponential function will be a good fit to a set of data? The answer is, Whenever the ratios of consecutive u-values are about the same. Let's consider some real data.

I dropped a ball from a height of 79 cm and recorded the height it bounced for each of the next 5 bounces. I recorded the ball bouncing using a video camera and repeatedly watched the replay of the video until I got fairly good estimates for the height the ball bounced. Even here, I rounded to the nearest centimeter. The data I collected is given in Table 3.6. The height is measured from the bottom of the ball from the floor.

TABLE 3.6 **Height of bouncing ball.**

bounce number, $n =$	0	1	2	3	4	5
height of bounce, $u(n) =$	79	63	47	35	27	20

These data are seen as points in Figure 3.21. I drew the line of best fit on this graph. From this line, we can see that the points appear to have a curved pattern that is not reflected by the line. If the points were truly linear, we would expect to see the points appear randomly above and below the line. Something else seems to be happening.

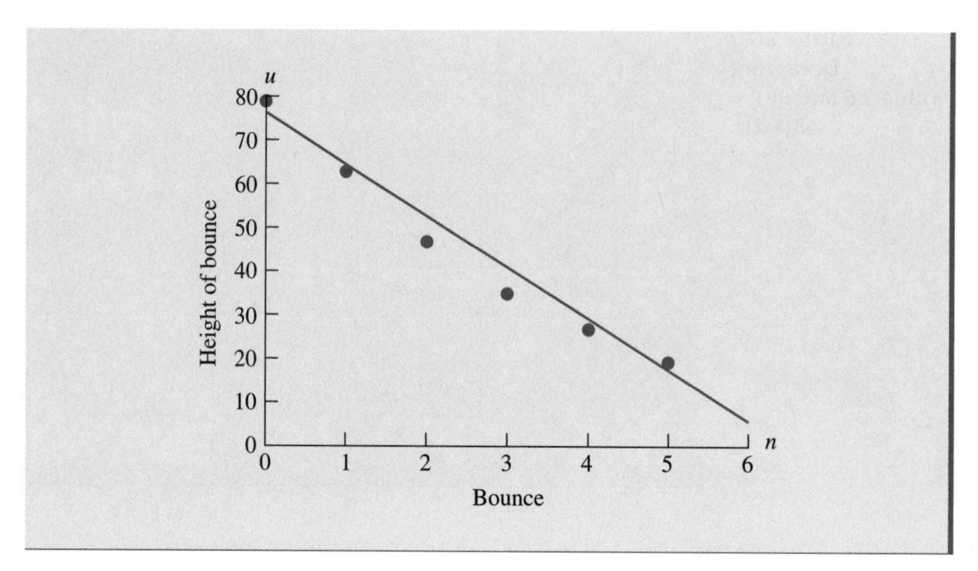

FIGURE 3.21 Heights of bounces and linear best fit.

If the points aren't linear, maybe an exponential function will fit them. To see if an exponential function is reasonable, I computed the ratios of consecutive heights in Table 3.7. Because these ratios are approximately the same, we expect that an exponential function will be a reasonable fit.

TABLE 3.7 Ratios of heights of consecutive bounces.

$u(1)/u(0)$	$u(2)/u(1)$	$u(3)/u(2)$	$u(4)/u(3)$	$u(5)/u(4)$
0.797	0.746	0.745	0.771	0.741

My calculator gave the exponential regression function $u(n) = 81.032(0.758)^n$, with numbers rounded to 3 decimal places, as being the best fit to the original set of data. This function is graphed along with the points from Table 3.6 in Figure 3.22. If you do computations with this function, you get fairly good predictions for the heights of the bounces.

We now know that my bouncing ball satisfies the dynamical system

$$u(n) = 0.757u(n-1)$$

This means that each bounce of the ball is about 76% of the height of the previous bounce, or that the ball seems to bounce about 24% less high on each bounce. From this we have discovered a property of this bouncing ball. Within reasonable ranges, each bounce is a certain fixed percentage less than the previous bounce. This actually means that the ball loses a fixed percentage of its energy on each bounce. The number $a = 0.757$ is called the **coefficient of restitution.**

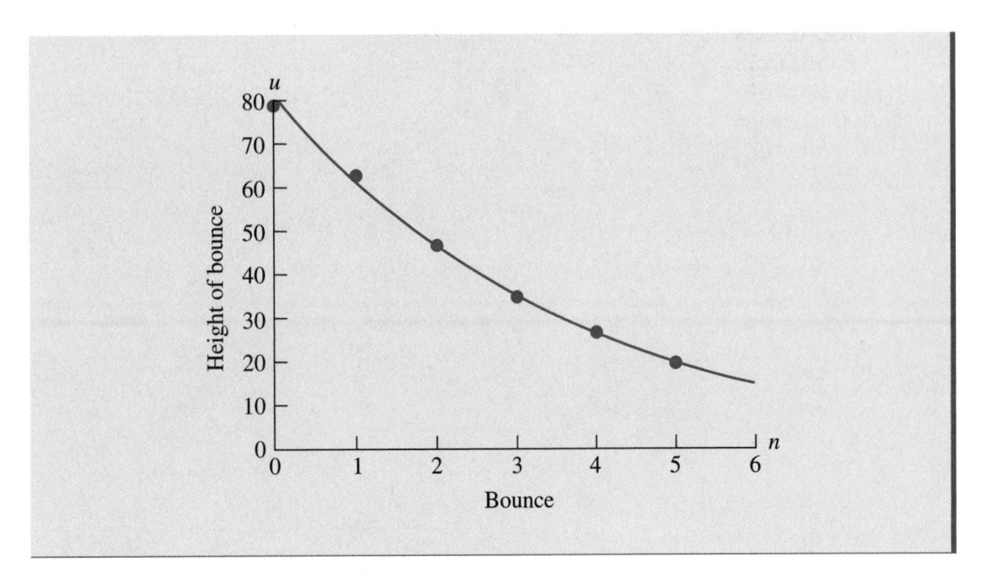

FIGURE 3.22 Heights of bounces and exponential fit.

REMARK Let me discuss how the exponential expression of best fit is found. For the ball bouncing experiment, the function we found was $u(n) = 81.032(0.758)^n$. We take logs of both sides of the equation to get

$$\log(u(n)) = \log(81.032 \times 0.758^n)$$

We rewrite the right side using properties of logarithms to get

$$\log(u(n)) = \log(81.032) + n \log(0.758)$$

We let $\log(u(n)) = y$. We find that $\log(81.032) = 1.91$ and $\log(0.758) = -0.12$, both to 2 decimal places. Making these substitutions, we get

$$y = -0.12n + 1.91$$

We take the logarithm of the $u(n)$ values in Table 3.6 to get the values in Table 3.8.

TABLE 3.8 **Logarithm of heights of bouncing ball.**

bounce number, n =	0	1	2	3	4	5
$\log(u(n))$ =	1.90	1.80	1.67	1.54	1.43	1.30

Plotting the data in Table 3.8 on an axis, as in Figure 3.23, we see that they are nearly linear. The line $y = -0.12n + 1.91$ is actually the least squares line of best fit for this data. This line can be converted to an exponential by computing

$$10^{-0.12n+1.91}$$

which simplifies to 81.032×0.758^n using properties of exponentials and a calculator.

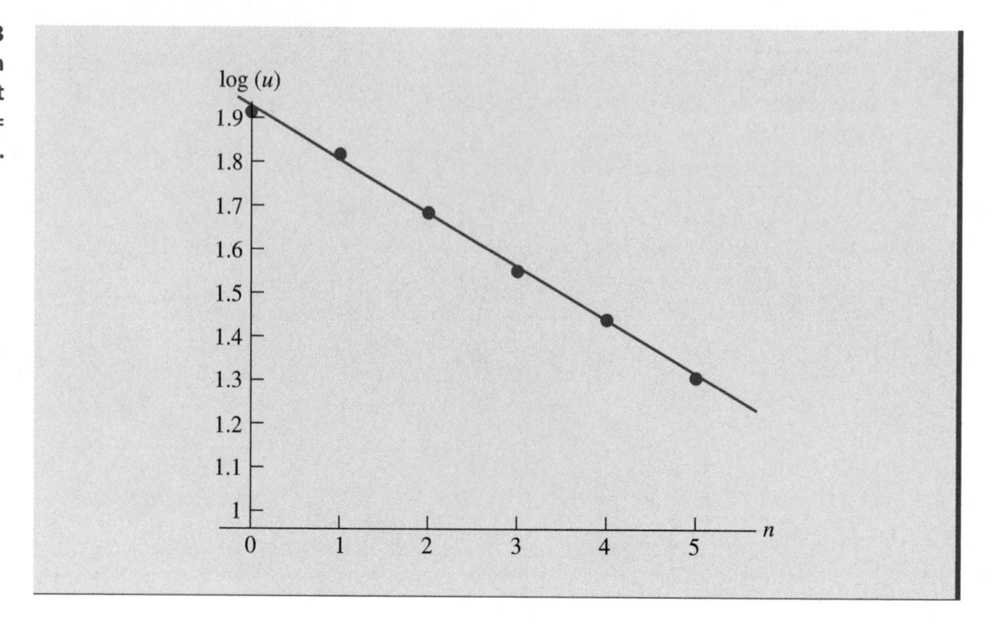

FIGURE 3.23 Points from Table 3.8 are fit with the line $y = -0.12n + 1.91$.

Thus, the data can be fit reasonably well with an exponential if the logarithms of values of the dependent variable are relatively linear. Some people determine this by plotting the values in Table 3.6 on log paper. Again, if the points are linear, then $u(n)$ is exponential.

3.4 Problems

1. Suppose that

$$u(n) = 100(0.93)^n$$

Note that $u(0) = 100$. Find the value of n, accurate to two decimal places, for which $u(n) = 50$.

2. Suppose that

$$u(n) = c(0.1)^n$$

Find the value of n, accurate to two decimal places, for which $u(n) = \frac{1}{2}u(0)$.

3. Suppose the amount of a chemical in the body decreases by 5% each day. What is the half-life of this chemical in the body?

4. Recall that each day about 1.5% of the lead in a child's blood system is removed.
   a. What is the half-life of lead in the blood system of a child?
   b. Suppose a girl has 2 mg of lead in her blood, which is dangerously high. Use part (a) to estimate how long will it take until the level of lead drops to the more desirable level of 0.25 mg.

5. Suppose you have a certain amount of a radioactive material. It is discovered that each minute, 8% of this material decays. Let n be the number of minutes and $u(n)$ the amount of this material remaining after n minutes. Find the half-life of this radioactive material.

6. Carbon-14 is a radioactive material that is present in animals and plants. When an animal or plant dies, its carbon-14 begins to decay. Let n represent the number of centuries since the organism died. Each century, a fixed proportion of the remaining carbon-14 will decay and a fixed proportion will remain. Let d represent the fraction of one century's carbon-14 that decays during the next century. A dynamical system that represents the decay of carbon-14 is

$$u(n) = u(n-1) - d\,u(n-1) \quad \text{or} \quad u(n) = (1-d)\,u(n-1).$$

   a. The half-life of carbon-14 is about 5700 years, meaning that if $u(0) = 1$ gram, then $u(57) = 0.5$ gram. Use this information to find d.
   b. When an organism died, it contained 1 gram of carbon-14. When it was discovered this year, only 0.8 grams of the carbon-14 remained. How old is this organism?

7. Doubling time is analogous to half-life in that it is the amount of time it takes for some quantity to double. One area in which doubling time is used is finance. In fact, to evaluate different investments, their doubling times are often compared. You invest $1000 into an account that earns 4.8% annual interest, compounded monthly. What is the doubling time for this investment; that is, how long will it take for your investment to double to $2000? Figure 3.24 is a visualization of this question.

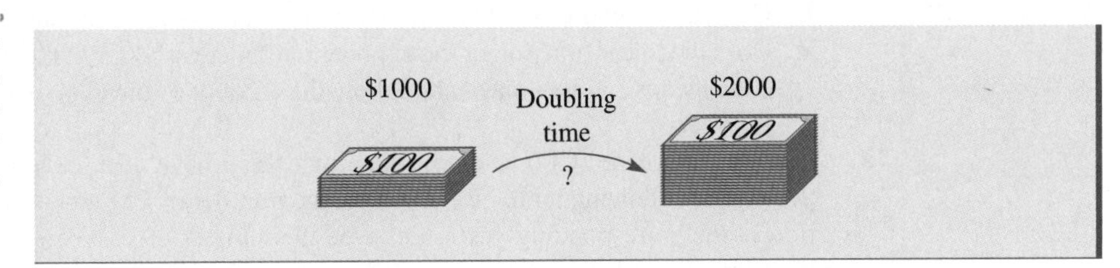

FIGURE 3.24 The unknown is time it takes to double your money.

8. The growth rate for a population is 3% per year, meaning that the size of the population increases by 3% each year. What is the doubling time for this population, that is, how long does it take for this population to double in size?

9. Mitochondrial DNA is passed from mother to daughter. In each generation, a certain fraction of the mother's mitochondrial DNA mutates as it is passed to her daughter. This fact has allowed researchers to estimate relative ages of different racial and ethnic groups. Suppose that the mutation rate is 0.00001, meaning that about $\frac{1}{100{,}000}$ of each mother's mitochondrial genes mutate when passed to her daughter. What is the half-life of mitochondrial DNA, meaning how many generations will it take for about half of the genes in mitochondrial DNA to mutate?

10. Consider the data given in the following table.

$n =$	1	2	3	4
$u(n) =$	4.00	3.20	2.56	2.05

a. Plot the points $(n, u(n))$ on an axis and fit those points to a line of best fit. Does the line give a good representation of the points?
b. Take the ratios of consecutive $u(n)$-values and observe if these ratios are all about the same. Find an exponential regression function that fits the data. Graph this function along with the points. Does this function give a good representation of the points?
c. Write a dynamical system that $u(n)$ satisfies. What conclusion can you make about the rate at which the u-values are changing?

11. The following table gives the trillions of vehicle-miles traveled on our nation's roads over the past 35 years. Let n be the number of years since 1960 and let $m(n)$ be the miles traveled that year. (U.S. Department of Transportation)

year	1960	1965	1970	1975	1980	1985	1990	1995
Trillions of miles	0.70	0.95	1.05	1.30	1.50	1.70	2.15	2.35

a. Plot these points and fit them with a line of best fit. Graph the line and plot the points on the same axis.
b. Generate an exponential function to fit this data using your calculator or computer.
c. Does the linear function or the exponential function better fit the data in the table? What conclusions can you make about how the amount of travel is increasing?

12. Carbon Monoxide (CO) levels in the atmosphere have been decreasing over the past 20 years. The following table gives the average number of CO violations per monitoring station in the corresponding years. Let n be the number of years since 1975 and let $v(n)$ be the average number of CO violations per monitoring station that year. (U.S. Environmental Protection Agency)

year	1975	1980	1985	1990	1995
violations	34	9	2.5	0.7	0.3

a. Plot the points $(n, m(n))$ on an axis.
b. Approximate the data with an exponential function. Does the function fit the data well?

c. Write a dynamical system that $m(n)$ satisfies. What does that mean about the decrease in violations every 5 years?

13. The following table gives the amount of waste in the United States that has been recycled over the past 35 years. (U.S. Environmental Protection Agency)

year	1960	1970	1980	1990	1995
millions of tons recycled that year	5.6	8.0	14.4	32.9	56.2

Let n be the number of years since 1960, so $n = 0$ represents 1960. Let $u(n)$ be the number of millions of tons of material that was recycled in year n. The points represented by the data in this table appear to be increasing exponentially.

a. Use your calculator or computer to get an exponential function to fit the data. Does the exponential function fit the data well?

b. Compute the ratios

$$\frac{u(n+10)}{u(n)}$$

for $n = 0$, 10, 20, and 30. If the data can be fit by an exponential function, these ratios should also be about the same. Are they approximately the same? What assumptions can you make about how the amount of trash that is recycled has been increasing?

14. The amount of U.S. currency in circulation over the past century is given in the following table, where the currency is in billions of dollars. (Source: U.S. Treasury.)

year	1910	1920	1930	1940	1950	1960	1970	1980	1990	2000
currency	3	6	5	8	27	32	54	127	267	517

a. Find an exponential function that fits this data. Is it a reasonably good fit?

b. Give a dynamical system that is satisfied by the exponential function. What can you say about the growth of currency in circulation.

15. The following table gives predictions about the number of online users in Latin America, where the users are in millions. (Jupiter Communications. Different organizations are making different predictions, and the same organizations may change their predictions.)

year	1999	2000	2001	2002	2003	2004	2005
users	11	16	23	32	43	55	67

Find the linear regression function and the exponential regression function that fits this data. Does it appear that they are using a linear model, an exponential model, or some other model to predict the number of online users in Latin America? Depending on the model you picked, linear or exponential, what does it say about the expected growth in online users in Latin America?

16. The number of teeth on each gear of the rear cluster of my 21-speed bicycle are given in the following table.

gear	1	2	3	4	5	6	7
# teeth	30	24	21	18	15	13	11

a. Fit these data points with a linear function. What does this function imply about the reduction in the number of teeth? How many teeth would a 10th gear have from this model?

b. Fit these data points with an exponential function. What does this function imply about the reduction in the number of teeth? How many teeth would a 10th gear have from this model?

c. Comparing graphs, which model, linear or exponential, do you believe fits this data and why?

17. The following tables give the population of Mexico over a several year period. (Source: U.S. Census Bureau. Other sources give different data for Mexican population.)

Table 1						
year	1950	1960	1970	1980	1990	2000
millions	28.5	38.6	52.8	68.7	84.4	100.3

Table 2											
year	1990	1991	1992	1993	1994	1995	1996	1997	1998	1999	2000
millions	84.4	86.1	87.7	89.3	90.9	92.5	94.1	95.7	97.2	98.8	100.3

a. Fit a linear function to the data in Table 1. What does this function say about Mexico's population? Use this function to approximate Mexico's population in 2005. Looking at the graph of the points and line, does the line fit the data well?

b. Fit an exponential function to the data in Table 1. What does this function say about Mexico's population? Use this function to approximate Mexico's population in 2005. Looking at the graph of the function and the points, does this function fit the data well?

c. Fit a linear function to the data in Table 2. What does this function say about Mexico's population? Use this function to approximate Mexico's population in 2005. Looking at the graph of the points and line, does the line fit the data well?

d. Fit an exponential function to the data in Table 2. What does this function say about Mexico's population? Use this function to approximate Mexico's population in 2005. Looking at the graph of the function and the points, does this function fit the data well?

e. Which functions from the previous parts fit the data better? Why does this make sense? Which model do you suggest using to approximate Mexico's population in 2005 and why?

Project 1. **a.** Pick 2 balls of your choosing. Estimate the coefficient of restitution for each ball by dropping that ball from several different heights and recording the height of the bounce (see Figure 3.25). Make all measurements from the bottom of the ball. You should drop the ball several times from the same height to get an estimate for the rebound height. Then you should actually record the rebound height for several drops and use the average of the bounces. Collect the data and fit it to an exponential curve of best fit. How do the coefficients of restitution compare?

b. Pick one of your 2 balls from part (a). Put the ball in a freezer for at least an hour. Collect data for the heights of bounces of the frozen ball and fit this data with an exponential function. How does this coefficient of restitution compare to that of the room-temperature ball?

FIGURE 3.25 Record heights of bounces.

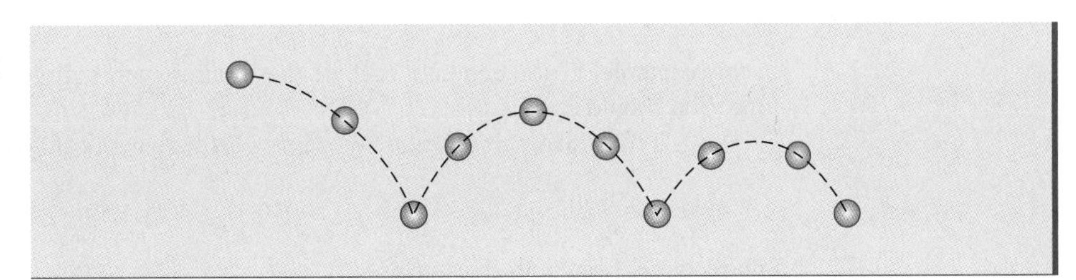

3.5 Translations of Exponential Functions

In this section, we will learn why the expression for the function $u(n)$ that satisfies the dynamical system

$$u(n) = au(n-1) + b \quad \text{with} \quad a \neq 1$$

is of the form

$$u(n) = ca^n + E$$

where E is the equilibrium value for the dynamical system and c is a constant that can be determined from the initial value. This function is just the exponential function ca^n translated by the equilibrium value E.

When $a = 1$, the dynamical system is

$$u(n) = u(n-1) + b$$

and you already know that the function can be written as

$$u(n) = nb + c$$

Consider the dynamical system

$$u(n) = 0.7u(n-1) + 1.2$$

with $u(0) = 14$. The equilibrium value is $E = 4$. Thus, the claim is that $u(n)$ can be written as

$$u(n) = c(0.7^n) + 4$$

for some value of c. Let's see why.

Recall that in Section 2.4, we analyzed the rate of change in the function $u(n)$ by comparing how far consecutive values were from equilibrium. In particular, we computed

$$\frac{u(1)-4}{u(0)-4}, \frac{u(2)-4}{u(1)-4}, \ldots, \frac{u(n)-4}{u(n-1)-4}$$

In this example, if you compute each of these ratios, you will get the constant 0.7 each time. You should try this.

Let's look at why this happens. When $n = 1$, the dynamical system gives

$$u(1) = 0.7u(0) + 1.2$$

Subtracting 4 from both sides gives

$$u(1) - 4 = 0.7u(0) - 2.8$$

Factoring 0.7 on the right side of the equation gives

$$u(1) - 4 = 0.7(u(0) - 4)$$

Dividing both sides by $u(0) - 4$ gives

$$\frac{u(1)-4}{u(0)-4} = 0.7$$

Similarly,

$$u(2) = 0.7u(1) + 1.2$$

so

$$u(2) - 4 = 0.7u(1) - 2.8 = 0.7(u(1) - 4)$$

and

$$\frac{u(2)-4}{u(1)-4} = 0.7$$

In fact, if we subtract 4 from both sides of the dynamical system, we get

$$u(n) - 4 = 0.7u(n-1) - 2.8$$

or

$$u(n) - 4 = 0.7(u(n-1) - 4)$$

The dynamical system written in the form $u(n) - 4 = 0.7(u(n-1) - 4)$ means that to get $u(1) - 4$, you multiply $u(0) - 4$ by 0.7. To get $u(2) - 4$, you multiply $u(1) - 4$ by 0.7, which is the same as multiplying $u(0) - 4$ by 0.7^2. Each time we increase n by one, we multiply $u(0) - 4$ by another 0.7. To get $u(n) - 4$, we multiply $u(0) - 4$ by 0.7 n times. This means that

$$u(n) - 4 = 0.7^n(u(0) - 4)$$

Adding 4 to both sides gives

$$u(n) = 0.7^n(u(0) - 4) + 4$$

Note that the value $c = u(0) - 4 = 14 - 4 = 10$, so

$$u(n) = 10(0.7^n) + 4$$

and we see that $u(n)$ is of the correct form.

Let's try this argument more formally. We have that $u(n) = 0.7u(n-1) + 1.2$ can be rewritten as

$$u(n) - 4 = 0.7(u(n-1) - 4)$$

by subtracting 4 from both sides then factoring out 0.7. We define a new function $v(n)$ as

$$v(n) = u(n) - 4$$

Note that the new function $v(n)$ is just the old function $u(n)$ translated by -4. This also means that

$$v(n-1) = u(n-1) - 4$$

Substitution of $v(n)$ and $v(n-1)$ for $u(n)$ and $u(n-1)$ in the dynamical system $u(n) - 4 = 0.7(u(n-1) - 4)$ gives

$$v(n) = 0.7v(n-1)$$

We know from Section 3.3 that

$$v(n) = c0.7^n$$

We now substitute $u(n) - 4$ for $v(n)$ to get $u(n) - 4 = c0.7^n$, or

$$u(n) = c0.7^n + 4$$

Geometrically, the solution $u(n) = c0.7^n + 4$ to the dynamical system

$$u(n) = 0.7u(n-1) + 1.2$$

is the solution $v(n) = c0.7^n$ to the dynamical system

$$v(n) = 0.7v(n-1)$$

translated by the equilibrium value. This can be seen in Figure 3.26, in which solutions to $u(n) = 0.7u(n-1) + 1.2$ and to $v(n) = 0.7v(n-1)$ are both graphed. Note that $u(0)$ is 4 more than $v(0)$.

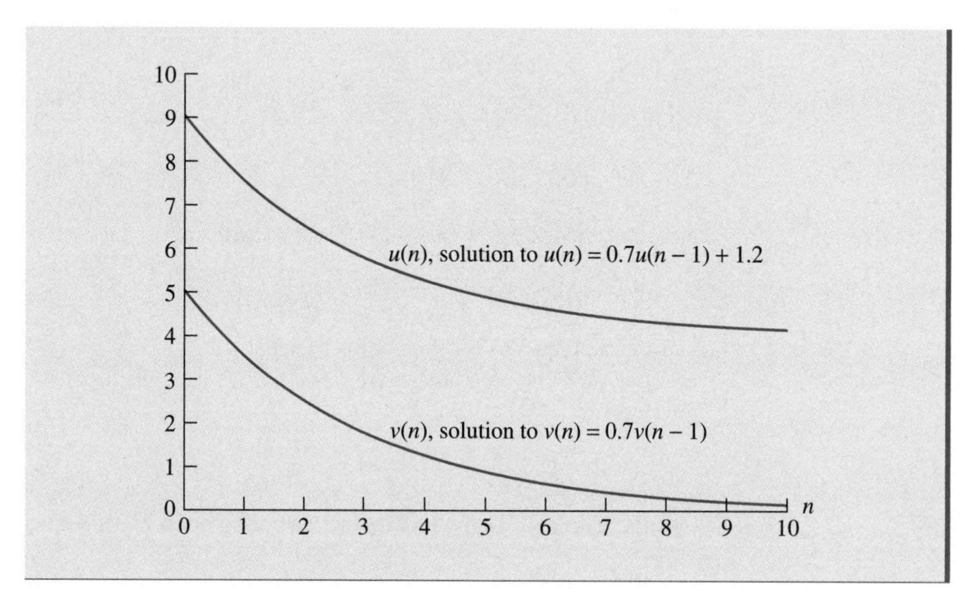

FIGURE 3.26 Comparing solution to $u(n) = 0.7u(n-1) + 12$ and $v(n) = 0.7v(n-1)$.

Using the same argument in the general case gives the following result.

THEOREM

The function that satisfies the dynamical system

$$u(n) = au(n-1) + b$$

where $a \neq 1$, can be written as

$$u(n) = ca^n + E$$

where E is the equilibrium value. The particular value for c depends on the given initial value.

Example 3.13 Consider the dynamical system

$$u(n) = 5u(n-1) + 2$$

The equilibrium value for this dynamical system is the solution to $E = 5E + 2$, which is $E = -0.5$. By the previous theorem,

$$u(n) = c5^n - 0.5$$

Suppose we are given that $u(0) = 3$. To find c, we substitute 0 for n into the expression to get

$$u(0) = c5^0 - 0.5 = c - 0.5$$

We substitute 3 for $u(0)$ and solve, getting

$$3 = c - 0.5 \quad \text{or} \quad c = 3.5$$

Thus

$$u(n) = 3.5\,(5)^n - 0.5$$

Instead of being given $u(0)$, suppose we were given that $u(2) = 12$. The method for finding c is the same. We substitute 2 for n into the expression $u(n) = c5^n - 0.5$, giving

$$u(2) = c5^2 - 0.5 = 25c - 0.5$$

Substituting 12 for $u(2)$ and solving gives

$$12 = 25c - 0.5 \quad \text{or} \quad c = 0.5$$

Thus the function can be written as

$$u(n) = 0.5(5)^n - 0.5.$$

As you see, the value of c is determined from the initial value you are given. The initial value that you are given does not need to be $u(0)$. You should try finding solutions given that $u(0) = -2$ and given that $u(1) = 15.5$. ■

Sometimes we know $u(n)$ for two different values of n, but we don't know the value for b in the dynamical system $u(n) = au(n-1) + b$. Using the explicit form of the function, $u(n)$, we can find the value for b algebraically.

Example 3.14

Suppose that we have a bank account that earns 5% interest, compounded annually. We deposit \$5000 into the account today and deposit b dollars at the end of each of the next 20 years. After 20 years, we want to have \$200,000 in our account. See Figure 3.27 for the flow diagram for this problem. What should our deposit be?

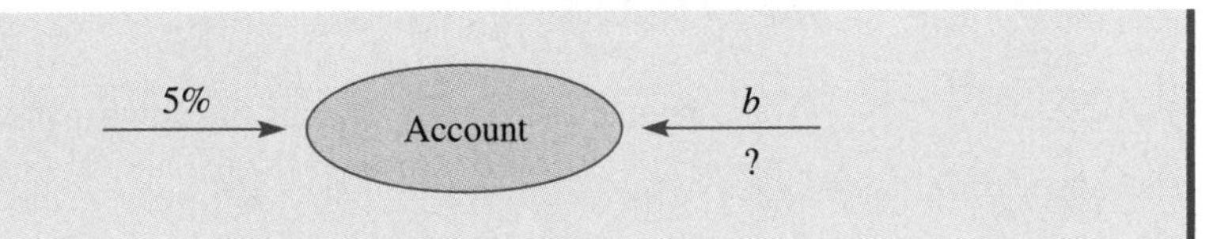

FIGURE 3.27 $u(0) = 5000$ and goal is $u(20) = 200{,}000$.

The dynamical system that models this account is

$$u(n) = 1.05u(n-1) + b$$

We also know that $u(0) = 5000$ and $u(20) = 200,000$. We could explore using our calculator or spreadsheet until we find a b-value that gives the desired result. Instead, we will find b directly.

The equilibrium value E is the solution to $E = 1.05E + b$ or $E = -20b$. Thus, $u(n)$ can be written as

$$u(n) = c(1.05)^n - 20b$$

We have

$$5000 = u(0) = c(1.05)^0 - 20b = c - 20b$$

This gives $c = 20b + 5000$. By substitution,

$$u(n) = (20b + 5000)(1.05)^n - 20b$$

We also have that

$$200,000 = u(20) = (20b + 5000)(1.05)^{20} - 20b$$

or

$$200,000 = 20b(1.05)^{20} + 5000(1.05)^{20} - 20b$$

You can solve this equation generally, or you can replace 1.05^{20} with its decimal equivalent (to 4 decimal places) 2.6533. This gives

$$200,000 = 20(2.6533)b + 5000(2.6533) - 20b$$

or, after multiplying,

$$200,000 = 53.066b + 13,266.5 - 20b$$

We subtract 13,266.5 from both sides, giving

$$186,733.5 = 53.066b - 20b = 33.06b$$

Dividing both sides by 33.06 gives $b = \$5648.32$, which is what we need to deposit each year to reach our goal of $u(20) = \$200,000$.

The more decimal places we keep when approximating 1.05^{20}, the more accurate will be our final answer. ■

We cannot usually solve for an unknown parameter a in a dynamical system $u(n) = au(n-1) + b$. Suppose we had the dynamical system

$$u(n) = au(n-1) - 8$$

and we knew that $u(0) = 9$ and $u(4) = 409$. We would find the equilibrium value by solving the equation $E = aE - 8$ for E. This gives $E = -8/(1-a)$. The function would then be written as

$$u(n) = ca^n - \frac{8}{1-a}$$

We substitute 0 for n, getting

$$9 = c - \frac{8}{1-a} \quad \text{or} \quad c = 9 + \frac{8}{1-a}$$

This gives the function

$$u(n) = \left(9 + \frac{8}{1-a}\right)a^n - \frac{8}{1-a}$$

Substitution of 4 for n gives

$$409 = \left(9 + \frac{8}{1-a}\right)a^4 - \frac{8}{1-a}$$

In general, we cannot solve equations such as this. There are 2 ways to get an accurate estimate for a. First, we could resort to the methods of Sections 1.3 and 1.4, in which we try different values for a in the dynamical system or the preceding equation until we get a value for a that gives an acceptable result. Second, we could graph the 2 functions

$$f(a) = 409 \quad \text{and} \quad g(a) = \left(9 + \frac{8}{1-a}\right)a^4 - \frac{8}{1-a}$$

which correspond to the 2 sides of the equation, as seen in Figure 3.28. We could then, using a calculator or computer, find the value for a where the functions intersect. Using either method, you would find that $a = 3$, exactly.

FIGURE 3.28 Functions $f(a)$ and $g(a)$ intersect at (3, 409).

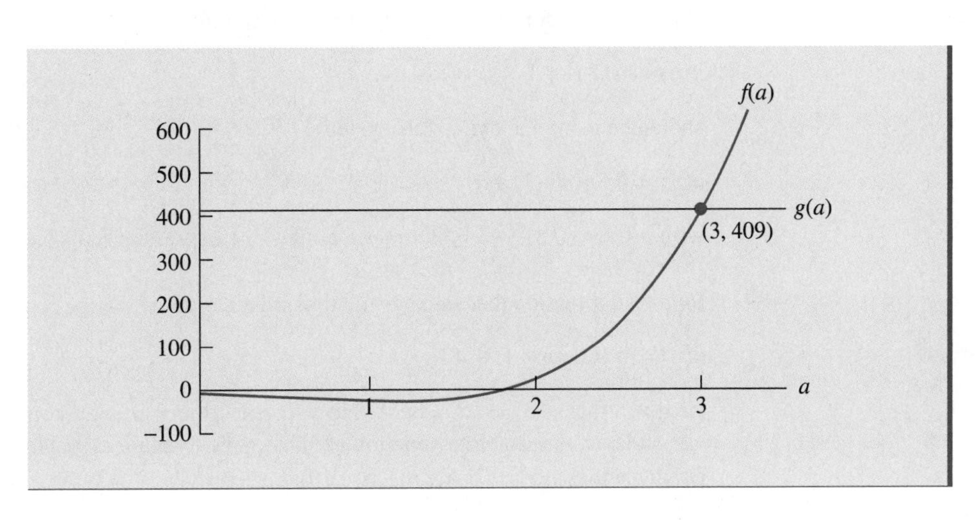

3.5 Problems

1. Compute $u(n)$ for several values of n using the function

$$u(n) = 15(0.8)^n + 5$$

and again using the dynamical system

$$u(n) = 0.8u(n-1) + 1$$

with $u(0) = 15(0.8)^0 + 5 = 20$, to determine if the expression satisfies the dynamical system.

2. Compute $u(n)$ for several values of n using the function

$$u(n) = -(0.8)^n + 5$$

and again using the dynamical system

$$u(n) = 0.8u(n-1) + 1$$

with $u(0) = -(0.8)^0 + 5 = 4$, to determine if the expression satisfies the dynamical system.

3. Compute $u(n)$ for several values of n using the function

$$u(n) = (0.2)^n + 5$$

and again using the dynamical system

$$u(n) = 0.8u(n-1) + 1$$

with $u(0) = (0.2)^0 + 5 = 6$, to determine if the expression satisfies the dynamical system.

4. Compute $u(n)$ for several values of n using the function

$$u(n) = (0.8)^n + 1$$

and again using the dynamical system

$$u(n) = 0.8u(n-1) + 1$$

with $u(0) = (0.8)^0 + 1 = 2$, to determine if the expression satisfies the dynamical system.

5. Find an expression that satisfies the dynamical system

$$u(n) = -0.5u(n-1) + 12$$

a. given that $u(0) = 18$. Check that your function works by computing $u(n)$ for several values of n using your function and using the dynamical system.
b. given that $u(5) = 7$. Again, check that it works.

6. Find an expression that satisfies the dynamical system

$$u(n) = -u(n-1) - 7$$

a. given that $u(0) = 10$. Check that it works.
b. given that $u(40) = 15.5$. Again, check that it works.

7. Find an expression that satisfies the dynamical system

$$u(n) = 0.5u(n-1) + 16$$

a. given that $u(0) = 5$. **b.** given that $u(8) = 35$.

8. Find an expression that satisfies the dynamical system

$$u(n) = 6u(n-1) - 20$$

a. given that $u(0) = 94$. **b.** given that $u(10) = 604{,}665.76$.

9. Find an expression that satisfies the dynamical system

$$u(n) = -2u(n-1) + 3 \quad \text{with} \quad u(0) = 4$$

10. Find an expression that satisfies the dynamical system

$$u(n) = u(n-1) + 0.321 \quad \text{with} \quad u(0) = -5.98$$

11. Find an expression that satisfies the dynamical system

$$u(n) = -0.5u(n-1) + 3 \quad \text{with} \quad u(4) = 1$$

12. Find the value of the parameter b algebraically, given that

$$u(n) = 2u(n-1) + b \quad u(0) = 3 \quad \text{and} \quad u(5) = 313$$

13. Find the value of the parameter b, algebraically, given that

$$u(n) = 0.5u(n-1) + b \quad u(0) = 106 \quad \text{and} \quad u(6) = -20$$

14. Suppose that a person burns 4% of his or her body weight each week through basic metabolism. Each week this person consumes 18,000 kcal, which is the equivalent of 5 pounds of weight.
a. Develop a dynamical system for how this person's weight changes.
b. Find an expression that satisfies that dynamical system given that this person weighs $w(0) = 140$ pounds now.

c. What will this person weigh in 1 year?

d. How long will it take until this person weighs 135 pounds? Solve for n algebraically.

15. Suppose you have $100,000 in a bank account that earns 10% interest, compounded annually. You will withdraw the same fixed amount from this bank account every year for 20 years, beginning next year. At that point, the amount in the account will be 0. Solve for this fixed withdrawal algebraically. Get your answer to the nearest cent. Check that it works.

16. Suppose that a person burns 4% of his or her body weight each week through basic metabolism. Each week this person consumes the same amount of food. Suppose this person weighs $w(0) = 140$ pounds this week and in 20 weeks weighs $w(20) = 135$ pounds. Find the amount of food, in pounds of weight, he or she consumes each week. How many kcal is this if there are 3600 kcal for each pound of weight?

3.6 Curve Fitting

In previous sections we have tried fitting data with linear and exponential functions, depending on whether the data "looked" linear or curved. In this section we discuss how to determine if it is reasonable to fit a translation of an exponential function

$$u(n) = ca^n + E$$

to the data. Let's consider a specific example.

Example 3.15 Consider the data in Table 3.9. (Source: Department of Transportation).

TABLE 3.9 Registered motorcycles in the United States.

year	1987	1988	1989	1990	1991
motorcycles in the United States	4,917,131	4,584,284	4,433,195	4,259,462	4,177,037

The corresponding points are plotted in Figure 3.29, with n representing the number of years since 1987. It is clear that a line does not fit these points very well.

Because of the "curved" pattern in the points, I tried fitting an exponential regression curve to the data. I got the function

$$u(n) = 4,838,440(0.96)^n$$

as the best exponential fit to these points. This function implies that the number of motorcycles is decreasing by about 4% per year. This function is seen in Figure 3.29. The coefficient of determination for this function, which is a measure of how well the curve fits the data, is

$$0.9560$$

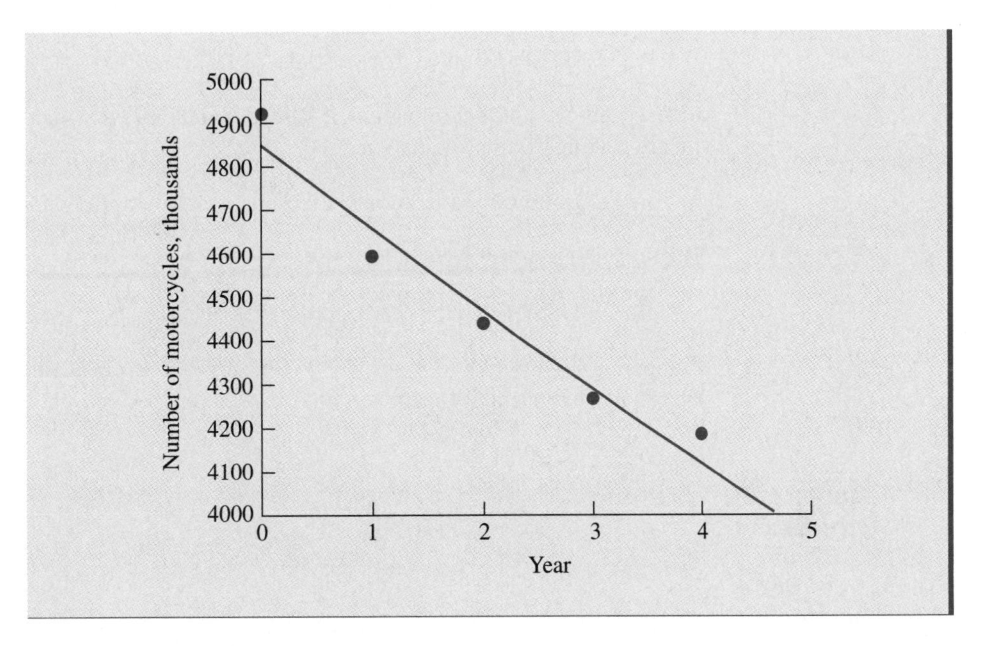

FIGURE 3.29 Motorcycles and exponential best fit.

which is close to 1. This suggests a pretty good fit. While the function fits the data fairly well, it appears in Figure 3.29 that there is more of a curve to the points than there is to the function.

Let's now try to fit the data with a function of the form

$$u(n) = ca^n + E$$

The constant E is the equilibrium value for the dynamical system and is a horizontal asymptote for the function. To obtain a crude estimate for E, we can sketch our own curve, by hand, on the graph and estimate where we think our curve has a horizontal asymptote. I chose $E = 3,500,000$, for no particular reason. Observe that fitting the data u-values with the function $u(n) = ca^n + 3,500,000$ is the same as fitting the values $u(n) - 3,500,000$ to an exponential ca^n, that is,

$$u(n) - 3,500,000 = ca^n$$

I computed the data values $u(n) - 3,500,000$, which are in Table 3.10.

TABLE 3.10 **Registered motorcycles in the United States.**

year	1987	1988	1989	1990	1991
(motorcycles − 3,500,000)	1,417,131	1,084,284	933,195	759,462	677,037

The exponential regression function that fits the values in Table 3.10 is

$$1,357,611(0.83)^n$$

with a coefficient of determination of 0.9825, which is better than before. This means that we are making the approximation

$$u(n) - 3,500,000 = 1,357,611(0.83)^n$$

This function can be rewritten as

$$u(n) = 1,357,611(0.83)^n + 3,500,000$$

Comparing Figure 3.29 to Figure 3.30 indicates that this new function is a better fit than the original exponential function.

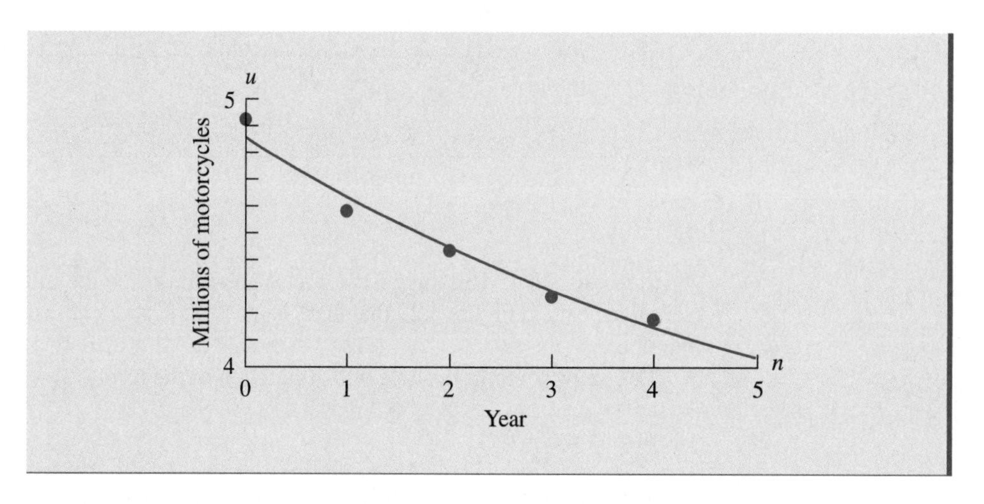

FIGURE 3.30 Motorcycles and $u(n) = 1,357,611(0.83)^n + 3,500,000$.

I tried this approach again to see if I could get an even better fit. This time I used $E = 3,700,000$. I got the function

$$u(n) = 1,167,273(0.792)^n + 3,700,000$$

with a coefficient of determination of 0.9883. This value is closer to 1 than the previous one, so we seem to have a better fit than before.

I kept repeating this process, increasing E by 100,000 each time, until the coefficient of determination moved away from 1. The best fit I found was when $E = 4,000,000$. In this case, the function was

$$u(n) = 915,867(0.66)^n + 4,000,000$$

with a coefficient of determination of 0.9955. The data points and function are seen in Figure 3.31.

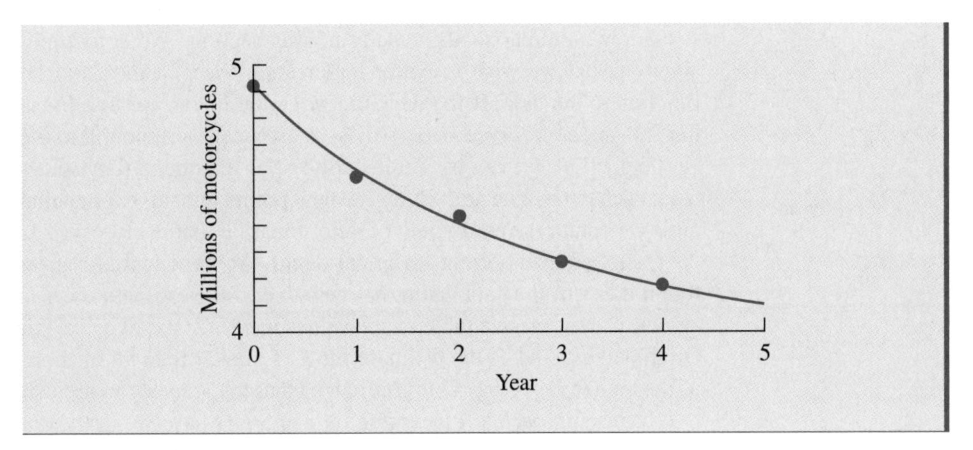

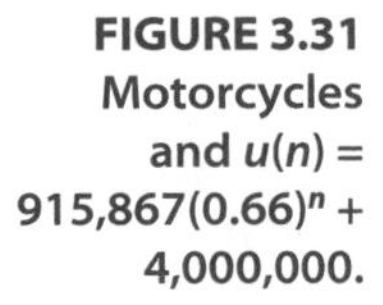

FIGURE 3.31 Motorcycles and $u(n) = 915{,}867(0.66)^n + 4{,}000{,}000$.

Recall that a function of the form $u(n) = ca^n + E$ satisfies a dynamical system of the form $u(n) = au(n-1) + b$. We conclude that the motorcycle data might be described by a dynamical system of the form

$$u(n) = 0.66u(n-1) + b$$

We also know that the equilibrium value is $E = 4{,}000{,}000$, so we solve

$$4,000,000 = 0.66(4,000,000) + b$$

to get $b \approx 1,360,000$. Thus the dynamical system is

$$u(n) = 0.66u(n-1) + 1,360,000$$

To interpret this dynamical system, we recall that the ratios of differences of u-values from equilibrium are constant and are equal to the coefficient of the $u(n-1)$-term, that is,

$$\frac{u(n) - 4,000,000}{u(n-1) - 4,000,000} = 0.66$$

This means that each year we expect that the number of motorcycles is about 66% as far from the equilibrium value as the previous year, or that each year the number of motorcycles is 34% closer to 4,000,000 than the previous year, and eventually we expect motorcycle ownership to level off at around 4,000,000.

I must note that there is nothing that we know about motorcycle ownership to suggest that this is true. A researcher would try to understand why motorcycle registration seems to be changing this way. Other factors might change and could result in a change in motorcycle registration (e.g., higher gasoline prices might result in more people switching from cars to motorcycles). The changing age structure of our population might also affect motorcycle ownership. We should not expect our model to remain true forever. ■

I now summarize the results of this section. We sometimes have data from a real situation that we wish to better understand. We plot the data on a graph and try to fit a function to the data. If the data appears to be linear, we fit a linear function to the data. If the data appears curved, we try to fit an exponential function to the data. If the exponential isn't a good fit, we can try a translation of an exponential by subtracting the same constant from each data value and fitting the new points with an exponential. We repeat this process until we obtain the coefficient of determination that is closest to 1.

Here is an important technical detail. An exponential regression function cannot be found if any of the data values are negative. If the estimate for E is less than all of the data values, u, then make a table of the values $u(n) - E$ and fit an exponential to this set of data. On the other hand, if the estimate for E is greater than all of the data values, make a table of the values $E - u(n)$. This guarantees that the values you are using will all be positive.

If this approach is successful, then not only have we fit the data to a function, we have an approximation for the equilibrium value for this system and can predict the rate at which we approach this equilibrium over time.

Sometimes no matter what number we subtract, the resulting exponential function is not a good fit with the data. This means that this set of data cannot be approximated well by an exponential or a translation of an exponential. In practice, we might try to fit the data to other functions, such as a polynomial. If this is successful, we would then try interpret what this means for the situation being studied.

3.6 Problems

1. Consider the following data.

$n =$	0	1	2	3	4
$u(n) =$	35	30	27.5	26.25	25.625

a. Plot these points on a graph.

b. The points are clearly not linear. Fit this data with an exponential function. Graph this function on the same axes as the data points. Is it a good fit? What is the coefficient of determination?

c. Fit the points $(n, u(n) - 5)$ with an exponential function and use this to find a function for $u(n)$. Graph this function on the same axes as the original set of points. Is it a good fit? What is the coefficient of determination?

$n =$	0	1	2	3	4
$u(n) - 5 =$	30	25	22.5	21.25	20.625

d. Repeat part (c) but subtract 15 instead of 5 from each $u(n)$ value.

e. Repeat part (c) but subtract 25 instead of 5 from each $u(n)$ value.

f. From parts (b), (c), (d), and (e), pick the one in which the function fits the data the best by seeing which coefficient of determination was closest to 1.

2. Consider the following data.

$n =$	0	1	2	3	4	5
$u(n) =$	130	126	122.8	120.24	118.19	116.55

a. Plot these points on a graph.
b. Fit this set of data with a function $u(n) = ca^n + E$, finding E accurate to the nearest integer.
c. Develop a dynamical system that is satisfied by the expression found in part (b).

3. Consider the following data.

$n =$	0	1	2	3	4
$u(n) =$	24200	23520	22894	22319	21789

a. Plot these points on a graph.
b. Fit this set of data with a function $u(n) = ca^n + E$, finding E accurate to the nearest hundred.
c. Develop a dynamical system that is satisfied by the expression found in part (b).

4. Consider the following data.

$n =$	0	1	3	6
$u(n) =$	158	174.74	195.88	210.62

a. Plot these points on a graph.
b. Fit this set of data with a function $u(n) = ca^n + E$, finding E accurate to the nearest integer. To do this, make a guess for E, then use the data $E - u(n)$ instead of $u(n) - E$. The data you use must be positive to be able to generate an exponential regression function.
c. Develop a dynamical system that is satisfied by the expression found in part (b).

5. The numbers of vehicle-miles traveled per capita in the United States from 1987 to 1991 are given in the following table. (Source: U.S. Department of Transportation)

year	1987	1988	1989	1990	1991
Vehicle-miles/capita = $m(n)$ =	7915	8266	8474	8580	8596

Let n be the number of years since 1987 and let $m(n)$ be the number of vehicle-miles per capita in year n.
a. Plot the points on a graph.

b. The points on the graph appear to be of the form

$$m(n) = ca^n + \text{"equilibrium value"}$$

From looking at your graph, make a guess for the equilibrium value E. Make a table similar to the following table. Note we are subtracting the values from the equilibrium so that the numbers in the table are positive.

$n =$	0	1	2	3	4
E – Vehicle-miles/capita = $E - m(n) =$					

Fit these points to an exponential function, and use that to find a regression function for $m(n)$. Keep repeating until you find a reasonably good fit. Graph this function on the same axis as the original data points.

c. What is the dynamical system that $m(n)$ satisfies? What is the equilibrium for the number of miles driven? What is the ratio

$$\frac{m(n) - E}{m(n-1) - E}$$

and what does this say about U.S. driving habits? (Note that this may not be a good estimate. We have no reason to believe the dynamical system fits data outside of the years used to develop the dynamical system. Things might change in U.S. driving habits to increase or decrease the equilibrium value and how we approach it. A dramatic increase in oil prices might considerably reduce the amount driven. Urban flight and an increase in roads might increase it.)

6. Let n be the number of years since 1956 and let $p(n)$ be the percentage of U.S. roads that were paved in year n. Data collected from 1956 through 1996 is given in the following table. (Source: U.S. Department of Transportation)

year	1956	1966	1976	1986	1996
percentage of U.S. roads that are paved	30.4	37.2	48.9	55.6	60.5

a. Plot the points on a graph.

b. The points on the graph appear to be of the form

$$p(n) = ca^n + E$$

From looking at your graph, make a guess for the equilibrium value, E. Complete a table similar to the following table

$n =$	0	10	20	30	40
$E - p(n) =$					

Fit these points to an exponential function and use that to find a regression function for $p(n)$. Keep repeating, finding E to the nearest integer. Graph this function on the same axis as the original data points.

c. What is the dynamical system that $p(n)$ satisfies? What is the equilibrium for the percentage of roads that are paved? How quickly is equilibrium being approached? (*Note:* As in the previous problem, it is rarely a good idea to predict the future based only a model developed from the data. It is better if you can justify the model, once it is developed. Also, the assumptions on which the model is based might change over time.)

4 Higher Order Dynamical Systems

4.1 Introduction to Higher Order Dynamical Systems

In this chapter, we are going to investigate situations in which $u(n)$ depends on two or more previous values, $u(n-1)$, $u(n-2)$, and so on. Consider the dynamical system

$$u(n) = 2u(n-1) - 3u(n-2) + 5$$

Substitution of 2 for n in this dynamical system gives

$$u(2) = 2u(1) - 3u(0) + 5$$

To use this equation, we need to be given the two starting values, $u(0)$ and $u(1)$. Suppose we are given that $u(0) = 1$ and $u(1) = 6$. Substitution gives

$$u(2) = 2(6) - 3(1) + 5 = 14$$

We could now substitute 3 for n in the dynamical system and use the values for $u(2)$ and $u(1)$ to find that $u(3) = 15$. I continued in this fashion, obtaining the values for $u(4)$ through $u(10)$, which are given in Table 4.1. You should see if you can obtain these same values.

TABLE 4.1 ***u*(*n*) – values where *u*(*n*) = 2*u*(*n* − 1) – 3*u*(*n* – 2) + 5, given that *u*(0) = 1 and *u*(1) = 6.**

$n =$	0	1	2	3	4	5	6	7	8	9	10
$u(n) =$	1	6	14	15	−7	−54	−82	3	257	510	254

We can make a time graph by plotting the points $(n, u(n))$, as seen in Figure 4.1. The points have been connected with lines so that the pattern is easier to see.

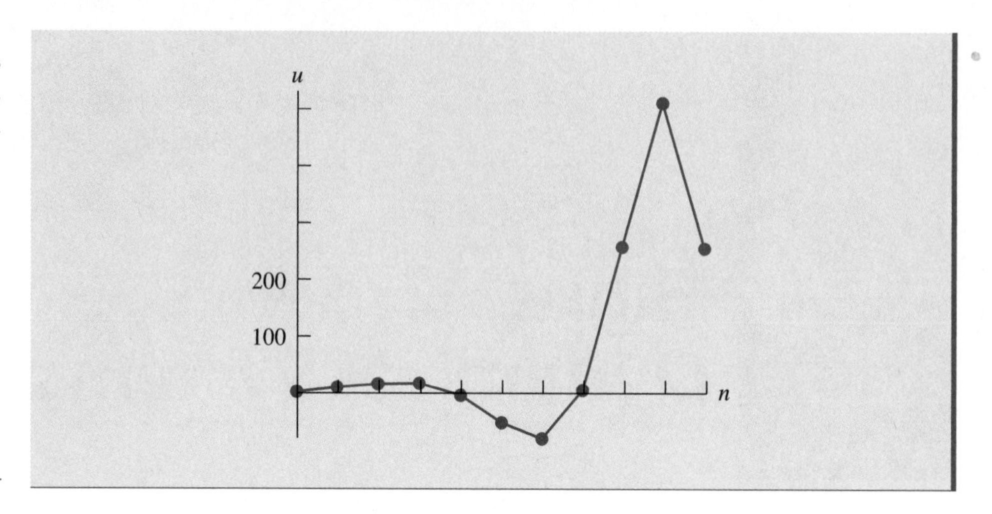

FIGURE 4.1 $(n, u(n))$ for $u(n) = 2u(n-1) - 3u(n-2) + 5$.

The dynamical system $u(n) = 2u(n-1) - 3u(n-2) + 5$ is called a second-order dynamical system because $u(n)$ depends on the previous two u-values.

A second-order dynamical system that you may already be familiar with is

$$u(n) = u(n-1) + u(n-2)$$

with $u(0) = 0$ and $u(1) = 1$. This gives the sequence of numbers in which each number is the sum of the previous two numbers. This sequence was studied by Leonardo Pisano Fibonacci, who lived in the late 12th and early 13th centuries. It is therefore called the **Fibonacci sequence.** You should use this dynamical system to check that $u(0)$ through $u(9)$ are

$$0, 1, 1, 2, 3, 5, 8, 13, 21, 34, \ldots$$

It is said that Fibonacci developed this sequence as a model for the growth of a population of rabbits. His assumptions were that each year, each rabbit gives birth to 1 rabbit and that each rabbit is alive for 2 years. Figure 4.2 demonstrates his assumptions starting with 1 adult rabbit. Note that the number of rabbits alive each year agrees with the second through fifth numbers in the above sequence. If you continued the diagram, you would continue getting consecutive numbers in this sequence. You should try thinking about why each number in this sequence is the sum of the previous 2 numbers, given Fibonacci's assumptions.

Fibonacci's assumptions are not very realistic and this is not a good model for the growth of populations of rabbits. But this sequence of numbers does occur in many different places in mathematics and in applications.

The dynamical systems in the previous chapters were first-order dynamical systems, since they depended on only the previous u-value. The dynamical system

$$u(n) = u(n-1) + u(n-2) - u(n-3)$$

is an example of a third-order dynamical system. To compute the u-values for a third-order dynamical system, we need to be given three initial values. For example, if we were given

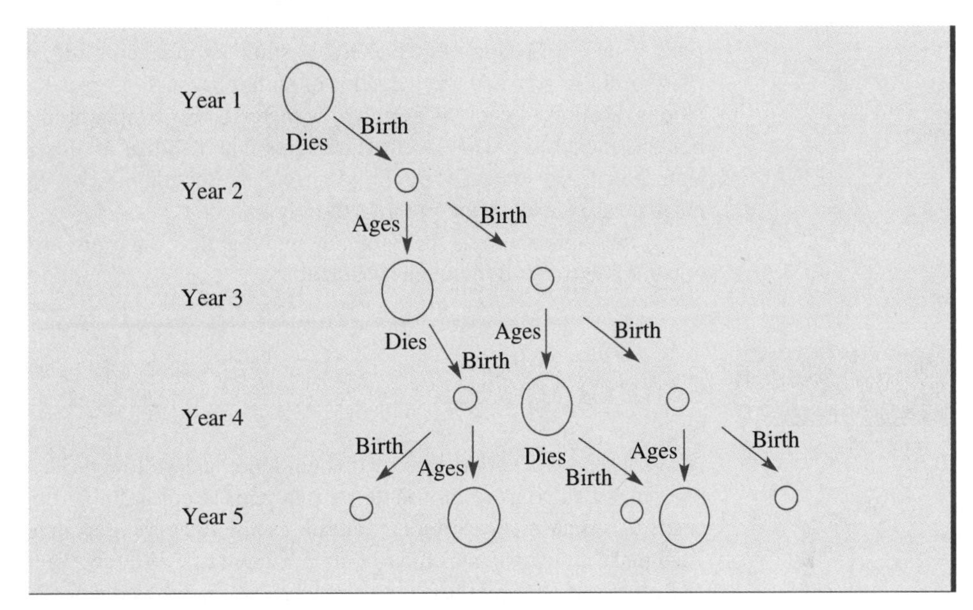

FIGURE 4.2 Fibonacci model for rabbit population. Each rabbit gives birth, baby rabbits age, and adult rabbits die.

that $u(0) = 1$, $u(1) = 3$, and $u(2) = 7$, then we would be able to compute that $u(3) = 7+3-1 = 9$, that $u(4) = 9+7-3 = 13$, and so on. The order of a system determines the number of initial values needed to compute $u(n)$-values. Dynamical systems that depend on more than the previous value of u are called **higher order dynamical systems.** It is intuitively easy to understand the order of a dynamical system. It is a little more complicated to define it formally, but we will do so anyway.

DEFINITION

Suppose we have a dynamical system for $u(n)$, given in terms of $u(n-1)$, $u(n-2)$, $\ldots$, $u(n-k)$. Then this system is called a ***k*th-order dynamical system.**

You do not actually need $u(n-1)$, $u(n-2)$, $\ldots$ as part of the dynamical system as long as $u(n-k)$ is an explicit part of the dynamical system. For example, the dynamical system

$$u(n) = 2u(n-2)$$

is a second-order dynamical system in which the term $u(n-1)$ is not part of the system. It is second order because $u(n-2)$ is part of the equation. Note that if we are given only $u(0)$, say $u(0) = 3$, then we can compute $u(2)$, $u(4)$, ... In fact, we can compute $u(n)$ for any even value of n. But we need to know $u(1)$ to compute $u(3)$, $u(5)$, ... This means that if we have a kth-order dynamical system and are given k initial values, $u(0)$, $u(1)$, $\ldots$, $u(k-1)$, then we can compute $u(k)$, $u(k+1)$, ...

In Section 4.2, you will see how higher order dynamical systems can be used to count complex collections of objects. Section 4.3 will discuss some background in probability that will be necessary in Section 4.4. In Section 4.4, higher order dynamical systems are used to investigate the effectiveness of some simple gambling systems. Section 4.5 discusses equilibrium and the behavior of solutions, such as growth rates, cycles, and oscillations. Sections 4.6 and 4.7 apply this material to a simple model of a national economy. For this model, we will develop strategies that stabilize an unstable economy as well as learn how these strategies might go wrong. In Section 4.8, we will discuss how $u(n)$ can be given as an expression in terms of only n.

A word of caution is that you may find that flow diagrams are not helpful when developing higher order dynamical systems.

4.2 Counting Sets

I have a renowned collection of orange, blue, and yellow beads. The yellow beads have been modified so as to be usable only in pairs. I enjoy lining up ordered sequences of n beads to form as many different patterns as possible, going from left to right. For example, I can make a lineup consisting of only 1 bead in two different ways, namely, 1 orange bead or 1 blue bead. I cannot use any yellow beads, since I would need to use at least 2 of them.

Let $u(n)$ represent the number of different lineups of n beads. So, $u(1) = 2$, the two ways being represented by o and b. Table 4.2 gives the number of lineups of $n = 1, 2$, and 3 beads, along with actually listing those lineups.

TABLE 4.2 Lineups of 1 to 3 beads.

$u(1) = 2$	*o*	*b*			
$u(2) = 5$	*oo*	*ob*	*bo*	*bb*	*yy*
	ooo	*oob*	*obo*	*obb*	*oyy*
$u(3) = 12$	*boo*	*bob*	*bbo*	*bbb*	*byy*
	yyo	*yyb*			

Note that the order of the beads is important. For example, the lineup *obb* is considered to be different from the lineup *bbo*.

Look at the system used in listing the 12 three-bead lineups in the third row of Table 4.2. The first row of 3-bead lineups consists of all the ways of making the lineup, beginning with an o, the second row consists of all lineups that begin with b, and the third row consists of all lineups beginning with yy. Since each lineup has to begin with either o, b, or yy, we know that we have them all.

The first row of the 3-bead lineups was obtained by putting o in front of each of the 2-bead lineups, that is,

$$ooo \quad oob \quad obo \quad obb \quad oyy = o(oo \quad ob \quad bo \quad bb \quad yy)$$

Similarly, the second row of 3-bead lineups is just the list of 2-bead lineups with a b in front of each.

I claim that $u(4) = 29$. You should try to list all 29 ways in a systematic manner. In particular, make a first row of 4-bead lineups in which each lineup begins with an o. You can easily make this list using the list of 3-bead lineups. You can similarly make a row of 4-bead lineups in which each lineup begins with a b. Make a third row of 4-bead lineups in which the first 2 beads are yellow. You can use the list of 2-bead lineups to make these.

We are now going to develop a systematic method for generating all lines of 5 beads using single orange beads, single blue beads, and double yellow beads. First, to generate all 5-bead lineups that begin with an orange, just put an "o" to the left of each 4-bead lineup. There are $u(4)$ lineups in this list. To generate all 5-bead lineups that begin with b, just put a "b" to the left of each 4-bead lineup. Again, there are $u(4)$ lineups in this list. Finally, to generate all 5-bead lineups that begin with a double-yellow bead, just put "yy" to the left of each 3-bead lineup. There are $u(3)$ lineups in this list. This will give all 5-bead lineups since each lineup must begin with o, b, or yy.

This approach makes it clear that there are

$$u(5) = u(4) + u(4) + u(3) = 29 + 29 + 12 = 70$$

5-bead lineups. Similar logic says that there are

1. $u(n-1)$ n-bead lineups that begin with o
2. $u(n-1)$ n-bead lineups that begin with b
3. $u(n-2)$ n-bead lineups that begin with yy

Thus, $u(n)$ satisfies the second-order dynamical system

$$u(n) = 2u(n-1) + u(n-2)$$

To use this dynamical system, we need 2 initial values. We can use $u(1) = 2$ and $u(2) = 5$, which we found by actually making the lists. Use either a spreadsheet or a calculator to compute the number of 15-bead and 28-bead lineups. Find about how many times more bead arrangements exist when using one more bead. For example, how many times bigger is $u(31)$ than $u(30)$. Figure 4.3 demonstrates how quickly the number of lineups is increasing.

FIGURE 4.3 Number of possible lineups using single orange and blue beads and double yellow beads.

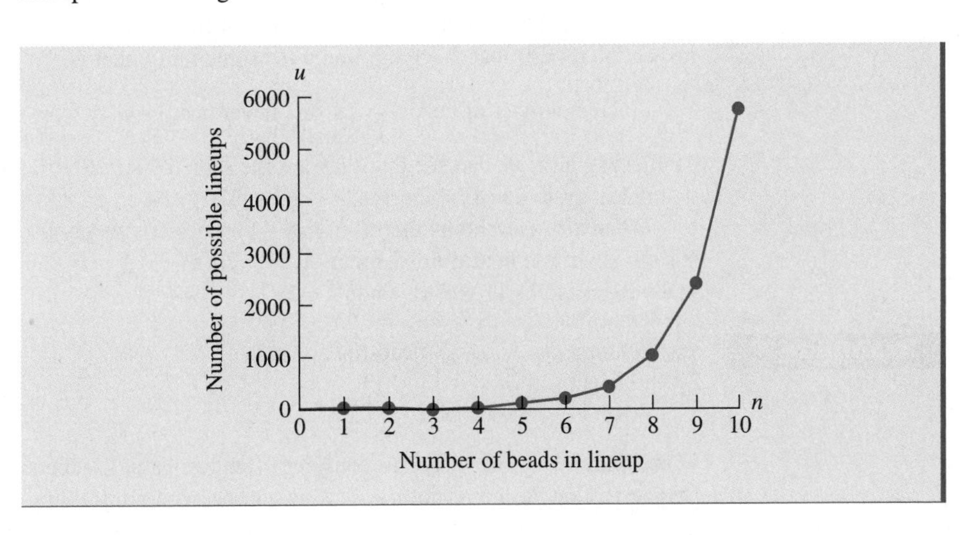

While flow diagrams are generally not helpful in developing higher order dynamical systems, tree diagrams can be very helpful. The tree diagram in Figure 4.4 summarizes the construction of the list of n beads.

FIGURE 4.4 Using tree to visualize construction of n-bead lineups.

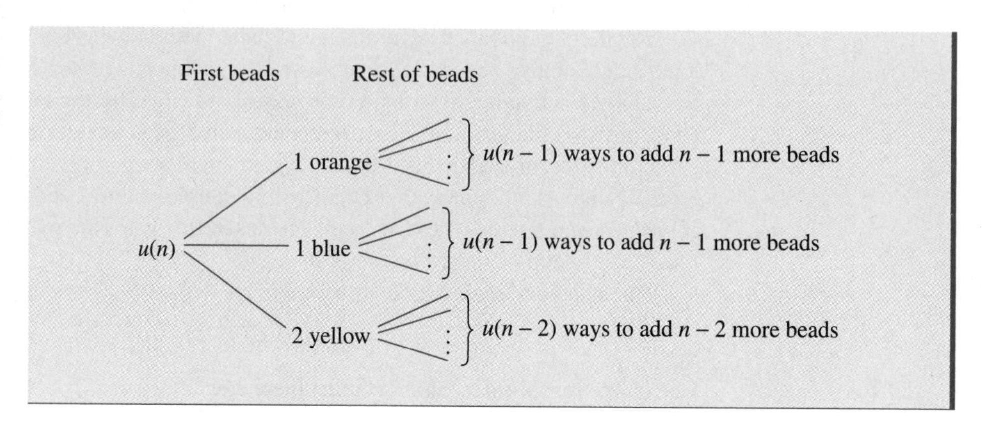

Let me summarize the method for computing the number of lineups of different lengths. Consider the first choice of beads in the lineup, then determine what remains to do. In this case, I assume I am constructing the $u(n)$ possible n-bead lineups. I put down an orange, a blue, or a double-yellow. I can complete the orange-starting lineups in $u(n-1)$ ways, since I still have $n-1$ beads to put into the lineup. Similarly, I can complete the blue-starting lineups in $u(n-1)$ ways and the yellow-starting lineups in $u(n-2)$ ways. Adding these ways together gives $u(n) = 2u(n-1) + u(n-2)$.

Sometimes higher order dynamical systems are written in the form

$$u(n+2) = 2u(n+1) + u(n)$$

This dynamical system appears to be different from $u(n) = 2u(n-1) + u(n-2)$. Further inspection reveals that they both imply the same thing, that is,

$$u(\text{some number of beads}) = 2u(\text{one fewer bead}) + u(\text{two fewer beads})$$

Mathematically, we can see that they are the same by substituting $n+2$ for every n in the dynamical system $u(n) = 2u(n-1) + u(n-2)$.

When $u(n)$ is written in terms of $u(n-1)$, $u(n-2)$, we say the second order dynamical system is written in **standard form.** $u(n) = 2u(n-1) + u(n-2)$ is in standard form, but $u(n+2) = 2u(n+1) + u(n)$ is not.

Example 4.1 Let's rewrite the dynamical system

$$u(n+1) = 3u(n) - 2u(n-1) + 7$$

in standard form. The idea is to consider $u(\text{highest value})$, which is $u(n+1)$. We want the "highest value" to be n, not $n+1$. This can be accomplished by substituting $n-1$ for n

throughout the dynamical system. This gives

$$u(n-1+1) = 3u(n-1) - 2u(n-1-1) + 7$$

which simplifies to

$$u(n) = 3u(n-1) - 2u(n-2) + 7$$

which is the dynamical system in standard form. ■

Example 4.2 Let's rewrite the dynamical system

$$u(n) = 2u(n+2) + 6u(n+1) - 10$$

in standard form. In this case, u(highest value) is $u(n+2)$. We now substitute $n-2$ for n throughout the dynamical system. This gives

$$u(n-2) = 2u(n+2-2) + 6u(n-2+1) - 10$$

which simplifies to

$$u(n-2) = 2u(n) + 6u(n-1) - 10$$

This is still not in standard form. We now solve for $u(n)$ by subtracting $6u(n-1) - 10$ from both sides, then dividing by 2. This gives

$$u(n) = -3u(n-1) + \tfrac{1}{2}u(n-2) + 5$$

which is the dynamical system in standard form. ■

Counting numbers of strands of beads does not seem to be a very important application, but problems similar to this can occur in many different areas. For example, consider deoxyribonucleic acid, or DNA, which makes up the human genome. DNA contains lineups of 4 different bases—adenine (A), thymine (T), cytosine (C), and guanine (G). The particular order of these bases determines specific genetic instructions. The number of lineups of these 4 bases determines the number of possible genetic traits. Molecules are made up of atoms that follow certain combination rules, such as the two yellows having to be together. The number of possible types of certain molecules can be determined by counting possible arrangements.

4.2 Problems

1. Rewrite the following dynamical systems in standard form.

a. $u(n+2) = 4u(n+1) - 3u(n)$
b. $u(n+1) = 1.3u(n) - 5$
c. $u(n+1) = 4.2u(n) - 2u(n-1)$
d. $u(n+1) = u(n) + 2.3u(n-1) + 3$
e. $u(n+3) - u(n+1) = 3u(n+2)$
f. $u(n-1) = 7u(n) - u(n+2)$

2. Suppose that orange beads come only as single beads, that blue beads come only in pairs, and that yellow beads come only in pairs. Let $u(n)$ be the number of different n-bead lineups, in which the order of the colors in the lineup is important.

a. Compute $u(1)$, $u(2)$, $u(3)$, and $u(4)$ by making lists of all possible 1-, 2-, 3-, and 4-bead lineups. Use the lists for 1- and 2-bead lineups to make the list of 3-bead lineups. Use the 2- and 3-bead lineups to make the list of 4-bead lineups.

b. Develop a second-order dynamical system for $u(n)$. Use this system and your calculator or spreadsheet to find $u(21)$, $u(22)$, $u(23)$, and $u(24)$.

c. Compute $u(22)/u(21)$, $u(23)/u(22)$, and $u(24)/u(23)$ to get a sense of about how many times bigger $u(n)$ is than $u(n-1)$. Check out $u(51)/u(50)$ to confirm your intuition. Note you are computing the ratios of consecutive values, which indicate the rate at which the size of the lists is growing.

3. Suppose you have blue and orange beads that come as singles, white beads that only come in pairs, and yellow beads that only come in pairs.

a. Find $u(1)$ and $u(2)$.

b. Develop a dynamical system for $u(n)$, the number of possible arrangements of n beads.

c. How many arrangements of 10 beads are there?

d. Find $u(10)/u(9)$, $u(16)/u(15)$, and $u(32)/u(31)$. What conclusion can you make?

4. Suppose you have 3 colored beads—yellow, orange, and blue—and they all come in singles. You want to make a line of n beads, except you never want to have 2 yellows together. Let $u(n)$ represent the number of ways to make this line of n beads.

a. What are $u(1)$, $u(2)$, $u(3)$, and $u(4)$?

b. Develop a dynamical system to model this problem. The following tree diagram may help.

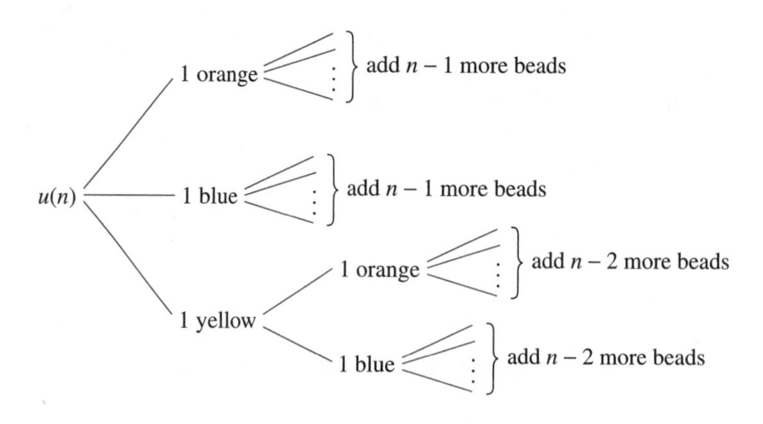

c. What is $u(10)$? What is the growth rate?

5. Suppose you have 4 colored beads, yellow, orange, green, and blue. Orange beads come in singles, blue beads and yellow beads come only as doubles, and green beads come only as triples. Let $u(n)$ represent the number of ways to make n-bead lineups, where the order is important; that is, *obb* is different from *bbo*.

a. What are $u(1)$, $u(2)$, and $u(3)$?

b. Develop a dynamical system to model this problem.

c. Find $u(4)$, $u(5)$, $u(6)$, and $u(7)$. (If you don't have access to a spreadsheet, you may have to do these computations by hand.)

6. Suppose you have a pile of pieces of bubble gum (g) that cost 1 cent each, a pile of licorice sticks (s) that cost 1 cent each, and a pile of breath mints (m) that cost 2 cents each. You pick n cents' worth of candy, one piece at a time. Let $u(n)$ be the number of ways you can select n cents' worth of candy, where the order in which the candy is chosen is important. Thus, $u(1) = 2$ ways, which are g and s, and $u(2) = 5$ ways, which are m, sg, ss, gg, and gs. Let $u(n)$ be the number of ways this can be done.

a. Find a dynamical system to model this problem.

b. Suppose that in addition to the bubble gum, licorice sticks, and breath mints, there is also a pile of candy mints that cost 3 cents each. Find a dynamical system to model this problem.

7. Suppose you have a pile of pieces of bubble gum (g), that cost 1 cent each, a pile of licorice sticks (s), that cost 1 cent each, and a pile of breath mints (m) that cost 2 cents each. You pick n cents' worth of candy, one piece at a time, with the rule that after you pick bubble gum, you always pick a licorice stick next (if the bubble gum wasn't the last pick). Thus, $u(1) = 2$ ways, which are g and s, and $u(2) = 4$ ways, which are m, sg, ss, and gs. Let $u(n)$ be the number of ways this can be done.

a. What are $u(1)$, $u(2)$, $u(3)$, and $u(4)$?

b. Develop a dynamical system to model this problem. (*Hint:* This is similar to Exercise 4.)

8. Suppose that orange beads come only as single beads and that yellow beads come only in pairs. Let $u(n)$ be the number of different n-bead lineups of orange and yellow beads in which the order of the colors in the lineups is important.

a. Make lists of the possible 1-, 2-, 3-, and 4-bead lineups.

b. Find a dynamical system for $u(n)$.

c. Suppose that in addition to the orange and yellow beads, you also have blue beads that come as singles. Let $b(n)$ represent the number of n-bead lineups that contain at least one blue bead. Make lists corresponding to $b(1)$, $b(2)$, $b(3)$, and $b(4)$. When making the list corresponding to $b(3)$, see if you can use the lists corresponding to $b(2)$, $b(1)$, and $u(2)$ to help you. Look for a similar pattern for the list corresponding to $b(4)$.

d. Develop a second-order equation for $b(n)$ in terms of $u(n-1)$, $b(n-1)$, and $b(n-2)$.

e. Find $b(20)$.

4.3 Introduction to Probability

Before proceeding to a study of gambling, we need to review a little probability that will be used in that section. Suppose we are performing some random process, such as drawing a card from a deck, rolling a die, or drawing a candy from a bag. Each possible result of this process is called an **event.** For example, one event might be that we get a red candy from the bag. For shorthand, we might name our event. For example we might let A represent the event in which an even number occurs on the roll of a die. The notation P(A) means "the probability of A occurring." For example,

$$\text{P(A)} = \text{probability of an even number on roll of die} = \frac{1}{2}$$

If we perform a process a large number of times, the fraction of the time we expect event A to occur is the probability of event A. We could actually perform a process a large number of times and observe the fraction of time A occurs. This will give us an estimate for P(A).

In this section, we are going to estimate the probability of events from our understanding of the situation. For example, when flipping a coin, we believe that heads and tails are equally likely to occur, so we assume $\text{P(H)} = \frac{1}{2}$ and $\text{P(T)} = \frac{1}{2}$. In some situations, it is impossible to determine the probabilities from the situation and we actually have to estimate the probability by conducting the experiment a large number of times. For example, suppose we drop a marshmallow. What is the probability of it landing on its side versus the probability it lands on an end? We doubt these are equally likely. We might get a bag of marshmallows, throw them in the air, then determine what fraction land on their sides and what fraction land on their ends. We could use these fractions to estimate

$$\text{P(landing on side)} \quad \text{and} \quad \text{P(landing on end)}$$

Many of the events we will be considering consist of 2 or more things occurring, such as drawing 3 face cards from a deck or rolling a total of 7 on the toss of a pair of dice. Such events are called **compound events.** Usually, we will consider the process as occurring in steps or stages. For example, we roll the first die, then we roll the second die. This process would then consist of 2 stages. Suppose we let event A be that we get a 6 on the roll of the first die and event B be that we get an even number on the roll of the second die. It should be clear that

$$\text{P(A)} = \frac{1}{6} \quad \text{and} \quad \text{P(B)} = \frac{1}{2}$$

We will define the compound event AB as the event in which a 6 occurs on the first die **and** an even number occurs on the second die.

Suppose that event A gives the result of the first step of a process and event B gives the result of the second step of the process. If event A occurring on the first step does not affect the probability of B occurring on the second step, then events A and B are called **independent events.** It should be clear that getting a 6 on the first die does not affect the probability that the second die is "even." Thus events A and B as defined earlier are independent events.

Suppose we have a bag with 2 red candies and 1 green candy. One person reaches into the bag and randomly picks a candy. A second person then reaches into the bag and draws one of the remaining candies. Let event C be that the first person draws a green candy. Let event D be that the second person draws a green candy. Clearly, event C occurring, the first person getting the green candy, effects the probability that event D will occur, that is, the second person will get a green candy. In this case, events C and D are called **dependent events.** Suppose events A and B are independent events. The **multiplication principle for probability** says that the probability of A happening on the first step of the process then B happening on the second step of the process is the product of the 2 probabilities, that is,

$$\mathrm{P(AB) = P(A)P(B)}$$

By the multiplication principle, the probability of rolling a 6 then rolling an "even" is

$$\mathrm{P(AB)} = \left(\tfrac{1}{6}\right)\left(\tfrac{1}{2}\right) = \tfrac{1}{12}$$

Sometimes we consider an event that is composed of a number of possible results. Suppose we consider event C, which is that we get a total of 3 on the roll of a pair of dice. Event C is composed of the results C_1 and C_2, where C_1 is the event in which we get a 1 on the first die and a 2 on the second die and C_2 is the event that we get a 2 on the first die and a 1 on the second die. Then the probability of C occurring is the probability of C_1 or C_2 occurring, which we write as

$$\mathrm{P(C) = P(C_1 \text{ or } C_2)}$$

When 2 events cannot both occur, as in the case of C_1 and C_2, then the 2 events are said to be **mutually exclusive.** The **addition principle for probability** states that if events A and B are mutually exclusive, then

$$\mathrm{P(A \text{ or } B) = P(A) + P(B)}$$

Since events C_1 and C_2 cannot both occur, by the addition principle,

$$\mathrm{P(C_1 \text{ or } C_2) = P(C_1) + P(C_2)}$$

To compute $\mathrm{P(C_1)}$, we use the multiplication principle

$$\mathrm{P(C_1) = P(1 \text{ on first die})\ P(2 \text{ on second die})} = \tfrac{1}{6} \times \tfrac{1}{6} = \tfrac{1}{36}$$

Similarly, $\mathrm{P(C_2)} = 1/36$. Combining the addition and multiplication principle gives

$$\mathrm{P(C) = P(C_1) + P(C_2)} = \tfrac{1}{36} + \tfrac{1}{36} = \tfrac{1}{18}$$

There is a geometric technique called **tree diagrams** that can help. A tree analyzing the probability of getting a total of 3 on the roll of a pair of dice is seen in Figure 4.5. We construct branches of a tree for different possible results of the first step of the process. We combined the results of a 3, 4, 5, and 6 into one result with a probability of 4/6. This gave 3 possible results of the first roll, so the tree begins with 3 branches. We then construct branches off of the first stage branches to indicate the possible results of the second step. Each branch is labeled with the probability of that result occurring. To get to the end of a branch, say the probability of rolling a 1, then a 2, we just multiply the probabilities of the 2 branches that give that result, which gives 1/36. This is just the multiplication principle. To get the total probability, we use the addition principle and add the results of each of the total branches, giving that the probability of getting a total of 3 on the dice is 1/18.

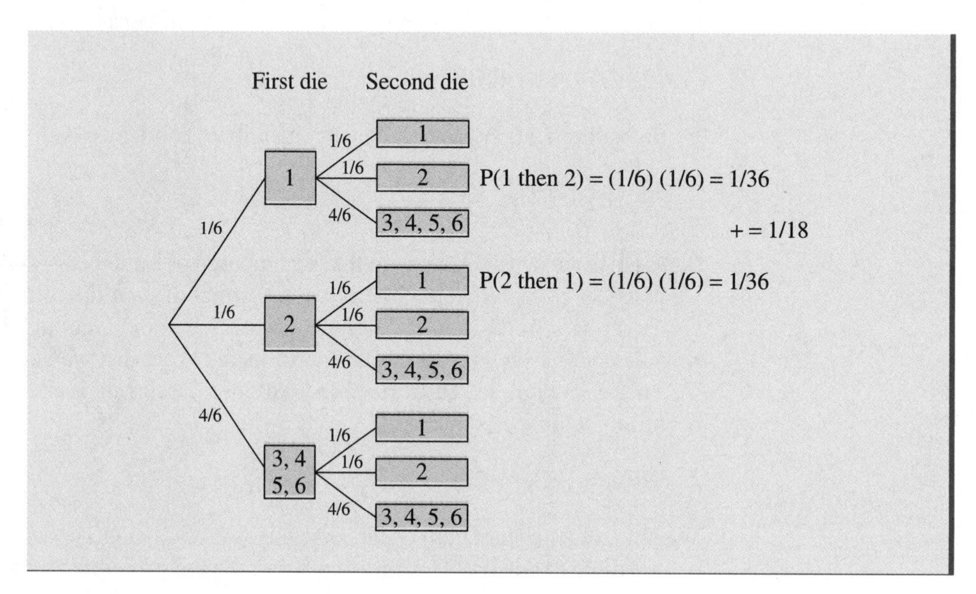

FIGURE 4.5 Tree diagram for computing that the probability of total of 3 on roll of pair of dice is (1/36) + (1/36) = 1/18.

Example 4.3

Suppose we have 2 bags. Bag 1 contains 3 white and 2 black beads. The second bag contains 1 white and 4 black beads. See Figure 4.6. We draw a bead from bag 1, then a bead from bag 2. Event A consists of getting 2 beads of the same color.

By the addition principle, we get that

$$P(A) = P(W_1W_2 \text{ or } B_1B_2) = P(W_1W_2) + P(B_1B_2)$$

By the multiplication principle, we get that

$$P(W_1W_2) = P(W_1)P(W_2) = \tfrac{3}{5} \cdot \tfrac{1}{5} \quad \text{and} \quad P(B_1B_2) = P(B_1)P(B_2) = \tfrac{2}{5} \cdot \tfrac{4}{5}$$

Putting this all together gives that

$$P(A) = \tfrac{3}{5} \cdot \tfrac{1}{5} + \tfrac{2}{5} \cdot \tfrac{4}{5} = \tfrac{11}{25} = 0.44$$

This can also be seen in the tree diagram in Figure 4.7. ■

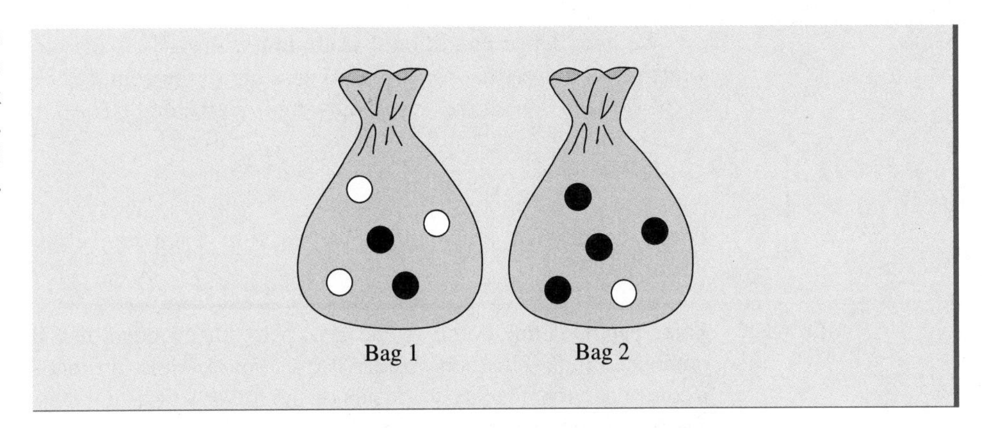

FIGURE 4.6 Two bags containing black and white beads. Draw one bead from each.

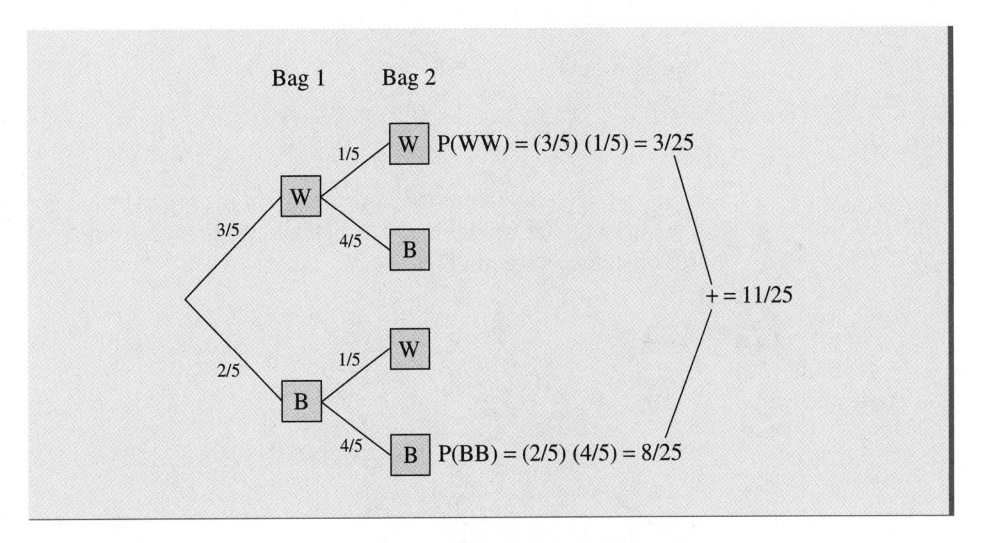

FIGURE 4.7 Tree diagram for drawing beads from bags.

We are now going to consider 2-step processes where the probability of the result of the second step is dependent on the result of the first step. For example, suppose that we flip a coin. If heads occurs, we draw a bead from bag 1 of Figure 4.6. If tails occurs, we draw a bead from bag 2. Let event H be that heads occurs on the flip of the coin and let W be the event we draw a white marble. Clearly, the result of the flip of the coin affects the probability of drawing a white marble, since the flip determines which bag we are drawing from. Therefore, H and W are dependent events. Since these events are dependent, we cannot use the multiplication principle in its present form.

We now define **conditional probability.** P(W|H) is the probability that we draw a white marble given that we get heads on a flip of the coin. But if we get heads on the flip of the coin, then we know we are drawing from the first bag, so

$$P(W|H) = \frac{3}{5}$$

Often, conditional probabilities are easy to compute because we are given extra information.

REMARK Some people define events A and B as being independent if P(B)=P(B|A). Usually it is obvious from the situation that this is true. For example, if event A is that heads occurs on a coin and event B is that a 1 occurs on the roll of a die, it is clear that $P(B) = P(B|A) = \frac{1}{6}$, since the flip of the coin is not going to affect the roll of the die.

Suppose we want to know P(HW), the probability that heads occurs on the flip of the coin and we draw a white bead from a bag. This is computed using the **generalized multiplication principle** for probability, which states that

$$P(HW) = P(H)P(W|H)$$

This is simply

$$P(HW) = P(H)P(W|H) = \frac{1}{2} \times \frac{3}{5} = \frac{3}{10}$$

Often, we are only interested in the result of the second step. For example, we may want a white bead, but we don't care which bag we drew it from. To do this, we make another tree diagram, as seen in Figure 4.8.

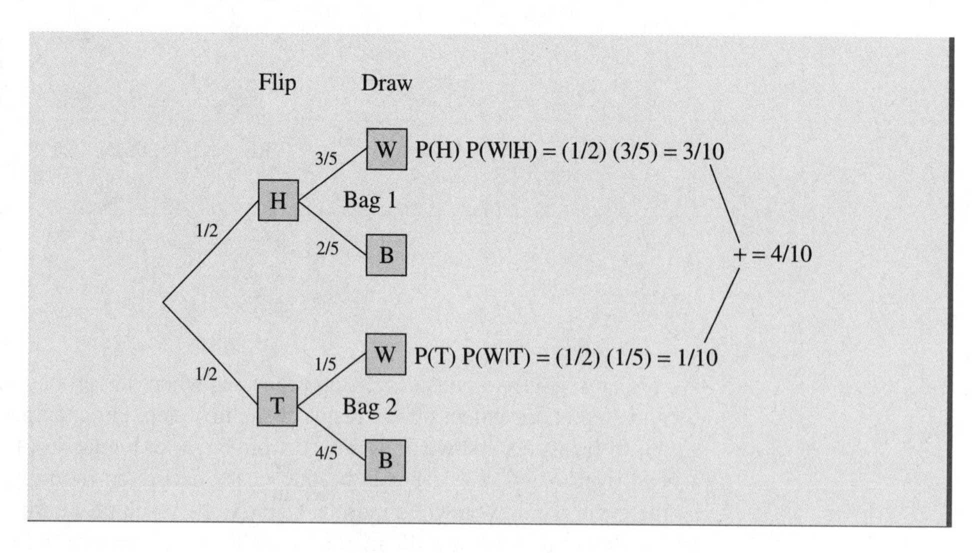

FIGURE 4.8 Probability of getting a white bead.

So P(W)= 0.4.

Example 4.4

Suppose we have a bag containing 4 red candies and 3 green candies, as in Figure 4.9. We randomly draw a candy and eat it. We then draw another candy and eat it. What is the probability that we get one candy of each color?

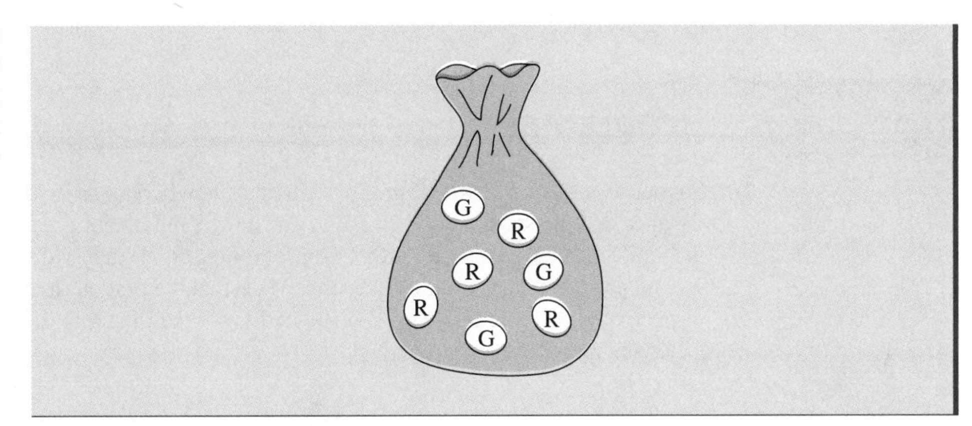

FIGURE 4.9 Representation of a bag containing 4 red and 3 green candies.

We let R_1 and G_1 be the events that the first candy is red and green, respectively. Similarly, we let R_2 and G_2 be the events that the second candy is red and green, respectively. Note that $P(R_2|G_1) = \frac{4}{6}$. This is because we know we drew a green, which leaves 2 green and 4 red in the bag when we make the second draw. Similarly, $P(G_2|R_1) = \frac{3}{6}$. From the tree diagram in Figure 4.10, we see that

$$P(\text{2 different color candies}) = \frac{3}{7} \times \frac{4}{6} + \frac{4}{7} \times \frac{3}{6} = \frac{24}{42} = \frac{4}{7}$$

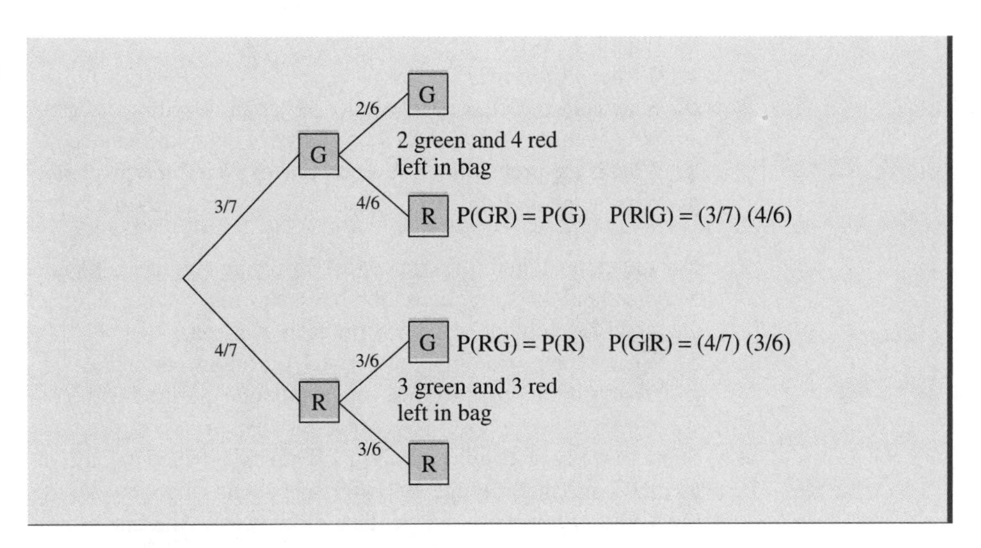

FIGURE 4.10 Probability of getting candies of 2 different colors, using tree diagram.

We can also find that the probability of 2 greens is

$$P(G_1G_2) = P(G_1)P(G_2|G_1) = \frac{3}{7} \times \frac{2}{6} = \frac{1}{7}$$

and the probability of 2 reds is $\frac{2}{7}$. ■

4.3 Problems

1. Suppose we have 2 bowls of m&m's. The first bowl contains 2 yellow, 3 red, and 5 green m&m's, and the second bowl contains 4 yellow, 2 red, and 4 green m&m's as represented in Figure 4.11. We draw an m&m from the first bowl, then we draw an m&m from the second bowl. Let Y_1, R_1, and G_1 be the events that the first m&m is yellow, red, and green, respectively, and similarly for Y_2, R_2, and G_2. Tree diagrams may help you answer the following questions.

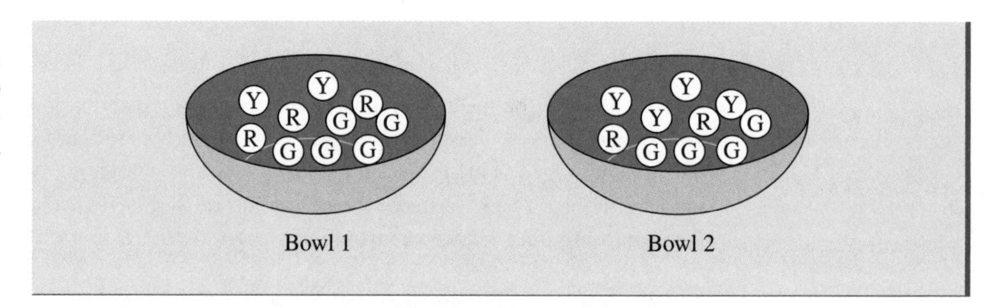

FIGURE 4.11 Representation of 2 bowls to be used with Problem 1.

a. What is $P(Y_1)$?
b. What is $P(Y_2)$?
c. What is $P(Y_1 \text{ or } G_1)$?
d. What is $P(Y_1Y_2)$, the probability of getting 2 yellow m&m's?
e. What is the probability of your 2 m&m's being the same color?
f. What is the probability of exactly one of your 2 m&m's being yellow?
g. What is the probability of at least one of your 2 m&m's being red?

2. You are taking a multiple choice test for which you haven't studied. The first question has 5 choices and the second question has 4 choices. Having no clue what the correct answers are, you randomly make a guess for each question.
a. What is the probability you get both questions correct?
b. What is the probability you get exactly one question correct?

3. Suppose we have 2 bowls of m&m's. The first bowl contains 2 yellow (Y), 3 red (R), and 5 green (G) m&m's, and the second bowl contains 4 yellow, 2 red, and 4 green m&m's. We flip a coin. If heads occurs, we draw a candy from the first bowl. If tails occurs, we draw a candy from the second bowl. Let H represent the event that we got heads on the flip of the coin. Let Y be the event that we got a yellow candy.

a. Find $P(Y|H)$. **b.** Find $P(G|T)$. **c.** Find $P(HR)$.

d. Find $P(Y)$. Drawing a tree diagram may help.

4. You are taking a multiple choice test for which you haven't studied. The first question has 5 choices, and the second question has 4 choices. You flip a coin. Heads, and you answer the first question, tails, and you answer the second question. Let R be the event you pick the correct answer and let W be the event you pick the wrong answer. As far as you're concerned, each answer is equally likely to be the correct one.

a. What is $P(R|H)$, that is, what is the probability you answer the question correctly, given that you flip heads?

b. What is $P(R)$, that is, what is the probability you correctly answer the one question you answer?

c. Instead of flipping a coin, you decide to answer the first question, then answer the second question. What is $P(RR)$, the probability you pick the right answer to both questions? What is the probability you correctly answer exactly one question?

5. The weather forecast states that there is a 30% chance of rain on Saturday and a 60% chance of rain on Sunday.

a. You decide to go camping this weekend. What is the probability that you will not get rained on?

b. You roll a die. If 1 or 2 occurs, you have your picnic on Saturday. If 3, 4, 5, or 6 occurs, you have your picnic on Sunday. What is the probability it will rain on your picnic?

6. Generally, customers tend to have brand loyalty, meaning that if they purchase a particular brand, then on their next purchase, they are more likely to purchase that same brand. Suppose there are 2 brands of a product, brand A and brand B. Suppose that 80% of customers who purchase brand A will purchase brand A on the following purchase, that is, $P(A|A) = 0.8$. This also means that $P(B|A) = 0.2$. Suppose that 60% of customers who purchase brand B will purchase brand B on the following purchase.

a. Suppose Roy purchased brand A. What is $P(AA)$; that is, what is the probability that Roy's next 2 purchases will also be brand A?

b. Suppose Sarah purchased brand B. What is the probability that Sarah will purchase brand A at least once in her next 2 purchases? A tree diagram may help.

4.4 The Gambler's Ruin

In 1986, Ann Landers published a letter that I had written her about a surprising result involving roulette. Quoting from my letter, "In simple roulette, if you bet a dollar on red, you will win a dollar about 48% of the time and will lose a dollar about 52% of the time because of the two greens. Suppose you decide to bet $1 at a time on red, until you have lost $50 or have won $50. You will lose your $50 more than 98 percent of the time, and will win $50 less than 2% of the time."

In this section, we are going to examine the claims that I made in that letter. For simplicity, I rounded off some of the probabilities. We will compute them more accurately in this section.

A standard U.S. roulette wheel has a loose ball and 38 slots; 18 for red numbers, 18 for black numbers, and 2 for green numbers as represented in Figure 4.12. Suppose we wager \$1 that a red number will occur on the next spin of the wheel. This means that if red occurs, we win \$1 and if black or green occurs, we lose \$1. The probability of winning this bet is $18/38 = 9/19$ since there are 18 red slots out of 38 possible slots and all slots are equally likely. This means that for every 19 times that we bet a red slot will come up, on average we will win 9 times and lose 10 times. That extra loss represents the fact that the casino should come out ahead over time.

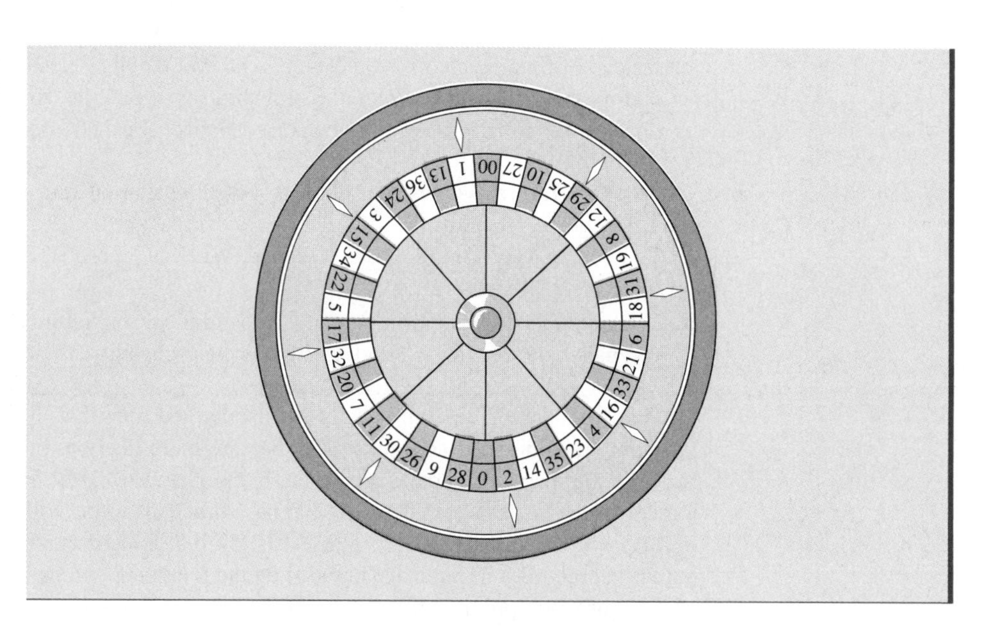

FIGURE 4.12 A standard U.S. roulette wheel.

In the letter to Ann Landers, I proposed the scenario that we come to the casino with \$50 and that we continue betting \$1 on red until we have a total of \$0 or \$100. We want to compute P(\$0) and P(\$100), that is, the probability that we will leave the casino with no money and the probability we will leave the casino with \$100. We will return to this scenario, but first we consider a simpler version of the roulette scenario.

Example 4.5

Suppose that you have \$4 and a friend has \$6. Someone flips a fair coin (equally likely for a heads or tails to occur). If heads comes up, you win a dollar from your friend, but if tails comes up, your friend wins a dollar from you. You continue this exciting game until one of you goes broke. What is the probability that you will eventually lose all of your hard-earned money?

There are several possible events that we could consider.

1. Events H and T are that heads and tails, respectively, occur on a flip of the coin.
2. Events $A_0, A_1, \ldots, A_9, A_{10}$, are that you have \$0, \$1, ..., \$9, \$10, respectively.
3. Event L is that you eventually go broke, and event W is that you eventually have \$10.

We wish to determine

$$P(L|A_4) \quad \text{and} \quad P(W|A_4)$$

that is, the probability you will end up with \$0, given that you now have \$4, and the probability you will end up with \$10, given that you now have \$4. We will denote

$$P(L\,|A_4) = P(4)$$

for short. Similarly,

$$P(L\,|A_0) = P(0),\ P(L|A_1) = P(1), \ldots,\ P(L|A_{10}) = P(10)$$

Clearly $P(0) = 1$, since you are certain to eventually have \$0 if you now have \$0. Also, $P(10) = 0$, since you can't possibly end up with \$0 if you now have \$10 because you are quitting the game when you have \$10.

Similarly, let

$$P(W|A_0) = Q(0),\ P(W|A_1) = Q(1), \ldots,\ P(W|A_{10}) = Q(10)$$

In this case, $Q(0) = 0$ and $Q(10) = 1$.

Let's try to understand all of this notation. Suppose you presently have \$4 and you want to compute the probability that you win the next bet, then eventually go broke (although you may win and lose quite a number of bets before you do). Then we are trying to compute

$$P(HL)$$

By the generalized multiplication principle, this gives

$$P(HL) = P(H)P(L|H) = \left(\tfrac{1}{2}\right)P(L|H)$$

Let's think about what $P(L|H)$ means. You had \$4. We are given that you won the next bet. This means that you have \$5. Thus, in this case

$$P(L|H) = P\text{ (you eventually go broke given you have \$5)} = P(L|A_5) = P(5)$$

Thus,

$$P(HL) = \left(\tfrac{1}{2}\right)P(5)$$

Let's compute a formula for $P(4)$. We are going to consider this process as a 2 step process. The first step is to flip a coin, winning or losing this bet. H corresponds to winning \$1 and T corresponds to losing \$1. The second step is to complete the game, either winning all of the \$10 or losing all of the money. These 2 steps are shown in Figure 4.13. There are many ways in which the second step can be completed.

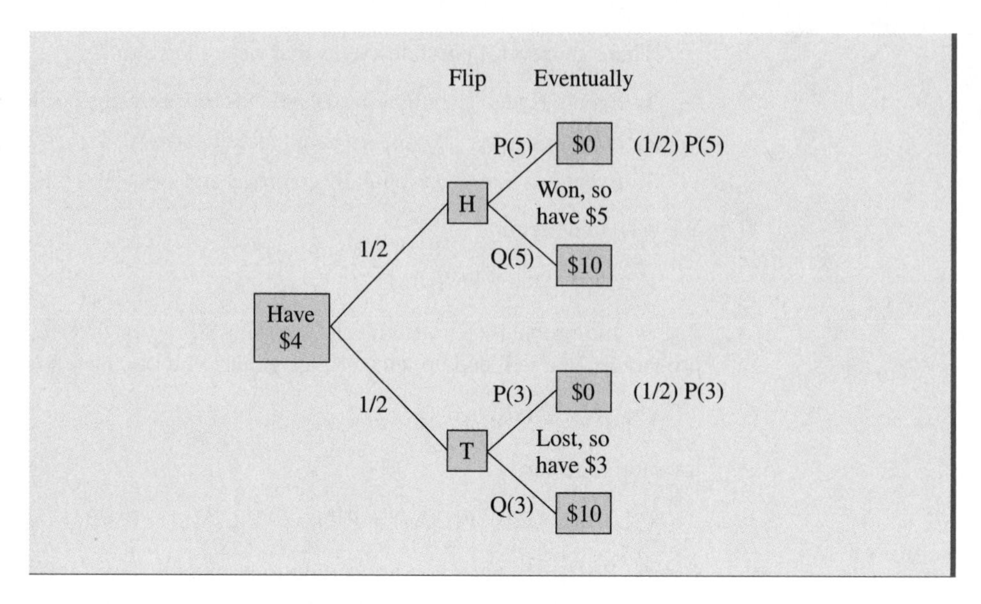

FIGURE 4.13 Probability of eventually going broke when starting with $4.

We can now apply the addition principle, getting

$$P(4) = \frac{1}{2}P(5) + \frac{1}{2}P(3)$$

Suppose you are at a point in the game where you have \$$n$. We now want to know the probability you will eventually lose all of your money; that is, we want to compute $P(n)$.

By Figure 4.14 and the addition principle, we get

$$P(n) = \frac{1}{2}P(n+1) + \frac{1}{2}P(n-1)$$

This is a dynamical system, but it is not in standard form. To get it in standard form, substitute $n-1$ for n, getting

$$P(n-1) = \frac{1}{2}P(n) + \frac{1}{2}P(n-2)$$

Next, subtract $\frac{1}{2}P(n-2)$ from both sides, then multiply by 2 to get

$$P(n) = 2P(n-1) - P(n-2)$$

The problem is that we know that $P(0) = 1$ and $P(10) = 0$. To use this dynamical system, we need to know $P(0)$ and $P(1)$. I will tell you that $P(1) = 0.9$. Using this value, you should be able to get the results in Table 4.3. What is important to note in Table 4.3 is that $P(10) = 0$, which agrees with what we already know. If $P(10)$ had not been 0, then you would have known that I gave you the wrong value for $P(1)$.

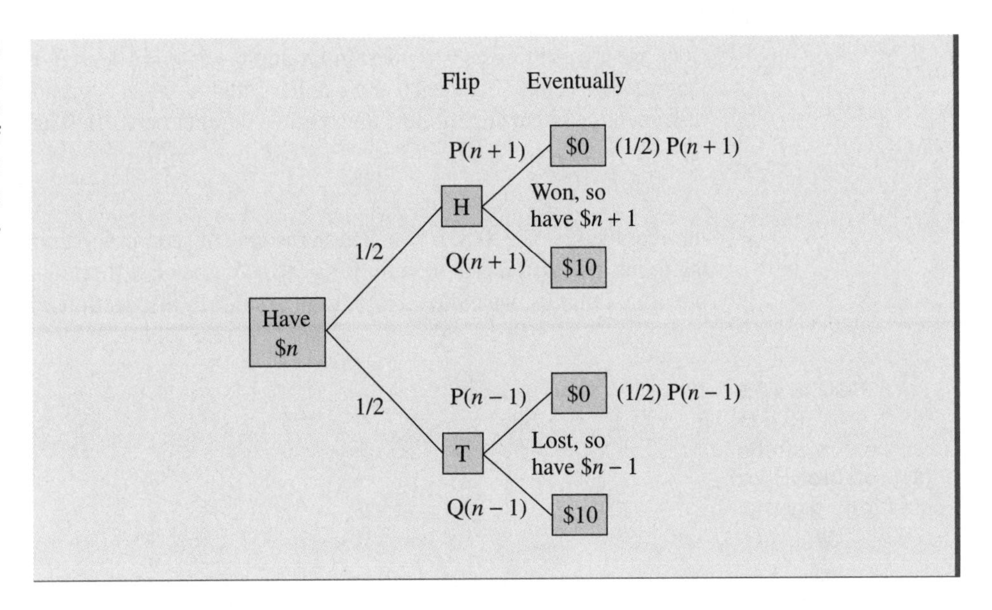

FIGURE 4.14 Tree diagram for finding P(*n*), the probability of eventually going broke when you have *n* dollars.

TABLE 4.3 **Probability of eventually losing and winning if you presently have *n* dollars.**

you have n =	0	1	2	3	4	5	6	7	8	9	10
eventually lose $P(n)$ =	1	0.9	0.8	0.7	0.6	0.5	0.4	0.3	0.2	0.1	0
win $1 - P(n)$ =	0	0.1	0.2	0.3	0.4	0.5	0.6	0.7	0.8	0.9	1

Suppose that I didn't give you the value for P(1). Then you would use $P(0) = 1$ and try different values for P(1) until you found a value for P(1) that resulted in $P(10) = 0$. It is important that you understand the notation introduced in this example. Reread this example until you understand it. ■

We are now going to use the approach of Example 4.5 to study slightly more general gambling situations. Let's make some assumptions about the gambling situations we are considering. n represents n dollars. $P(n)$ is the **probability** that we will eventually go broke given that we presently have n dollars, that is,

$$P(n) = P(L|A_n) = P(\text{eventually broke given now have } \$n)$$

Since we always quit playing the game when we have \$0, then

$$P(0) = 1$$

We always have a predetermined goal, g. When we have a total of g dollars, we quit playing. This means that

$$P(g) = 0$$

since we are quitting as a winner. In Example 4.5, $g = 10$, so $P(10) = 0$. This also means that we always have between 0 and g dollars, that is, $0 \leq n \leq g$. Since $P(n)$ is the probability of something occurring, it must always be a number between 0 and 1; that is,

$$0 \leq P(n) \leq 1$$

for every value of n, $0 \leq n \leq g$. These results are seen in the graph in Figure 4.15, where the points $(n, P(n))$ are plotted for $0 \leq n \leq 10$ in the coin-flipping game of Example 4.5. In previous chapters, we called this a "time graph." In this section n is not time but is money.

FIGURE 4.15 Points $(n, P(n)) =$ (\$, prob broke) for coin flipping game.

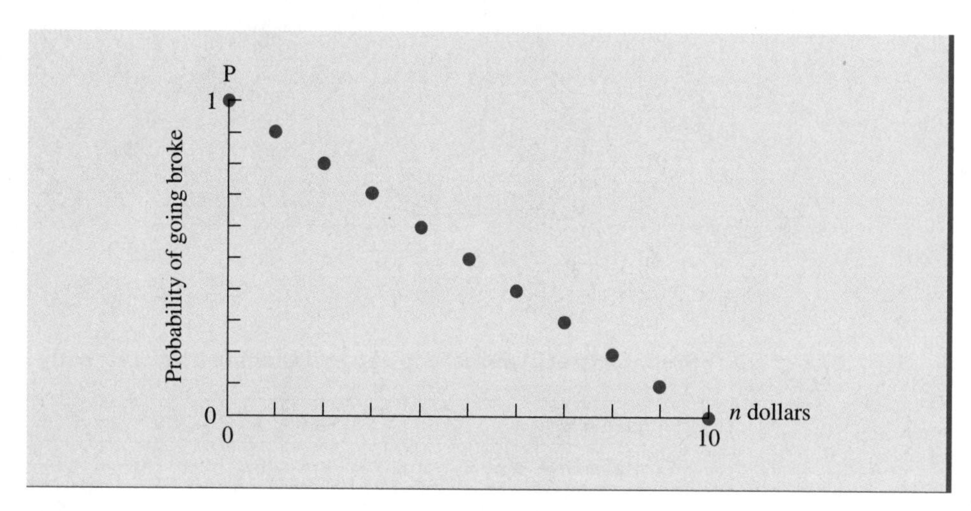

Let's consider another example.

Example 4.6

Consider a "roulette" wheel that has 4 red slots, 6 black slots and no green slots. It is equally likely that the ball will fall into any of the 10 slots. Suppose we decide to play this game until we either go broke or have a total of $g = \$20$, at which point we will quit playing. Let R and B represent the events that we get red and black, respectively, on the next bet. Let L and W represent the events that we quit the game with \$0 and \$20, respectively.

Assume we presently have \$10. We want to compute $P(10) = P(L|A_{10})$, which is the probability that we eventually go broke given we presently have \$10. We consider this game as a 2 step process. The first step is to make 1 bet on red. The second step, which is really a lot of bets considered together, is to complete playing the game until we have \$0 or \$20. These 2 steps can be visualized in Figure 4.16.

Figure 4.16 shows that we can lose our money 2 ways, RL or BL. This means that

$$P(L|A_{10}) = P(10) = P(\text{RL or BL})$$

By the addition principle,

$$P(10) = P(\text{RL}) + P(\text{BL})$$

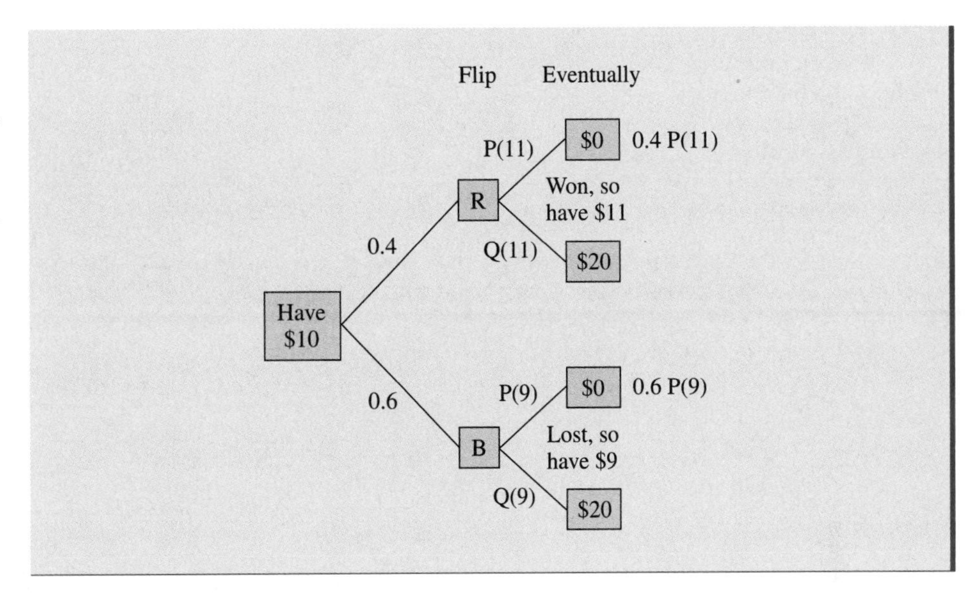

FIGURE 4.16 Tree diagram for finding P(10), the probability of eventually going broke when you have $10.

By the generalized multiplication principle,

$$P(10) = P(R)P(L|R) + P(B)P(L|B)$$

In this case, we know that $P(R) = 0.4$ and $P(B) = 0.6$, since there are 4 red and 6 black slots, so

$$P(10) = 0.4P(L|R) + 0.6P(L|B)$$

Remember that $P(L|R)$ is the probability we lose all our money if we win the next bet, giving us \$11. Thus, $P(L|R) = P(L|A_{11}) = P(11)$. Similarly, $P(L|B) = P(9)$, since if black occurs on the next spin, we lose a dollar and have \$9. Thus, if we have \$10, the equation becomes

$$P(10) = 0.4P(11) + 0.6P(9)$$

If we now have $\$n$, then $P(L|A_n) = P(n)$. In addition, $P(L|R) = P(n+1)$ and $P(L|B) = P(n-1)$, since if red occurs we have $\$n+1$, and if black occurs we have $\$n-1$. This can be seen in Figure 4.17, which is just a duplication of Figure 4.13 using the variable amount of money. By the addition principle, this gives

$$P(n) = 0.4P(n+1) + 0.6P(n-1)$$

This is a dynamical system, although it is not in standard form. By substituting $n-1$ for n, then solving for $P(n)$, we get the dynamical system

$$P(n) = 2.5P(n-1) - 1.5P(n-2)$$

which is in standard form.

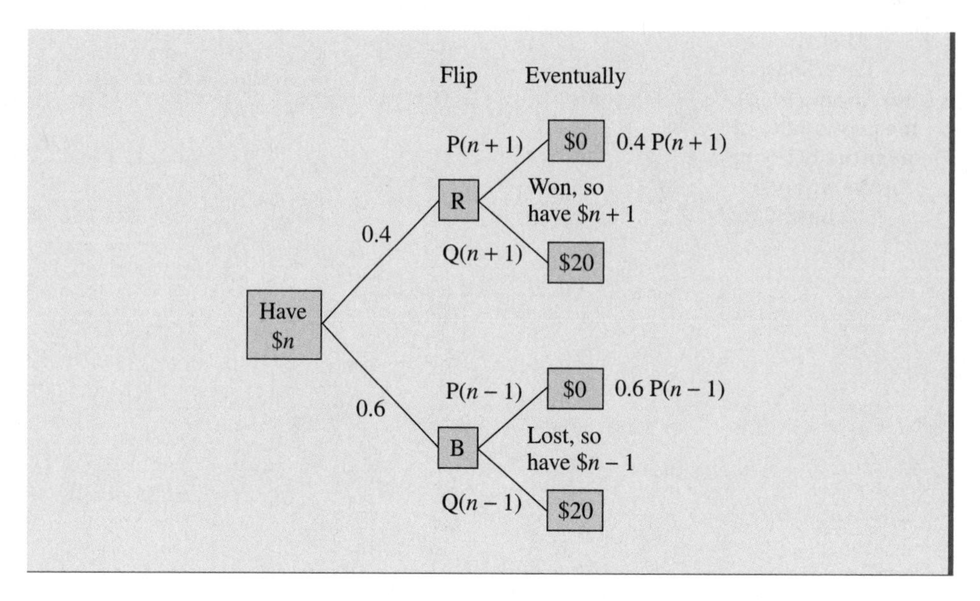

FIGURE 4.17 Tree diagram for finding P(n), the probability of eventually going broke when you have $$n$.

We will use this dynamical system to find the probability of going broke given that we presently have any amount of money between \$0 and \$20. We know that $P(0) = 1$ and $P(20) = 0$. In Table 4.4 are my guesses at P(1) and the resulting values for P(20). In making my guesses, I used the context of the problem. Since P(1) is the probability of eventually going broke, given that we have only \$1, I assumed that P(1) should be fairly close to 1. The question I had to answer was, How close to 1?

TABLE 4.4 Guesses for P(1) and resulting values for P(20).

P(1) =	0.9	0.99	0.999	0.9999
P(20) =	−664	−65.5	−5.65	0.335

IMPORTANT It might appear that $P(1) = 0.9999$ is pretty good, since $P(20) = 0.335$ is "close" to 0. But remember that P(20) is the probability of going broke given that we have \$20 and are quitting the game as a winner. $P(20) = 0.335$ implies that we have a 33.5% chance of going broke. Taken in this context, we see that 0.335 is not sufficiently close to 0. In Table 4.5, I continued making guesses for P(1) until I achieved $0 < P(20) < 0.01$, meaning less than 1%.

TABLE 4.5 More guesses for P(1) and resulting values for P(20).

P(1) =	0.9997	0.9998	0.99985
P(20) =	−0.995	−0.33	0.0027

Being somewhat compulsive, I continued looking and found that if $P(1) = 0.9998495904377$, then $P(20) = -0.00000000069$. I couldn't get P(20) to exactly equal 0. Using $P(0) = 1$ and $P(1) = 0.99985$, I found the values that are given by the points $(n, P(n))$ as seen in Figure 4.18.

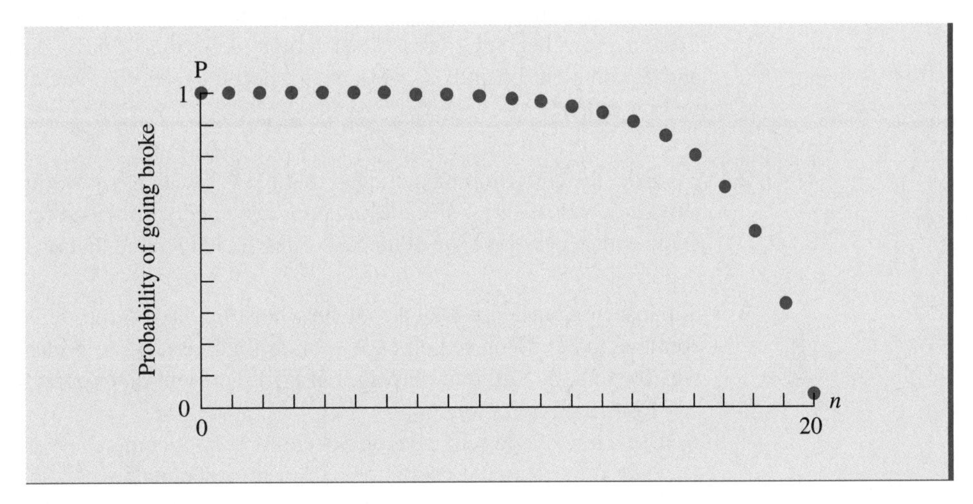

FIGURE 4.18 Points $(n, P(n))$, which give present amount of money and probability of eventually going broke given we have that amount of money.

As you can see, $P(n)$ is near 1 for most n values. Some of the actual probabilities of going broke, using $P(1) = 0.99985$ and rounding to the nearest percent, are given in Table 4.6.

TABLE 4.6 **Probability of losing all your money, given present amount of money.**

current money	8	10	14	15	16	17	18	19
% chance of going broke	99%	98%	91%	87%	80%	70%	56%	34%

Note what Table 4.6 implies. Given this "roulette" wheel, if you have \$10, there is a 98% chance that you will eventually lose all \$10, and only a 2% chance you will eventually have \$20, winning \$10. If you have \$19, there is a 34% chance you will lose all \$19, and there is only a 66% chance you will arrive at \$20, for a total gain of \$1. To rephrase, one-third of the time you lose \$19 and two-thirds of the time you win \$1. Not very good odds. ■

Recall my letter to Ann Landers. A person plays regular roulette, betting \$1 on red on each bet. He starts with \$50 and has a goal of $g = \$100$. I stated that the person would lose the \$50 more than 98% of the time. In fact, in this situation, $P(50) \approx 0.995$, meaning that the person should lose about 99.5% of the time. (You will compute this value in Problem 4.) This means he should lose, on average, 199 out of every 200 nights. Note that losing \$50 for each of 199 nights and winning \$50 on only one night means that over 200 nights, this person should expect to lose a total of \$9900. This averages out to losing \$49.50 per night.

4.4 Problems

1. Consider the coin-flipping game of Example 4.5. We found that $P(2) = 0.8$. This means that if you begin with \$2 and play until you have \$0 or \$10, you will lose your \$2 80% of the time and will win \$8 20% of the time, on average. Suppose you play this game on 10 days in a row, beginning with \$2 each time. After the 10 days, do you expect to come out ahead, come out behind, or break even? If you expect to come out ahead or behind, then by how much?

2. Consider the coin-flipping game in Example 4.5, which is governed by the dynamical system $P(n) = 2P(n-1) - P(n-2)$. Suppose you and your friend each begin with \$10, so the game ends when you have \$0 or \$20. Find the $P(1)$-value that makes $P(20) = 0$.

3. Suppose you have a modified roulette wheel that has 45 red, 55 black, and no green. You continue to bet \$1 on red on each spin until you either are broke or have a total of \$6, at which point you will quit playing. Let $P(n)$ represent the probability that you will eventually lose all of your money.

a. Use a tree diagram to develop a second-order dynamical system in standard format for $P(n)$. Use fractional coefficients instead of decimals when solving for $P(n)$. This will improve the accuracy of your answers in the next parts of this problem.

b. You know that $P(0) = 1$, since you are certain to go broke if you are already broke. You also know that $P(6) = 0$, since if you have \$6, you quit the game a winner. Using a table, make trial-and-error guesses for the correct value of $P(1)$, to 4 decimal places.

c. Find $P(3)$, the probability that you will eventually go broke given that you started the game with \$3, to 2-decimal-place accuracy.

4. Recall my letter to Ann Landers, in which a person bets \$1 at a time on red in roulette until he has a total of either \$0 or \$100. The probability of winning any one bet is $18/38 = 9/19$, and the probability of losing any one bet is $20/38 = 10/19$.

a. Develop a second-order dynamical system for $P(n)$ in standard format. Making a tree-diagram may help you. Using fractions instead of decimals for your system's coefficients will improve accuracy of your answers in the next parts.

b. You know that $P(0) = 1$. You also know that $P(100) = 0$, since if he ever has \$100, he quits a winner. Make trial-and-error guesses for the correct value of $P(1)$, to 7 decimal places.

c. Find $P(50)$ to 3 decimal places. Does it agree with my claim?

d. Suppose that this person starts with \$95 instead of \$50. Just five more dollars and he can go home happy! What is the probability, to 2 decimal places, that he will, from this tantalizing point, eventually win for the night?

5. You are playing roulette with a wheel that has 11 black slots, 10 red slots, and 29 green slots. You bet \$1 on each spin of the wheel. If red occurs, you win \$1; if black occurs, you lose your \$1; and if green occurs, you get your dollar back. In this situation, there are 3 results of the first step of the process, R, B, and G. Let $P(n)$ represent the probability you quit the game with nothing, given that you presently have n dollars.

a. Develop a tree diagram to model this situation. There will be 3 branches on the first step, R, B, and G.

b. Use your tree-diagram to develop a second-order dynamical system in standard format.

c. You will play this game until you have \$0, in which case you quit as a loser, or you have \$20, at which point you quit as a winner. Find $P(1)$ to 5 decimal places.

d. What percentage of the time will you quit as a winner given you start with \$10?

6. You are playing roulette with a wheel that has 1 black slot, 5 red slots, and 4 green slots. If red occurs, you win \$1; if black occurs, you lose \$2; and if green occurs, you lose \$1. In this situation, there are 3 results of the first step of the process—R, B, and G. You will play this game until you have \$1 or \$0, in which case you quit as a loser, or you have \$50, at which point you quit as a winner. Let $P(n)$ represent the probability you quit the game with \$1 or less, given that you presently have n dollars. (Note that you must quit when you have \$1, since you might lose \$2 on the next play and it might be dangerous not to pay your gambling debts.)

a. Develop a tree diagram to model this situation.

b. Use your tree diagram to develop a third-order dynamical system in standard format.

c. To analyze your third-order system, you need 3 initial conditions—$P(0), P(1)$, and $P(2)$. Note that $P(0) = 1$ and $P(1) = 1$. If you have a spreadsheet, make guesses for $P(2)$ until you get $P(50)$ is close to zero. Find $P(2)$ to 6 decimal places.

Project 1. Suppose you are playing a fair game in which the probability that you win \$1 on any particular bet is 0.5 and the probability that you lose \$1 on any particular bet is also 0.5. You will quit playing this game when you have a total of \$0 or a predetermined goal. Let $e(n)$ represent the number of bets you expect to make before quitting, given you now have $\$n$. $e(n)$ is an average number of bets. For example, suppose you always start with \$10. One night you might make a total of 30 bets before quitting. Another night you might make 53 bets before quitting. Your average number of bets is 41.5. $e(10)$ represents the average number of bets you would expect to make each night if you played this game repeatedly for a large number of nights.

a. Suppose you will quit playing this game if you have \$0 or \$4. Estimate $e(2)$ the following way. You are starting with \$2. Flip a coin. If heads occurs, then you lose \$1. If tails occurs, you win \$1. If you won \$1, you then have a total of \$3. If you lost \$1, you have a total of \$1. Flip again using the same win/lose rules. Keep repeating this process until you reach \$0 or \$4. Record the number of bets you made, which is the number of flips you made. Repeat this entire process a total of 10 times. You now have 10 estimates for $e(2)$. Compute their average. This is an even better estimate for $e(2)$. Repeat this entire process, but instead start with \$1 to get an estimate for $e(1)$.

b. Develop an equation for $e(n)$ in terms of $e(n-1)$ and $e(n+1)$. Note that $e(n)$ is an average number of bets you make given that you now have n dollars. You make one bet, so $e(n)$ must include that 1. After making that bet, on 5 nights out of 10, you make $e(n+1)$ more bets, since the probability of winning that bet is 0.5. On the other 5 nights out of 10, you make $e(n-1)$ more bets. To find $e(n)$, count the one bet you just made and add to it the average number of additional bets you will make. Use this equation to find a second order dynamical system for the expected number of bets made. Suppose you will quit with \$0 or with \$4, so you know that $e(0) = 0$ and $e(4) = 0$. Find $e(1)$. Find $e(2)$.

c. My letter to Ann Landers also included the following paragraph. "You will make on the average more than 1,200 bets before you lose your \$50. Therefore, if you view your losses as entertainment expenses, you may feel it is worth it. I personally do not find that throwing my money away is entertaining." A person starts playing roulette with a standard wheel, so on any bet, the probability of winning \$1 is 9/19 and the probability of losing \$1 is 10/19. The person begins with \$50 and plays until he has \$0 or \$100. Evaluate my statement in Ann Landers' column by developing a dynamical system for the expected number of bets he will make on a typical night and use that to find $e(50)$. Assume this person has a good night playing roulette and presently has \$99. How many more bets does he make on average?

4.5 Analyzing Higher Order Dynamical Systems

In Sections 4.2 and 4.4, we used higher order dynamical systems **quantitatively,** meaning that we used them to compute actual numerical values in situations we were studying, such as the number of arrangement of beads or the probability of going broke. In this section, we are less interested in exact values and more interested in the general, **qualitative** behavior of the functions that satisfy a dynamical system. We are going to apply the methods we learned in Chapter 2 to higher order dynamical systems. In particular, we will find equilibrium values for higher order dynamical systems, determine the stability of the equilibrium values, compute the rates at which consecutive values go toward or away from equilibrium, and describe cyclic behavior when it occurs.

Recall that a number E is an equilibrium value for a dynamical system if the constant function $u(n) = E$ satisfies the dynamical system. Consider the dynamical system

$$u(n) = 2u(n-1) + u(n-2) - 6$$

If $u(0) = 3$ and $u(1) = 3$, then

$$u(2) = 2u(1) + u(0) - 6 = 2(3) + 3 - 6 = 3$$

Similarly, $u(3) = 3$, $u(4) = 3$, ... In this case, $u(n) = 3$ for every value of n. So for this dynamical system, $E = 3$ is an equilibrium value. Notice that if $u(0) = 3$ and $u(1) = 2$, then

$$u(2) = 2u(1) + u(0) - 6 = 2(2) + 3 - 6 = 1$$

and $u(3) = -2$. Even though $u(0)$ is equal to the equilibrium value, the other $u(n)$-values are not equal to equilibrium. If all initial values equal the equilibrium value, then $u(n)$ remains at equilibrium. Otherwise, $u(n)$ will not remain at equilibrium. Figure 4.19 shows the graph of $u(n)$ with $u(0) = u(1) = 3$ and again when $u(0) = 3$ and $u(1) = 2$. In both cases, u started in equilibrium with $u(0) = 3$. But in 1 case, u did not remain in equilibrium.

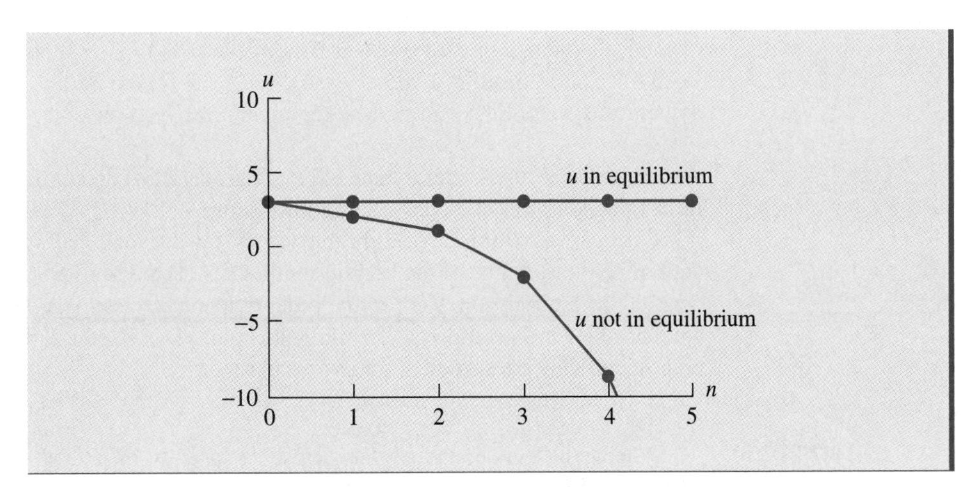

FIGURE 4.19 Graph of $u(n)$ for 2 different sets of initial values.

Equilibrium values for first-order dynamical systems were found by substituting a variable, say E, into the dynamical system for $u(n)$ and $u(n-1)$, and solving. The method is the same for second order dynamical systems. To find the equilibrium values for

$$u(n) = au(n-1) + bu(n-2) + c$$

just solve the equation

$$E = aE + bE + c$$

for E. If no solution exists, then the dynamical system does not have an equilibrium value. To find the equilibrium value for $u(n) = 2u(n-1) + u(n-2) - 6$, solve the equation

$$E = 2E + E - 6$$

The solution to this equation is $E = 3$, as expected.

Example 4.7

To find the equilibrium values for the dynamical system

$$u(n) = 5u(n-1) - 4u(n-2) + 7$$

we solve the equation $E = 5E - 4E + 7$. This equation simplifies to $E = E + 7$. Subtracting E from both sides gives $0 = 7$ which has no solution. So this dynamical system has no equilibrium values. ■

Example 4.8

To find the equilibrium values for the dynamical system

$$u(n) = -u(n-1) + 2u(n-2)$$

we solve the equation $E = -E + 2E$, which simplifies to $E = E$. This equation is balanced for E equal to any real number. Therefore, every real number is an equilibrium value for this dynamical system. For example, if $u(0) = 3.5$ and $u(1) = 3.5$, then $u(2) = -3.5 + 2(3.5) = -3.5 + 7 = 3.5$. Similarly, $u(3) = u(4) = \cdots = 3.5$ and $u(n) = 3.5$ is a solution to this dynamical system. The same is true for any number you pick. ■

As you see, a second-order dynamical system may have no equilibrium values, 1 equilibrium value, or infinitely many equilibrium values.

Once we have found equilibrium values for a dynamical system, we should try to determine the stability of the equilibrium values. The definitions for stable, unstable and neutral equilibrium values for higher order dynamical systems are similar to those for stable, unstable, and neutral equilibrium points given in Chapter 2. These definitions are not precise, but they are sufficient for our purposes.

DEFINITION

If the $u(n)$-values eventually get "close" to an equilibrium value E and remain close to it, no matter what the initial values, then the equilibrium value is **globally stable.** Geometrically, this means that for any pair of horizontal lines drawn on either side of the line $u = E$, the time graph $(n, u(n))$ will eventually be between those lines and stay between them.

Example 4.9

Consider the dynamical system

$$u(n) = 0.5u(n-1) - 0.9u(n-2) + 14$$

This system has the equilibrium value $E = 10$. A time graph of the function is seen in Figure 4.20, with $u(0) = 15$ and $u(1) = 5$. It oscillates around $u = 10$. Some of the time, the $u(n)$-values are moving away from 10, other times the $u(n)$-values are moving toward 10. But the equilibrium value is stable, because all of the points $(n, u(n))$ will eventually be close to the line $u = 10$. The line $u = 10$ is a horizontal asymptote for the function $u(n)$. ■

The **amplitude** of an oscillation is the distance of the peak of the oscillation from the equilibrium value, $u = 10$ in the case of Example 4.9. We say that this function oscillates with **decreasing amplitude.**

When functions oscillate, as in Figure 4.20, you should estimate the length of the oscillations. The length of the oscillations is called the **period** of the function. You can estimate the period of a function by estimating the horizontal distance between consecutive peaks of the function. In Figure 4.20, vertical lines go through each peak point. It is not clear if $u(0)$ is a peak or not, since we don't know $u(-1)$, so we will consider the first complete cycle as the one that goes from $n = 4$ to $n = 9$. Vertical lines are drawn on Figure 4.20 to indicate these points. This cycle has a length of $9 - 4 = 5$. The second complete cycle goes from $n = 9$ to $n = 14$ and also has a length of 5. The third, fourth, and fifth cycles have lengths 5, 4, and 5, respectively. We can then say that the period of this function is between 4 and 5 units.

It happens that the period of a function is usually not an integer value. At this point we can only give a range for the period, between 4 and 5. A better estimate of the period can

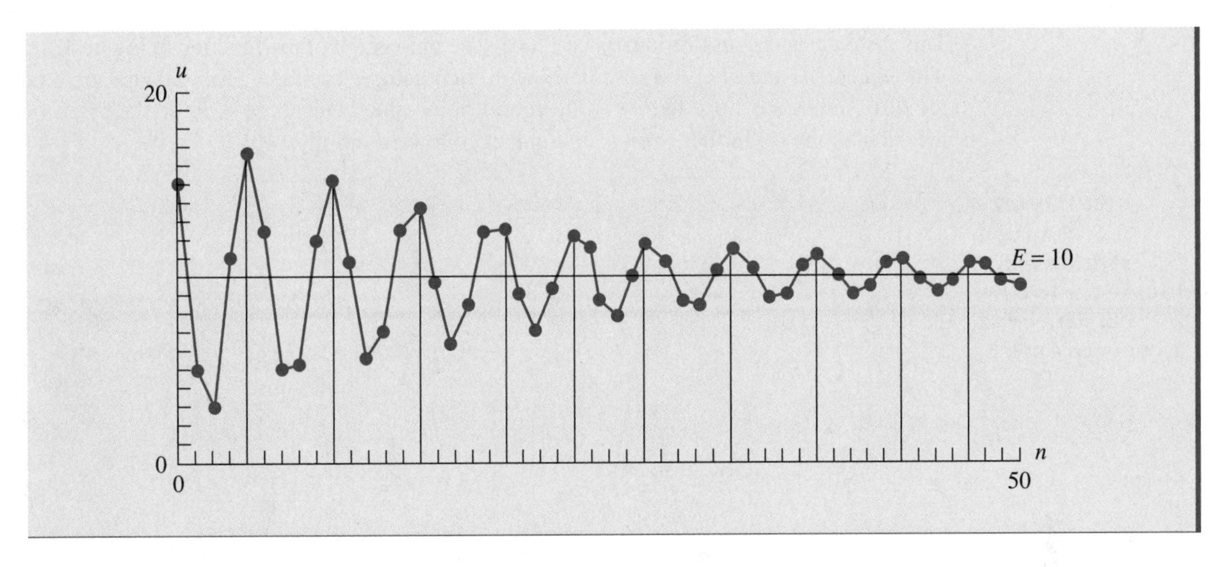

FIGURE 4.20 Function that satisfies $u(n) = 0.5u(n) - 1)) - 0.9u(n) - 2) + 14$ oscillates to equilibrium with a period of between 4 and 5.

be found by using an average. Since 9 complete cycles take place as n goes from 4 to 47, the average is

$$\frac{47-4}{9} \approx 4.8$$

We can now state that for the dynamical system of Example 4.9, the equilibrium value 10 is stable and that solutions oscillate toward equilibrium with a period of about 4.8.

With a little trigonometry, we would be able to compute the exact period for functions that satisfy second-order dynamical systems.

In the case of Example 4.9, you should try other initial values to see that different initial values result in functions with the same period. This means that the period of the functions is intrinsic to the dynamical system and does not depend on the initial values.

DEFINITION

Suppose that for any set of initial values, not all equal to the equilibrium value E, some of the $u(n)$-values go toward positive or negative infinity. Then the equilibrium value is said to be **globally unstable.** This means that for any pair of lines drawn on either side of the line $u = E$, some of the points $(n, u(n))$ will be outside of those lines, as long as the initial values are not all equal to E.

Example 4.10 Consider the dynamical system

$$u(n) = 0.5u(n-1) - 1.1u(n-2) + 8$$

This system has the equilibrium value $E = 5$. The time graph of the function in Figure 4.21, with $u(0) = 10$ and $u(1) = 7$, oscillates with **increasing** amplitude. Even though some of the $u(n)$-values are close to $u = 5$, the equilibrium value is unstable because the peaks of the oscillations get farther from 5, and in fact go toward infinity.

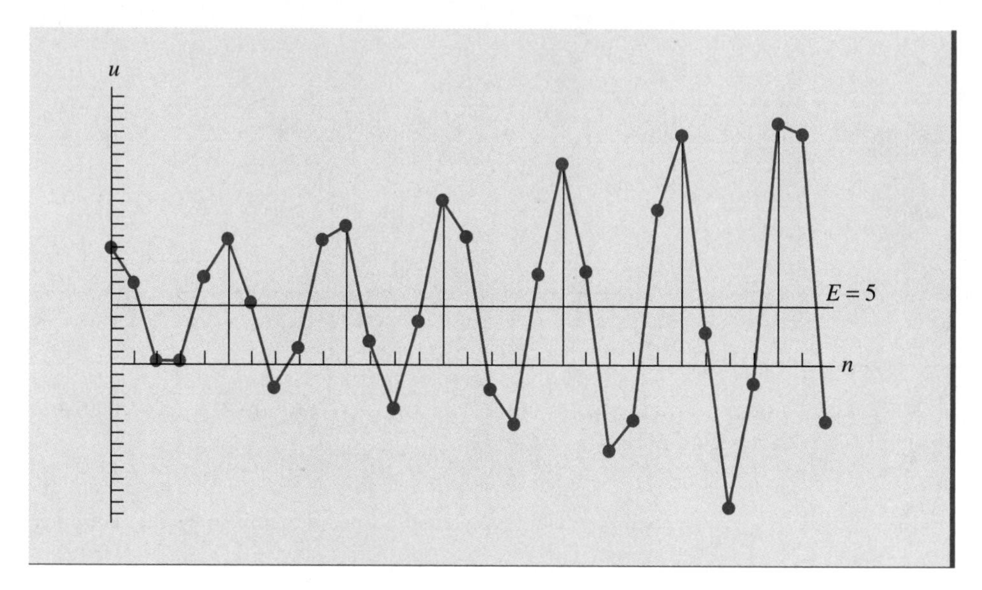

FIGURE 4.21 Function satisfies $u(n) = 0.5u(n-1) - 1.1u(n-2) + 8$. Has period between 4 and 5.

You will find that no matter what initial values you pick (as long as they are not both equal to 5), the solution will oscillate and the oscillations will increase in size.

In Figure 4.21, vertical lines go through each peak point of the time graph. The first complete cycle goes from $n = 5$ to $n = 10$. This cycle has a length of 5. The second cycle goes from $n = 10$ to $n = 14$ and has a length of 4. The third, fourth, and fifth cycles have lengths 5, 5, and 4, respectively. We can then say that the length of a cycle for this function is between 4 and 5 units long. We note that the function completes 5 cycles when n goes from 5 to 28, so a better estimate for the period is

$$\frac{28-5}{5} = 4.6$$

You should try other initial values to see that different initial values result in functions with the same period.

We can now state that $u(n) = 0.5u(n-1) - 1.1u(n-2) + 8$ has the unstable equilibrium value of 5 and that solutions oscillate with increasing amplitude and a period of about 4.6. ■

The definition of globally unstable is slightly different from the one in Chapter 2, due to the different form of the dynamical systems. We could have defined globally stable by saying that the points $(u(n-1), u(n))$ go away from the point (E, E), that is, eventually lie outside any circle centered at (E, E). Then the definition would be similar to the definition in Chapter 2. But plotting points $(u(n-1), u(n))$ doesn't seem to make contextual sense.

DEFINITION

> If the time graph oscillates $(n, u(n))$ but the amplitude of the oscillations neither goes toward 0 nor toward infinity, then the equilibrium value is **neutral.**

It is often difficult to know if the size of oscillations is decreasing slightly, increasing slightly, or remaining approximately constant. In such cases, you may have to generate time graphs using a relatively large window for the n-axis.

Example 4.11 Consider the dynamical system

$$u(n) = 0.8u(n-1) - u(n-2) + 9$$

which has the equilibrium value $E = 7.5$. The function $u(n)$ with $u(0) = 8$ and $u(1) = 11$ is seen in Figure 4.22. It oscillates with approximately constant amplitude around $E = 7.5$. The peaks are not all the exact same size, but if you kept generating values, you would see that the amplitude of the oscillations does not decrease or increase. In this case, we say that the function oscillates with **constant amplitude.** You will find that the particular amplitude for any one function depends on the particular initial values. Such behavior indicates that the equilibrium value is neutral.

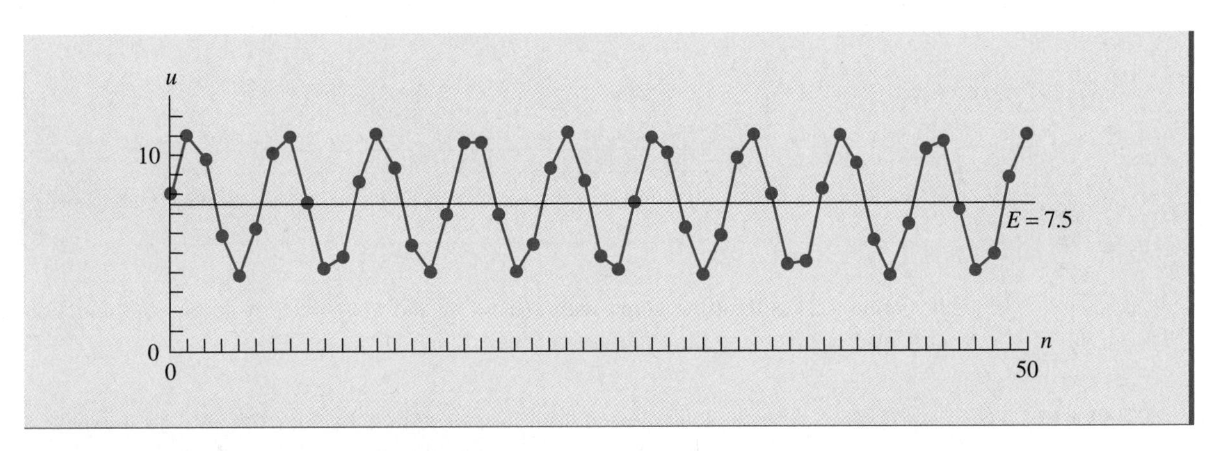

FIGURE 4.22 Function satisfying $u(n) = 0.8u(n-1) - u(n-2) + 9$ oscillates with constant amplitude and has a period of between 5 and 6.

You should check that this function, and functions with other initial values, have periods of between 5 and 6. In the case of Figure 4.22, the function completes 8 cycles as n goes from 1 to 45, so we can state that the period is approximately

$$\frac{48-1}{8} \approx 5.9$$

In this case, we state that the equilibrium value 7.5 is neutral and that solutions have constant amplitude and a period of nearly 6. ■

Sometimes it is difficult to determine if the amplitude of the oscillations is slowly increasing, slowly decreasing, or remaining constant. In this case, it is advisable to generate a graph of the solution for n going from 0 to a relatively large number so that you can compare many of the oscillations.

We now discuss another type of behavior that can occur.

Example 4.12 Consider the dynamical system

$$u(n) = 2.5u(n-1) - u(n-2) - 3$$

You should be able to determine that the equilibrium value is $E = 6$. In Figure 4.23 is the time graph when $u(0) = 16.1$ and $u(1) = 11.2$. Note that the solution initially goes toward the equilibrium value of 6, which is seen as the horizontal line $u = 6$, but then the values go away from $u = 6$ and toward positive infinity.

FIGURE 4.23 Points $(n, u(n))$ for $u(n) = 2.5u(n-1) - u(n) - 2) - 3$.

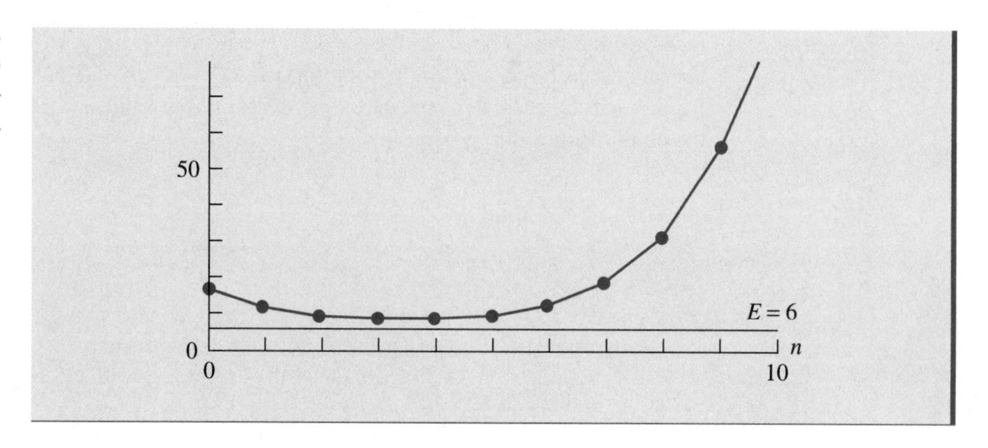

In Figure 4.24 is the time graph with $u(0) = 26$ and $u(1) = 16$. Note that this function continues to go toward the equilibrium value. ■

DEFINITION

Suppose $u(n)$ goes toward equilibrium for some initial values and away from equilibrium for other initial values. Then this equilibrium is called a **saddle.**

In the case of a saddle equilibrium value, you will find that $u(n)$ goes away from equilibrium for "most" initial values. Let me describe what I mean by "most." For $u(n) = 2.5u(n-1) - u(n-2) - 3$, functions go toward the equilibrium $E = 6$ if $u(1) = 0.5u(0) + 3$. You should check that this statement is true by trying several different pairs of initial values that satisfy this equation, such as $u(0) = 20$ and $u(1) = 13$. If $u(1) \neq 0.5u(0) + 3$, then $u(n)$ goes toward infinity. You should check the validity of this statement by trying several pairs of initial values that satisfy this inequality, such as $u(0) = 2$ and $u(1) = 1$. To say this another way, the only pairs of initial values that result in $u(n)$ going toward equilibrium in this example are those in which $u(0) = x$, $u(1) = y$ and $y = 0.5x + 3$. This consists only of pairs of values on the line in Figure 4.25.

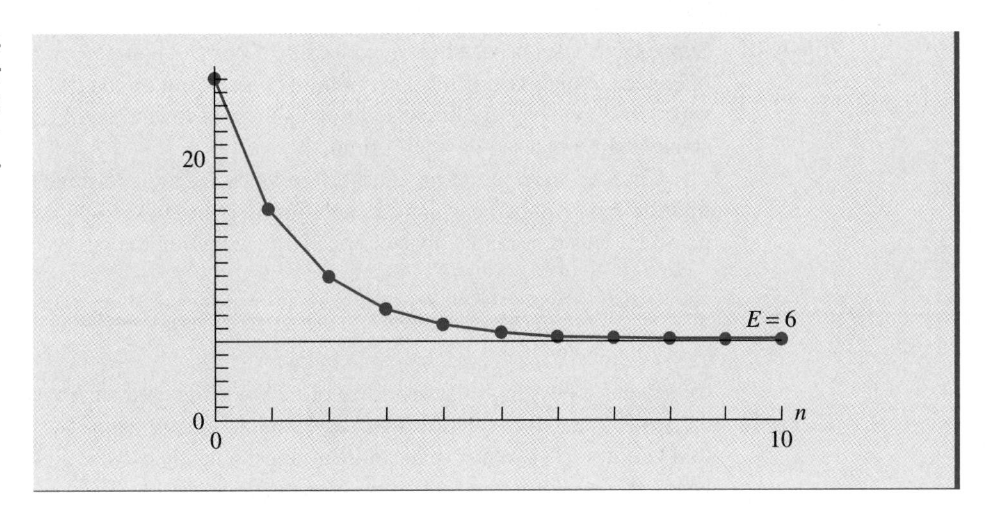

FIGURE 4.24 Another function satisfying $u(n) = 2.5u(n-1) - u(n-2) - 3$.

REMARK In case of a saddle equilibrium, the set of initial values for which $u(n)$ goes toward equilibrium is "small." It may be hard to find a set of initial values so that $u(n)$ goes toward equilibrium. Therefore, it is usually difficult to decide if an equilibrium value is globally unstable or a saddle. For this reason, we will label saddle equilibrium values as unstable for the purpose of the exercises.

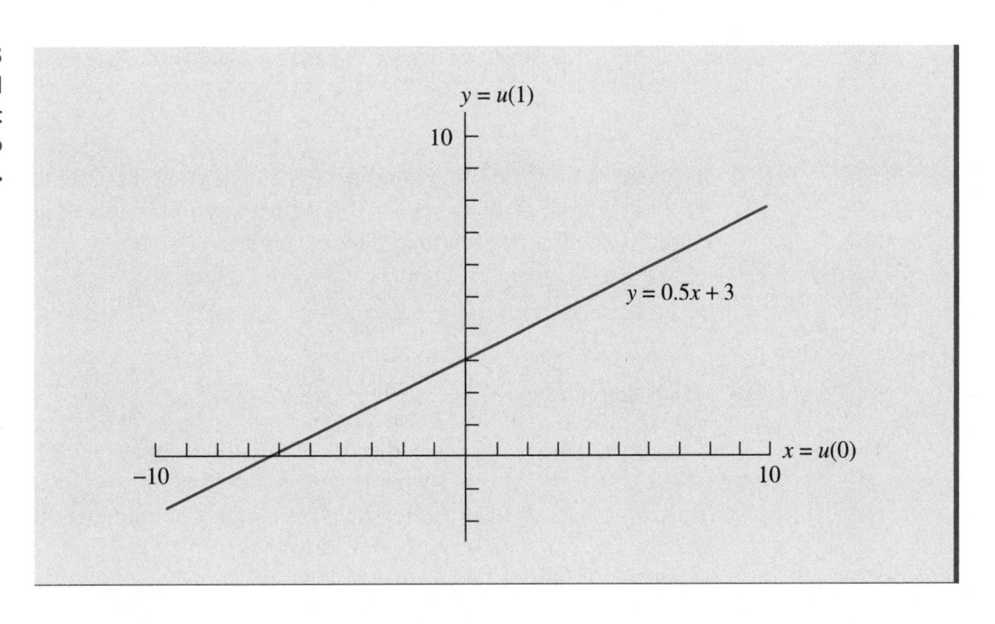

FIGURE 4.25 Pairs of initial values on line result in $u(n)$ going to equilibrium.

REMARK Suppose you pick a set of initial values $u(0)$ and $u(1)$, and $u(n)$ appears to be stable. You only need to pick 1 additional set of initial values, not on the line going through the points $(u(0), u(1))$ and (E, E). If the solution again goes toward equilibrium, then E is definitely stable and is not a saddle equilibrium.

Once we have found an equilibrium value E and determined its stability, we should find the rate, if any, at which the solution is going toward or away from equilibrium, if possible. This rate can be approximated by computing the ratios

$$\frac{u(n)-E}{u(n-1)-E}$$

for several increasingly large values of n. You computed such ratios in Chapter 2. In Exercises 2, 3, and 4 of Section 4.2, you found the rate at which $u(n)$ went to infinity, which told you about how many times more arrangements there were with n beads than with $n-1$ beads. Thus, computing rates is nothing new.

Example 4.13 Consider the dynamical system

$$u(n) = 0.1u(n-1) + 0.72u(n-2) + 9$$

The equilibrium value is $E = 50$, and it is stable. Using $u(0) = 30$ and $u(1) = 37$, I obtained the results in Table 4.7.

TABLE 4.7 Ratios for functions satisfying $u(n) = 0.1u(n-1) + 0.72u(n-2) + 9$.

$n =$	5	10	20	30	50	100
$\frac{u(n)-50}{u(n-1)-50} =$	1.25	0.75	0.85	0.88	0.899	0.899996

It appears that the rate is approximately 0.9, meaning that eventually, each $u(n)$ is about 90% as far from 50 as is $u(n-1)$. A better way to say this is that $u(n)$ gets about 10% closer to equilibrium each time. This means that

$$\frac{u(n)-50}{u(n-1)-50} \approx 0.9$$

which simplifies to

$$u(n) \approx 0.9u(n-1) + 5$$

Thus, $u(n)$ behaves like functions that satisfy the much simpler dynamical system

$$u(n) = 0.9u(n-1) + 5$$

From Chapter 3, we know that functions that satisfy this first-order dynamical system are of the form

$$u(n) \approx c0.9^n + 50$$

This tells us that $u(n)$ approaches 50 approximately exponentially, like 0.9^n. Again, we wouldn't use this expression to actually find values for $u(n)$. It gives the general behavior.

You should try other initial values and see if you get the same results. ■

Sometimes, a different set of initial values gives a different rate. For second-order dynamical systems, it is possible to get up to 2 different rates. The rate we are looking for is the larger rate. Therefore, you should always try several sets of initial values when approximating the rate. In Example 4.13, it happens that if you pick $u(0)$ and $u(1)$ so that $u(1) = 90 - 0.8u(0)$, then you will get ratios equal to -0.8. Otherwise, you will get ratios approaching 0.9.

In some cases, you won't be able to see a pattern when computing ratios. Consider $u(n) = 0.5u(n-1) - 1.1u(n-2) + 8$ in Example 4.10. In this case, the equilibrium value $E = 5$ was unstable and the function oscillated. Using $u(0) = 10$ and $u(1) = 7$, I obtained the results in Table 4.8.

TABLE 4.8 **Ratios for $u(n) = 0.5u(n-1) - 1.1u(n-2) + 8$.**

$n =$	5	10	20	30	50	100
$\frac{u(n)-5}{u(n-1)-5} =$	2.3	1.2	0.27	-0.66	1.7	0.7

Notice that there is no apparent pattern in the ratios. This is because the function is oscillating, as was seen in Figure 4.21. This means that sometimes $u(n)$ is further from equilibrium than $u(n-1)$ but at other times, it is closer. If the function oscillates, there is no need to compute the ratios.

4.5 Problems

1. Consider the second-order dynamical system

$$u(n) = -0.4u(n-1) + 0.45u(n-2) + 2.85$$

a. Find the equilibrium value.

b. Figure 4.26 is a graph of u with $u(0) = 1$ and $u(1) = 2$. Graph the solution to this dynamical system with $u(0) = 2$ and $u(1) = 1$. Your horizontal axis should go to a larger value of n than is seen in Figure 4.26 so you can be sure about the behavior of the graph. If the graph oscillates, give the approximate period of the oscillations.

c. Determine the stability of the equilibrium values, if there is one. Use at least 3 pairs of initial values, including the ones given in part (b).

d. If the $u(n)$-values are oscillating, what is the period? If they are not oscillating, at what rate are they going toward infinity or toward equilibrium?

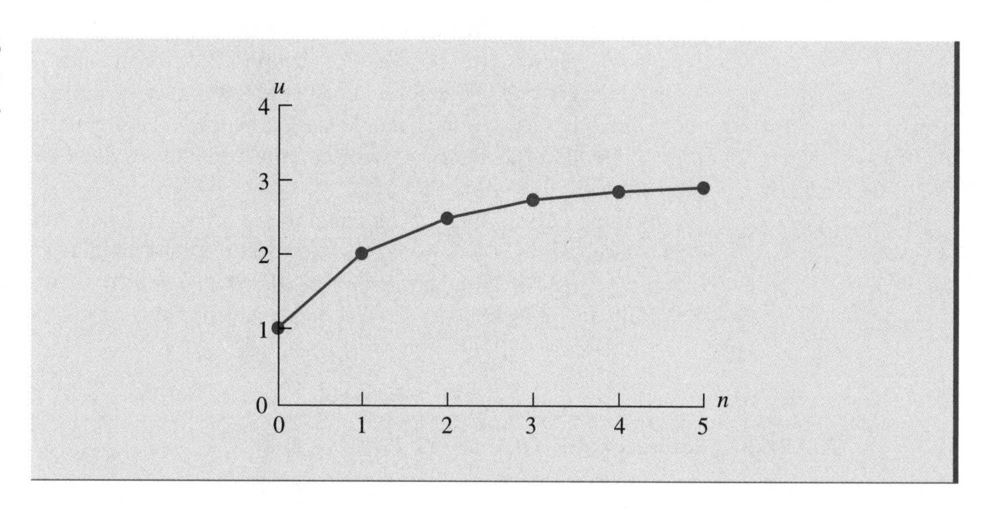

FIGURE 4.26 Graph of $u(n)$ for Problem 1b.

2. Consider the second-order dynamical system

$$u(n) = -1.5u(n-1) + u(n-2) + 12$$

a. Find the equilibrium value.

b. Graph the solution to this dynamical system with $u(0) = 5$ and $u(1) = 7$. If the graph oscillates, give the approximate period of the oscillations.

c. Determine the stability of the equilibrium values, if there is one. Use at least 3 pairs of initial values, including the ones given in part (b).

d. If the $u(n)$-values are oscillating, what is the period? If they are not oscillating, at what rate are they going toward infinity or toward equilibrium?

3. Consider the second-order dynamical system

$$u(n) = 1.7u(n-1) - 0.72u(n-2) + 0.5$$

a. Find the equilibrium value.

b. Graph the solution to this dynamical system with $u(0) = 3$ and $u(1) = 2$.

c. Determine the stability of the equilibrium values, if there is one. Use at least 3 pairs of initial values, including the ones given in part (b).

d. Find the rate at which $u(n)$ goes toward infinity or toward equilibrium.
e. For the dynamical system $u(n) = 1.7u(n-1) - 0.72u(n-2) - 0.6$, find the equilibrium value, give the stability, and determine the rate at which $u(n)$ goes toward or away from equilibrium. How does the rate seem to be affected by the constant added or subtracted at the end of the dynamical system?

4. Consider the second-order dynamical system

$$u(n) = 4u(n-1) - 4u(n-2) + 3.1$$

a. Find the equilibrium value.
b. Graph the solution to this dynamical system with $u(0) = 0$ and $u(1) = 0$.
c. Determine the stability of the equilibrium value, if there is one. Use at least 3 pairs of initial values, including the ones given in part (b).
d. Find the rate at which $u(n)$ goes toward infinity or toward equilibrium.
e. For the dynamical system $u(n) = 4u(n-1) - 4u(n-2) + 6$, find the equilibrium value, give the stability, and determine the rate at which $u(n)$ goes toward or away from equilibrium. How does the rate seem to be effected by the constant added or subtracted at the end of the dynamical system?

5. Consider the second-order dynamical system

$$u(n) = u(n-1) - 0.5u(n-2) + 4$$

a. Find the equilibrium value.
b. Graph the solution to this dynamical system with $u(0) = 20$ and $u(1) = 12$. If the graph oscillates, give the approximate period of the oscillations.
c. Determine the stability of the equilibrium value, if there is one. Use at least 3 pairs of initial values, including the ones given in part (b).
d. Find the rate at which $u(n)$ goes toward infinity or toward equilibrium if $u(n)$ does not oscillate.

6. Consider the second-order dynamical system

$$u(n) = 2u(n-1) - 1.02u(n-2) - 0.2$$

a. Find the equilibrium value.
b. Graph the solution to this dynamical system with $u(0) = -3$ and $u(1) = 1$. If the graph oscillates, give the approximate period of the oscillations.
c. Determine the stability of the equilibrium value, if there is one. Use at least 3 pairs of initial values, including the ones given in part (b).
d. Find the rate at which $u(n)$ goes toward infinity or toward equilibrium if $u(n)$ does not oscillate.

7. Consider the second-order dynamical system

$$v(n) = v(n-1) - v(n-2) + 7$$

a. Find the equilibrium value.

b. Graph the solution to this dynamical system with $v(0) = 2$ and $v(1) = -4$. The graph oscillates. Give the approximate period of the oscillations. Graph the function using other initial values. How do the initial values effect the length of the period?

c. Determine the stability of the equilibrium value, if there is one. Use at least 3 pairs of initial values, including the ones given in part (b). How do the initial values effect the stability?

d. For the dynamical system $v(n) = v(n-1) - v(n-2) - 3$, find the equilibrium value, determine the stability, and give the approximate period of the oscillations. How does the period seem to be effected by the constant added or subtracted at the end of the dynamical system?

8. Consider the second-order dynamical system

$$w(n) = 1.2w(n-1) - w(n-2) + 16$$

a. Find the equilibrium value.

b. Graph the solution to this dynamical system with $w(0) = 0$ and $w(1) = 1$. The graph oscillates. Give the approximate period of the oscillations. Graph the solution using different initial values. How do the initial values effect the length of the period?

c. Determine the stability of the equilibrium value, if there is one. Use at least 3 pairs of initial values, including the ones given in part (b). How do the initial values effect the stability?

d. For the dynamical system $w(n) = 1.2w(n-1) - w(n-2) + 4$, find the equilibrium value, determine the stability, and give the approximate period of the oscillations. How does the period seem to be effected by the constant added or subtracted at the end of the dynamical system?

9. Consider the second-order dynamical system

$$u(n) = 1.8u(n-1) - 0.8u(n-1) + 3$$

a. Find the equilibrium value.

b. Graph the solution to this dynamical system with $u(0) = 2$ and $u(1) = 2$. If the graph oscillates, give the approximate period of the oscillations.

c. Determine the stability of the equilibrium value, if there is one. Use at least 3 pairs of initial values, including the ones given in part (b).

d. Describe how the $u(n)$-values are changing. (Constant difference, constant rate, oscillating, etc.)

10. Consider the second-order dynamical system

$$u(n) = -0.4u(n-1) + 1.4u(n-2)$$

a. Find equilibrium.
b. Graph the solution to this dynamical system with $u(0) = 5$ and $u(1) = 2$. If the graph oscillates, give the approximate period of the oscillations.
c. Determine the stability of the equilibrium values, if there is one. Use at least 3 pairs of initial values, including the ones given in part (b).
d. Describe how the $u(n)$-values are behaving.

11. Consider the second-order dynamical system

$$u(n) = 0.4u(n-1) + 0.6u(n-2)$$

a. Find equilibrium.
b. Graph the solution to this dynamical system with $u(0) = 0$ and $u(1) = -1$. If the graph oscillates, give the approximate period of the oscillations.
c. Try different initial values, not both equaling the same number. Describe what happens to the solutions.

12. In Section 4.2, we developed the dynamical system

$$u(n) = 2u(n-1) + u(n-2)$$

in which $u(n)$ is the number of arrangements of n beads using single orange and blue beads and double yellow beads. The equilibrium value is 0, which is unstable. Find the rate at which $u(n)$ grows. This rate tells about how many times more arrangements there are with n beads then with $n-1$ beads.

4.6 An Economic Model

In this section, we are going to make several simple assumptions about a nation's economy. From those assumptions, we are going to develop a second-order affine dynamical system that models the economy. From that model, we are going to try to understand how each of the components of the model affects the long-term behavior of the economy. While we do not expect this dynamical system to perfectly describe anything as complex as a nation's economy, we do hope that this model will give us some insight into how major components of an economy affect its behavior.

Once we understand how our assumptions affect the behavior of the economy, we will add new components to the economy and modify the components we have already included. In Section 4.7, we modify one particular assumption, government spending, and investigate how alternate approaches to governmental spending can positively and negatively affect this economy.

A word of caution is that our conclusions only apply to an economy satisfying our assumptions and no others. Therefore, any conclusions we make must be considered in this light.

When modeling, it helps to begin with a simple model, then add components and assumptions about these components as we gain understanding of the situation. For this reason, we assume our nation's economy consists of only three components: consumer expenditures, private investment, and government expenditures, denoted by c, p, and g, respectively, represented in Figure 4.27. The total national economy, denoted by t, is the sum of these parts; that is

$$t = c + p + g$$

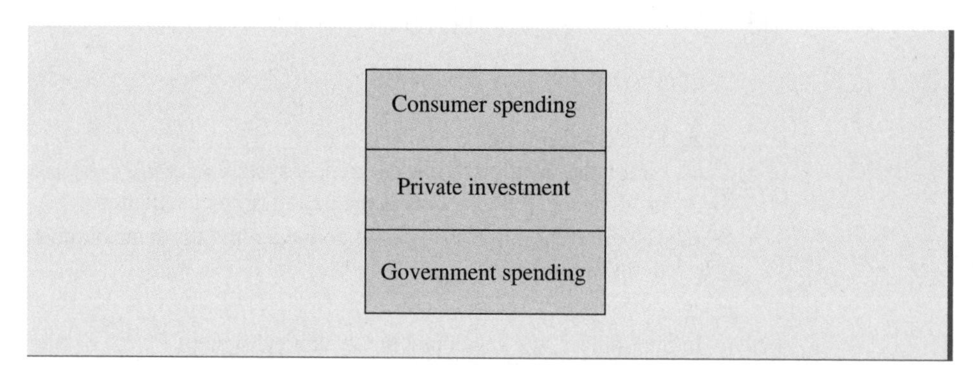

FIGURE 4.27 Total economy.

While consumption, investment, and government expenditures are being made continuously, they are generally known only at discrete points in time, for example, quarterly reports issued by the government. We shall assume that these components depend on a time period, n. Thus, the total national economy during time period n is

$$t(n) = c(n) + p(n) + g(n)$$

where $c(n)$ is the amount consumers spend, $p(n)$ is the amount of additional investment made, and $g(n)$ is the amount the government spends, all during the nth time period. These components could be subdivided into more components, and other components could be added to give a more complex model of the economy.

We now make one assumption about each component of the economy—$c(n)$, $p(n)$, and $g(n)$.

Consumer Spending Assumption We assume that individuals spend a certain fixed proportion of their income and their income is a fixed proportion of the total economy. This money is spent in the time period after it is earned. This means that consumption in time period n is proportional to the total income in time period $n-1$. In short,

$$c(n) = at(n-1)$$

The constant a is known as the **marginal propensity to consume,** or MPC. Since what is not spent is saved, $1 - a = 1-\text{MPC}$ is called the **marginal propensity to save,** or MPS.

We assume that $a < 1$. If $a \geq 1$, then consumer spending would exceed the economy, which would lead to some unrealistic results. Before making the Investment Assumption, let me describe what is meant by investment. $p(n)$ is the increase or decrease in private investment made during the nth time period. For example, buying new machines or building a new factory would be a positive investment. Closing a factory would be a negative investment. When people buy stock, they are making additional, positive investments. When people sell stock, they are making negative investments. When the value of a stock goes up, that is considered additional investment, since the total value of all investments has increased. Similarly, if the value of a stock goes down, that would be considered a negative investment.

Let $p(n)$ be the total investment made during time period n. If more money is invested during time period n, then $p(n)$ will be positive. If total investment decreases through closing factories or stock prices declining, then $p(n)$ is negative.

Investment Assumption Investment depends on the **change** in consumption; that is, if consumption increases, more factories will be needed to produce the additional material being consumed, so more investment must be made and $p(n)$ will be positive. People will be encouraged by the increase and will buy more stocks, causing stock prices to rise. If consumption decreases, then existing factories are sufficient to produce materials, so additional investment will be small or even negative because of closing factories or laying off employees. Investors will be discouraged and will sell stock, causing stock prices to decrease. This is summarized in Figure 4.28. Thus, an investor, in time period n, compares the present consumption, $c(n)$, to the previous consumption, $c(n-1)$, and looks at the change in consumption, $c(n)-c(n-1)$. If this is positive people are spending more, so the additional investment, $p(n)$, should be positive. The larger this change is, the larger is $p(n)$. If $c(n)-c(n-1)$ is negative, then investment is reduced and $p(n)$ is negative. In short, additional investment in time period n is proportional to the **change** in consumption from time period $n-1$ to time period n, that is

$$p(n) = b(c(n) - c(n-1))$$

We will call b the **marginal propensity to investment** or MPI.

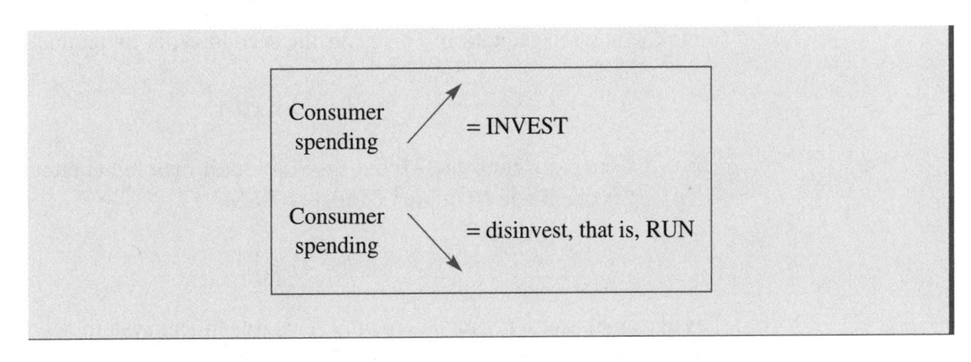

FIGURE 4.28 Investment advice.

Government Spending Assumption

For simplicity, we assume that government expenditures are constant. Whatever this amount, we let it represent one unit of money; that is, for every n,

$$g(n) = 1$$

While this assumption may seem unrealistic, it is more reasonable if we are assuming that all units of money are based on today's value of money, that is, the government's increases in expenditures equal the inflation rate. Thus, the **value** of the government's expenditures remains constant. In Section 4.7, we will investigate how varying government expenditures can effect the economy.

A word of caution is that the results obtained in this section only apply to an economy that satisfies our assumptions and no other assumptions. The model that will be developed is therefore only a starting point to understanding something as complex as a real nation's economy.

Summary of Assumptions about an Economy

We recall that our economy consists of 3 parts; that is,

$$t(n) = c(n) + p(n) + g(n) \qquad \text{(Economy Equation)}$$

Our 3 assumptions are summarized by the 3 equations

$$c(n) = at(n-1) \qquad \text{(Consumption Equation)}$$

$$p(n) = b(c(n) - c(n-1)) \qquad \text{(Investment Equation)}$$

$$g(n) = 1 \qquad \text{(Government Equation)}$$

The Investment Equation allows substitution of $b(c(n) - c(n-1))$ for $p(n)$ into the Economy Equation, giving $t(n) = c(n) + b(c(n) - c(n-1)) + g(n)$ or, after simplification,

$$t(n) = (1+b)c(n) - bc(n-1) + g(n)$$

The Consumption Equation allows the substitutions

$$c(n) = at(n-1) \quad \text{and} \quad c(n-1) = at(n-2)$$

into the previous equation. This gives the second-order dynamical system

$$t(n) = (1+b)at(n-1) - bat(n-2) + g(n) \qquad \textbf{(GEM)}$$

as our **General Economic Model** or GEM for a national economy. Substitution of 1 for $g(n)$ gives our **Basic Economy Model** or BEM

$$t(n) = (1+b)at(n-1) - bat(n-2) + 1 \qquad \textbf{(BEM)}$$

as our model for a national economy. This dynamical system is a variation of the Samuelson accelerator–multiplier model, which was first developed in 1939.

Example 4.14 Consider the case in which $a = 0.5$ and $b = 1.8$ in the Basic Economy Model. This means that people invest an amount equal to 180% of the change in consumption and that people spend 50% of the previous economy. Substitution of these values into the Basic Economy Model gives

$$t(n) = 1.4t(n-1) - 0.9t(n-2) + 1$$

The equilibrium value for this dynamical system is the solution to $E = 1.4E - 0.9E + 1$, which is $E = 2$. This means that if the total economy at time $n = 0$ and $n = 1$ is 2 units of money, it will remain there. To investigate the stability of the equilibrium value, we pick an arbitrary set of initial values, say $t(0) = 1$ and $t(1) = 1.5$. The graph of the function $t(n)$ is seen in Figure 4.29.

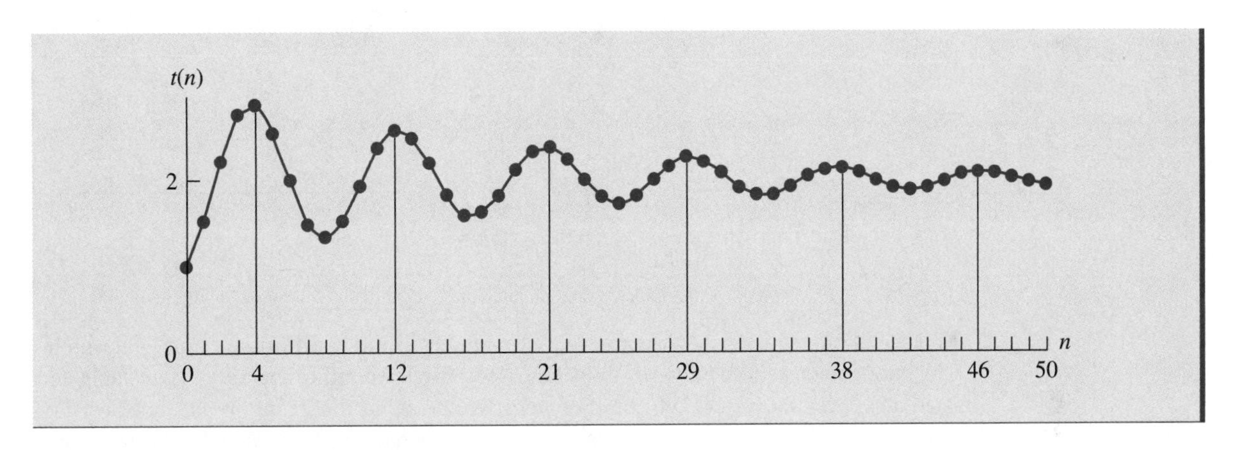

FIGURE 4.29 An oscillating economy, where $a = 0.5$ and $b = 1.8$.

Note that the function is oscillating with decreasing amplitude, so the equilibrium value is stable. Each oscillation is either 8 or 9 units long. Averaging the 5 complete oscillations in Figure 4.29, we get that the period is about 8.4 units long. This means that the economy makes a complete cycle about every 8.4 time units. If one time unit is a quarter of a year, then the economy tends to cycle about every 2 years, or a little more.

You should try other initial values and check that the functions oscillate toward equilibrium and the period of the oscillations is about 8.4 units long. (More detailed study will show that the period is closer to 8.5 units long.) ■

Example 4.15 Consider the case in which $a = 0.5$ and $b = 2.2$. This gives the dynamical system

$$t(n) = 1.6t(n-1) - 1.1t(n-2) + 1$$

The equilibrium value for this dynamical system is also $E = 2$. To investigate the stability of the equilibrium value, I picked the same initial values as before, $t(0) = 1$ and $t(1) = 1.5$. The graph of the function is seen in Figure 4.30.

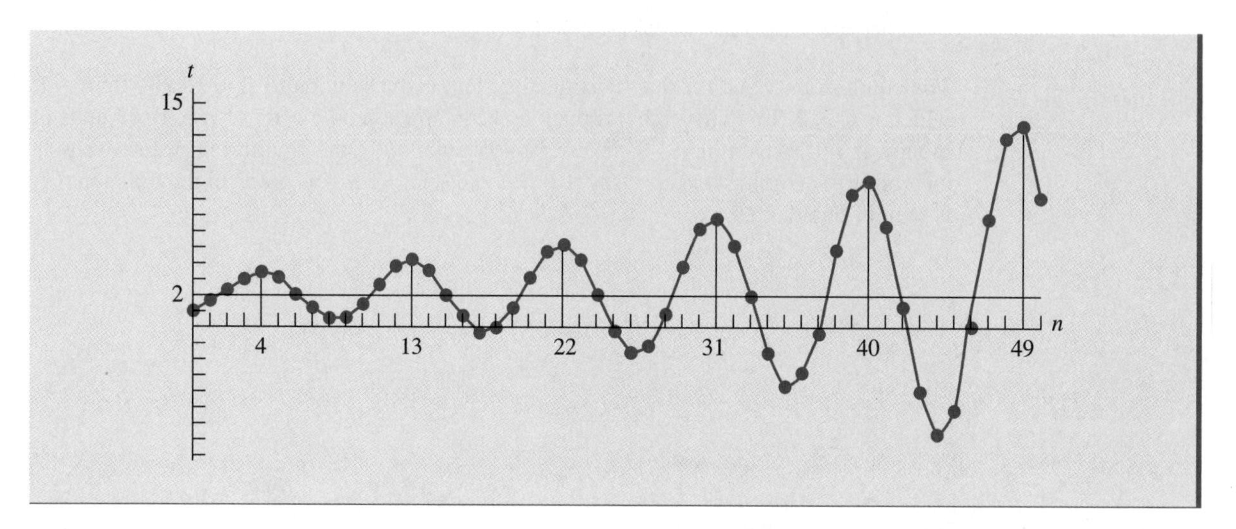

FIGURE 4.30 An unstable, oscillating economy where $a = 0.5$ and $b = 2.2$.

This function oscillates with increasing amplitude, so the equilibrium value is unstable. The period appears to be about 9 units long. Because the equilibrium value is unstable and oscillates, eventually $t(n)$ will be 0 or negative, meaning the economy has crashed. For our initial values, the economy crashed at $n = 17$, which is the n-value of the first point that falls below the n-axis. I found that $t(17) = -0.4$. This is an unstable economy and is undesirable. You should check that the same behavior occurs for any initial values, other than both being equal to 2. ■

Example 4.16 Consider the case in which $a = 0.5$ and $b = 2.0$. This gives the dynamical system

$$t(n) = 1.5t(n-1) - 1t(n-2) + 1$$

The equilibrium value for this dynamical system is still $E = 2$. To investigate the stability of the equilibrium value, I picked the same initial values as before, $t(0) = 1$ and $t(1) = 1.5$. The time graph of the function is seen in Figure 4.31.

This function oscillates with constant amplitude, so the equilibrium value is neutral. The period appears to be about 8.6 units long. This is a neutral economy. It will have good times at the peaks and bad times at low points.

You should check that the same behavior occurs for any initial values other than both being equal to 2. You will find that the amplitude of the cycles depends on the initial values.

■

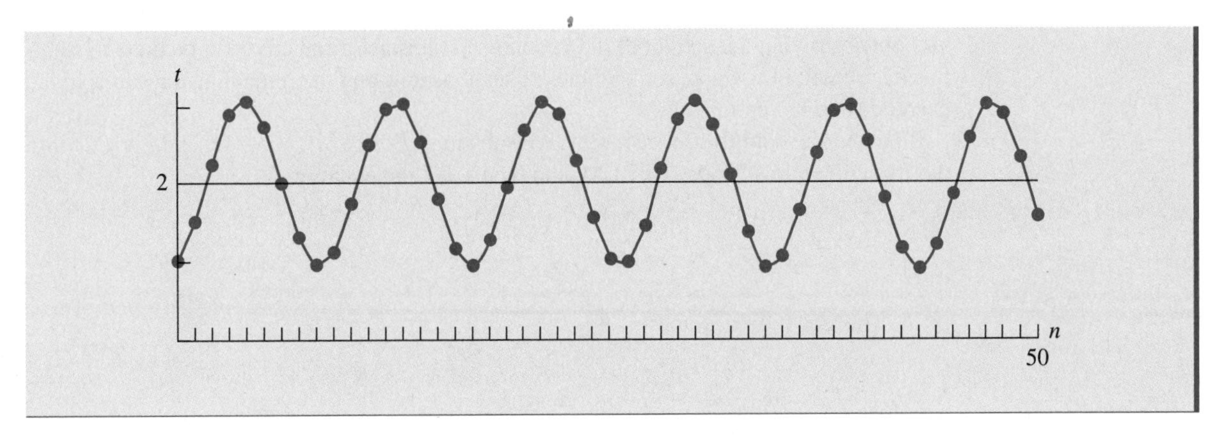

FIGURE 4.31 A neutral, oscillating economy where $a = 0.5$ and $b = 2$.

In Examples 4.14, 4.15, and 4.16, we fixed the marginal propensity to spend at $a = 0.5$ and investigated the behavior for different values of b, the marginal propensity to invest. We found that the equilibrium value was the same in each case, $t = 2$. Thus, it appears that the b-value does not affect the equilibrium value. It appears that the equilibrium is stable for $b < 2$, is unstable for $b > 2$, and is neutral for $b = 2$ as seen in Figure 4.32. If you try other b-values, you will see that these results appear to be true. The value $b = 2$ is a cutoff point between stable and unstable economies. This means that when people spend 50% of the economy, then the marginal propensity to invest must be less than $b = 2$ for the economy to remain stable.

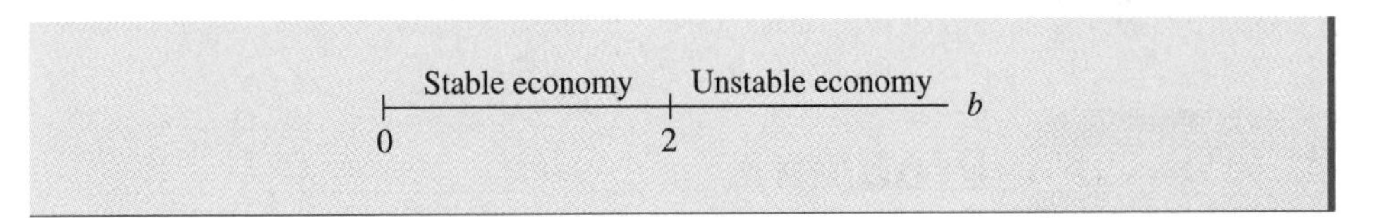

FIGURE 4.32 Summary of results when marginal propensity to spend is $a = 0.5$.

In the exercises, you will discover that for each value of a, a different b-value is the cutoff point between a stable and unstable economy. One question a government might have is, Given that an economy is unstable, what could be done to lower the value of b to a point where the economy becomes stable? Since b is a measure of how people invest, this means the government would need to discourage investment, which might be accomplished through tax increases on investments. The government could also try to influence the value for a so that it is at a value where the cutoff point is above the present value for b. This could be accomplished by encouraging or discouraging consumer spending.

Other questions we might want to answer are, For any given a and b, what is the equilibrium economy? For what values of a and b will the economy be stable? If we determine the equilibrium economy is too small, can anything be done to increase the equilibrium without destroying the stability? If an economy is unstable, can anything be done to stabilize it? We will investigate some of these questions now and the remaining questions in the exercises and in Section 4.7.

To find the equilibrium economy, we substitute E for $t(n)$, $t(n-1)$, and $t(n-2)$ into the Basic Economy Model, or BEM, and then solve the equation

$$E = (1+b)aE - abE + 1$$

for E. This equation becomes

$$E = aE + abE - abE + 1 \quad \text{or} \quad E = aE + 1$$

We subtract aE from both sides and divide by $(1-a)$ to get that the equilibrium value is

$$E = \frac{1}{1-a} = \frac{1}{\text{MPS}}$$

since $1-a$ is the marginal propensity to save. Thus, the equilibrium value is the multiplicative inverse of the marginal propensity to save. This means that the higher the savings rate is, the lower is the equilibrium economy, and the lower the savings rate is, the higher is the equilibrium economy. Thus, if the government wants to raise the equilibrium economy, it must encourage people to spend more and/or save less. In the exercises, you will investigate the affect on stability caused by increasing the equilibrium economy. You will also investigate the relationship among a, b, and stability.

4.6 Problems

1. Assume that $a = 0.9$ and $b = 1$ for the Basic Economy Model.

a. Find the equilibrium economy.

b. Determine the stability of the economy, that is, the stability of the equilibrium value.

c. Approximate the period of the economy.

2. Assume that $a = 0.8$ and $b = 1.25$ for the Basic Economy Model.

a. Find the equilibrium economy.

b. Determine the stability of the economy, that is, the stability of the equilibrium value.

c. Approximate the period of the economy.

d. Suppose that an election is coming at time period $n = 5$. The present economy is $t(0) = 1$. As head of the government, what is your goal for $t(1)$ so that the economy will be at the desirable level of $9.9 < t(5) < 10.1$? Refer to Figure 4.33. What will happen to the economy from $n = 6$ to $n = 12$?

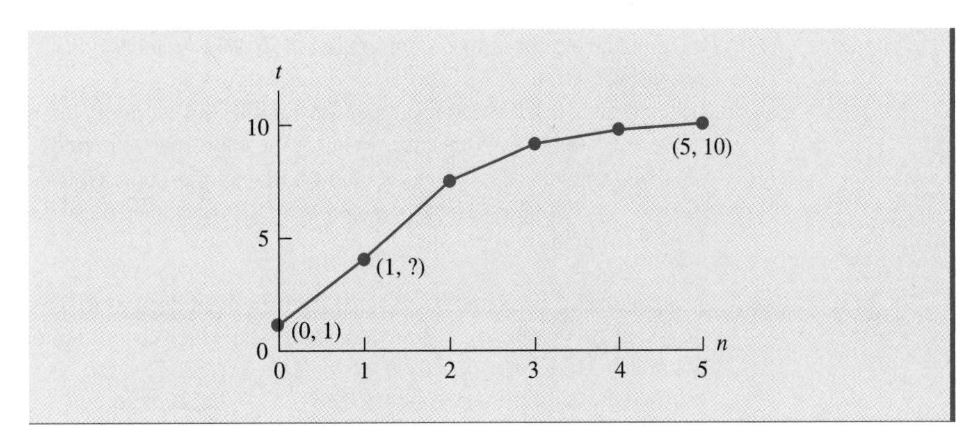

FIGURE 4.33 This figure indicates what the question is for Problems 2d and 3d.

3. Assume that $a = 0.7$ and $b = 1.6$ for the Basic Economy Model.

a. Find the equilibrium economy.

b. Determine the stability of the economy, that is, the stability of the equilibrium value.

c. Approximate the period of the economy.

d. Suppose that an election is coming at time period $n = 5$. The present economy is $t(0) = 1$. As head of the government, what is your goal for $t(1)$ so that the economy will be at the desirable level of $9.9 < t(5) < 10.1$? How long will it be before the economy crashes; that is, $t(n) \leq 0$ (see Figure 4.33)?

4. Assume that $a = 0.25$. Find, to the nearest tenth, the value of b that makes the equilibrium economy neutral. For what positive values of b is the economy stable? Unstable?

5. Assume that $a = 0.4$. Find, to the nearest tenth, the value of b that makes the equilibrium economy neutral. For what positive values of b is the economy stable? Unstable?

6. Assume that $a = 0.75$ and $b = 1.2$.

a. Determine the equilibrium economy and find its stability.

b. The government wants the equilibrium economy to equal 8. To do this, it needs for a to equal what number? If this is accomplished, what will happen to the stability of the equilibrium economy, assuming b remains at 1.2?

7. Assume that $a = 0.5$ and $b = 1.4$.

a. Determine the equilibrium economy and find its stability.

b. The government wants the equilibrium economy to equal 5. To do this, it needs for a to equal what number? If this is accomplished, what will happen to the stability of the equilibrium economy, assuming b remains at 1.4?

8. Assume that $a = 0.75$ and $b = 1.2$. The government decides to have a new spending policy; that is, $g(n) = c$ where c is some constant. Make the substitution of c for $g(n)$ in dynamical system (GEM), $t(n) = (1 + b)at(n - 1) - bat(n - 2) + g(n)$.

a. Find the equilibrium economy if $c = 1.1$. What is its stability? What is the approximate period?

b. Find a value for c that results in the equilibrium economy being 8. What is the stability of the economy in this case? What is the approximate period?

c. Investigate this dynamical system for other values of c. How does the value for government spending, c, affect the stability of the economy? How does it affect the period of the functions $t(n)$?

9. Assume that $a = 0.5$ and $b = 1.4$. The government decides to have a new spending policy, that is, $g(n) = c$ where c is some constant. Make the substitution of c for $g(n)$ into General Economic Model, $t(n) = (1+b)at(n-1) - bat(n-2) + g(n)$.

a. Find the equilibrium economy if $c = 1.2$. What is its stability? What is the approximate period?

b. Find a value for c that results in the equilibrium economy being 5. What is the stability of the economy in this case? What is the approximate period?

c. Investigate this dynamical system for other values of c. How does the value for government spending, c, affect the stability of the economy? How does it affect the period of the functions $t(n)$?

4.7 Controlling an Economy

In the previous section, we began exploring the behavior of a nation's economy through the General Economic Model

$$t(n) = (1+b)at(n-1) - bat(n-2) + g(n) \qquad \textbf{(GEM)}$$

where a is the marginal propensity to consume, b is the marginal propensity to invest, and $g(n)$ is government spending. We discovered that when government spending was equal to 1, then the equilibrium economy was

$$E = \frac{1}{1-a} = \frac{1}{\text{MPS}}$$

where MPS is the marginal propensity to save. In fact, if government spending is any fixed number c, then the equilibrium economy is

$$E = \frac{c}{\text{MPS}}$$

This means that the equilibrium economy is proportional to government spending and is inversely proportional to the marginal propensity to save. The marginal propensity to invest b has no effect on the value of the equilibrium economy.

The value for b does affect the stability of the economy. In Problems 4 and 5 of Section 4.6, you found the value of b that was the cutoff between stable and unstable economies for particular values of a. You might have noticed that the larger the value for a, the smaller the cutoff value for b. In fact, if

$$ab < 1$$

then the equilibrium economy is stable; if $ab > 1$, then the equilibrium economy is unstable; and if $ab = 1$, then the equilibrium economy is neutral. This can be seen in Examples 4.14, 4.15, and 4.16. In Example 4.14, $a = 0.5$, $b = 1.8$, and $ab = 0.9 < 1$. The equilibrium economy is stable. In Example 4.15, $a = 0.5$, $b = 2.2$, $ab = 1.1 > 1$, and the equilibrium economy is unstable. Finally, in Example 4.16, $a = 0.5$, $b = 2$, $ab = 1$, and the equilibrium economy is neutral. These examples are not proof of the claim about ab, but they are consistent with that claim. The actual proof of this statement requires a bit more algebra than we wish to do.

Since the equilibrium economy is inversely proportional to the marginal propensity to save, $E = c/\text{MPS}$, decreasing the MPS results in increasing the equilibrium economy. Since $\text{MPS} = 1 - \text{MPC} = 1 - a$, the smaller the MPS is, the larger is the value for $\text{MPC} = a$. Thus, if a government wanted to increase the equilibrium economy, it could institute policies that encourage consumer spending or discourage saving, which would increase a. But if the value of a is increased so that $ab > 1$, then a stable economy might become unstable from these policies. This was seen in Problems 6 and 7 of Section 4.6. In an attempt to improve one facet of the economy, increasing the equilibrium economy, another facet of the economy can be harmed, the equilibrium becoming unstable.

In Problems 8 and 9 of Section 4.6, the stability of the equilibrium economy is not dependent on the government's fixed spending. In fact, the government's spending does not affect the stability of the equilibrium value as long as the government spending is the same every time period, that is, as long as $g(n) = c$.

We are now going to investigate what happens if the government varies its spending from one time to another in an attempt to stabilize an unstable economy. To do this, we make several alternate government spending assumptions and determine how each of these assumptions affect the stability of an economy.

The idea behind the new government spending assumptions is the following. If the total economy $t(n)$ at time n is below the equilibrium economy $E = 1/(1 - a)$, then the government will spend more to help bring it up toward equilibrium. If the total economy $t(n)$ is greater than the equilibrium economy, the government will spend less to help bring it down (see Figure 4.34). The main difference in these assumptions will be in how additional spending or spending cuts are determined.

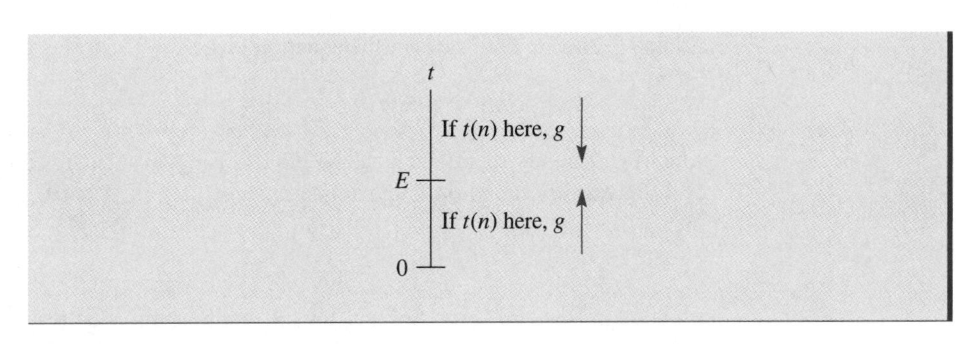

FIGURE 4.34 Government increases or decreases spending depending on value of economy *t*.

Immediate Government Response

The equilibrium economy is $E = 1/(1-a)$, and it is now time period n. The normal governmental spending is $g(n) = 1$. The government determines the amount the present economy is below equilibrium, $E - t(n)$, and increases its spending by 20% of that amount in an attempt to improve the economy. If the economy $t(n)$ is above equilibrium, then the government reduces spending by 20% of that amount in an attempt to slow down the economy. This means that government spending will be

$$g(n) = 1 + 0.2\left(\frac{1}{1-a} - t(n)\right) \quad \text{or} \quad g(n) = 1 + 0.2(E - t(n))$$

Note that if $E - t(n)$ is positive, then the current economy is less than the equilibrium economy, so $0.2(E - t(n))$ is positive and the government increases spending; that is, $g(n) > 1$. If the current economy exceeds the equilibrium economy, then $E - t(n)$ is negative and $g(n) < 1$, meaning that government spending has been reduced.

Example 4.17

In Example 4.14, $a = 0.5$ and $b = 1.8$, resulting in the dynamical system

$$t(n) = 1.4t(n-1) - 0.9t(n-2) + g(n)$$

In this case, the equilibrium economy is

$$\frac{1}{1-0.5} = 2$$

so government spending is

$$g(n) = 1 + 0.2(2 - t(n)) = 1.4 - 0.2t(n)$$

Notice that if the economy is at $t(n) = 1.5$ at time n, then it is $E - t(n) = 2 - 1.5 = 0.5$ under equilibrium. This means that the government will spend $0.2(0.5) = 0.1$ units more in money than usual; that is, $g(n) = 1.1$ at time n.

On the other hand, if the economy is $t(n) = 3$ at time n, then $E - t(n) = 2 - 3 = -1$. This means that the government will spend $0.2(-1) = -0.2$, or 0.2 units less money than usual. So $g(n) = 1 - 0.2 = 0.8$.

Substitution of $1.4 - 0.2t(n)$ for $g(n)$ into the GEM gives

$$t(n) = 1.4t(n-1) - 0.9t(n-2) + 1.4 - 0.2t(n)$$

Adding $0.2t(n)$ to both sides of the equation gives

$$1.2t(n) = 1.4t(n-1) - 0.9t(n-2) + 1.4$$

which is a second-order dynamical system that is not in standard form. Dividing both sides by 1.2 gives the second-order dynamical system

$$t(n) = \tfrac{7}{6}t(n-1) - \tfrac{3}{4}t(n-2) + \tfrac{7}{6}$$

which is in standard form. Solving for the equilibrium value results in $E = 2$, as before. Using the same initial values as in Example 4.14, $t(0) = 1$ and $t(1) = 1.5$, results in the time graph in Figure 4.35. As in Example 4.14, the equilibrium economy is stable, but a comparison of Figures 4.29 and 4.35 indicates that this change in government spending has resulted in $t(n)$ approaching equilibrium even more quickly. Thus, this new government policy has made the equilibrium economy even more stable than before. Also notice that the period has decreased from about 8.5 to about 7.5 in this case. ■

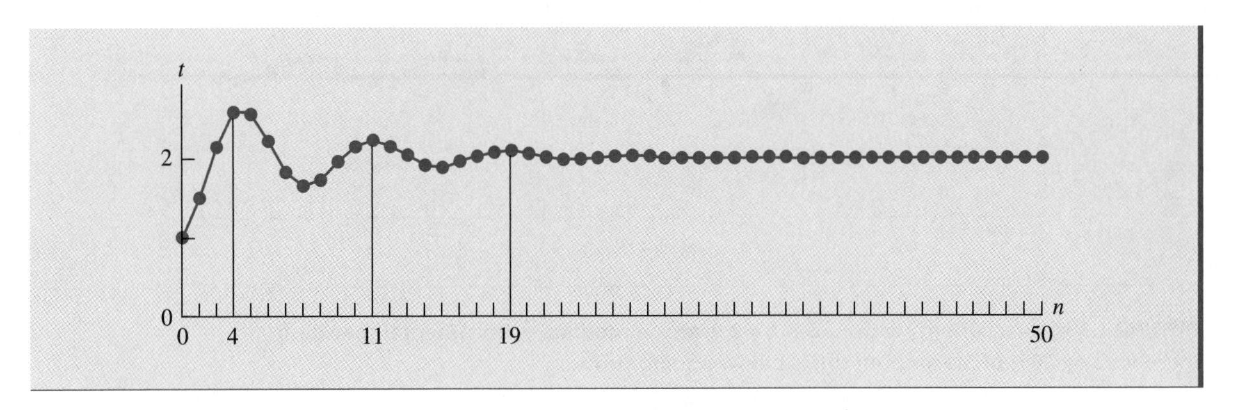

FIGURE 4.35 An economy with $a = 0.5$, $b = 1.8$, and increased government spending that is 20% of the difference of the economy from equilibrium.

In previous chapters, we have learned to develop and analyze dynamical systems that model real situations. This approach to adjusting government spending to control the economy is called **feedback control.** We are taking feedback from the economy, $t(n)$, and using it to control the stability of the economy.

Example 4.18 In Example 4.15, $a = 0.5$ and $b = 2.2$, resulting in the dynamical system

$$t(n) = 1.6t(n-1) - 1.1t(n-2) + g(n)$$

The equilibrium economy was still $E = 2$, but it was unstable when government spending was $g(n) = 1$. Since the equilibrium economy is again $E = 2$, we have the same government spending as in Example 4.17,

$$g(n) = 1 + 0.2(2 - t(n)) = 1.4 - 0.2t(n)$$

Substitution for $g(n)$ in the dynamical system gives

$$t(n) = 1.6t(n-1) - 1.1t(n-2) + 1.4 - 0.2t(n)$$

Adding $0.2t(n)$ to both sides of the equation and dividing both sides by 1.2 gives the second-order dynamical system

$$t(n) = \frac{4}{3}t(n-1) - \frac{11}{12}t(n-2) + \frac{7}{6}$$

The equilibrium value is $E = 2$, as before. Using the initial values $t(0) = 1$ and $t(1) = 1.5$ results in the time graph in Figure 4.36.

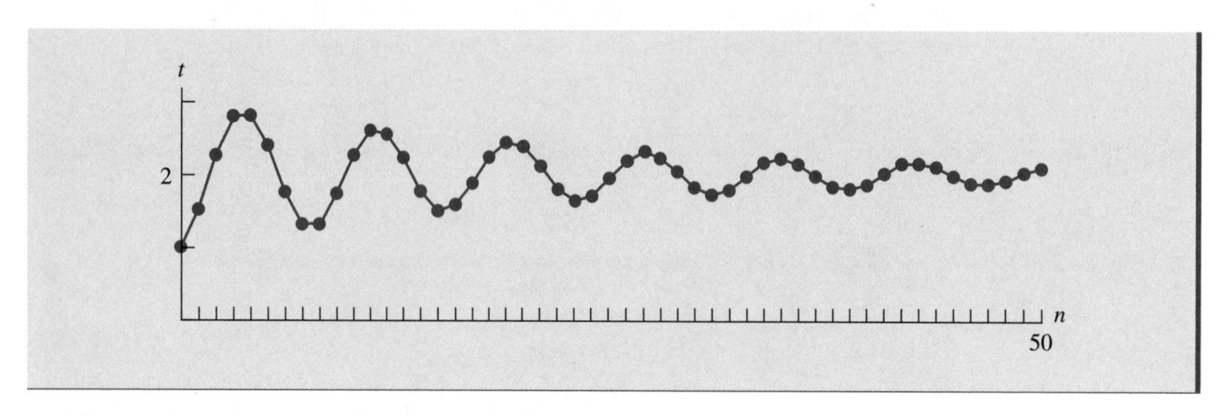

FIGURE 4.36 **An economy with $a = 0.5$, $b = 2.2$, and immediate governmental spending increased by 20% of the amount $t(n)$ is below equilibrium.**

Comparison of Figures 4.30 and 4.36 indicates that this change in government spending has caused the previously unstable economy to become stable. Also notice that the period has decreased from about 9 to about 8. ■

It appears that this change in government spending can result in stabilizing an unstable economy. In the **Immediate Government Response,** 20% was chosen at random. The government could have decided to increase spending by 5%, 10% or some other percentage. In the Problems, you will look for the smallest percentage by which the government can increase spending to stabilize an unstable economy.

It could be argued that the government cannot know the present value of the economy, $t(n)$, and must therefore use the value of the previous economy, $t(n-1)$. This would result in the following assumption.

Short Delay in Government Response The government spending at time n is increased by 20% of the amount the previous economy was below equilibrium, that is,

$$g(n) = 1 + 0.2\left(\frac{1}{1-a} - t(n-1)\right) \quad \text{or} \quad g(n) = 1 + 0.2(E - t(n-1))$$

It might seem that this assumption should have much the same positive results as the Immediate Government Response. But this is not the case. Using the Short Delay in Government Response method, stable economies remain stable and unstable economies remain unstable. This approach to government spending has no effect on the stability of the equilibrium economy. All it does is change the period of the oscillations without affecting the rate at which the oscillations approach or diverge from equilibrium. In this case, where $g(n)$ is given in terms of the previous economy, $t(n-1)$, we are said to have a **delay** in the control.

Example 4.19 As in Examples 4.15 and 4.18, $a = 0.5$ and $b = 2.2$ gives

$$t(n) = 1.6t(n-1) - 1.1t(n-2) + g(n)$$

Let government spending be

$$g(n) = 1 + 0.2(2 - t(n-1)) = 1.4 - 0.2t(n-1)$$

Substitution of $1.4 - 0.2t(n-1)$ for $g(n)$ gives

$$t(n) = 1.6t(n-1) - 1.1t(n-2) + 1.4 - 0.2t(n-1)$$

which simplifies to

$$t(n) = 1.4t(n-1) - 1.1t(n-2) + 1.4$$

The equilibrium value is still $t = 2$. The initial values $t(0) = 1$ and $t(1) = 1.5$ result in the function in Figure 4.37.

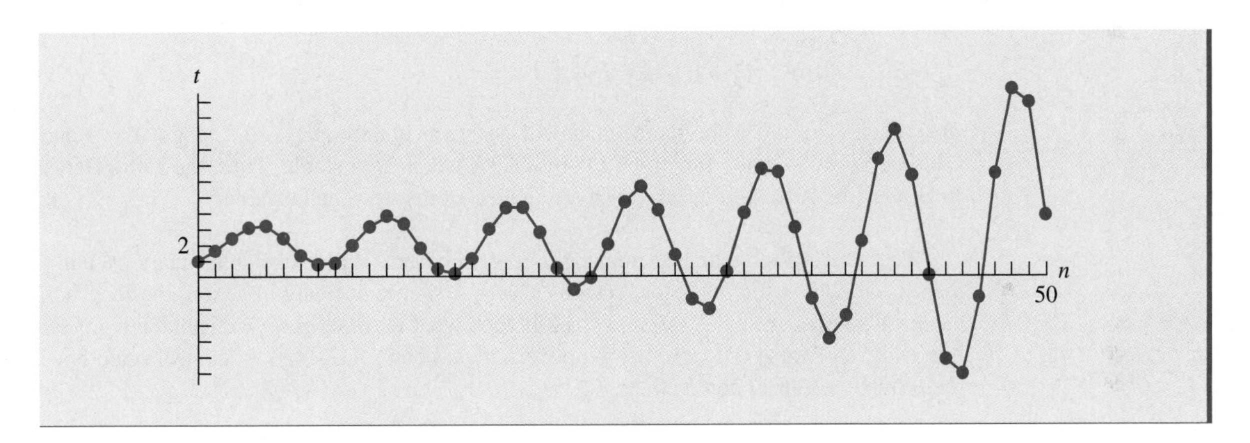

FIGURE 4.37 Economy with $a = 0.5$, $b = 2.2$, and a 1-unit delay in government spending.

Comparison of Figures 4.30 and 4.37 indicates that this change in government spending has not resulted in an improvement in the unstable economy, but the period has decreased from about 9 to about 7.5. ■

This implies that a delay in the control can negate the positive affect of that control. In fact, worse things can happen under similar conditions. Suppose that by the time the change in government spending has been approved by congress, it is 2 time units later. Thus, government spending at time n is based on the economy at time $n-2$.

Long Delay in Government Response Government spending at time n is increased by 20% of the amount the economy is below equilibrium in time period $n-2$, that is,

$$g(n) = 1 + 0.2\left(\frac{1}{1-a} - t(n-2)\right) \quad \text{or} \quad g(n) = 1 + 0.2(E - t(n-2))$$

In this case, not only does the government spending not improve an unstable economy, it can destabilize a stable economy.

Example 4.20 In Examples 4.14 and 4.17, we assumed that $a = 0.5$, $b = 1.8$. This resulted in the dynamical system

$$t(n) = 1.4t(n-1) - 0.9t(n-2) + g(n)$$

This dynamical system has a stable equilibrium of $E = 2$ when there is fixed government spending of $g(n) = 1$. Suppose government spending is

$$g(n) = 1 + 0.2(2 - t(n-2)) = 1.4 - 0.2t(n-2)$$

Substitution of $1.4 - 0.2t(n-2)$ for $g(n)$ gives

$$t(n) = 1.4t(n-1) - 0.9t(n-2) + 1.4 - 0.2t(n-2)$$

which simplifies to

$$t(n) = 1.4t(n-1) - 1.1t(n-2) + 1.4$$

By coincidence, this is the same dynamical system as in Example 4.19. Using $t(0) = 1$ and $t(1) = 1.5$ results in the function in Figure 4.37, which is unstable. Thus, the **Long Delay in Government Response** has made a stable economy become unstable. ■

These results indicate that immediate feedback can stabilize unstable situations, but if there is a delay in the feedback, such as a 1- or 2-time-unit delay in action, the feedback can have no effect or a deleterious effect. In fact, what we have observed is that the greater the delay, the worse the effect of the governmental control. Delays in control often have undesirable results in applications.

4.7 Problems

1. Consider an economy that is modeled by the General Economic Model

$$t(n) = (1+b)at(n-1) - bat(n-2) + g(n) \qquad \text{(GEM)}$$

with $a = 0.75$ and $b = 1.4$. Suppose a government spends 1 unit per time period. The equilibrium economy is therefore $E = 4$.

a. Suppose the government decides to spend an additional 10% of the amount $t(n)$ is below equilibrium. If $t(n) = 2.5$, what will be the government spend $g(n)$?

b. Suppose the government decides to cut spending by 10% of the amount that $t(n)$ is above equilibrium. If $t(n) = 7$, then what will $g(n)$ be?

c. Repeat parts (a) and (b), but assume the increase or decrease in spending is 30%, not 10%.

2. Consider an economy that is modeled by the GEM with $a = 0.4$ and $b = 1.8$. Suppose a government spends 3 units per time period. The equilibrium economy is therefore $E = 5$.

a. Suppose the government decides to spend an additional 5% of the amount $t(n)$ is below equilibrium. If $t(n) = 2$, what will the government spend, $g(n)$?

b. Suppose the government decides to cut spending by 5% of the amount that $t(n)$ is above equilibrium. If $t(n) = 5.5$, then what will $g(n)$ be?

c. Repeat parts (a) and (b), but assume the increase or decrease in spending is 40%, not 5%.

3. Consider an economy that is modeled by the GEM with $a = 0.8$ and $b = 1.2$. Suppose a government spends 1 unit per time period. The equilibrium economy is therefore $E = 5$. Suppose that $t(0) = 3$, $t(1) = 3.5$, and $t(2) = 4.28$.

a. Suppose the government decides to spend an additional 10% of the amount $t(n)$ is below equilibrium, that is, $g(n) = 1 + 0.1(5 - t(n))$. What is $g(2)$?

b. Suppose the government decides to spend an additional 10% of the amount the previous economy $t(n-1)$ is below equilibrium; that is, $g(n) = 1 + 0.1(5 - t(n-1))$. What is $g(2)$?

c. Suppose the government at time n decides to spend an additional 10% of the amount the economy at time $n-2$ is below equilibrium; that is, $g(n) = 1 + 0.1(5 - t(n-2))$. What is $g(2)$?

4. Consider an economy that is modeled by the GEM with $a = 0.4$ and $b = 1.5$. Suppose a government spends 3 units per time period. The equilibrium economy is then $E = 5$. Suppose that $t(0) = 4$, $t(1) = 5$, and $t(2) = 5.6$.

a. Suppose the government decides to spend an additional 20% of the amount $t(n)$ is below equilibrium or reduce spending by 20% of the amount that $t(n)$ is above equilibrium. What is $g(2)$?

b. Suppose the government decides to spend an additional 20% of the amount the previous economy is below equilibrium or reduce spending by 20% of the amount that it is above equilibrium. What is $g(2)$?

c. Suppose the government at time n decides to spend an additional 20% of the amount the economy at time $n-2$ is below equilibrium or reduce spending by 20% of the amount it is above equilibrium. What is $g(2)$?

5. Consider an economy that is modeled by the GEM with $a = 0.8$ and $b = 1.5$. Suppose a government spends 1 unit per time period. Then there is an unstable equilibrium economy of $E = 5$.

a. Suppose the government decides to increase (or decrease) spending by an additional 10% of the amount $t(n)$ is below (or above) equilibrium. Find the corresponding dynamical system, and determine the stability of the equilibrium economy.

b. Suppose the government decides to increase (or decrease) spending by an additional 20% of the amount $t(n)$ is below (or above) equilibrium. Find the corresponding dynamical system, and determine the stability of the equilibrium economy.

c. Suppose the government decides to increase (or decrease) spending by an additional 30% of the amount $t(n)$ is below (or above) equilibrium. Find the corresponding dynamical system, and determine the stability of the equilibrium economy.

6. Consider an economy that is modeled by the GEM with $a = 0.25$ and $b = 5$. Suppose a government spends 1 unit per time period. Suppose the government decides to increase (or decrease) spending by an additional I% of the amount $t(n)$ is below (or above) equilibrium. By trying different values for I, determine the smallest value for I that results in the equilibrium value being stable or neutral. (Find I to the nearest percent.)

7. Consider an economy that is modeled by the GEM with $a = 0.6$ and $b = 3$. Suppose a government spends 1 unit per time period. Suppose the government decides to increase (or decrease) spending by an additional I% of the amount $t(n)$ is below (or above) equilibrium. By trying different values for I, determine the smallest value for I that results in the equilibrium value being unstable, stable, or neutral. (Find I to the nearest percent.)

8. Consider an economy that is modeled by the GEM with $a = 0.8$ and $b = 0.9$. Suppose a government spends 1 unit per time period. The stable equilibrium economy is $E = 5$.

a. Suppose the government decides to increase (or decrease) spending by an additional 20% of the amount $t(n-2)$ is below (or above) equilibrium. Find the corresponding dynamical system and determine the stability of the equilibrium economy.

b. Suppose the government decides to increase (or decrease) spending by an additional 40% of the amount $t(n-2)$ is below (or above) equilibrium. Find the corresponding dynamical system and determine the stability of the equilibrium economy.

c. Suppose the government decides to increase (or decrease) spending by an additional I% of the amount $t(n-2)$ is below (or above) equilibrium. As the value for I increases, the economy becomes less stable and eventually becomes unstable. Find the cutoff value for I that results in the equilibrium economy becoming neutral.

9. Consider an economy that is modeled by the GEM with $a = 0.25$ and $b = 3$. Suppose a government spends 1 unit per time period. Suppose the government decides to increase (or decrease) spending by an additional I% of the amount $t(n-2)$ is below (or above) equilibrium. By trying different values for I, determine the smallest value for I that results in the equilibrium value being unstable or neutral. (Find I to the nearest percent.)

10. Consider an economy that is modeled by the GEM with $a = 0.4$ and $b = 1.5$. Suppose a government spends 3 units per time period. Suppose the government decides to increase (or decrease) spending by an additional I% of the amount $t(n-2)$ is below (or above) equilibrium. By trying different values for I, determine the smallest value for I that results in the equilibrium value being unstable or neutral. (Find I to the nearest percent.)

11. Consider an economy that is modeled by the GEM with $a = 0.5$ and $b = 1.8$. Suppose a government spends 1 unit per time period. The equilibrium economy is $E = 2$, which is stable. In an attempt to increase the equilibrium economy, the government has decided to adopt the following policy. If the current economy $t(n)$ is below 3, then the government will increase its spending by 25% of the amount that $t(n)$ is below 3. If the current economy is above 3, then the government will decrease spending by 25% of the amount that $t(n)$ is above 3.

a. Develop a second-order dynamical system in standard form to model this scenario.
b. Find the equilibrium value for this dynamical system and determine its stability.
c. Instead of increasing or decreasing spending by 25%, the government wants to increase or decrease spending by a percentage I that will result in an equilibrium economy of 2.5. What should I be, exactly? Will the economy still be stable?

12. Consider an economy that is modeled by the GEM with $a = 0.8$ and $b = 1.5$. Suppose a government spends 1 unit per time period. The equilibrium economy is $E = 5$, which is unstable. In an attempt to increase the equilibrium economy and improve its stability, the government has decided to adopt the following policy. If the current economy $t(n)$ is below 8, then the government will increase its spending by 10% of the amount that $t(n)$ is below 8. If the current economy is above 8, then the government will decrease spending by 10% of the amount that $t(n)$ is above 8.

a. Develop a second-order dynamical system in standard form to model this scenario.
b. Find the equilibrium value for this dynamical system and determine its stability.
c. Instead of increasing or decreasing spending by 10%, the government wants to increase or decrease spending by a percentage I of the difference between $t(n)$ and 8, so that the equilibrium economy will be 7. What should I be, exactly? Will the economy be stable?
d. Instead of increasing or decreasing spending by 10% of the difference of the economy from 8, the government wants to increase or decrease spending by I percent, where I is the smallest percentage for which the economy will no longer be unstable. What is I?

4.8 Exponential and Trigonometric Functions

Functions satisfying second-order affine dynamical systems have displayed quite a variety of behaviors. In this section, I am going to explain why the different behaviors occur. Recall that when we had a first-order affine dynamical system

$$u(n) = au(n-1) + b$$

we could express $u(n)$ explicitly as

$$u(n) = ca^n + E$$

where E is the equilibrium value and c could be found from the given initial value. This was assuming there was an equilibrium value. When there was no equilibrium value, $u(n)$ could be written as a linear function.

Some functions satisfying some second-order affine dynamical systems can be written explicitly in a similar form, except that there are 2 exponentials, not 1. In particular, suppose we have the dynamical system

$$u(n) = a_1u(n-1) + a_2u(n-2) + b$$

and it has the single equilibrium value E. Suppose that the quadratic

$$x^2 = a_1x + a_2$$

has 2 real, distinct solutions, r_1 and r_2. Then $u(n)$ has the explicit form

$$u(n) = c_1 r_1^n + c_2 r_2^n + E$$

where the values for c_1 and c_2 can be found from the 2 given initial values, $u(0)$ and $u(1)$.

You can sometimes find the solutions to $x^2 = a_1 x + a_2$ by factoring. Other times, you may need to apply the quadratic formula to the equation $x^2 - a_1 x - a_2 = 0$. This would give the 2 solutions

$$r_1 = \frac{a_1 + \sqrt{a_1^2 + 4a_2}}{2} \quad \text{and} \quad r_2 = \frac{a_1 - \sqrt{a_1^2 + 4a_2}}{2}$$

Example 4.21 Consider the dynamical system

$$u(n) = 2.5u(n-1) - u(n-2) - 3$$

from Example 4.12, with $u(0) = 17$ and $u(1) = 16$. The time graph for $u(n)$ is seen in Figure 4.38.

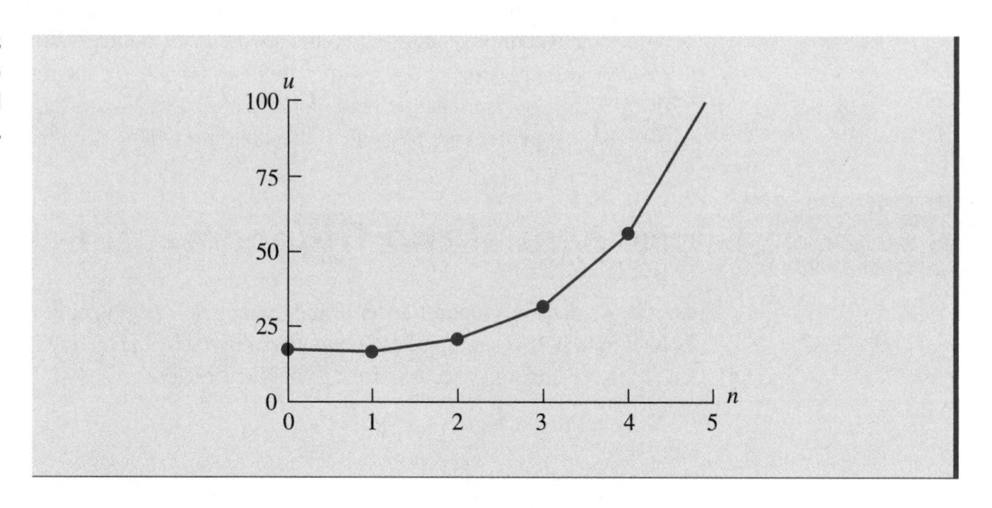

FIGURE 4.38 Points $(n,u(n))$ with $u(0) = 17$ and $u(1) = 16$.

The equilibrium value is $E = 6$. The roots of the equation

$$x^2 = 2.5x - 1$$

are $x = 2$ and $x = 0.5$. These can be found by applying the quadratic formula to

$$x^2 - 2.5x + 1 = 0$$

giving

$$r_1 = \frac{2.5 + \sqrt{2.5^2 - 4}}{2} = 2 \quad \text{and} \quad r_2 = \frac{2.5 - \sqrt{2.5^2 - 4}}{2} = 0.5$$

This means that

$$u(n) = c_1 2^n + c_2(0.5)^n + 6$$

Since $u(0) = 17$ and $u(1) = 16$, we can solve the 2 equations in 2 unknowns

$$u(0) = c_1 2^0 + c_2(0.5)^0 + 6$$
$$u(1) = c_1 2^1 + c_2(0.5)^1 + 6$$

which, after substitution of 17 for $u(0)$ and 16 for $u(1)$, simplify to

$$17 = c_1 + c_2 + 6$$

$$16 = 2c_1 + 0.5c_2 + 6$$

The solution is $c_1 = 3$ and $c_2 = 8$. In this case,

$$u(n) = 3(2)^n + 8(0.5)^n + 6$$

You should check that you get the same values for $u(2)$, $u(3)$, and so on using this expression and using the dynamical system.

In Section 4.5, we computed

$$\frac{u(n) - 6}{u(n-1) - 6}$$

for large values of n and found that the result was close to 2. Let's see why this happens. For the function $u(n) = 3(2)^n + 8(0.5)^n + 6$, this would be

$$\frac{u(n) - 6}{u(n-1) - 6} = \frac{3(2)^n + 8(0.5)^n}{3(2)^{n-1} + 8(0.5)^{n-1}}$$

Note that 0.5^n is a lot less than 2^n, so

$$\frac{u(n) - 6}{u(n-1) - 6} = \frac{3(2)^n + 8(0.5)^n}{3(2)^{n-1} + 8(0.5)^{n-1}} \approx \frac{3(2)^n}{3(2)^{n-1}} = 2$$

When $u(0) = 26$ and $u(1) = 16$, we found that the ratios were 0.5, not 2. The reason is that $c_1 = 0$ and $c_2 = 20$ in this case. Thus,

$$u(n) = 20(0.5)^n + 6$$

and

$$\frac{u(n) - 6}{u(n-1) - 6} = \frac{20(0.5)^n}{20(0.5)^{n-1}} = 0.5$$

When we were finding ratios, we were trying to find the largest of the bases that form the expression for $u(n)$. This shows why we get the smaller of the bases in some cases. ■

We note that if $|r_1| < 1$ and $|r_2| < 1$, then the equilibrium value is stable. If $|r_1| > 1$ and $|r_2| > 1$, then the equilibrium value is unstable. If $|r_1| > 1$ and $|r_2| < 1$ or vice versa, as in Example 4.21, the equilibrium value is a saddle. Sometimes the equation

$$x^2 = a_1 x + a_2$$

has no real roots. This happens when we apply the quadratic formula to $x^2 - a_1 x - a_2 = 0$ and end up with a negative under the square root. In this case, $u(n)$ is given explicitly as

$$u(n) = r^n(c_1 \cos(n\theta) + c_2 \sin(n\theta)) + E$$

where E is the equilibrium value. The numbers c_1 and c_2 can be found using the initial values, and $r = \sqrt{|a_2|}$. The value for the angle θ is

$$\theta = \tan^{-1}\left(\frac{\sqrt{|a_1^2 + 4a_2|}}{a_1}\right)$$

For this reason these functions

1. Oscillate because of the trigonometric functions;
2. Oscillate toward equilibrium (stable) if $r < 1$, oscillate away from equilibrium (unstable) if $r > 1$, and oscillate with constant amplitude (neutral) if $r = 1$; and
3. The period of the oscillations is $360/\theta$, if θ is given in degrees, or $2\pi/\theta$, if θ is given in radians.

The actual derivation of this explicit expression and the value for θ is beyond the scope of this book. There are more advanced books on dynamical systems in which the derivation can be found.

Example 4.22 Consider the dynamical system

$$t(n) = 1.4t(n-1) - 0.9t(n-2) + 1$$

of Example 4.14. The equilibrium economy $E = 2$ is stable. In Figure 4.29, you could see a time graph of the solution with $t(0) = 1$ and $t(1) = 1.5$. This function oscillated with decreasing amplitude toward 2 as can be seen in Figure 4.39. The period of each oscillation was about 8.5. Let's see why this happened.

To find an explicit expression for this function, we first note that the solutions to the quadratic

$$x^2 = 1.4x - 0.9$$

are, by the quadratic formula,

$$x = \frac{1.4 \pm \sqrt{-1.64}}{2}$$

which are not real. Therefore, the explicit expression is given as

$$t(n) = r^n(c_1 \cos(n\theta) + c_2 \sin(n\theta)) + E$$

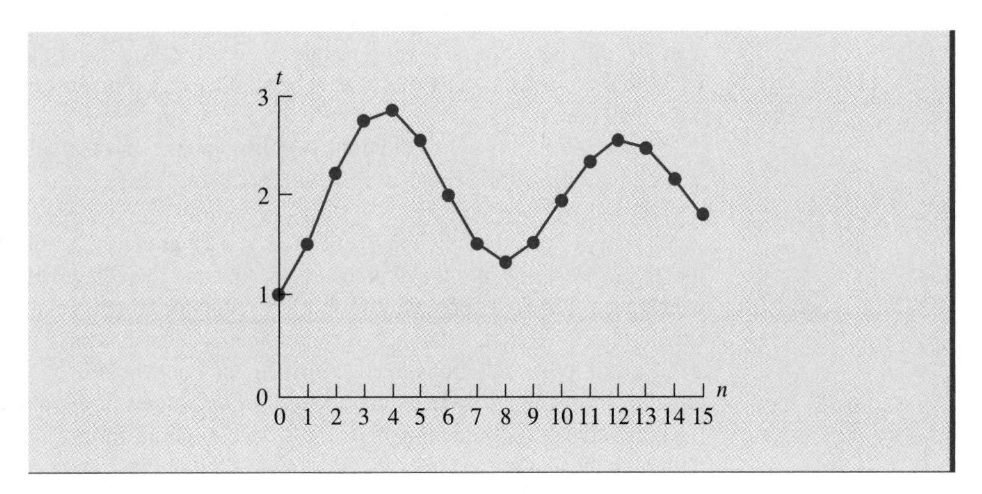

FIGURE 4.39 Oscillating economy with stable equilibrium.

where

$$r = \sqrt{0.9} < 1$$

The angle θ is

$$\theta = \tan^{-1}\left(\frac{\sqrt{|1.4^2 + 4(-0.9)|}}{1.4}\right) = \tan^{-1}\left(\frac{\sqrt{1.64}}{1.4}\right) \approx 42.45 \text{ degrees} \quad \text{or} \quad 0.74 \text{ radians}$$

This means that

$$t(n) = \left(\sqrt{0.9}\right)^n \left(c_1 \cos(42.45n) + c_2 \sin(42.45n)\right) + 2$$

We need to solve

$$t(0) = \left(\sqrt{0.9}\right)^0 \left(c_1 \cos(0) + c_2 \sin(0)\right) + 2 = 1$$

$$t(1) = \left(\sqrt{0.9}\right)^1 \left(c_1 \cos(42.45) + c_2 \sin(42.45)\right) + 2 = 1.5$$

The first equation simplifies to

$$c_1 + 2 = 1$$

or $c_1 = -1$. The second equation becomes

$$\sqrt{0.9}\left(-\cos(42.45) + c_2 \sin(42.45)\right) = 1.5$$

Since $\cos(42.45) \approx 0.738$ and $\sin(42.45) = 0.675$, we get $c_2 = 0.31235$, to five decimal places. Therefore,

$$t(n) = \left(\sqrt{0.9}\right)^n \left(-\cos(42.45n) + 0.31235 \sin(42.45n)\right) + 2$$

You should check that you get the same values for $t(n)$ using the dynamical system and using this expression. (The values may be slightly different, since I rounded off 0.31235 to 5 decimal places.)

Since $\sqrt{0.9} < 1$, the oscillations decrease in size and the equilibrium value is stable. The period of the oscillations is $360/42.45 = 8.48$. ■

The proof that the function $u(n)$ can be written as claimed requires only that we substitute the expressions into the dynamical systems and then check that the resulting equation is balanced. For the exponential form of the function, this is relatively easy to do. For the trigonometric version, a number of trigonometric identities need to be used.

Several other situations may occur. One or both roots of $x^2 - a_1x - a_2 = 0$ could be equal to 1, in which case there are no equilibrium values or there are infinitely many equilibrium values. The equation $x^2 - a_1x - a_2 = 0$ could have a double root other than 1. Although the explicit expressions for $u(n)$ in each of these cases are not complicated, they add little to our understanding, so I omit them.

4.8 Problems

1. Consider the dynamical system

$$u(n) = 5u(n-1) - 6u(n-2) + 8$$

The equilibrium value is $E = 4$, and the solutions to the equation $x^2 = 5x - 6$ are $r_1 = 2$ and $r_2 = 3$. The claim is that

$$u(n) = c_1 2^n + c_2 3^n + 4.$$

a. Find c_1 and c_2, given that $u(0) = 7$ and $u(1) = 12$. Check that you get the same values for $u(2)$ and $u(3)$ using your expression and using the dynamical system.

b. Find c_1 and c_2, given that $u(0) = 1$ and $u(1) = -2$. Check that you get the same values for $u(2)$ and $u(3)$ using your expression and using the dynamical system.

2. Consider the dynamical system

$$u(n) = 2u(n-1) + 8u(n-2) + 18$$

The equilibrium value is $E = -2$, and the solutions to the equation $x^2 = 2x + 8$ are $r_1 = -2$ and $r_2 = 4$. The claim is that

$$u(n) = c_1(-2)^n + c_2 4^n - 2.$$

a. Find c_1 and c_2, given that $u(0) = 6$ and $u(1) = 0$. Check that you get the same values for $u(2)$ and $u(3)$ using your expression and using the dynamical system.

b. Find c_1 and c_2, given that $u(0) = 0$ and $u(1) = 0$. Check that you get the same values for $u(2)$ and $u(3)$ using your expression and using the dynamical system.

3. Consider the dynamical system

$$u(n) = 2.5u(n-1) - u(n-2) + 3$$

The equilibrium value is $E = -6$.

a. Find the solutions to $x^2 = 2.5x - 1$, and write the general form of the expression for $u(n)$.

b. Write the explicit form of the function by finding c_1 and c_2, given that $u(0) = -3$ and $u(1) = -3$. Check that you get the same values for $u(2)$ and $u(3)$ using your expression and using the dynamical system.

4. Consider the dynamical system

$$u(n) = 4.2u(n-1) - 0.8u(n-2) - 12$$

The equilibrium value is $E = 5$.

a. Find the solutions to $x^2 = 4.2x - 0.8$, and write the general form of the expression for $u(n)$.

b. Write the explicit form of the function by finding c_1 and c_2, given that $u(0) = 1$ and $u(1) = 8$. Check that you get the same values for $u(2)$ and $u(3)$ using your expression and using the dynamical system.

5. Consider the dynamical system

$$u(n) = -3u(n-1) + 10u(n-2) + 12$$

Write the explicit form of the function $u(n)$, given that $u(0) = 3$ and $u(1) = -6$.

6. Consider the dynamical system

$$u(n) = -u(n-1) + 12u(n-2) - 20$$

Write the explicit form of the function $u(n)$, given that $u(0) = 3$ and $u(1) = -2$.

7. Consider the dynamical system

$$u(n) = u(n-1) - 0.6u(n-2) + 6$$

The equilibrium value is $E = 10$, and it is stable. The equation $x^2 = x - 0.6$ has no solutions, so the function $u(n)$ has the form

$$u(n) = r^n(c_1 \cos(n\theta) + c_2 \sin(n\theta)) + 10$$

Find r to 2 decimal places and find θ to the nearest degree. What is the period of $u(n)$ to 1 decimal place?

8. Consider the dynamical system

$$u(n) = 1.2u(n-1) - u(n-2) + 16$$

The equilibrium value is $E = 20$ and is neutral. The equation $x^2 = 1.2x - 1$ has no solutions, so the function $u(n)$ has the form

$$u(n) = r^n(c_1 \cos(n\theta) + c_2 \sin(n\theta)) + 20$$

Find r to 2 decimal places and find θ to the nearest degree. What is the period of $u(n)$, to 1 decimal place?

9. Consider the dynamical system

$$u(n) = 2u(n-1) - 4u(n-2) + 6$$

The equilibrium value is $E = 2$, and it is unstable. The equation $x^2 = 2x - 4$ has no solutions, so the function $u(n)$ has the form

$$u(n) = r^n(c_1 \cos(n\theta) + c_2 \sin(n\theta)) + 2$$

a. Find r to 2 decimal places, and find θ to the nearest degree. What is the period of $u(n)$, to 1 decimal place?
b. Find c_1 and c_2, given that $u(0) = 3$ and $u(1) = 3$

10. Consider the dynamical system

$$u(n) = 6u(n-1) - 25u(n-2)$$

The equilibrium value is $E = 0$, and it is unstable. The equation $x^2 = 6x - 25$ has no solutions, so the function $u(n)$ has the form

$$u(n) = r^n(c_1 \cos(n\theta) + c_2 \sin(n\theta))$$

a. Find r to 2 decimal places, and find θ to 2 decimal places.
b. Find c_1 and c_2, given that $u(0) = 5$ and $u(1) = 11$.

5 Nonlinear Dynamical Systems

5.1 Introduction

We have developed affine dynamical systems to model many situations. Affine dynamical systems involve constant multiples of the dependent values plus the addition or subtraction of constants, such as

$$u(n) = 5u(n-1) - 3$$

or

$$\begin{aligned} u(n) &= 2u(n-1) + 4v(n-1) + 5 \\ v(n) &= -3.5u(n-1) + v(n-1) - 3 \end{aligned}$$

or

$$u(n) = 2u(n-1) - 5u(n-2) + 9$$

The reason the dynamical systems were so simple is that we were often simplifying real world situations. When we make our models more realistic by considering other aspects that might affect the model, we may need to use nonlinear dynamical systems, such as

$$u(n) = 3[u(n-1)]^2 - 2 \tag{5.1}$$

and $$u(n) = \frac{2u(n-1) - 1}{3u(n-1) + 2} \tag{5.2}$$

Dynamical system (5.1) is nonlinear because it includes a square of the dependent variable, $(u(n-1))^2$, which is usually written as $u^2(n-1)$. Dynamical system (5.2) is nonlinear because it includes a quotient involving $u(n-1)$. In Section 5.2, we develop a nonlinear dynamical system involving a quotient to study the elimination of alcohol from a person.

To be able to understand a situation being modeled, we need to be able to analyze a nonlinear dynamical system. One time graph of a function satisfying dynamical system (5.2) can be seen in Figure 5.1. As seen from this strange graph, functions satisfying nonlinear dynamical systems can be extremely complex, which often makes their analysis quite difficult. Another reason it is difficult to analyze nonlinear systems is that it is **impossible** to find expressions for the function $u(n)$ for most nonlinear systems, even though the function clearly exists, since we can find $u(1)$, $u(2)$, and so forth. Therefore other techniques must be used to study them.

FIGURE 5.1 Time graph for dynamical system (5.2) with $u(0) = 1$.

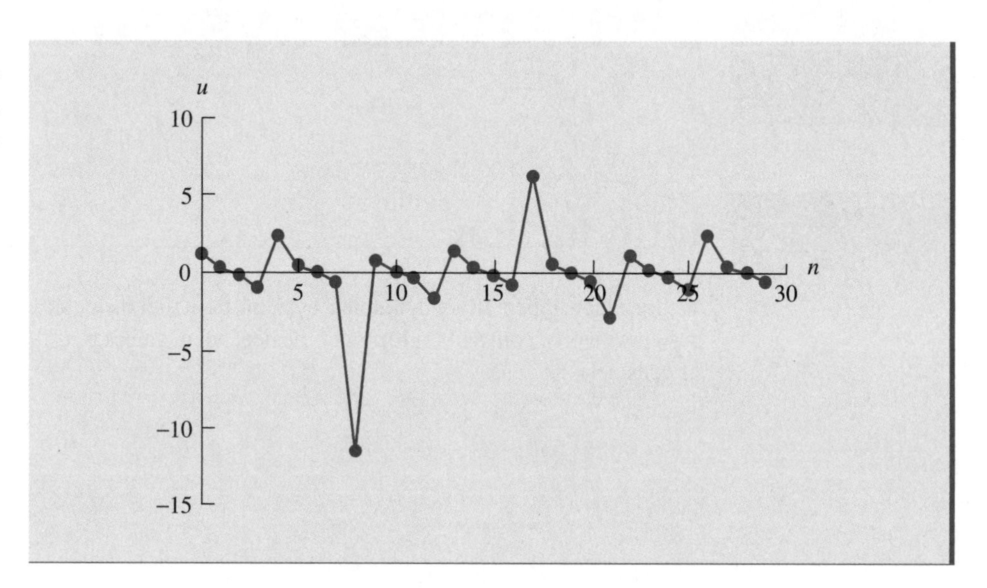

You are already familiar with several techniques that are often used to analyze a nonlinear dynamical system. These are (1) examining a table of values, (2) making time graphs, (3) finding equilibrium values, (4) determining the stability of the equilibrium values, and (5) approximating the rate at which the function goes toward or away from equilibrium.

The method for finding equilibrium values is the same as before. We substitute E for the dependent variable in the dynamical system and then solve the equation. For dynamical system (5.1), we would need to solve the equation

$$E = 3E^2 - 2$$

To do this, we subtract E from both sides, giving $0 = 3E^2 - E - 2$. We then factor, getting

$$0 = (3E + 2)(E - 1)$$

Setting each factor equal to 0 gives the two equilibrium values $E = 1$ and $E = -\frac{2}{3}$. You can find the equilibrium values for (5.2) by multiplying both sides of the equation

$$E = \frac{2E - 1}{3E + 2}$$

by the denominator, $3E+2$, and then solving. After simplifying, you get the equation

$$3E^2+1=0$$

which has no solutions. So dynamical system (5.2) has no equilibrium values.

In Section 5.3, you will learn to analyze the stability of the equilibrium values. In Section 5.4, you will learn a new technique, graphical analysis, for analyzing a nonlinear dynamical system.

In Chapter 6, you will use the techniques developed in this chapter to analyze nonlinear dynamical systems arising in the study of population growth and harvesting strategies. This analysis will help explain issues related to survival of species, optimal harvesting, and sustainable harvesting. In Chapter 7, you will analyze nonlinear dynamical systems that arise in the study of genetics. These systems will help explain the prevalence of certain genetic diseases and the difficulty in eliminating harmful genetic traits.

5.2 The Dynamics of Alcohol

In this section, we are going to study the dynamics involved in the buildup and elimination of alcohol in the blood system. We study this for 2 reasons. First, the dynamics of alcohol in the body is quite interesting from a mathematical point of view. Second, for people to make intelligent choices about drinking, they should understand the risks involved in alcohol consumption.

Some students believe that moderate drinking enhances the quality of social interactions, and it can become an integral part of their social life. Nevertheless, the misuse of alcohol exacts a heavy toll. It has been estimated that the abuse of alcohol in this country costs in excess of \$100 billion a year in lost wages and medical care. The personal costs in terms of emotional pain, destroyed relationships, and death of loved ones cannot be calculated. Alcohol is a contributing factor in 40% to 55% of all traffic fatalities, 50% of homicides, and 30% of suicides every year. Death from binge drinking, the rapid consumption of 5 or more drinks (4 or more for women) for the purpose of getting drunk, has been increasing at an alarming rate on college campuses.

There has been a considerable amount of scientific research on the effects of alcohol on memory, reflexes, coordination, and depth perception as well as a host of other cognitive and psychomotor processes. In these studies, as well as in the legal arena, blood alcohol levels are presented as a percentage of alcohol per volume of blood.

A blood alcohol level (BAC) of 0.10 is defined as 1 gram of alcohol per kg of blood, meaning that alcohol is 0.10% or $\frac{1}{1000}$ of the blood. While some states have set 0.10 as the cutoff for the legal definition of Driving While Intoxicated (DWI), others have reduced the BAC level for DWI to 0.08. The BAC of an individual is determined by four primary factors, which include body weight, amount of alcohol consumed, gender, and amount of elapsed time from first drink until a breath or blood sample is taken. For example, 140-pound male who rapidly consumes 3 alcoholic drinks will have an 0.08 BAC. One drink is defined as 1 ounce of 100-proof liquor, 5 ounces of wine, or one 12-ounce beer. A drink consists of approximately 14 grams or half an ounce of alcohol. Alcohol is absorbed into the body primarily through the lining of the stomach. For this reason, a BAC peak is reached within about 20 minutes of the last drink.

Table 5.1 gives approximate results for drinking for a 140-pound male. A summary of this data is seen in Figure 5.2. Depending on the weight of a male, the BAC numbers are proportionately larger or smaller than the numbers in Table 5.1. Females will reach the BAC benchmarks with fewer grams of alcohol than males of the same weight. The actual number of grams of alcohol that can cause alcohol poisoning depends on many factors.

TABLE 5.1 Benchmarks for alcohol consumption.

	140-pound male		
grams of alcohol in body	10	40	200–250
BAC	0.02	0.08	0.4–0.5
affect on person	impairment begins	possible DWI	risk of death or coma

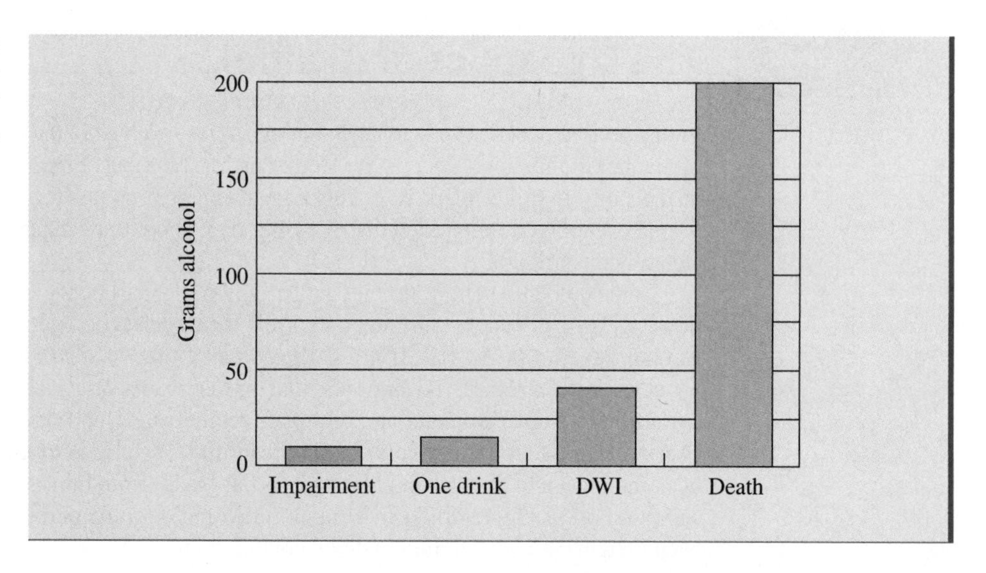

FIGURE 5.2 Alcohol in bloodstream for 140 pound male.

There are 2 main ways the body deals with chemicals in the blood stream. The first is filtration by the kidneys. The second is breaking down the chemicals by enzymes from the liver. As discussed in Section 1.3, when the kidneys filter out a chemical, they tend to eliminate a fixed proportion each time period. For example, the average person eliminates about 13% of the caffeine in his or her body each hour. Thus, if the person has 100 mg of caffeine in the body (about the amount from one cup of coffee) at the beginning of an hour, then the body will eliminate about 13 mg over the next hour. If the person has 200 mg in the body, about 26 mg will be eliminated in the next hour. Let $a(n)$ equal the amount of caffeine in the person's body at the beginning of hour n. Then the amount of caffeine eliminated during the $n-1$th hour is

$$0.13a(n-1)$$

This is true because $a(n-1)$ is the amount of caffeine in the body at the beginning of hour $n-1$. The percentage of caffeine removed from a person's body per hour can vary greatly from person to person. The number we used, 13%, is an average.

For chemicals, such as caffeine, in which a fixed percentage is eliminated each time period, the amount of the chemical in the body is modeled by the dynamical system

$$a(n) = a(n-1) - ra(n-1) + d$$

where r is the fraction of the chemical eliminated by the kidneys during each time period and d is the amount of the chemical consumed in each time period. For caffeine, $r = 0.13$.

The liver eliminates chemicals by breaking them down with enzymes. However, the liver generally does not break down a constant proportion each hour. Instead, the fraction of the chemical being broken down can depend on the amount of the chemical that is in the body, which is the case for alcohol. What happens is that as the amount of alcohol in the body increases, the fraction of alcohol the liver can break down decreases toward 0.

For alcohol, we have 3 variables. n represents the number of hours after beginning the consumption of alcohol, $a(n)$ represents the amount of alcohol in the body at the beginning of the nth hour, and $r(a(n))$ represents the fraction of the alcohol in the body that is eliminated during the nth hour. Note that the amount of alcohol a depends on the time n but the rate r depends on the amount of alcohol a.

For simplicity, we begin working with only 2 variables at a time. If we are considering time and amount of alcohol, then n is the independent variable and a is the dependent variable, and we will make time graphs of the points $(n, a(n))$. But if we are considering only the variables a and r, then we consider a, the amount of alcohol, to be the independent variable and r, the fraction of the alcohol that is eliminated, to be the dependent variable. In this case, we graph the function $r(a)$, which compares the fraction or rate at which alcohol is eliminated to the amount of alcohol. For caffeine, the rate was constant, so the function would be $r(a) = 0.13$, which would be a horizontal line on the ra-axis. For alcohol, our assumption is that r goes to 0 as a gets larger. This means that the graph of r as a function of a should look something like the graph in Figure 5.3.

FIGURE 5.3 **Fraction of alcohol broken down as function of alcohol in body.**

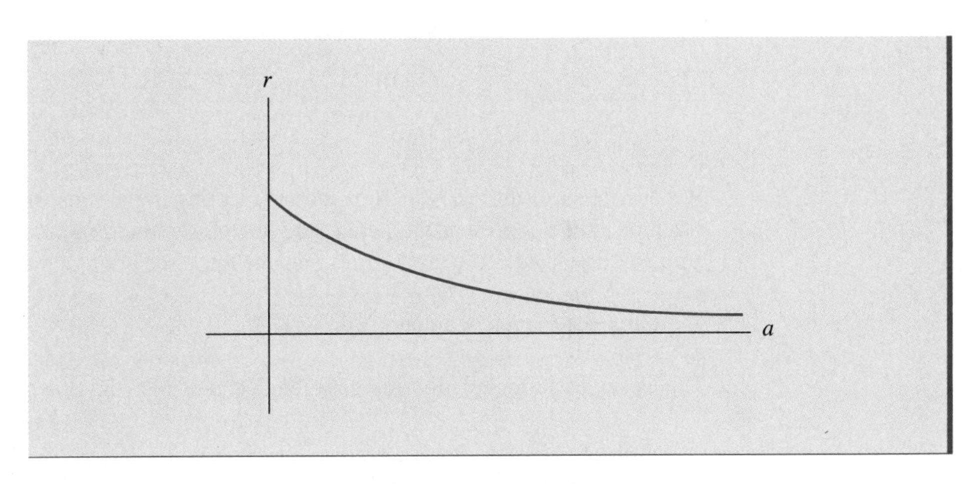

To model this situation, we need to develop a function that has the general shape of the function in Figure 5.3. One possible choice is a rational function, such as

$$r = \frac{1}{1+a}$$

Another choice is an exponential function, such as

$$r = \frac{1}{2^a}$$

The graphs of both of these functions are seen in Figure 5.4. They both have vertical intercepts at $(0, 1)$, and the a-axis is the horizontal asymptote for both. The exponential function approaches the a-axis much faster than the rational function.

FIGURE 5.4 Rational and exponential function used for elimination rate function.

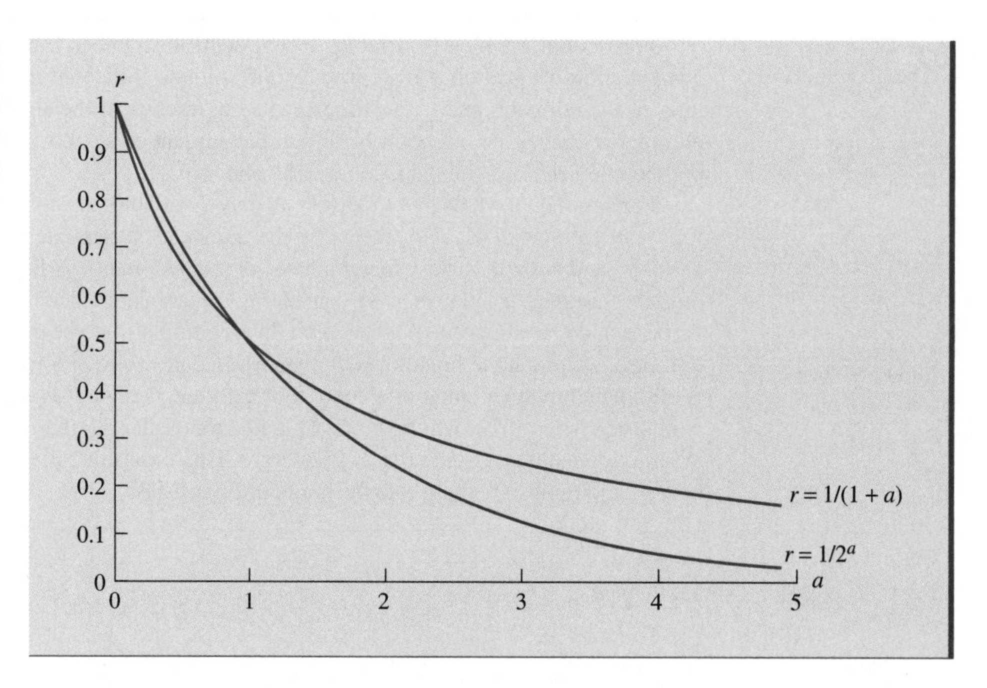

We now need to discuss which of these functions is the most reasonable to model this situation. Let e represent the actual amount of alcohol eliminated during an hour. The amount eliminated is the rate r times the amount in the body a; that is,

$$\text{eliminated} = \text{rate} \times \text{amount} \quad \text{or} \quad e = ra$$

This is similar to the relationship between distance, rate and time, that is,

$$\text{distance} = \text{rate} \times \text{time}$$

When the rate is given by the rational function, the amount of alcohol eliminated is given by the function

$$e = ra = \left(\frac{1}{1+a}\right)a = \frac{a}{1+a}$$

The graph of this function is seen in Figure 5.5. Note that this function implies that as the amount of alcohol in the body increases, the amount eliminated increases, and that the amount eliminated levels off at a maximum of 1 gram per hour, the horizontal asymptote of the function $e = a/(1+a)$.

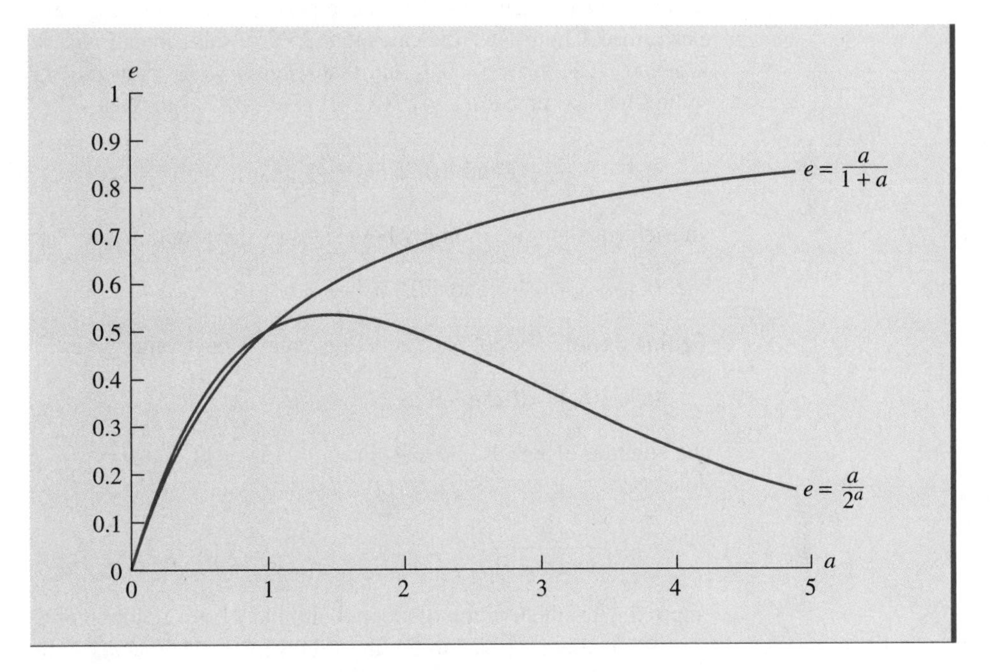

FIGURE 5.5 Comparing amounts of chemical eliminated using a rational function and an exponential function.

When the rate is given by the exponential, the amount eliminated is given by

$$e = ra = \frac{a}{2^a}$$

This function is also seen in Figure 5.5. The exponential function suggests that as the amount of alcohol in the body increases, the actual amount eliminated decreases. Experience and studies indicate that the rational function gives a more reasonable result. There is no reason to think that the actual amount of alcohol eliminated will decrease as we consume more alcohol.

From this discussion, we are going to assume that the **fraction** r of alcohol eliminated during an hour is given by a rational function of the form

$$r = \frac{b}{c+a}$$

The constants b and c depend on the particular person. Since a depends on time, then r depends on time. This means that the fraction of the alcohol eliminated during the $n-1$th hour is approximated by the formula

$$r = \frac{b}{c + a(n-1)}$$

where $a(n-1)$ is the number of grams of alcohol in the body at the beginning of hour $n-1$.

We now discuss how we might be able to approximate b and c. Suppose a person consumes 21 grams of alcohol. A blood test performed 1 hour later indicates that 40% of this alcohol was removed. The same person consumes 36 grams of alcohol. A blood test performed 1 hour later indicates that 25% of this alcohol was removed. This means that when $a = 21$, then $r = 0.4$, and that when $a = 36$, then $r = 0.25$. Substitution of these values into the function $r = b/(c+a)$ gives the 2 equations

$$0.4 = \frac{b}{c+21} \quad \text{and} \quad 0.25 = \frac{b}{c+36}$$

In each equation, we multiply both sides by the denominator. This gives the 2 equations

$$0.4c + 8.4 = b \quad \text{and} \quad 0.25c + 9 = b$$

Setting the left sides of each equation equal to each other gives

$$0.4c + 8.4 = 0.25c + 9$$

the solution of which is $c = 4$. From this, we get that $b = 10$. Thus, for this person, the function

$$r = \frac{10}{4 + a(n-1)}$$

approximates the fraction of alcohol eliminated as a function of the amount of alcohol.

While this approach to finding values for c and b seems somewhat simplistic, it is not far from what is actually done. In fact, the numbers $c = 4$ and $b = 10$ are approximately equal to the values that are actually used for the elimination rate for males. The number for b is generally lower for women.

Recall that the dynamical system for the change in chemicals in the body is

$$a(n) = a(n-1) - ra(n-1) + d$$

where r is the fraction eliminated and d is the constant amount consumed each time period. Substitution of our function for r into this dynamical system gives

$$a(n) = a(n-1) - \left(\frac{10}{4 + a(n-1)}\right) a(n-1) + d \tag{5.3}$$

(If the amount consumed each hour varied, we would have to modify this dynamical system.) This dynamical system is a **nonlinear dynamical system** because there is a quotient involving $a(n-1)$.

Alcohol is an example of what is called **capacity-limited metabolism,** in which the fraction of the chemical metabolized depends on the amount of the chemical in the body. Note that the dynamical system is an average, and the particular system will vary somewhat from one individual to another.

Let's use (5.3) to investigate several different drinking scenarios.

Example 5.1

Suppose someone has half a drink every hour. This means that $d = 7$ grams of alcohol. The dynamical system is

$$a(n) = a(n-1) - \left(\frac{10}{4 + a(n-1)}\right) a(n-1) + 7$$

Assuming the first half a drink is at time $n = 0$, then $a(0) = 7$. Using this starting value and the dynamical system, you should be able to obtain the numbers in Table 5.2.

TABLE 5.2 Amount of alcohol in body after *n* hours, consuming half a drink per hour.

$n =$	0	1	2	3	4	10	24	48
$a(n) =$	7	7.64	8.07	8.39	8.62	9.19	9.33	9.33

These results can be seen in the time graph in Figure 5.6. Note that even if a person could continue this consumption for 2 days, the amount of alcohol in his body would be under 10 grams. If this person was a 140-pound male, he would not have reached the threshold amount for feeling the effects of the alcohol.

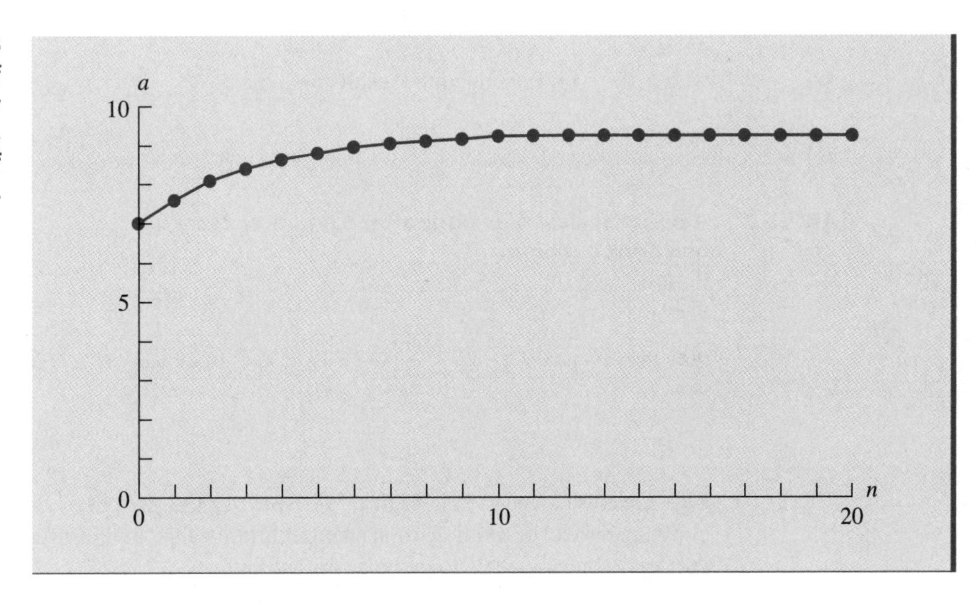

FIGURE 5.6 Amount of alcohol in body after *n* hours, drinking 7 grams of alcohol per hour.

The amount of alcohol $a(n)$ has approached an equilibrium amount. To find this equilibrium amount, we substitute E for $a(n)$ and $a(n-1)$ into the dynamical system, giving the equation

$$E = E - \frac{10E}{4+E} + 7$$

We subtract E from both sides and multiply through by the denominator, $4+E$, to get

$$0 = -10E + 7(4+E) \quad \text{or} \quad 0 = -10E + 28 + 7E$$

We solve this equation, getting that the equilibrium value is

$$E = \frac{28}{3} = 9\tfrac{1}{3} \approx 9.33$$

which we might have expected after seeing the results in Table 5.2 and Figure 5.6.

If you try other reasonable starting values for $a(0)$, you will see that $a(n)$ goes toward 9.33 in each case. Thus, $E = 9.33$ appears to be a stable equilibrium value. ■

Notice that consuming half of a drink per hour for a prolonged period of time keeps the level of alcohol in the body at about the level at which impairment begins, mildly intoxicated.

Example 5.2 Suppose a person consumes one drink every hour, starting at time $n = 0$. The dynamical system that models the amount of alcohol in this person's body is

$$a(n) = a(n-1) - \left(\frac{10}{4+a(n-1)}\right)a(n-1) + 14$$

with $a(0) = 14$. I obtained the results in Table 5.3.

TABLE 5.3 Amount of alcohol in body after n hours, consuming one drink per hour.

$n =$	0	1	2	3	4	10	24	48
$a(n) =$	14	20.22	25.87	31.21	36.35	65.04	126.95	228.43

These results can be seen in Figure 5.7. Note that the amount of alcohol seems to continue climbing instead of leveling off at an equilibrium value. In fact, after the first several points, the data "almost" seems to lie on a straight line.

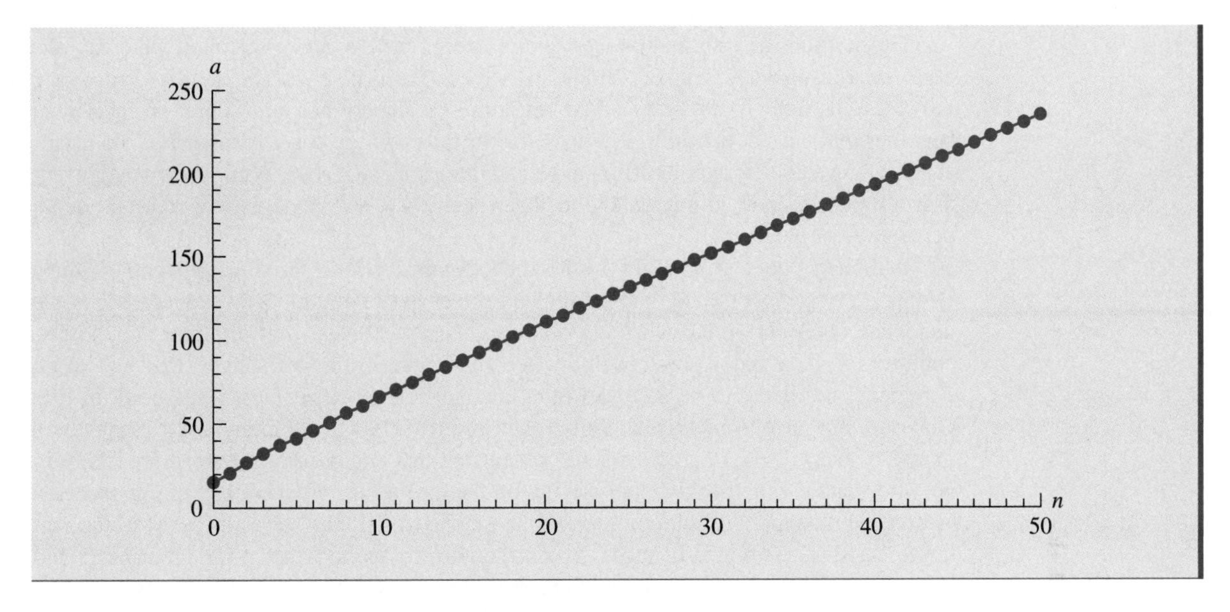

FIGURE 5.7 **Amount of alcohol in body after consuming 14 grams per hour.**

From Figure 5.7, we can see that a 140-pound male could be arrested for DWI after about 5 hours, and would be at risk of alcohol poisoning after about 40 hours. Of course, consuming one drink per hour for 40 hours is a highly artificial scenario.

To find the equilibrium value for this dynamical system, we solve the equation

$$E = E - \frac{10E}{4+E} + 14$$

The solution is

$$E = -14$$

This equilibrium value has no physical significance. ■

Note that if a person consumes 1 drink per hour, the level of alcohol in that person's blood will continue to rise, indefinitely. This is in stark contrast to a person consuming half a drink per hour.

In Example 5.2, we did not test the stability of the equilibrium value. You will find that if you pick $a(0)$ close to -14, say $-20 < a(0) < -5$, then $a(n)$ will approach -14. We noticed in Figure 5.7 that if $a(0) = 14$, then $a(n)$ goes toward infinity. This means that the equilibrium value is considered **stable,** but is not **globally stable.** We will discuss this in more detail in Section 5.3.

In the problems, you are going to investigate several drinking scenarios. One result of these problems is to see that it is practically impossible to maintain a moderate to high degree of intoxication. To achieve certain equilibrium amounts, you have to be very precise in the amount of alcohol you consume. Considering that the dynamical system is only an approximation for each person, you cannot be confident of achieving these results.

Given all of the information we have regarding the dangers of alcohol, what can one do to be a responsible drinker? Moderation for social drinkers is the key. As Problem 6 indicates, the body oxidizes less than ten grams of alcohol per hour. Once an individual becomes intoxicated, the only significant factor in becoming sober is time. The old myths about taking a cold shower or drinking black coffee are dangerous. While they may prevent a drunk person from falling asleep at the wheel, they will not improve reflexes, depth perception, or judgment.

Although public campaigns by Mothers Against Drunk Driving, Students Against Drunk Driving, Sober Ride, and designated driver awareness have all helped reduce the incidence of alcohol-related driving fatalities, it is very clear that drinking and driving continues to be a major problem in our society. Recent estimates suggest that half of all Americans are likely to be involved in an alcohol-related accident at some point in life. One does not have to be an alcoholic to get into trouble with drinking and driving. As a matter of fact, there is some evidence to suggest that social drinkers are more likely to get into trouble with drinking and driving on the rare occasion when they might exceed a reasonable alcohol consumption level. This increased risk has been attributed to the fact that social drinkers typically have a lower tolerance for alcohol than problem drinkers and alcoholics, and are thus more likely to be impaired at lower blood alcohol levels.

I would like to thank Michael Kaiser, Ph.D., for supplying much of the information in this section. Dr. Kaiser is a clinical psychologist who is a certified addictions counselor.

5.2 Problems

1. Suppose that a person who consumes 21 grams of alcohol eliminates $r = \frac{1}{2}$ of it in the next hour. If this same person consumes 33 grams of alcohol, $r = \frac{1}{3}$ is eliminated. Determine the values for b and c in the function $r = b/(c + a)$ that approximates the elimination rate for this person.

2. Suppose a person with $a(0) = 100$ grams of alcohol in the body stops drinking, so $d = 0$. How many hours will it take until the amount of alcohol drops below 40 grams, the level at which the person can "safely" drive? How many hours will it take until the amount of alcohol drops below 10 grams, the level at which this person might no longer feel its affect? This is assuming $b = 10$ and $c = 4$.

3. Assume $b = 10$ and $c = 4$. To 2-decimal places, how many grams of alcohol can a person consume each hour so that the resulting dynamical system will have a positive equilibrium value of

a. 50 grams of alcohol?

b. 100 grams?

c. 200 grams?

d. In each part, determine, to 1 decimal place, the number of ounces of beer that a person would consume to reach the desired equilibrium value, remembering that 12 ounces of beer contains about 14 grams of alcohol.

4. **a.** Find the amount of alcohol in your body that would qualify you as legally intoxicated by solving the proportion

$$\frac{40}{140} = \frac{x}{\text{your weight}}$$

for x.

b. Suppose you wished to maintain an alcohol level that is 20 grams above your legal limit x. How many grams of alcohol d would you consume each hour to maintain that amount as your equilibrium level?

c. What would be the result if you consumed 1 gram of alcohol more each hour than the amount found in part (b) (which is about the amount of alcohol in 2 tablespoons of beer)?

5. Suppose the metabolism of some person is such that the dynamical system modeling the elimination of alcohol was

$$a(n) = a(n-1) - \frac{9a(n-1)}{4.2 + a(n-1)} + d$$

a. Find the equilibrium value, given that this person drinks half a drink per hour? What happens to the amount of alcohol in this person's body?

b. Find the equilibrium value, given that this person drinks 8 grams of alcohol per hour, about 7 ounces of a beer? What happens to the amount of alcohol in this person's body?

c. Find the equilibrium value, given that this person drinks 8.9 grams of alcohol per hour. What happens to the amount of alcohol in this person's body?

d. Find the equilibrium value, given that this person drinks 9 grams of alcohol per hour. What happens to the amount of alcohol in this person's body?

6. The amount of alcohol eliminated from the body each hour depends on the amount of alcohol in the body and is estimated to be

$$e = \frac{10a}{4 + a}$$

where a is the amount of alcohol in the body. Approximate the amount of alcohol eliminated from the body given that $a = 20, 50, 100, 200, 500$ grams of alcohol. Graph the function e. Does there appear to be a limit on the total amount of alcohol eliminated from the body in any hour? What does that limit appear to be?

7. The function

$$r = \frac{b}{c + a^2}$$

also has the shape of Figure 5.3. Again, assume $b = 2$ and $c = 3$. Graph the function

$$ra = \frac{2a}{3 + a^2}$$

and describe what happens to the amount of alcohol eliminated as a function of a, the amount of alcohol in the body.

8. Suppose that the kidneys eliminate 10% of a chemical every hour and that the fraction of the chemical broken down by the liver every hour is given by

$$r = \frac{5}{10+a}$$

where a is the number of milligrams of the drug in the body. Assume a person consumes 8 mg of the chemical every hour.

a. Develop a dynamical system to model this situation.

b. What is the positive equilibrium value for this dynamical system? Is it stable?

5.3 Stability

Let's informally recall what we already know about stability. An equilibrium value E is globally stable if for any initial value, $u(n)$ gets close to E. An equilibrium value is globally unstable if for any initial value, $u(n)$ goes away from the equilibrium value. An equilibrium value is neutral if for any initial value, $u(n)$ stays about the same distance from E. For affine dynamical systems of the form

$$u(n) = ru(n-1) + b$$

the equilibrium value is globally stable when $|r| < 1$, is globally unstable when $|r| > 1$, and is neutral when $r = -1$. There was no equilibrium value if $r = 1$.

If the dynamical system is nonlinear, then $u(n)$ may go to E for only a limited range of initial values but not all initial values. This is a major difference between affine and nonlinear dynamical systems.

Example 5.3 Consider the dynamical system

$$u(n) = -0.5u^2(n-1) + 1.5u(n-1)$$

The equilibrium values are the solutions to

$$E = -0.5E^2 + 1.5E$$

Subtract E from both sides of the equation, then factor out $0.5E$, giving

$$0 = 0.5E(-E+1)$$

The equilibrium values can be found by setting each factor equal to 0. The solutions are $E = 0$ and $E = 1$.

Let's see how the functions $u(n)$ behave by making time graphs for several different initial values. I picked initial values above and below each of the equilibrium values $u(0) = -0.1$, $u(0) = 0.1$, $u(0) = 0.5$, and $u(0) = 1.5$. In Figure 5.8 are time graphs corresponding to each of those values.

Note that for some initial values, $u(n)$ goes toward 1, indicating it might be a stable equilibrium value, but that for $u(0) = -0.1$, the values for $u(n)$ go away from $E = 1$.

We might ask ourselves, For what initial values will $u(n)$ approach 1? You should make time graphs using other initial values and see if you observe the following.

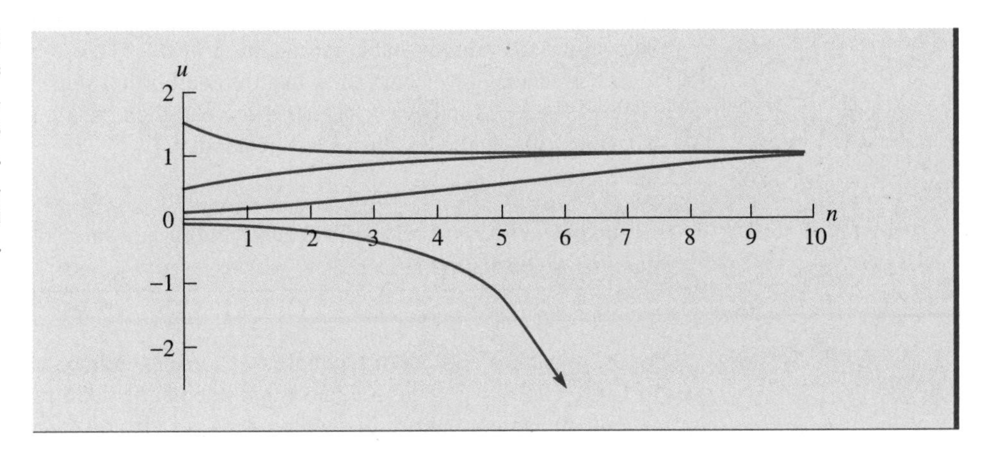

FIGURE 5.8 When $u(0) =$ 0.1, 0.5, or 1.5, $(n, u(n))$ goes toward line $u = 1$. When $u(0) = -0.1$, $u(n)$ goes toward negative infinity.

1. If $0 < u(0) < 3$, then $u(n)$ goes toward 1.
2. If $u(0) < 0$ or $u(0) > 3$, then $u(n)$ goes toward negative infinity.
3. If $u(0) = 0$ or $u(0) = 3$, then $u(n) = 0$ for $n = 1, 2, \ldots$.

Thus, we note that sometimes $u(n)$ goes toward equilibrium and sometimes it goes away. We need to expand our concepts of stability. ■

DEFINITION

An equilibrium value E is **stable** if there are numbers a and b such that $a < E < b$ and whenever $a < u(0) < b$, then $u(n)$ goes to E.

Figure 5.9 gives a depiction of what it means to be stable. If $u(0)$ starts in the indicated interval, then $u(n)$ goes to the equilibrium value. The difference between stable and globally stable is that for globally stable, it must be true for every choice of a and b; while for stable, it is true for some choice of a and b, but not every choice.

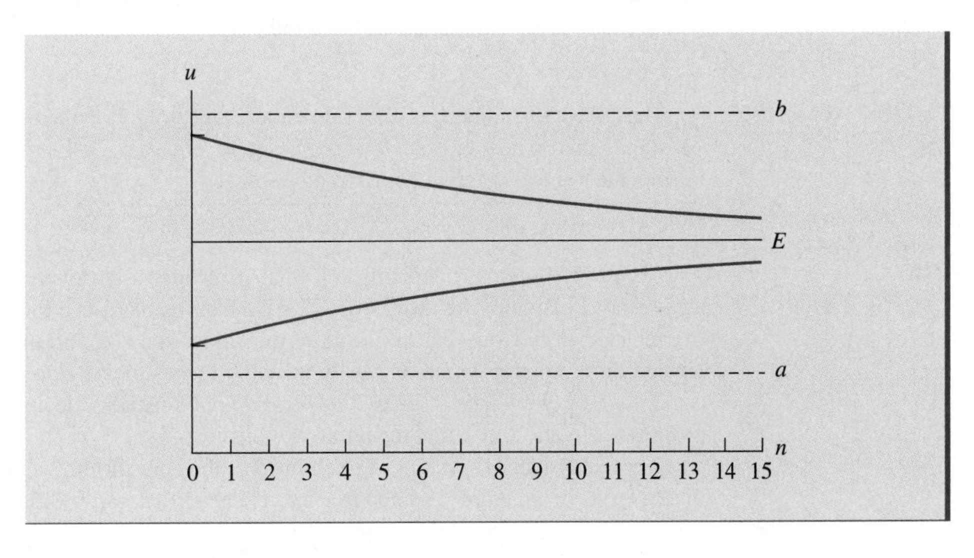

FIGURE 5.9 If $a < u(0) < b$ for some numbers a and b, then the points $(n, u(n))$ approach the line $u = E$.

An equilibrium value is stable means that if $u(0)$ is "close enough" to E, then $u(n)$ goes to E. For example, we could show that the equilibrium value $E = 1$ in Example 5.3 is stable by picking $a = 0$ and $b = 3$. Then $0 < 1 = E < 3$ and if $0 < u(0) < 3$, then $u(n)$ goes to 1. Note that $u(0)$ can be less than E or greater than E.

DEFINITION

The **maximum interval of stability** for an equilibrium value E is the set of all initial values for which $u(n)$ goes to E.

For the dynamical system of Example 5.3, some experimentation will show that $u(n)$ goes to 1 if $0 < u(0) < 3$. Thus the maximum interval of stability is the interval $(0, 3)$. In this case, the maximum interval of stability does not include the end points, 0 and 3. The length of the maximum interval of stability, $3 - 0 = 3$ for $E = 1$ of Example 5.3, gives some indication about how stable the equilibrium value is.

In this section, we will consider nonlinear dynamical systems in which it is relatively easy to find the maximum interval of stability through experimentation. In Section 5.4, we will learn to algebraically find the maximum interval of stability for these dynamical systems.

You don't need to find the maximum interval of stability to show an equilibrium value is stable. You only need to find some interval that satisfies the definition. In Example 5.3, we could let $a = 0.5$ and $b = 1.5$. Again, $0.5 < E = 1 < 1.5$, and if $0.5 < u(0) < 1.5$, then $u(n)$ goes to 1. Fortunately we don't need to find the maximum interval of stability, because the maximum interval of stability is often quite difficult to determine.

When we determine that an equilibrium value is stable by testing $u(0)$-values on both sides of E, we are not actually proving E is stable, since we haven't checked every value close to E. We could prove an equilibrium value is stable using a little calculus, but will not do so here.

You need to be careful when determining stability, because the maximum interval of stability can be quite small. When this happens, a and b might need to be very close to the equilibrium value, as Example 5.4 will indicate.

DEFINITION

An equilibrium value E is **unstable** if there are numbers a and b, such that $a < E < b$, and no matter how close $u(0)$ is to E (but not equal to E), eventually $u(n)$ is outside the interval (a, b) for some value of n.

Figure 5.10 gives a depiction of what it means to be unstable; that is, $u(n)$ moves away from E. By moving away from E, we mean that $u(n)$ is either eventually less than a or greater than b. Note that in one case, the values for $u(n)$ did not go off to infinity, but leveled off at another value. Globally unstable means that the points $(n, u(n))$ eventually are outside of every pair of lines $u = a$ and $u = b$. Unstable means that there is **some** pair of lines $u = a$ and $u = b$ that the points get outside of.

For the dynamical system of Example 5.3, the equilibrium value $E = 0$ is unstable. In this case, we could let $a = -1$ and $b = 0.5$. Then $-1 < E = 0 < 0.5$. You should experiment

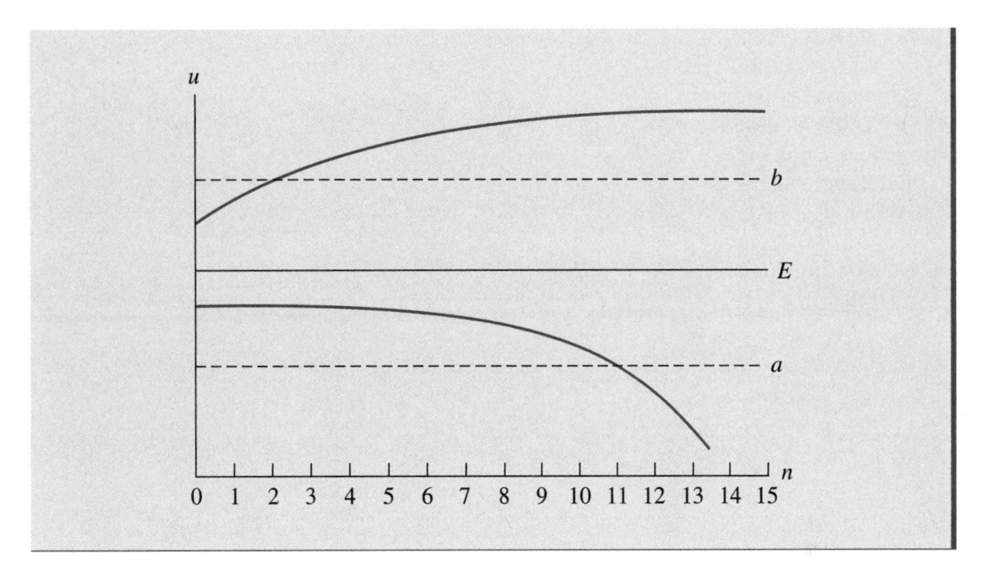

FIGURE 5.10 *E* is unstable if, whenever $u(0)$ is between some a and b, the points $(n, u(n))$ are eventually outside the lines $u = a$ and $u = b$.

to see that if $-1 < u(0) < 0$, then $u(n)$ goes to negative infinity and is thus less than -1. Because $E = 1$ is stable, we know that if $0 < u(0) < 0.5$, then $u(n)$ goes to 1. But then $u(n)$ is eventually bigger than 0.5. In either case, $u(n)$ is eventually outside of the interval $(-1, 0.5)$. There was no magical reason for picking -1 and 0.5. We could have picked any negative number and any positive number less than 1.

The definition of unstable may seem a little confusing. In fact, defining unstable is somewhat difficult because of all of the strange types of behavior that can occur for nonlinear dynamical systems. For our purposes, an equilibrium value is unstable if, no matter how close $u(0)$ is to E, $u(n)$ will get further away from E. The values for $u(n)$ do not have to get "far away" from E, as Example 5.4 will indicate.

Example 5.4 Consider the dynamical system

$$u(n) = 3u^2(n-1) + 0.7u(n-1)$$

The equilibrium values are the solutions to

$$E = 3E^2 + 0.7E$$

which are $E = 0$ and $E = 0.1$. We test the equilibrium values by plotting the points $(n, u(n))$ when $u(0) = -0.1$, $u(0) = 0.03$, $u(0) = 0.08$, and $u(0) = 0.11$. I let $a = -0.12$, $b = 0.05$, and $c = 0.13$. We note that $E = 0$ is stable, since the functions in which $-0.12 < u(0) < 0.05$ go toward 0. The equilibrium point $E = 0.1$ is unstable, since the functions in which $0.05 < u(0) < 0.13$ eventually get outside the lines $u = 0.05$ and $u = 0.13$. Even though $E = 0.1$ is unstable, some of the functions $u(n)$ get no further away than to 0.

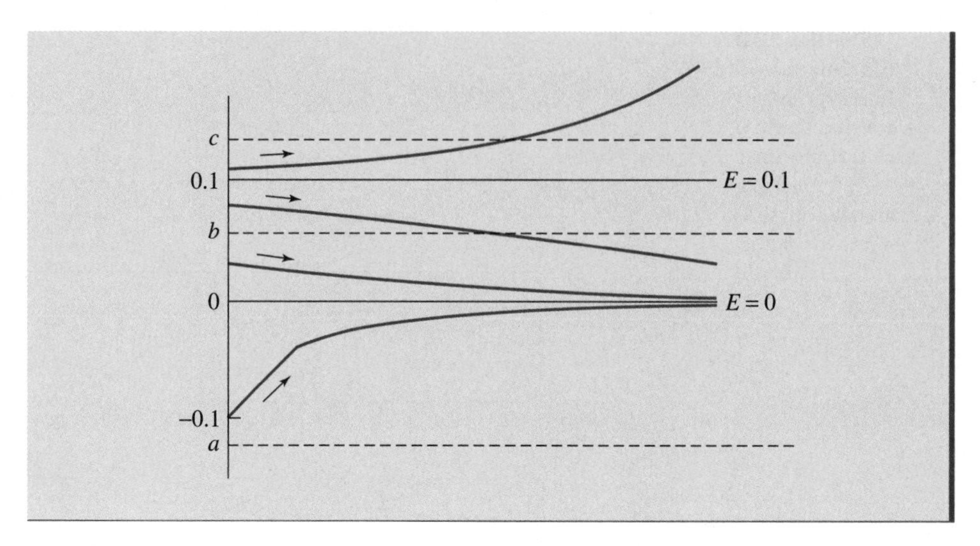

FIGURE 5.11 $E = 0$ is stable because when $a < u(0) < b$, $u(n)$ goes to 0. $E = 0.1$ is unstable because when $b < u(0) < c$, $(n, u(n))$ gets outside the lines $u = b$ and $u = c$.

There are many other choices that could be made for a, b, and c. The maximum interval of stability is $(-\frac{1}{3}, 0.1)$. Notice how small this interval is. ■

Example 5.5 Consider the dynamical system

$$u(n) = u^2(n-1) - 3u(n-1) + 4$$

The equilibrium values are solutions to

$$E = E^2 - 3E + 4$$

Subtracting E from both sides of the equation and factoring gives

$$0 = (E-2)^2$$

The only solution to this equation is $E = 2$. To test the stability of 2, I picked $u(0) = 1.5$ and $u(0) = 2.2$ which are "close" to 2 and on either side of 2. The time graphs of the 2 functions are seen in Figure 5.12. Note that one solution goes toward equilibrium and the other goes away.

Some experimentation will indicate that the maximum interval of stability for $E = 2$ is $1 \leq u(0) \leq 2$. For any other value for $u(0)$, the function $u(n)$ goes toward positive infinity. This equilibrium value is not stable, since whenever $u(0)$ is greater than 2, $u(n)$ goes away from 2. This equilibrium value is not unstable since for some $u(0)$ less than 2, $u(n)$ goes toward 2. ■

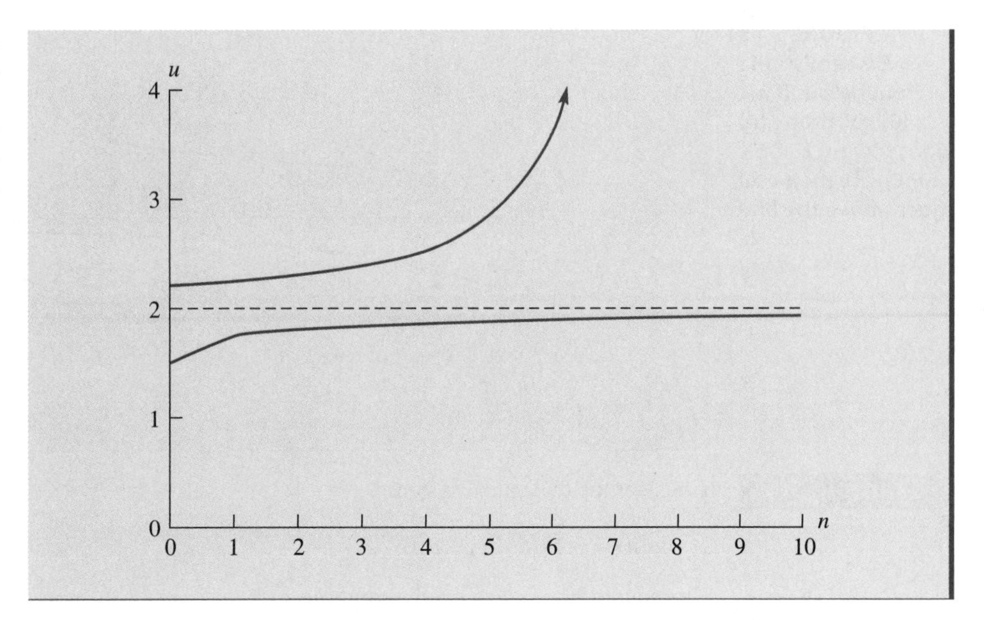

FIGURE 5.12 When $u(0) = 1.5$, $u(n)$ goes toward $E = 2$, but when $u(0) = 2.2$, $u(n)$ goes toward infinity.

DEFINITION

An equilibrium value is **semistable** if there are numbers a and b, such that $a < E < b$, and either (1) when $a < u(0) \leq E$, then $u(n)$ goes to E and when $E < u(0) < b$, then $b < u(n)$ for some value of n or (2) when $a < u(0) < E$, then $u(n) < a$ for some value of n and when $E \leq u(0) < b$, then $u(n)$ goes to E. In case 1, where the stable interval is below the equilibrium value, we say that E is **semistable from below.** In case 2, where the stable interval is above E, we say that E is **semistable from above.**

Figure 5.13 shows what it means for an equilibrium value to be semistable from below. Semistable from above would be the reverse of this figure. Essentially, a semistable equilibrium value is stable on one side and unstable on the other side. Note that Figure 5.13 is like Figure 5.9 below E and is like Figure 5.10 above E. The equilibrium value acts like a wall, preventing the $u(n)$ values from passing.

To determine if an equilibrium value is stable, unstable, or semistable, you need to generate time graphs using initial values both below and above equilibrium.

One difficulty with determining stability is that the meaning of "close" depends on the particular dynamical system. Thus, before making a conclusion, be sure you have picked $u(0)$-values that are "close enough" to the equilibrium value you are analyzing. If you think about the equilibrium values and experiment, you will soon develop a "feel" for whether you are close enough or not. The graphical analysis that we will learn in Section 5.4 will help clarify what is happening.

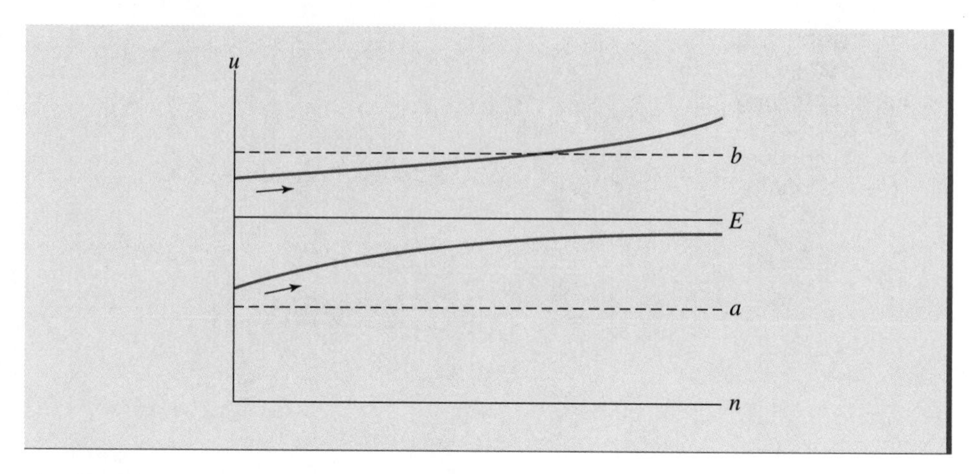

FIGURE 5.13 E is semistable from below. If $a < u(0) < E$, then $u(n)$ goes toward E. If $E < u(0) < b$, then $u(n)$ goes above the line $u = b$.

Example 5.6 Consider the dynamical system

$$u(n) = 3.1u(n-1) - u^2(n-1)$$

The equilibrium values are the solutions to

$$E = 3.1E - E^2$$

which are $E = 0$ and $E = 2.1$.

Picking $u(0) = 2.0$, which is close to $E = 2.1$, gives the time graph seen in Figure 5.14. Note that the function oscillates away from the equilibrium value $E = 2.1$, eventually getting outside the lines $u = a$ and $u = b$. The values do not get far away from 2.1, though. This indicates that $E = 2.1$ is unstable. Picking other initial values close to 2.1 gives the same result, time graphs oscillating away from equilibrium.

Figure 5.15 gives time graphs for $u(n)$ using $u(0) = -0.1$ and $u(0) = 0.1$, initial values on each side of the equilibrium value $E = 0$. In both cases, $u(n)$ goes away from $E = 0$ indicating that $E = 0$ is also unstable. Note that when $u(0) = 0.1$, then $u(n)$ eventually begins oscillating similarly to the time graph in Figure 5.14.

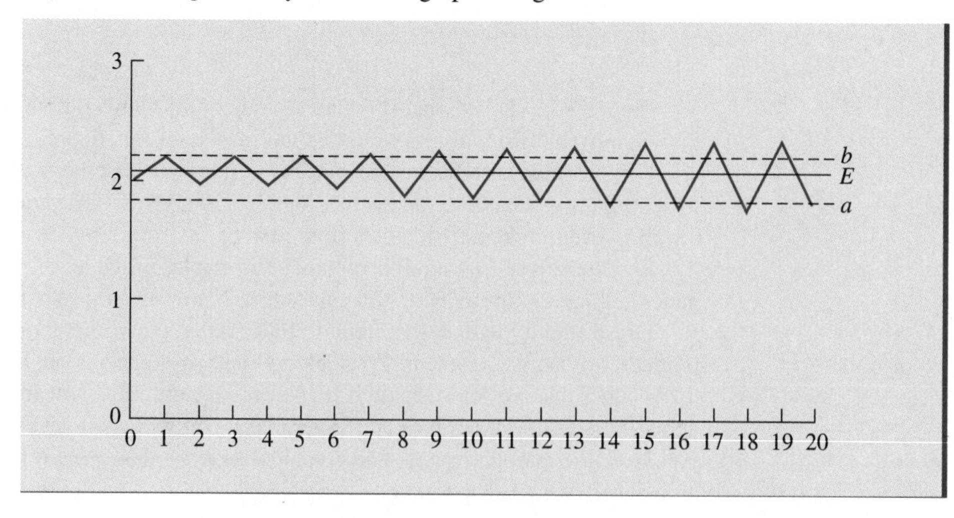

FIGURE 5.14 When $u(0) = 2$, the function $u(n)$ oscillates with period 2 away from the equilibrium, $E = 2.1$.

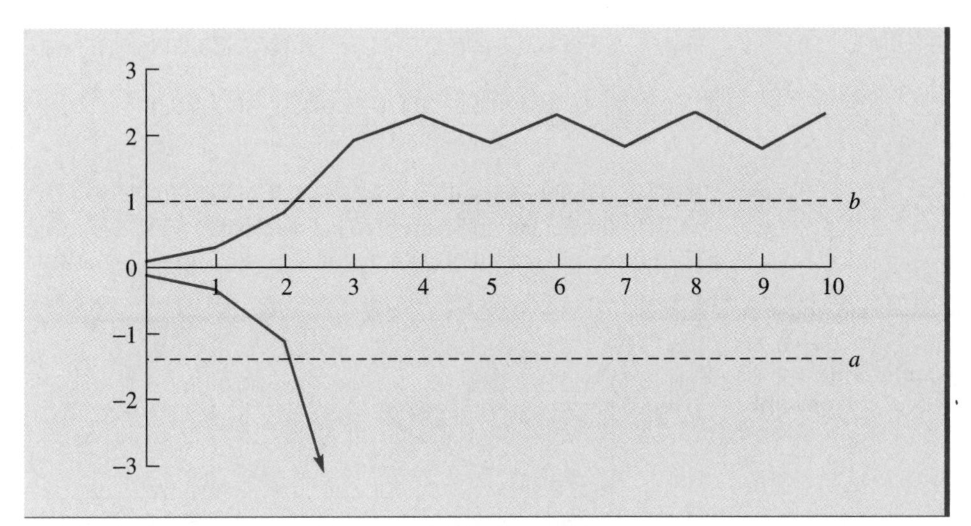

FIGURE 5.15 When $u(0) = 0.1$ or -0.1, $u(n)$ goes away from $E = 0$.

Since both equilibrium values are unstable, you may wonder what happens to functions with different starting values. If $u(0) < 0$ or $u(0) > 3.1$, then $u(n)$ goes to negative infinity. You should try several values for $u(0)$ to see if this happens.

You should also check that if $0 < u(0) < 3.1$ (except for $u(0) = 2.1$), then, eventually, $u(n)$ repeats the same 2 values, approximately 2.37 and 1.73. This can be seen in Figures 5.14 and 5.15. The 2 values, $E_1 = 2.37$ and $E_2 = 1.73$, form what is called a **stable 2-cycle.** It is a 2-cycle since the same 2 values keep repeating. It is stable in that functions with nearby initial values eventually repeat these 2 values. I will not define a stable 2-cycle, but I wanted you to have some experience with the different types of behavior that can occur with nonlinear dynamical systems. ■

Example 5.7 Consider the dynamical system

$$u(n) = u^2(n-1) + 1.5u(n-1) - 0.14$$

The equilibrium values are the solutions to

$$E = E^2 + 1.5E - 0.14$$

Subtracting E from both sides gives the equation

$$0 = E^2 + 0.5E - 0.14$$

The expression on the right is not easy to factor, so we will use the quadratic formula. As you remember, the solutions to an equation

$$0 = ax^2 + bx + c$$

are

$$x = \frac{-b \pm \sqrt{b^2 - 4ac}}{2a}$$

In this case, $a = 1$, $b = 0.5$, and $c = -0.14$. The solutions are therefore

$$E = \frac{-0.5 \pm \sqrt{0.5^2 - 4(1)(-0.14)}}{2(1)} = \frac{-0.5 \pm 0.9}{2}$$

Thus, the equilibrium values are $E = -0.7$ and $E = 0.2$.

The time graphs in Figure 5.16 indicate that $E = -0.7$ is stable and $E = 0.2$ is unstable. The maximum interval of stability for $E = -0.7$ is $-1.7 < u(0) < 0.2$. ■

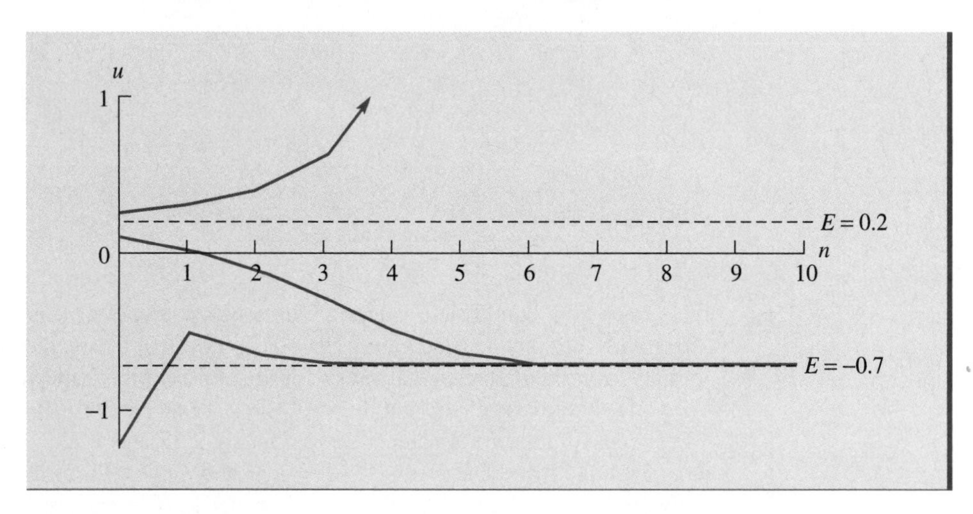

FIGURE 5.16 $E = -0.7$ **is stable, and** $E = 0.2$ **is unstable.**

Suppose equilibrium value E is stable. As in Sections 2.4 and 4.5, we compute the ratios

$$\frac{u(n) - E}{u(n-1) - E}$$

for several n values to determine the rate at which functions approach equilibrium. For example, the dynamical system $u(n) = -0.5u^2(n-1) + 1.5u(n-1)$ has a stable equilibrium value of $E = 1$. Using $u(0) = 0.5$, since it is in the interval of stability for $E = 1$, I obtained the results in Table 5.4. It appears that, eventually, each $u(n)$-value is about half as far from 1 as the previous value, or about 50% closer to 1. If you try computing the ratio for larger values of n, the values may be so close to zero that you get a division by zero error or the wrong answer on your calculator or computer, so be careful.

As another example, the dynamical system $u(n) = 3u^2(n-1) + 0.7u(n-1)$ has the stable equilibrium value $E = 0$. Using $u(0) = 0.05$, which is in the interval of stability, I obtained the results in Table 5.5. In this case, it appears that $u(n)$ is 70% as far from 0 as $u(n-1)$, or is 30% closer.

TABLE 5.4 Ratios of *u*(*n*)-values for dynamical system (Example 5.3).

$n =$	1	2	5	10	20	30
$\frac{u(n)-1}{u(n-1)-1} =$	0.75	0.6875	0.547	0.502	0.5	0.5

TABLE 5.5 **Ratios of $u(n)$-values for dynamical system (Example 5.4).**

$n =$	1	2	5	10	20	30	50
$\frac{u(n)-0}{u(n-0)-0} =$	0.85	0.8275	0.767	0.715	0.7005	0.70001	0.7

We will only compute ratios if the equilibrium value is stable and if $u(0)$ is in the maximum interval of stability.

CAUTION We do not classify dynamical systems as stable or unstable. We do not classify initial values as stable or unstable. We only classify equilibrium values as stable or unstable.

5.3 Problems

1. Consider the dynamical system

$$u(n) = 0.1u^2(n-1) + 0.5u(n-1)$$

a. Determine the two equilibrium values for this dynamical system by substituting E for $u(n-1)$ and $u(n)$ and solving.

b. The time graph for u with $u(0) = 4.5$ can be seen in the Figure 5.17. Pick several more $u(0)$-values close to each equilibrium value and on both sides of them, and use a table of $u(n)$-values and graphs of the functions to determine the stability of each equilibrium value.

c. Approximate the maximum interval of stability of all stable equilibrium values by observing the behavior of functions using a variety of initial values.

d. Approximate the rate at which $u(n)$ goes toward each stable equilibrium value, to 1-decimal-place accuracy.

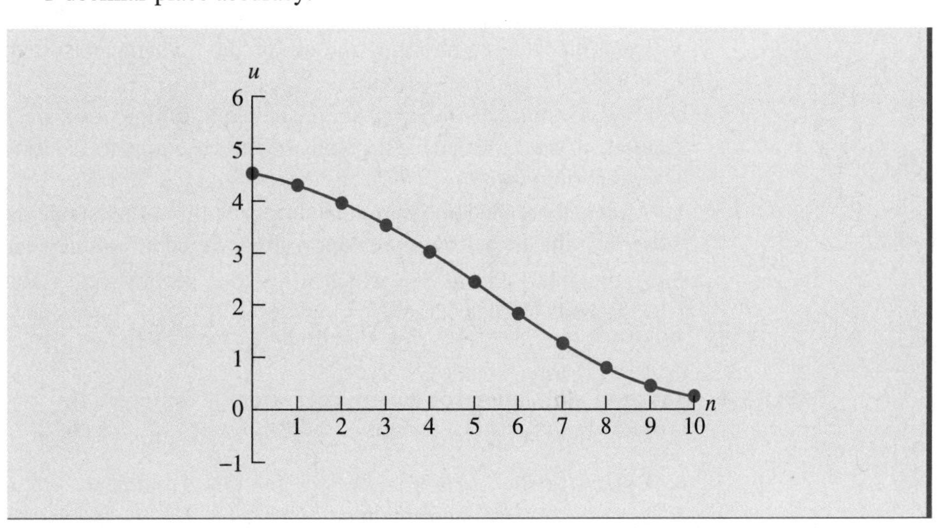

FIGURE 5.17 One possible time graph for Problem 1B.

2. Consider the dynamical system

$$u(n) = 0.5u^2(n-1) - 0.5u(n-1) + 1$$

a. Determine the equilibrium values for this dynamical system by substituting E for $u(n-1)$ and $u(n)$ and solving.

b. Pick several $u(0)$-values close to each equilibrium value and on both sides of it. Use a table of $u(n)$-values and graphs of the functions to determine the stability of each equilibrium value.

c. Approximate the maximum interval of stability of all stable equilibrium values by observing the behavior of functions using a variety of initial values.

d. Approximate the rate at which $u(n)$ goes toward each stable equilibrium value, to 1-decimal-place accuracy.

3. Consider the dynamical system

$$u(n) = 0.3u^2(n-1) + 0.4u(n-1) - 0.9$$

a. Determine the equilibrium values for this dynamical system by substituting E for $u(n-1)$ and $u(n)$ and solving.

b. Pick several $u(0)$-values close to each equilibrium value and on both sides of it. Use a table of $u(n)$-values and graphs of the functions to determine the stability of each equilibrium value.

c. Approximate the maximum interval of stability of all stable equilibrium values by observing the behavior of functions using a variety of initial values.

d. Approximate the rate at which $u(n)$ goes toward each stable equilibrium value, to 1-decimal-place accuracy.

4. Consider the dynamical system

$$u(n) = 0.5u^2(n-1) + 2u(n-1) + 0.5$$

a. Determine the equilibrium values for this dynamical system by substituting E for $u(n-1)$ and $u(n)$ and solving.

b. Pick several $u(0)$-values close to each equilibrium value and on both sides of it. Use a table of $u(n)$-values and graphs of the functions to determine the stability of each equilibrium value.

c. Approximate the maximum interval of stability of all stable equilibrium values by observing the behavior of functions using a variety of initial values.

d. Approximate the rate at which $u(n)$ goes toward each stable equilibrium value, to 1-decimal-place accuracy.

5. Consider the dynamical system

$$u(n) = -0.4u^2(n-1) + 3.4u(n-1) - 2$$

a. Determine the equilibrium values for this dynamical system by substituting E for $u(n-1)$ and $u(n)$ and solving.

b. Pick several $u(0)$-values close to each equilibrium value and on both sides of it. Use a table of $u(n)$-values and graphs of the functions to determine the stability of each equilibrium value.

c. Approximate the maximum interval of stability of all stable equilibrium values by observing the behavior of functions using a variety of initial values.

d. Approximate the rate at which $u(n)$ goes toward each stable equilibrium value, to 1-decimal-place accuracy.

6. Consider the dynamical system

$$u(n) = -2.2u^2(n-1) + 7.6u(n-1) - 4.4$$

a. Determine the equilibrium values for this dynamical system by substituting E for $u(n-1)$ and $u(n)$ and solving.

b. Pick several $u(0)$-values close to each equilibrium value and on both sides of it. Use a table of $u(n)$-values and graphs of the functions to determine the stability of each equilibrium value.

c. Approximate the maximum interval of stability of all stable equilibrium values by observing the behavior of functions using a variety of initial values.

d. Approximate the rate at which $u(n)$ goes toward each stable equilibrium value, to 1-decimal-place accuracy.

7. Consider the dynamical system

$$u(n) = -u^2(n-1) + 2u(n-1) - 0.25$$

a. Determine the equilibrium values for this dynamical system by substituting E for $u(n-1)$ and $u(n)$ and solving.

b. Pick several $u(0)$-values close to each equilibrium value and on both sides of it. Use a table of $u(n)$-values and graphs of the functions to determine the stability of each equilibrium value.

c. Approximate the maximum interval of stability of all stable equilibrium values by observing the behavior of functions using a variety of initial values.

d. Approximate the rate at which $u(n)$ goes toward each stable equilibrium value, to 1-decimal-place accuracy.

8. Consider the dynamical system

$$u(n) = \frac{3}{4 - u(n-1)}$$

a. Determine the equilibrium values for this dynamical system by substituting E for $u(n-1)$ and $u(n)$ and solving.

b. Pick several $u(0)$-values close to each equilibrium value and on both sides of it. Use a table of $u(n)$-values and graphs of the functions to determine the stability of each equilibrium value.

c. Approximate the maximum interval of stability of all stable equilibrium values by observing the behavior of functions using a variety of initial values.

d. Approximate the rate at which $u(n)$ goes toward each stable equilibrium value, to 2-decimal-place accuracy.

9. Consider the dynamical system

$$u(n) = \frac{4 - u(n-1)}{2 + u(n-1)}$$

a. Determine the equilibrium values for this dynamical system by substituting E for $u(n-1)$ and $u(n)$ and solving.

b. Pick several $u(0)$-values close to each equilibrium value and on both sides of it. Use a table of $u(n)$-values and graphs of the functions to determine the stability of each equilibrium value.

c. Approximate the rate at which $u(n)$ goes toward each stable equilibrium value, to 2-decimal-place accuracy.

10. Consider the dynamical system

$$u(n) = \frac{16}{u^{\frac{1}{3}}(n-1)} \quad \text{or} \quad u(n) = \frac{16}{\sqrt[3]{u(n-1)}}$$

a. Determine the equilibrium values for this dynamical system by substituting E for $u(n-1)$ and $u(n)$ and solving.

b. Pick several $u(0)$-values close to each equilibrium value and on both sides of it. Use a table of $u(n)$-values and graphs of the functions to determine the stability of each equilibrium value.

c. Approximate the rate at which $u(n)$ goes toward each stable equilibrium value, to 2-decimal-place accuracy.

d. Approximate the maximum interval of stability of all stable equilibrium values by observing the behavior of functions using a variety of initial values.

11. Consider the dynamical system

$$u(n) = -0.2u^3(n-1) + 0.8u^2(n-1) + 0.4u(n-1)$$

a. Determine the equilibrium values for this dynamical system by substituting E for $u(n-1)$ and $u(n)$ and solving.

b. Pick several $u(0)$-values close to each equilibrium value and on both sides of it. Use a table of $u(n)$-values and graphs of the functions to determine the stability of each equilibrium value.

c. Approximate the rate at which $u(n)$ goes toward each stable equilibrium value, to 2-decimal-place accuracy.

12. Consider the dynamical system

$$u(n) = 0.3u^3(n-1) + 0.3u^2(n-1) + 0.4u(n-1)$$

a. Determine the equilibrium values for this dynamical system by substituting E for $u(n-1)$ and $u(n)$ and solving.
b. Pick several $u(0)$-values close to each equilibrium value and on both sides of it. Use a table of $u(n)$-values and graphs of the functions to determine the stability of each equilibrium value.
c. Approximate the rate at which $u(n)$ goes toward each stable equilibrium value, to 2-decimal-place accuracy.

13. Consider the dynamical system

$$u(n) = -0.2u^3(n-1) + 0.8u^2(n-1) + 0.2u(n-1)$$

a. Determine the equilibrium values for this dynamical system by substituting E for $u(n-1)$ and $u(n)$ and solving.
b. Pick several $u(0)$-values close to each equilibrium value and on both sides of it. Use a table of $u(n)$-values and graphs of the functions to determine the stability of each equilibrium value.
c. Approximate the rate at which $u(n)$ goes toward each stable equilibrium value, to 2-decimal-place accuracy.

5.4 Web Analysis

In this section we are going to explore a new graphical technique, called **web analysis,** that will aid in understanding the behavior of nonlinear dynamical systems. Let's begin by returning to the study of the dynamical system

$$u(n) = -0.5u^2(n-1) + 1.5u(n-1)$$

of Example 5.3.

You know that a dynamical system is an equation that uses a current value, $u(n-1)$, to obtain a new value, $u(n)$. As such, we could use "now" for $u(n-1)$ and "next" for $u(n)$. The dynamical system could then be written as

$$\text{next} = -0.5\text{now}^2 + 1.5\text{now}$$

or, using x for "now" and y for "next",

$$y = -0.5x^2 + 1.5x$$

Thus, depending on our notation, we can write the dynamical system 3 equivalent ways. A graph of the function $y = -0.5x^2 + 1.5x$, along with the line $y = x$ is seen in Figure 5.18.

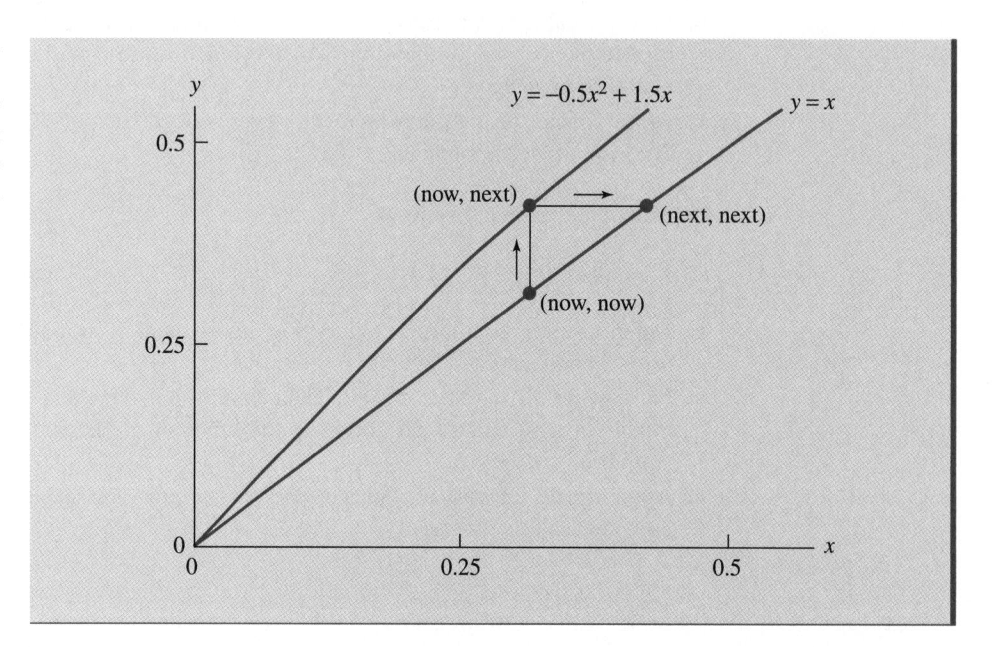

FIGURE 5.18 Graphical method for converting current value "now" into "next" value.

Suppose we have a value for $u(n-1)$ or "now." We plot the point (now, now) on the line $y = x$. Graphically, we can substitute "now" into the function by going vertically from the point (now, now) to the graph of the function. The y-coordinate of that point is $u(n)$ or "next." By going horizontally back to $y = x$, we then get the point (next, next).

The values of the coordinates of (now, now) are $u(n-1)$, and the values of the coordinates of the point (next, next) are $u(n)$. Thus, the combination of the vertical then horizontal moves is a graphical approach of moving from $u(n-1)$ to $u(n)$.

We can repeat this process of moving horizontally and vertically to get a sequence of points on $y = x$ corresponding to a sequence of values $u(0), u(1), \ldots$.

In Figure 5.19 are the graphs of the function $y = -0.5x^2 + 1.5x$ and the line $y = x$ and four repetitions of the vertical and horizontal moves, giving the points $(u(0), u(0))$ through $(u(4), u(4))$.

By continuing to move

1. vertically from the line $y = x$ to the function $y = -0.5x^2 + 1.5x$, then
2. horizontally from the function back to the line $y = x$

we continue to find the points $(u(5), u(5))$, $(u(6), u(6))$, $(u(7), u(7))$, and so on.

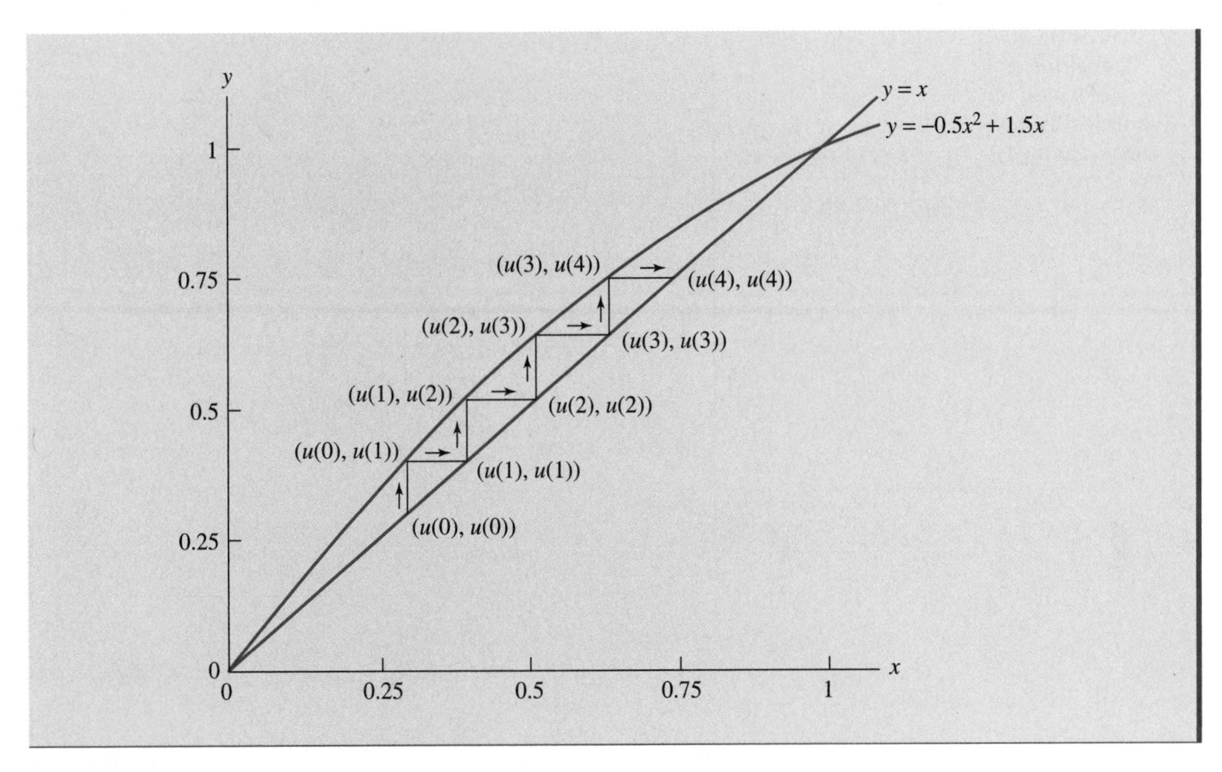

FIGURE 5.19 Horizontal and vertical movements giving sequence of points ($u(0), u(0)$) through ($u(4), u(4)$).

In Figure 5.19, you can see that the points are moving toward the point $(1,1)$, which is the intersection of the curve and the line. This means that the values for $u(n)$ are approaching the equilibrium value $E = 1$. This is a graphical indication that $E = 1$ is a stable equilibrium value.

The points are also moving away from the point $(0,0)$. This is a graphical indication that $E = 0$ is an unstable equilibrium value.

Let me summarize what we have seen. We have a dynamical system

$$u(n) = f(u(n-1))$$

meaning that $u(n)$ is given in terms of $u(n-1)$. We have an equilibrium value E for this dynamical system, meaning that $E = f(E)$. We draw the graphs of $y = f(x)$ and $y = x$. The point (E,E) is on both graphs, so it is a point of intersection of the 2 graphs. We plot the point $(u(0), u(0))$ on the $y = x$. From there, we go vertically to the point $(u(0), u(1))$ on the graph of the function, as seen in Figure 5.19. We then go horizontally from that point to the point $(u(1), u(1))$ on the line. We keep repeating this process, getting pairs of points $(u(n), u(n))$ on the graph of $y = x$ and pairs of points $(u(n-1), u(n))$ on the graph of $y = f(x)$. This process is called **web analysis.** The actual horizontal and vertical lines, as seen in Figure 5.20, are called a **web.**

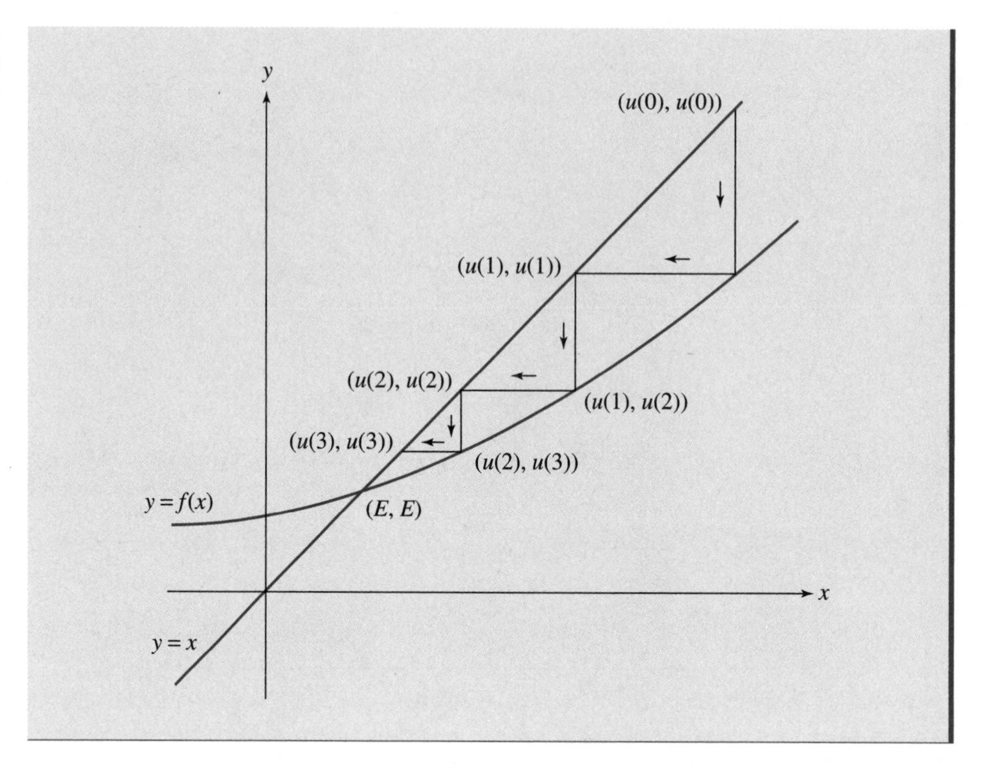

FIGURE 5.20 Development of a web for dynamical system $u(n) = f(u(n-1))$.

Let's look at some examples to see how web analysis can shed light on the stability of equilibrium values.

Example 5.8 Consider the dynamical system

$$u(n) = 0.1u^2(n-1) + 0.7u(n-1) - 0.4$$

You should be able to find that the solutions to the equation

$$E = 0.1E^2 + 0.7E - 0.4$$

are $E = -1$ and $E = 4$. If you try values for $u(0)$ close to -1, you will find that it is a stable equilibrium value. If you try values for $u(0)$ close to 4, you will find that it is an unstable equilibrium value.

The graph of

$$y = 0.1x^2 + 0.7x - 0.4$$

is seen in Figure 5.21, along with the graph of $y = x$. Note that the graphs of the function and the line intersect at $(-1, -1)$ and $(4, 4)$. These 2 points correspond to the 2 equilibrium values.

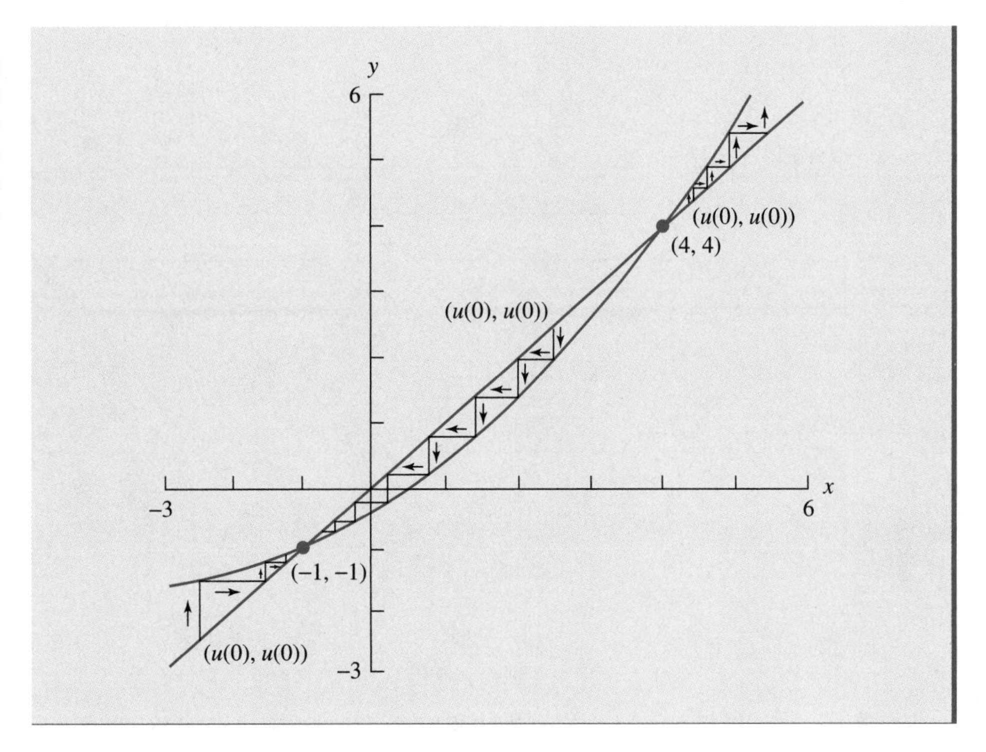

FIGURE 5.21 Three different webs showing that -1 is a stable equilibrium value and 4 is an unstable equilibrium value.

Figure 5.21 also shows 3 different webs. One web starts with $u(0) < -1$. The points on this web go toward the point $(-1, -1)$. Another web starts with $u(0)$ between -1 and 4. This web goes toward the point $(-1, -1)$ and away from the point $(4, 4)$. The final web starts with $u(0) > 4$. This web goes away from $(4, 4)$. As the webs indicate, if $u(0)$ is close to -1, then $u(n)$ goes toward -1, which is therefore a stable equilibrium value. If $u(0)$ starts close to 4, then $u(n)$ goes away from 4, indicating that it is an unstable equilibrium value.

From this web analysis, it appears that the manner in which the graph of $y = f(x)$ intersects the line $y = x$ determines the stability of the equilibrium value. This is in fact true. You will investigate this relationship in the Problems. ■

Note that in the examples, we have not used an actual number for $u(0)$. Web analysis is not used to find actual values. If we want the actual value for $u(n)$ for some n, we will compute it. Web analysis is used to determine the general behavior of the values for $u(n)$.

In the examples, we began at the point $(u(0), u(0))$. The actual starting point is not relevant. We could have started at the point $(u(0), u(1))$ and moved horizontally first. Some calculators begin at the point $(u(0), 0)$ and move vertically to $(u(0), u(1))$. From there, the calculator repeats the same steps as described here. The important thing to observe is not the starting point, but the behavior of the points after you begin.

Sometimes it is not convenient to begin at the point $(u(0), u(0))$. Figure 5.22 shows 3 webs, each beginning with $u(0)$ on or near -11. The purpose of these webs is to shed light on the maximum interval of stability. If we began at the point $(u(0), u(0))$, we would need

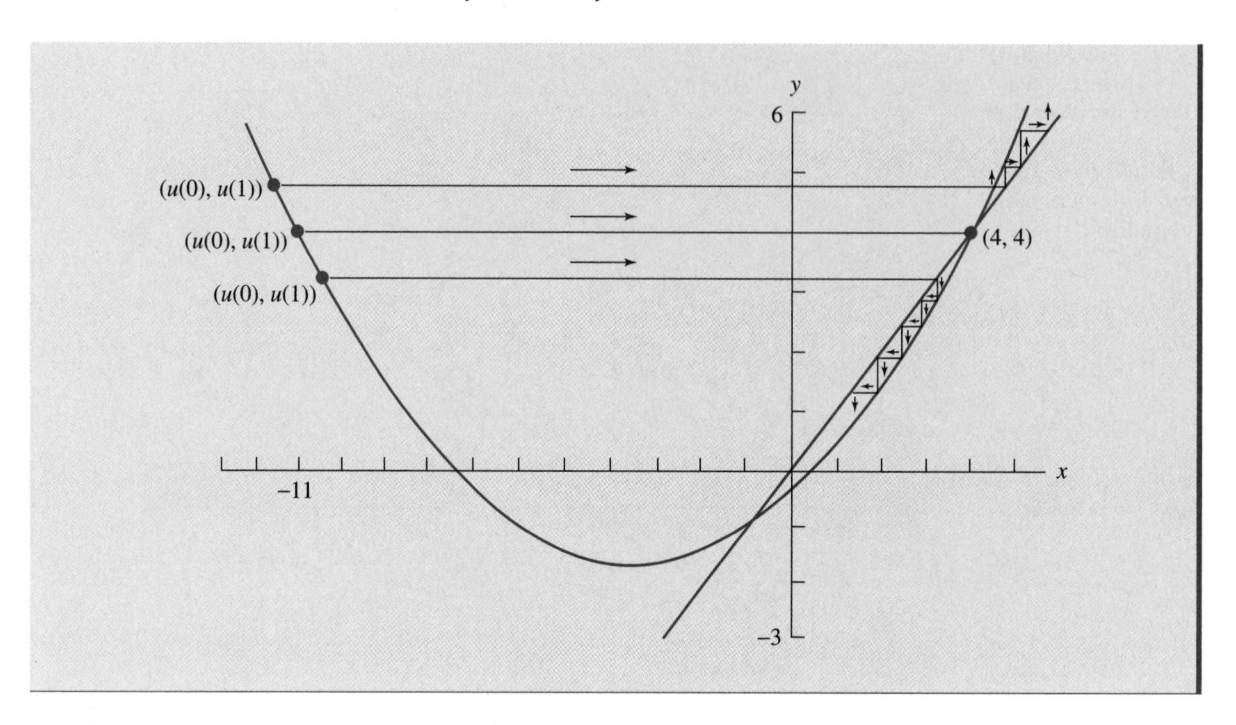

FIGURE 5.22 Graphs of $y = 0.1x^2 + 0.7x - 0.4$ and $y = x$. Webs indicate that the maximum interval of stability is $-11 < u(0) < 4$.

to make the figure much bigger to see what we want. It is simpler to begin the webs at the points $(u(0), u(1))$.

The web that begins with $u(0) < -11$ goes toward infinity. The web beginning with $-11 < u(0)$ goes toward the point $(-1, -1)$. The web beginning with $u(0) = -11$ goes directly to the point $(4, 4)$ and remains there. You should try drawing other webs until you understand why the ones with $-11 < u(0) < 4$ result in $u(n)$ going toward $E = -1$. Thus, the maximum interval of stability is $(-11, 4)$.

Note that for the function $y = 0.1x^2 + 0.7x - 0.4$, if $x = -11$ or $x = 4$, then $y = 4$. So to find the maximum interval of stability in this case, we could have solved the equation

$$4 = 0.1x^2 + 0.7x - 0.4$$

for x.

Let's consider an example that sheds some light on semistable equilibrium values.

Example 5.9

Let's reconsider the dynamical system

$$u(n) = u^2(n-1) - 3u(n-1) + 4$$

that we first studied in Section 5.3. This dynamical system has only $E = 2$ as an equilibrium value. Figure 5.23 shows 2 webs for the function

$$y = x^2 - 3x + 4$$

Note that the web with $u(0) < 2$ goes toward $(2, 2)$ but the web with $2 < u(0)$ goes away from $(2, 2)$. This is because the line $y = x$ is tangent to the function $y = x^2 - 3x + 4$ at the point $(2, 2)$. You can also use the horizontal line $y = 2$ to determine that the maximum interval of stability for $E = 2$ is $1 \leq u(0) \leq 2$. ■

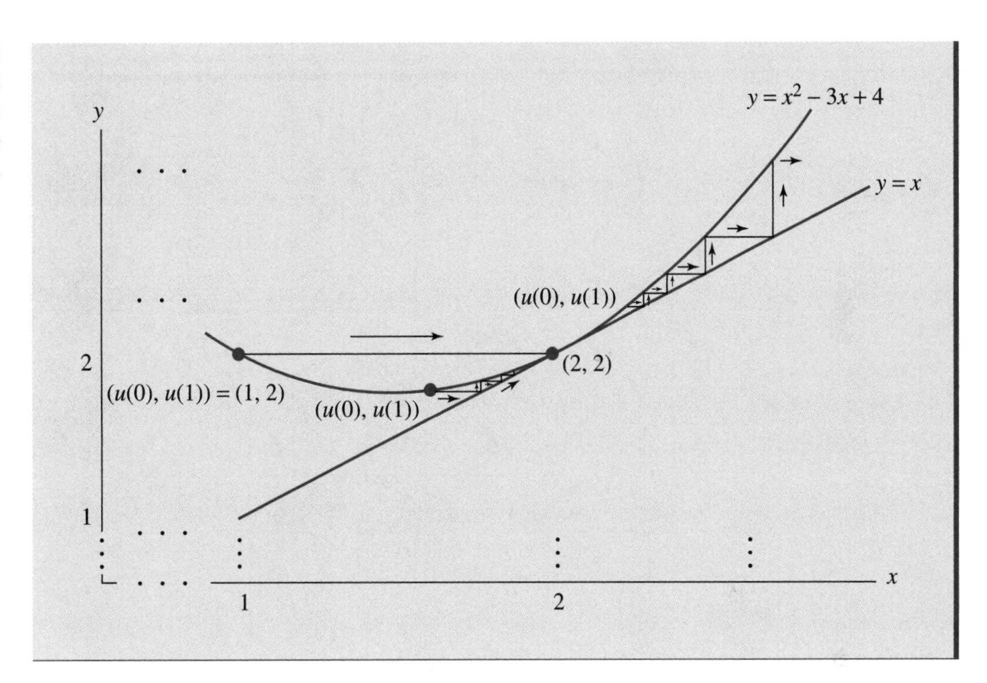

FIGURE 5.23 Web analysis showing a semistable equilibrium.

When the nonlinear dynamical system is more complex, it can be quite difficult to find the maximum interval of stability, even using web analysis.

Let's consider an example of web analysis to see what can cause both equilibrium values to be unstable.

Example 5.10 Consider dynamical system

$$u(n) = 3.1u(n-1) - u^2(n-1)$$

of Example 5.6. This dynamical system has 2 unstable equilibrium values, $E = 0$ and $E = 2.1$. Figure 5.24 displays a web starting at P near $(2.1, 2.1)$.

Note how the web spirals away from the point of intersection toward the square bounded by the four points $(1.73, 1.73)$, $(1.73, 2.37)$, $(2.37, 2.37)$, and $(2.37, 1.73)$. This indicates that the 2 values $E_1 = 1.73$ and $E_2 = 2.37$ form a 2-cycle.

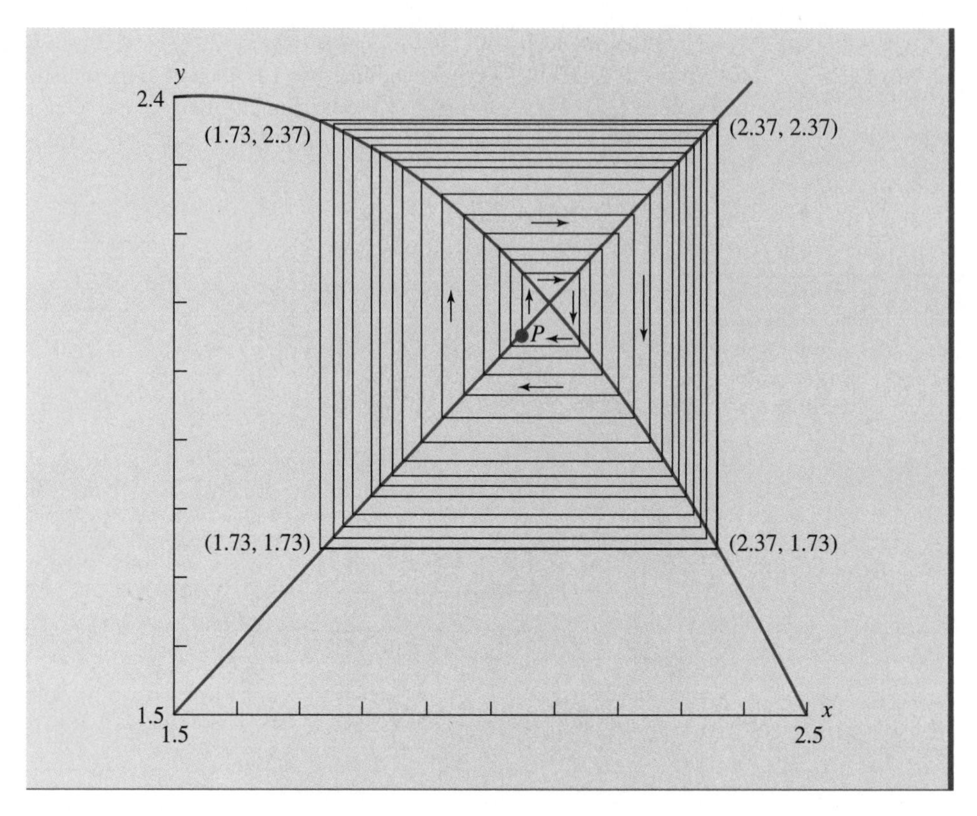

FIGURE 5.24 Web beginning near equilibrium point (2.1, 2.1) spirals out toward a square, eventually repeating same values.

A web beginning with $u(0)$ close to 0, but positive, will eventually spiral out toward the same square as seen in Figure 5.24. ■

Figure 5.24 indicates why the horizontal and vertical lines between the function and $y = x$ are called "webs." Spirals such as these occur near equilibrium values where the function is decreasing. Steps, such as those in earlier examples, occur when the function is increasing near the equilibrium value.

For most purposes, the web is only to indicate a general pattern in the behavior of the $u(n)$ values, so it is sufficient to draw the webs by hand on a graph that was generated by a computer or calculator. In fact, I encourage you to draw them by hand so that you develop a better insight into how webs work and how they relate to solutions to dynamical systems.

Suppose you are using web analysis to study the stability of an equilibrium value E. It is often helpful to at first generate only a portion of the graph that is near the point (E, E) and draw webs on that portion of the graph. Then, to get a more global picture of how solutions behave for a variety of initial values, you can generate graphs that have a larger viewing window. This is what I did in Figures 5.21 and 5.22. You may need to generate several different graphs for the same function to get a complete idea of how solutions behave. Don't be afraid to explore.

5.4 Problems

1. A web is displayed on the left of the following graphs.

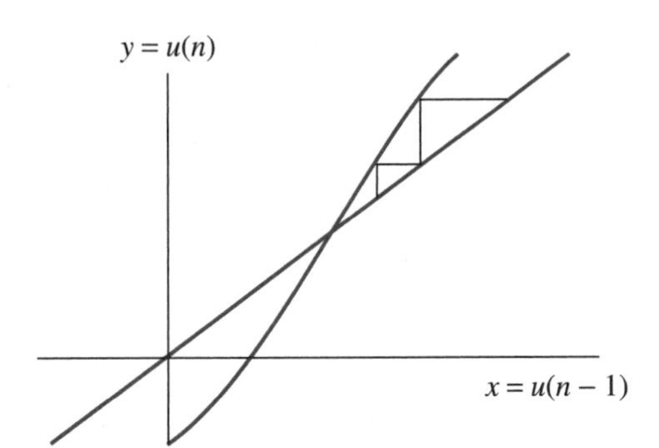

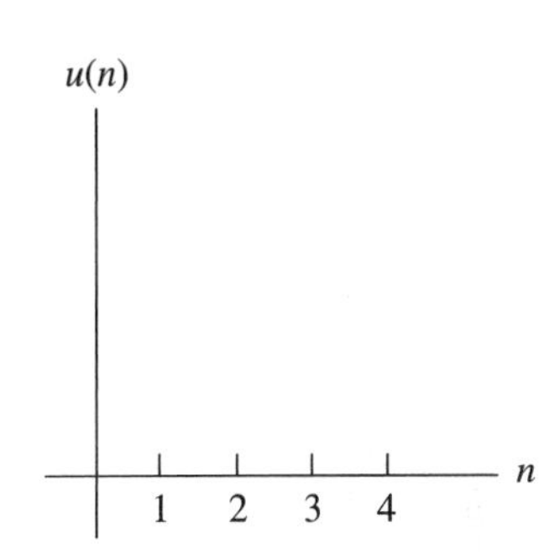

a. Draw arrows on the left graph indicating the direction of the web.
b. Assuming the web on the left graph began at the point $(u(0), u(0))$, plot the points $(0, u(0))$, $(1, u(1))$ and $(2, u(2))$ on the right graph.

2. A web is displayed on the left of the following graphs.

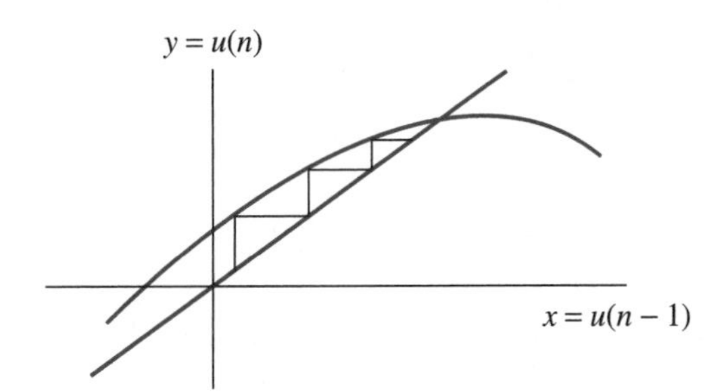

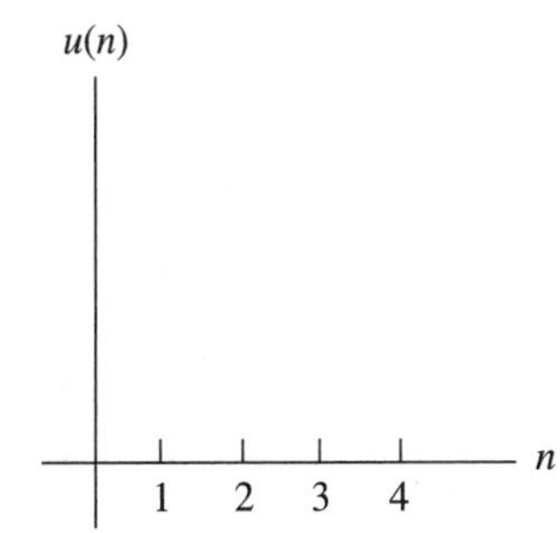

a. Draw arrows on the left graph indicating the direction of the web.
b. Assuming the web on the left graph began at the point $(u(0), u(0))$, plot the points $(0, u(0))$, $(1, u(1))$, $(2, u(2))$, and $(3, u(3))$ on the right graph.

3. A web is displayed on the left of the following graphs.

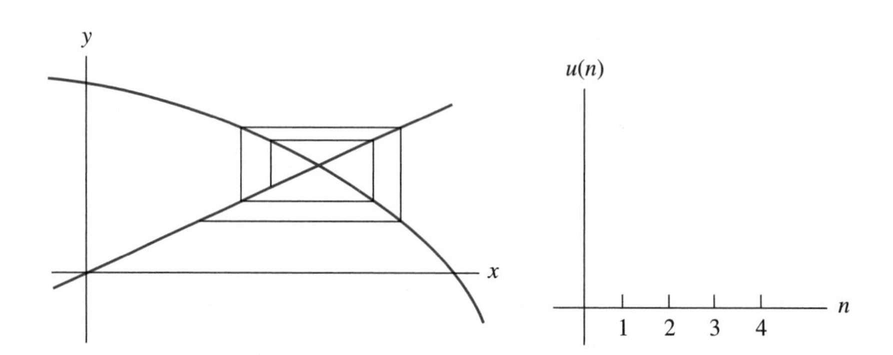

a. Draw arrows on the left graph indicating the direction of the web.

b. Assuming the web on the left graph began at the point $(u(0), u(0))$, plot the points $(0, u(0))$, $(1, u(1))$, $(2, u(2))$, $(3, u(3))$, and $(4, u(4))$ on the right graph.

4. A web is displayed on the left of the following graphs.

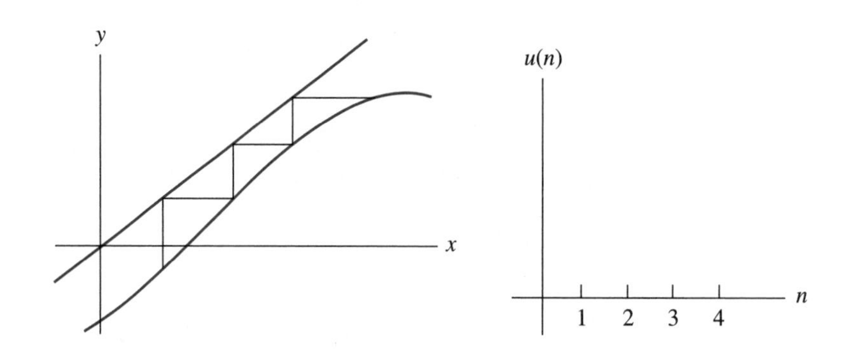

a. Draw arrows on the left graph indicating the direction of the web.

b. Assuming the web on the left graph began at the point $(u(0), u(1))$, plot the points $(1, u(1))$, $(2, u(2))$, $(3, u(3))$, and $(4, u(4))$ on the right graph.

5. Sketch a web on the left of the following graphs, starting at the indicated point. Does it indicate that the equilibrium value is stable or unstable? Use the web to plot the points $(0, u(0))$, $(1, u(1))$, $(2, u(2))$, and $(3, u(3))$ on the right graph.

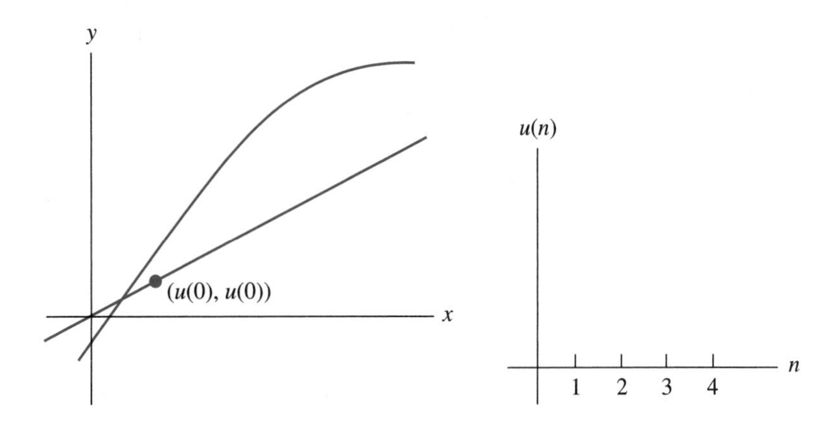

6. Sketch a web on the left of the following graphs, starting at the indicated point. Does it indicate that the equilibrium value is stable or unstable? Use the web to plot the points $(0, u(0))$, $(1, u(1))$, $(2, u(2))$, and $(3, u(3))$ on the right graph.

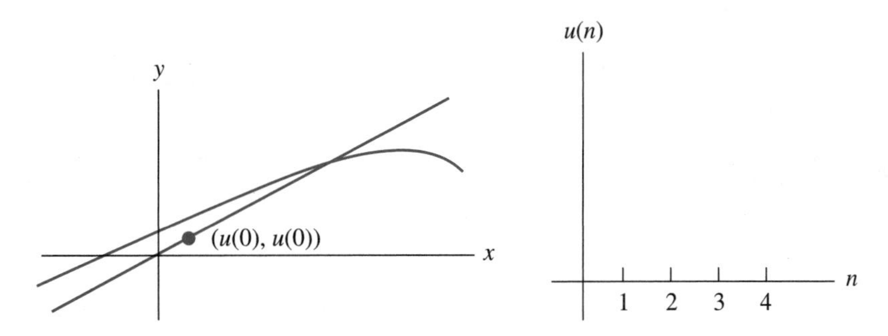

7. Sketch a web on the left of the following graphs, starting at the indicated point. Use the web to plot the points $(1, u(1))$, $(2, u(2))$, $(3, u(3))$, and $(4, u(4))$ on the right graph.

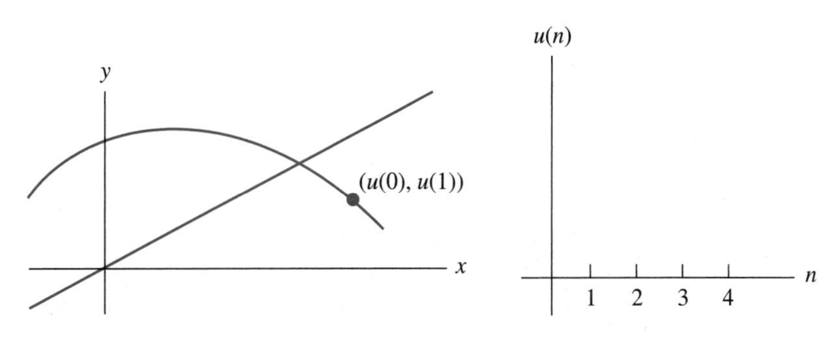

8. Sketch a web on the left of the following graphs, starting at the indicated point. Use the web to plot the points $(0, u(0))$, $(1, u(1))$, $(2, u(2))$, and $(3, u(3))$ on the right graph.

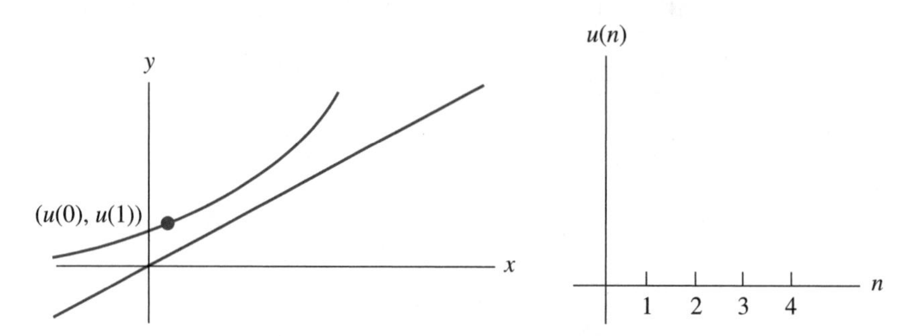

9. Consider the dynamical system

$$u(n) = 0.1u^2(n-1) + 0.5u(n-1).$$

You found 1 stable and 1 unstable equilibrium value for this dynamical system in Problem 1 of Section 5.3.

a. Draw webs near each equilibrium value to confirm your results.

b. Use web analysis to find the maximum interval of stability for the stable equilibrium value. This will also involve solving a quadratic to find the end points of the interval.

10. Consider the dynamical system

$$u(n) = 0.5u^2(n-1) - 0.5u(n-1) + 1.$$

You found 1 stable and 1 unstable equilibrium value for this dynamical system in Problem 2 of Section 5.3.

a. Draw webs near each equilibrium value to confirm your results.

b. Use web analysis to find the maximum interval of stability for the stable equilibrium value. This will also involve solving a quadratic to find the end points of the interval.

11. Consider the dynamical system

$$u(n) = 0.3u^2(n-1) + 0.4u(n-1) - 0.9.$$

You found 1 stable and 1 unstable equilibrium value for this dynamical system in Problem 3 of Section 5.3.

a. Draw webs near each equilibrium value to confirm your results.

b. Use web analysis to find the maximum interval of stability for the stable equilibrium value. This will also involve solving a quadratic to find the end points of the interval.

12. Consider the dynamical system

$$u(n) = 0.5u^2(n-1) + 2u(n-1) + 0.5.$$

You found 1 equilibrium value for this dynamical system in Problem 4 of Section 5.3.

a. Draw webs near that equilibrium value to confirm it is semistable from below.

b. Use web analysis to find the maximum interval of stability for this equilibrium value. This will also involve solving a quadratic to find the end points of the interval.

13. Consider the dynamical system

$$u(n) = -0.4u^2(n-1) + 3.4u(n-1) - 2.$$

You found 1 stable and 1 unstable equilibrium value for this dynamical system in Problem 5 of Section 5.3.

a. Draw webs near each equilibrium value to confirm your results.

b. Use web analysis to find the maximum interval of stability for the stable equilibrium value.

14. Consider the dynamical system

$$u(n) = -2.2u^2(n-1) + 7.6u(n-1) - 4.4.$$

You found 2 unstable equilibrium values for this dynamical system in Problem 6 of Section 5.3.

a. Draw webs near each equilibrium value to confirm your results.

b. Use web analysis to help predict the behavior of the solution $u(n)$ for any $u(0)$-value.

15. Consider the dynamical system

$$u(n) = -u^2(n-1) + 2u(n-1) - 0.25.$$

You found 1 equilibrium value for this dynamical system in Problem 7 of Section 5.3.

a. Draw webs near that equilibrium value to confirm it is semistable from above.

b. Use web analysis to find the maximum interval of stability for this equilibrium value. This will also involve solving a quadratic to find the end points of the interval.

16. Consider the dynamical system

$$u(n) = \frac{3}{4 - u(n-1)}.$$

You found 1 stable and 1 unstable equilibrium value for this dynamical system in Problem 8 of Section 5.3.

a. Draw webs near each equilibrium value to confirm your results.

b. Draw webs for $u(0) = -1, 2.5, 3.5,$ and 7. In each case, what happens to the points $(u(n-1), u(n))$?

c. Try drawing a web for $u(0) = 3.25$. What problem occurs? [It might help to try computing the values for $(u(1), u(2))$, and so on.]

17. Consider the dynamical system

$$u(n) = \frac{4 - u(n-1)}{2 + u(n-1)}.$$

You found 1 stable and 1 unstable equilibrium value for this dynamical system in Problem 9 of Section 5.3.

a. Draw webs near each equilibrium value to confirm your results.

b. Draw a web with $u(0) = -3$ and use it to see what eventually happens to $u(n)$. What is the behavior of $u(n)$ for "most" values of $u(0)$?

c. Try drawing a web using $u(0) = -8$. It might help to try computing $u(1)$, $u(2)$, and so on. What problem occurs? What happens to the web using $u(0) = -20/7$?

18. Consider the dynamical system

$$u(n) = \frac{16}{u^{\frac{1}{3}}(n-1)} \quad \text{or} \quad u(n) = \frac{16}{\sqrt[3]{u(n-1)}}.$$

You found 2 stable equilibrium values for this dynamical system in Problem 10 of Section 5.3.

a. Draw webs near each equilibrium value to confirm your results.

b. Use web analysis to find the maximum interval of stability for each stable equilibrium value.

19. Consider the dynamical system

$$u(n) = 0.3u^3(n-1) + 0.3u^2(n-1) + 0.4u(n-1).$$

You found 1 stable and 2 unstable equilibrium values for this dynamical system in Problem 12 of Section 5.3.

a. Draw webs near each equilibrium value to confirm your results.

b. Use web analysis to find the maximum interval of stability for the stable equilibrium value.

20. Consider the dynamical system

$$u(n) = -0.2u^3(n-1) + 0.8u^2(n-1) + 0.2u(n-1).$$

You found 1 stable and 1 semistable equilibrium value for this dynamical system in Exercise 13 of Section 5.3. Draw webs near each equilibrium value to confirm your results.

21. Consider the dynamical system

$$a(n) = a(n-1) - \frac{10a(n-1)}{4.2 + a(n-1)} + d$$

that determined the grams of alcohol in the body after n hours, assuming the person drinks d grams of alcohol each hour.

a. Construct a web for this dynamical system using $d = 7$ and $u(0) = 7$ to see that 9.8 is a stable equilibrium value.

b. Construct a web for this dynamical system using $d = 14$ and $u(0) = 14$ to see why $u(n)$ goes to infinity in this case.

Project 1. Consider the dynamical system

$$u(n) = -0.2u^3(n-1) + 0.8u^2(n-1) + 0.4u(n-1).$$

You found 2 stable equilibrium values and 1 unstable equilibrium value for this dynamical system in Problem 11 of Section 5.3.

a. Draw webs near each equilibrium value to confirm your results.

b. Draw a web using $u(0) = -2$. Why does this web indicate that there might be some difficulty finding all of the $u(0)$-values where $u(n)$ goes to the equilibrium value $E = 3$? Experiment by drawing webs for a variety of $u(0)$-values. What patterns do you notice? Can you find a value for $u(0)$ for which $u(n)$ goes to infinity? Can you find a 2-cycle?

Project 2. Consider the dynamical system

$$u(n) = 0.1u^2(n-1) + 0.5u(n-1).$$

In Problem 1 of Section 5.3, you found that the equilibrium value $E = 0$ is stable and that the rate at which $u(n)$ approaches 0 is 0.5, meaning that

$$\frac{u(n)}{u(n-1)} \approx 0.5$$

for moderately large values of n.

a. Draw a web using the function

$$y = 0.1x^2 + 0.5x$$

and $y = x$ with a viewing window of $-0.1 < x < 0.1$ and $-0.1 < y < 0.1$, and using $u(0) = 0.08$.

b. Estimate the "slope" of the parabola $y = 0.1x^2 + 0.5x$ at $(0, 0)$ by finding the slope of the line connecting $(0, 0)$ to some other point on the parabola that is close to the origin. You might try finding the estimate using several different points close to the origin. What do you think the slope of the parabola is at the origin? How does this compare to the rate?

Project 3. Consider the dynamical system

$$u(n) = 0.3u^2(n-1) + 0.4u(n-1) - 0.9.$$

In Problem 3 of Section 5.3, you found that the equilibrium value $E = -1$ is stable and that the rate at which $u(n)$ approaches -1 is -0.2, meaning that

$$\frac{u(n) - (-1)}{u(n-1) - (-1)} \approx -0.2$$

for moderately large values of n.

a. Draw a web using the function

$$y = 0.3x^2 + 0.4x - 0.9$$

and $y = x$ with a viewing window of $-1.1 < x < -0.9$ and $-1.1 < y < -0.9$, and using $u(0) = -0.91$.

b. Estimate the "slope" of the parabola $y = 0.3x^2 + 0.4x - 0.9$ at $(-1, -1)$ by finding the slope of the line connecting $(-1, -1)$ to some other point on the parabola that is close to that point. You might try finding the estimate using several different points. What do you think the slope of the parabola is at $x = -1$? How does this compare to the rate?

6 Population Dynamics

6.1 Introduction to Population Growth

Mismanagement of renewable resources, especially in the fishing industry, has resulted in serious problems. One well-known problem was the collapse of the Peruvian anchovy fishery in the 1970s. As can be seen from Figure 6.1, there was a major reduction in anchovy catch which caused financial problems for Peru, and could have led to a major shortage of anchovy pizzas. All kidding aside, it is now recognized that without some controls, there will be major shortages of many important renewable resources and extinction of numerous species of animals and plants.

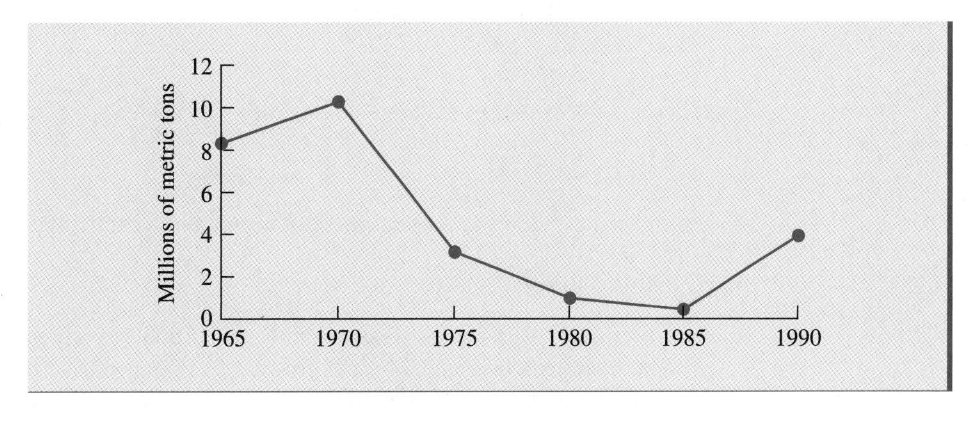

FIGURE 6.1 Peruvian anchovy catch for selected years.

In this chapter, we are going to apply our knowledge of dynamical systems to study models of population growth. We will start in Section 6.2 with a few simple assumptions about how the size of a population changes and analyze the consequences of these assumptions. In Section 6.3, we will analyze new systems that are developed from improving our assumptions about the growth of populations. In Section 6.4, we will graphically study how the effects of hunting and harvesting can effect the size of a population. This will include a study of sustainable harvesting of renewable resources, an important topic in resource management. Sustainable harvesting is also related to many environmental issues, such as the risk of extinction for species. In Section 6.5, we will algebraically study the effects of harvesting. In Section 6.6, we will investigate how economic aspects can make sustainable harvesting difficult to obtain.

6.2 The Logistic Model for Population Growth

We begin with a very simplistic model for the growth of a population. Consider a population of some species, say snipes. We will assume that the reproductive rate for snipes is $b = 2.2$, meaning that on average each snipe alive at the beginning of the year gives birth to $2\frac{2}{10}$ new snipes each year, or that for every 10 snipes, 22 births occur. Let's assume the death rate for snipes is $d = 0.2$, meaning that 20% of the snipes that were alive at the beginning of the year die during the year. Let n represent the beginning of the nth year. Let $u(n-1)$ be the number of snipes alive at the beginning of year $n-1$. To find the new population of snipes after one year, $u(n)$, add the number of births during year $n-1$ to $u(n-1)$ and subtract the number of deaths. This can be seen in the flow diagram in Figure 6.2.

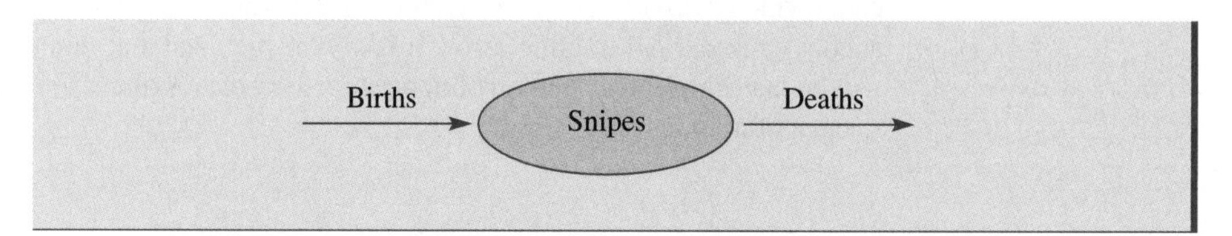

FIGURE 6.2 Flow diagram for population of snipes.

Thus, $u(n) = u(n-1) + 2.2u(n-1) - 0.2u(n-1)$, or

$$u(n) = 3u(n-1)$$

From Section 3.3, we know that $u(n)$ can be written as explicitly as

$$u(n) = 3^n u(0)$$

and that the population grows exponentially, with a 200% yearly increase in size each year.

In developing this model for the growth of snipes, we have assumed that the number of births during year $n-1$ is a **fixed proportion of the population size** at the beginning of that year. Likewise, we have assumed that the number of deaths during year $n-1$ is a fixed proportion of $u(n-1)$. To find the new population, we add the number of births to and subtract the number of deaths from $u(n-1)$. This gives $u(n) = u(n-1) + bu(n-1) - du(n-1)$, or

$$u(n) = u(n-1) + ru(n-1)$$

where $r = b - d$. This can be visualized in the flow diagram of Figure 6.3.

FIGURE 6.3
Simplified flow diagram for snipe population.

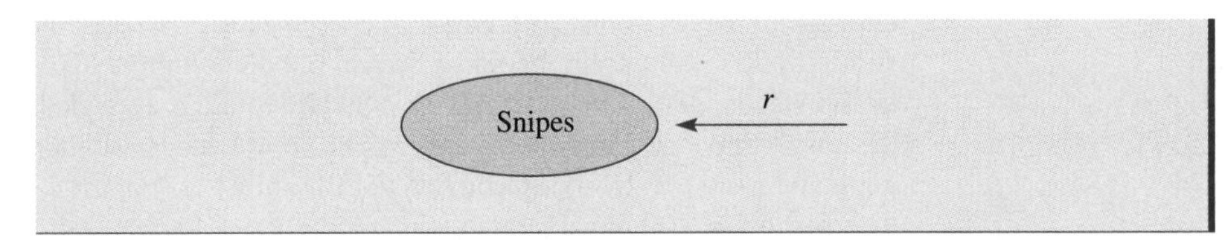

DEFINITION

The fraction by which a population increases in 1 unit of time is the **growth rate** for the population.

We will denote the growth rate for a population by the variable r. If the size of the population is u and the growth rate is r, then the actual change or **growth** g in the population is given by the equation

$$\text{change} = \text{rate} \times \text{population} \quad \text{or} \quad g = ru$$

This is similar to the relationship

$$\text{distance} = \text{rate} \times \text{time} \quad \text{or} \quad d = rt$$

as well as the relationship

$$\text{elimination} = \text{rate} \times \text{amount} \quad \text{or} \quad e = ra$$

which was used in Section 5.2 to model the change in amounts of chemicals in a person's body.

This means that the dynamical system for population growth is

$$u(n) = u(n-1) + ru(n-1)$$

For example, if $r = 0.1$, then the population grows by 10% per year and the dynamical system is

$$u(n) = u(n-1) + 0.1u(n-1) \text{ or } u(n) = 1.1u(n-1)$$

If $u(0) = 1000$, then $u(1) = 1000 + 0.1(1000) = 1100$ and $g = 100$ individuals the first year. Similarly, $u(2) = 1100 + 0.1(1100) = 1210$ and $g = 110$ the second year. Note that while the growth rate is a constant 10%, the actual growth or change in population size is not constant, but is increasing as the population size increases.

The solution to the dynamical system $u(n) = (1+r)u(n-1)$, you remember, is

$$u(n) = (1+r)^n u(0) \quad \text{or} \quad u(n) = 1.1^n u(0)$$

in this case. Then $u(10) = 2594$, $u(40) = 45{,}259$, and $u(100) = 13{,}780{,}612$. While the model seems to make sense, it predicts that the size of the population eventually gets unrealistically large. In fact, it predicts that $u(n)$ goes exponentially toward infinity. This is similar to the Malthusian model of human population growth. This model predicted that the human population would grow exponentially until it exceeded the carrying capacity of the environment, at which point catastrophe would occur. Although it is unclear if human population will eventually exceed the world's ability to support it, it has become clear that the growth rate for human populations in different countries does not remain constant, but can depend on a variety of circumstances.

In Section 2.4, we considered a similar population growth model for a population of Bison, except that we assumed the population could be broken into several distinct age groups, and that there was a separate fixed, constant birth and death rate for each age group. As we saw in that section, the population size of each age group eventually grew at the same rate; that is, the growth rate became constant for each age group. This results again in the population growing exponentially. But we know that no population can sustain exponential growth forever.

Does this mean these models are wrong? The answer depends on what information we want from the model and the particular circumstances of the population. When the population size is relatively small, population sizes often appear to grow exponentially and growth rates often remain approximately the same. But a problem with the model is that it assumes that birth and death rates remain constant even though an increased size in the population may put a strain on the available resources.

There are 3 variables, time n, population size u, and growth rate r. Population size depends on time. In Section 5.2, we assumed that the elimination rate r depended on the amount of chemical a. Similarly, in modeling populations, we assume the growth rate r depends on the population size u. A graphical representation of the constant growth rate model is seen in Figure 6.4. It assumes that the growth rate is the same for every population size. Different species would have different values for r, which would be represented by different horizontal lines.

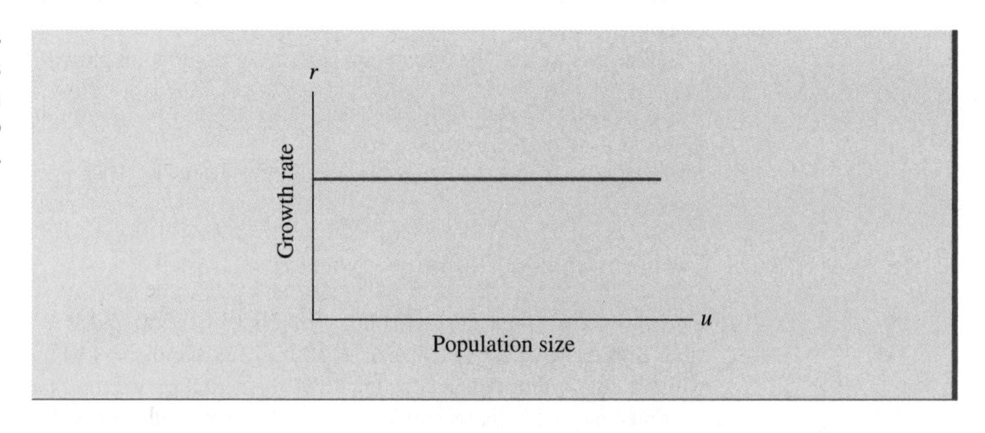

FIGURE 6.4 Growth rate is constant with respect to population size.

As we have discussed, a problem with this model of population growth is that **growth rates do not remain constant but change as the population size changes.** So a better model would be to replace the constant r with a variable growth rate $r(u)$. Here, we assume that r depends on the population size u. In other words, the growth rate is a function of the population size.

One reasonable assumption is that the **growth rate decreases as the population size increases.** There are several arguments for this assumption. As the population grows, overcrowding might cause the birth rate to go down. As the population grows, a lack of resources might lead to a higher death rate, which would decrease the growth rate. This is seen in the growth rate function shown in Figure 6.5.

FIGURE 6.5 Growth rate decreases as population size increases.

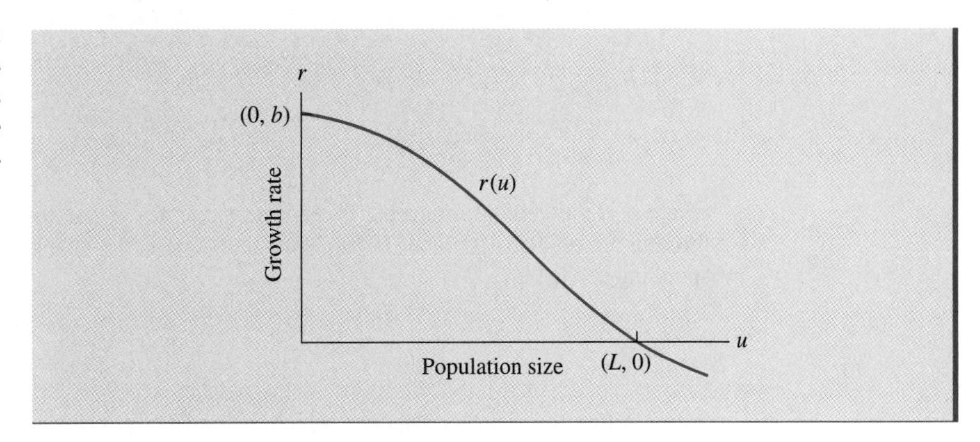

DEFINITION

The u-coordinate of the growth rate function where it decreases to 0 is called the **carrying capacity** of the environment.

The carrying capacity, L in Figure 6.5, is the population size that results in a 0 growth rate, meaning the population neither increases in size nor decreases in size but stays at a size of L. If the size of the population exceeds L, the growth rate is negative and the population must decrease in size because of a lack of resources.

DEFINITION

The largest r-value of the growth rate function is called the **intrinsic growth rate** for the population.

The intrinsic growth rate is the optimal growth rate for the species. If the growth rate function is decreasing, then the intrinsic growth rate is just the r-coordinate of the vertical intercept, b in Figure 6.5. This means that the intrinsic growth rate occurs when the population size is 0. While this is unrealistic, it implies that the optimal growth rate occurs when the population size is small. This seems reasonable, since there would be plenty of food and space available.

For simplicity, let's assume that $r(u)$ is a straight line, as seen in Figure 6.6. The equation for a line, given in slope-intercept form, is

$$r = mu + b$$

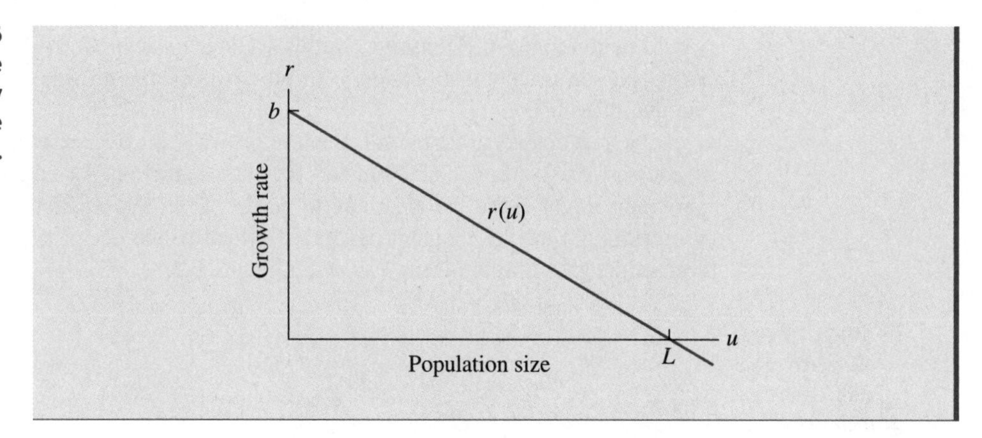

FIGURE 6.6 Growth rate decreases linearly as population size increases.

where b is the vertical intercept. To find the slope m, we use the fact that the line goes through the points $(L, 0)$ and $(0, b)$. Thus, the slope is the change in "y" over the corresponding change in "x," or

$$m = \frac{0 - b}{L - 0} = -\frac{b}{L}$$

We note that the slope is negative, which is implied from the fact that the growth rate decreases as the population size increases. Thus, the growth rate is given by

$$r = b - \frac{b}{L}u$$

Since the population size u depends on time n, then the growth rate r also depends on time, that is,

$$r(u(n)) = b - \frac{b}{L}u(n)$$

The dynamical system for population growth is

$$u(n) = u(n-1) + ru(n-1)$$

which, after substitution, becomes

$$u(n) = u(n-1) + \left(b - \frac{b}{L}u(n-1)\right)u(n-1)$$

Simplification gives the dynamical system

$$u(n) = (1+b)u(n-1) - \frac{b}{L}u^2(n-1).$$

which is often called the **logistic equation** for population growth.

Example 6.1

Suppose we have estimated that the intrinsic growth rate for a population is $b = 0.2$ and that the carrying capacity is $L = 8000$ of the species. Then the logistic equation is, after substitution and simplification,

$$u(n) = 1.2u(n-1) - 0.000025u^2(n-1)$$

To find the equilibrium values for this system, we solve

$$E = 1.2E - 0.000025E^2$$

The solutions are, naturally, $E = 0$ and $E = 8000$. These answers make sense. If the size of the population is 0, then the species is extinct and its population size remains at 0. If the size of the population is 8000, the carrying capacity, then the growth rate is 0, meaning that the population neither increases nor decreases in size. Thus, it is at equilibrium.

Time graphs for the population size are given in Figure 6.7. These graphs correspond to initial population sizes of $u(0) = 10$, $u(0) = 100$, $u(0) = 7000$, and $u(0) = 10,000$. They indicate that $E = 8000$ is a stable equilibrium population size and $E = 0$ is an unstable equilibrium population size.

FIGURE 6.7 Graphs of $(n,u(n))$ for several initial values, $u(0)$.

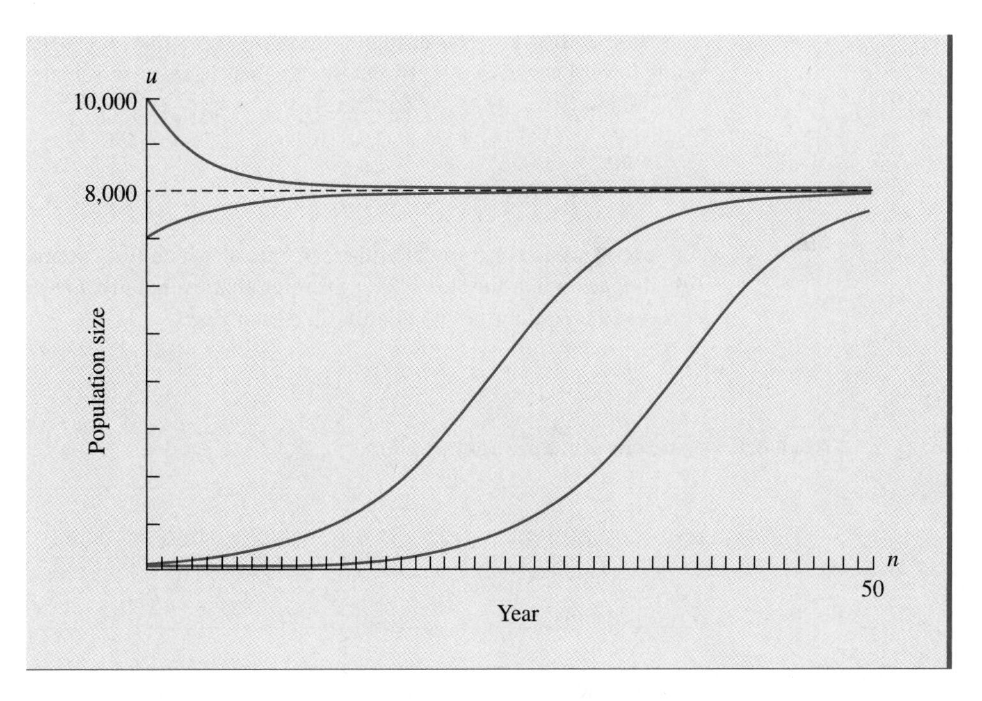

Web analysis for this logistic equation is seen in Figure 6.8. The initial population size is $u(0) = 5000$. The web goes toward the point $(8000, 8000)$, implying that $E = 8000$ is stable. The web also goes away from $(0, 0)$, implying that $E = 0$ is unstable.

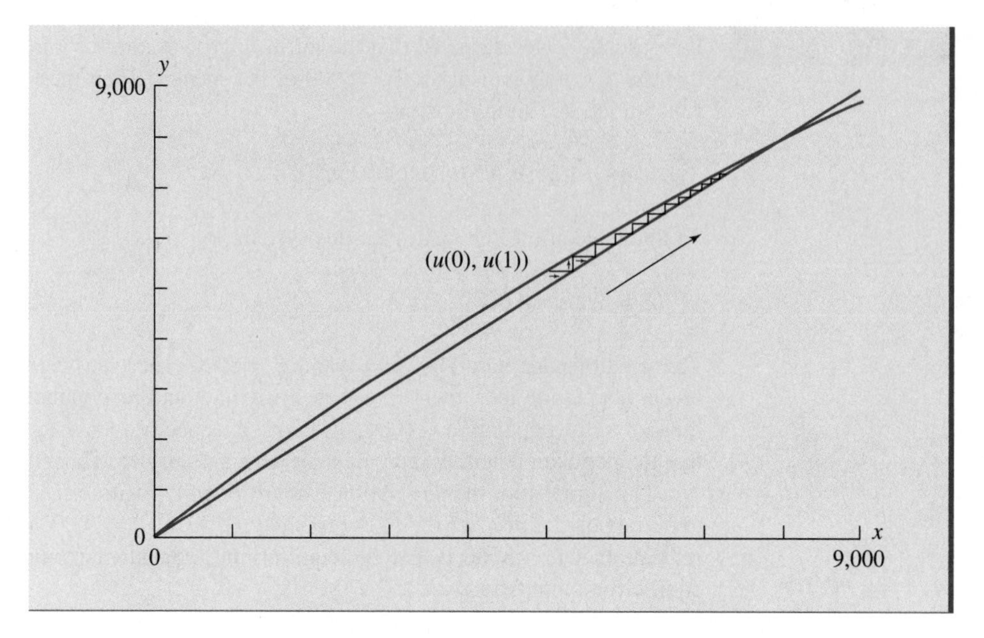

FIGURE 6.8 Web analysis for a population satisfying the logistic equation, with $b = 0.2$ and $L = 8000$.

To really see that $E = 0$ is unstable, we should try values for $u(0) < 0$. This results in $u(n)$ going toward negative infinity, but these values have no physical significance.

Using $u(0) = 1000$, we compute

$$\frac{u(n) - 8000}{u(n-1) - 8000}$$

for several values of n to determine the rate at which $u(n)$ approaches 8000. From Table 6.1, it appears that the rate is 0.8, meaning that eventually, $u(n)$ gets approximately 20% closer to the equilibrium population size each year. ■

TABLE 6.1 **Ratios as $u(n)$ approaches 8000.**

$n =$	1	2	5	20	50	70
$\frac{u(n)-8000}{u(n-1)-8000} =$	0.975	0.971	0.954	0.827	0.800	0.800

There are several questions to be asked. The first question is, What are the equilibrium values for the logistic equation? This question is easy to answer. We just solve the equation

$$E = (1+b)E - \frac{b}{L}E^2$$

for E. Subtracting E from both sides of the equation and factoring out E and b gives

$$0 = bE\left(1 - \frac{E}{L}\right)$$

Setting each factor equal to 0 gives the two solutions $E = 0$ and $E = L$. These should not be a surprise, considering the context of the equation.

Among the other questions we could ask are, What is an expression for the function $u(n)$ in terms of n? For what values of b and L is the equilibrium value $E = L$ stable? For what values of b and L is the equilibrium value $E = 0$ stable? When L is a stable equilibrium value, what is the maximum interval of stability; that is, for which initial population sizes $u(0)$ does $u(n)$ go to that equilibrium value?

Let me answer the first question. It is usually impossible to find an algebraic expression for the function that satisfies a nonlinear dynamical system. This includes the relatively simple logistic equation. We can compute $u(0)$, $u(1)$, ..., $u(n)$ for large values of n, so the function exists. But a "simple" algebraic expression for the function $u(n)$ in terms of n does not exist except in a few special cases.

We will investigate the stability of the equilibrium values experimentally in the exercises, including the relationship between stability and the values for b and L. In particular, we will discover that when b is reasonably small, then L will be stable.

REMARK A calculus technique for determining if an equilibrium value is stable or unstable involves finding the derivative of the function

$$y = (1+b)x - \frac{b}{L}x^2$$

and evaluating the derivative at the equilibrium value. If that derivative is between -1 and 1, then the equilibrium value is stable. If you know how to compute derivatives, you are encouraged to try this approach and compare the results to the results found using exploration.

While it is generally difficult to find the maximum interval of stability for a nonlinear dynamical system, it is not difficult to find this interval for the logistic equation, especially using the web analysis of Section 5.4. You will deal with this in the problems.

6.2 Problems

1. Let $b = 0.1$. Then the logistic equation becomes

$$u(n) = 1.1u(n-1) - \frac{0.1}{L}u^2(n-1)$$

a. Suppose that $L = 1000$. Then the equilibrium values are 0, which is unstable, and 1000, which is stable. Find the rate at which $u(n)$ approaches 1000.

b. Suppose that somehow, the carrying capacity has increased to $L = 2000$. Rewrite the logistic equation using $b = 0.1$ and $L = 2000$. Find the rate at which $u(n)$ approaches the stable equilibrium value 2000.

c. Pick your own positive value for L. Rewrite the logistic equation using your value for L. Find the rate at which $u(n)$ approaches this value. How does the value for L effect the stability of the equilibrium value L?

2. Let $b = 0.2$. Then the logistic equation becomes

$$u(n) = 1.2u(n-1) - \frac{0.2}{L}u^2(n-1)$$

a. Suppose that $L = 1000$. Then the equilibrium values are 0, which is unstable, and 1000, which is stable. Find the rate at which $u(n)$ approaches 1000.

b. Suppose that somehow, the carrying capacity has increased to $L = 2000$. Rewrite the logistic equation using $b = 0.2$ and $L = 2000$. Find the rate at which $u(n)$ approaches the stable equilibrium value 2000.

c. Pick your own positive value for L. You can pick as small or as large a number as you want for L. Rewrite the logistic equation using your value for L. Find the rate at which $u(n)$ approaches this value. How does the value for L effect the stability of the equilibrium value L?

3. Let $b = 0.3$. Then the logistic equation becomes

$$u(n) = 1.3u(n-1) - \frac{0.3}{L}u^2(n-1)$$

a. Suppose that $L = 1000$. Then the equilibrium values are 0, which is unstable, and 1000, which is stable. Find the rate at which $u(n)$ approaches 1000.

b. Suppose that somehow, the carrying capacity has increased to $L = 2000$. Rewrite the logistic equation using $b = 0.3$ and $L = 2000$. Find the rate at which $u(n)$ approaches the stable equilibrium value 2000.

c. Pick your own positive value for L. Rewrite the logistic equation using your value for L. Find the rate at which $u(n)$ approaches this value.

4. Suppose that $0 < b < 1$. From working Problems 1 through 3, make an estimate for the rate at which $u(n)$ approaches the stable equilibrium value L for the logistic equation. Test your hypothesis by picking your own value between 0 and 1 for b and your own value for L. Find the rate at which $u(n)$ approaches L. Does this rate agree with your hypothesis?

5. Let $b = 1.4$. Suppose that $L = 1000$. Then the equilibrium values are 0, which is unstable, and 1000, which is stable. Find the rate at which $u(n)$ approaches 1000. Does this rate agree with the rate predicted from Problem 4?

6. Let $b = 2.4$ and $L = 1000$. Find the stability of the equilibrium value 1000. Make a graph of the values $(n, u(n))$ for $u(0) = 900$. What is the behavior of this graph? Make a graph using other values for $u(0)$. What is the behavior? Draw a web for the dynamical system. What does it indicate?

7. Let $L = 1000$ in the logistic equation. Experimentally find the largest value you can for b in which 1000 appears to be a stable equilibrium value. Get b to the nearest tenth.

8. Let $b = 0.1$ and $L = 1000$ so that the logistic equation becomes

$$u(n) = 1.1u(n-1) - 0.0001u^2(n-1)$$

We know that 1000 is a stable equilibrium value. Find the maximum interval of stability for 1000. It will help to draw several webs for this dynamical system, using different values for $u(0)$. What is $u(1)$ when $u(0)$ equals the right end point of the maximum interval of stability?

9. Let $b = 0.2$ and $L = 1000$. We know that 1000 is a stable equilibrium value. Find the maximum interval of stability for 1000. It will help to draw several webs for this dynamical system, using different values for $u(0)$. What is $u(1)$ when $u(0)$ equals the right end point of the maximum interval of stability?

10. Let $b = 0.5$ and $L = 1000$. We know that 1000 is a stable equilibrium value. Find the maximum interval of stability for 1000. It will help to draw several webs for this dynamical system, using different values for $u(0)$. What is $u(1)$ when $u(0)$ equals the right end point of the maximum interval of stability?

11. From working Problems 8, 9, and 10, you should have determined that when $u(0)$ equals the value of the right end of the maximum interval of stability, $u(1) = 0$. Solve the equation

$$u(1) = 0 = (1+b)u(0) - \frac{b}{1000}u^2(0)$$

for $u(0)$ in terms of b to get a formula for the value of the right end of the maximum interval of stability. Does this formula give the same values that you found in Problems 8, 9, and 10? Test your formula when $b = 1$.

6.3 Nonlinear Growth Rates

In Section 6.2, we assumed that the growth rate for a population remained constant. Then we assumed that the growth rate decreased linearly as the size of the population increased. This is more realistic than a constant growth rate, but it ignores other factors that may affect the population. In this section, we are going to consider a variety of possible growth rate functions that can help model other factors affecting population growth.

First, we consider growth rates that decrease, but not linearly. We might expect the growth rate to remain close to the intrinsic growth rate for small population sizes, but then decrease more rapidly as the size of the population approaches the carrying capacity. Such a growth rate is seen in Figure 6.9. When we use a nonlinear function for our growth rate, we will call the resulting dynamical system a **generalized logistic model** for population growth.

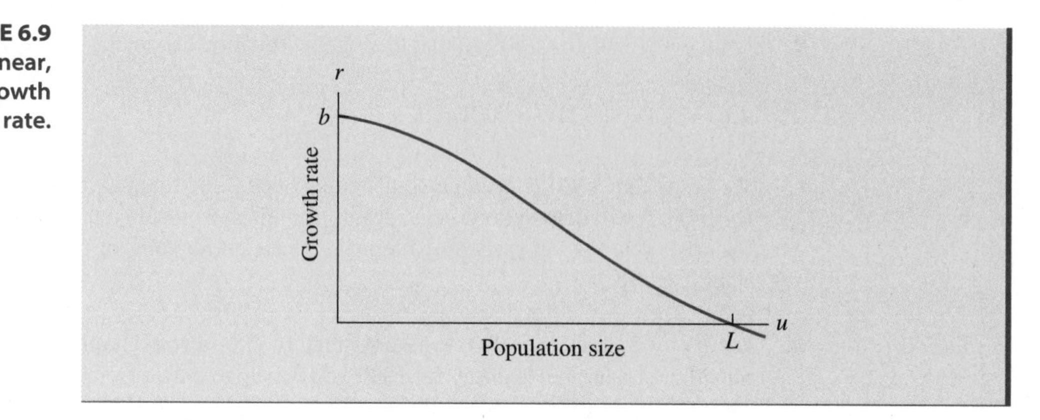

FIGURE 6.9 Nonlinear, decreasing growth rate.

We are going to begin by using parabolas for the growth rate functions. Recall that the equation for a parabola with vertex at the point (h, k) is of the form

$$y - k = a(x - h)^2$$

If $a > 0$, then the parabola opens upward and if $a < 0$, then the parabola opens downward. The size of $|a|$ determines whether the parabola is narrow or wide. In fact, the smaller the value of $|a|$, the wider the parabola.

We will use r instead of y for the dependent variable and u instead of x for the independent variable. This means the equation for the parabola will be of the form

$$r - k = a(u - h)^2$$

Example 6.2

Suppose we determine that the intrinsic growth rate for a population is $b = 0.3$, its carrying capacity is $L = 1000$, and the growth rate decreases as the population size increases. We want to develop a function for the growth rate $r(u)$ that has a shape similar to the graph in Figure 6.10. This can be accomplished by using one branch of a parabola with vertex at $(0, 0.3)$. This is a parabola with equation

$$r - 0.3 = a\,(u - 0)^2$$

which simplifies to

$$r = au^2 + 0.3$$

We want the parabola to go through the point $(1000, 0)$, so we substitute 1000 for u and 0 for r into this equation and solve for a. This gives

$$0 = 0.3 + a(1000)^2$$

FIGURE 6.10 Intrinsic growth rate is 0.3 and carrying capacity is 1000.

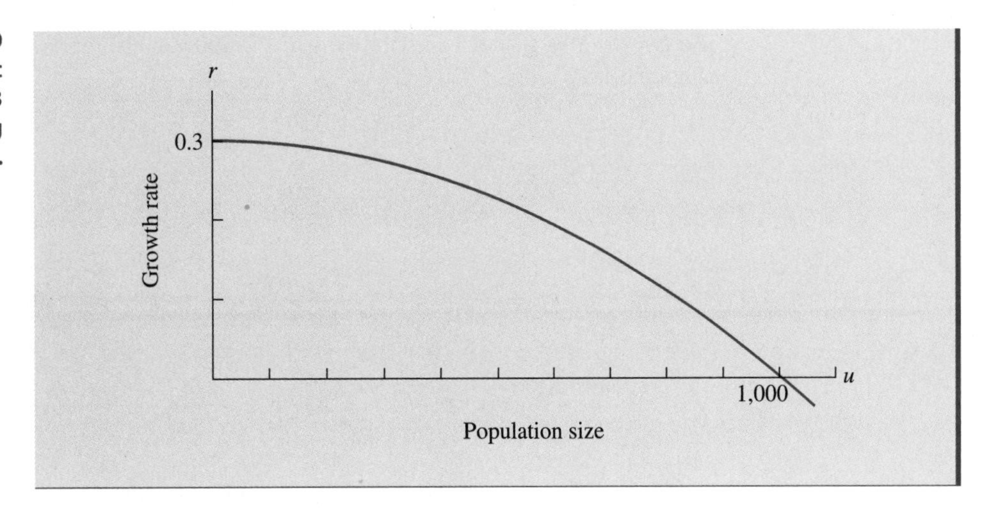

or $a = -0.0000003$. Our growth rate is now given by the function

$$r = 0.3 - 0.0000003u^2$$

This means that the growth rate for the $n-1$th year is

$$r = 0.3 - 0.0000003u^2(n-1)$$

Substitution into the dynamical system

$$u(n) = u(n-1) + ru(n-1)$$

gives the generalized logistic equation

$$u(n) = u(n-1) + \left(0.3 - 0.0000003u^2(n-1)\right)u(n-1)$$

which can be rewritten as

$$u(n) = 1.3u(n-1) - 0.0000003u^3(n-1)$$

The equilibrium population sizes are the solutions to the equation

$$E = 1.3E - 0.0000003E^3$$

which are $E = -1000$, 0, and 1000. We will ignore -1000, since we can't have a population of negative size. If you generate a time graph for $(n, u(n))$ using a variety of $u(0)$-values, you will see that $u(n)$ approaches 1000 for reasonable initial population sizes.

So $E = 1000$ is stable. The stability of $E = 1000$ can also be seen in the web analysis of Figure 6.11.

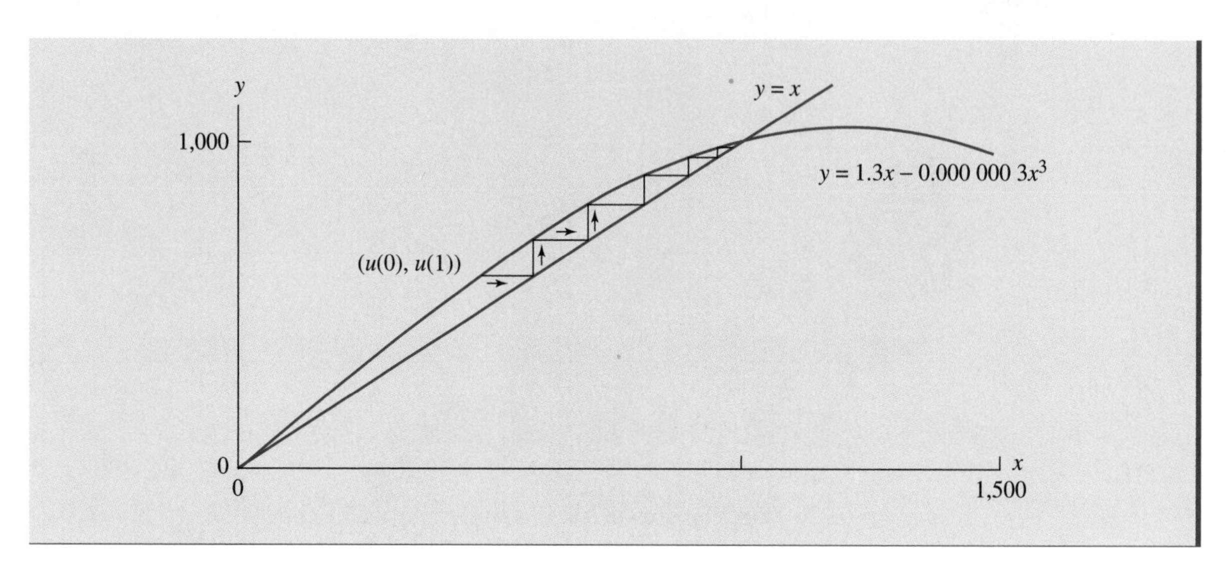

FIGURE 6.11 **Web analysis for the dynamical system of Example 6.2. $E = 1000$ is a stable equilibrium, $E = 0$ is unstable.**

If you compute

$$\frac{u(n) - 1000}{u(n-1) - 1000}$$

for several values of n, you will see that the rate at which $u(n)$ approaches 1000 is 0.4, meaning that the population size eventually gets 60% closer to 1000 each year. ■

Whenever the growth rate for a population strictly decreases, such as the growth rate in Figure 6.10, then the model is called a **compensation model.**

Suppose we are developing a model for the size of a population of whales. Even though a whale is a large animal, when a population of whales is spread throughout the Atlantic, Pacific, Arctic and Antarctic Oceans, they may have problems finding each other. Thus, if the size of the population size is small, the whales may have problems finding a mate. This would result in a small growth rate. Whales are just 1 of many species in which (1) the growth rate may be small for small population sizes, (2) the growth rate increases as the population size increases because it is easier to find a mate, but (3) when the population size gets large, the growth rate decreases again because of a lack of sufficient resources. This can be summarized by the growth rate graph in Figure 6.12. Recall that the intrinsic growth rate is the largest r-value, which occurs at the peak of the graph, point (m, b) in Figure 6.12.

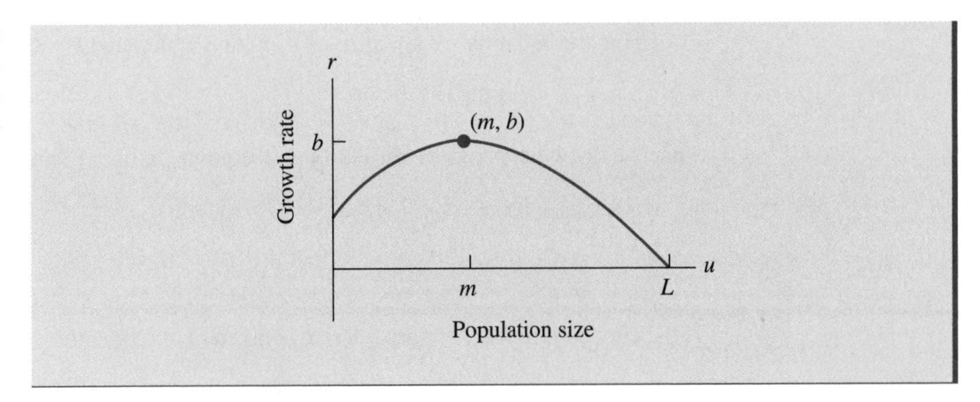

FIGURE 6.12 Increasing, then decreasing, growth rate.

Note that as the population size increases from near 0 to m, the growth rate increases. But after the population exceeds a size of m, the growth rate begins decreasing due to a lack of sufficient resources; m is the optimal population size and b is the optimal growth rate. Once the population exceeds L, the growth rate becomes negative. Again, L is the carrying capacity of the environment because that is the point at which the population's growth rate decreases to 0.

Example 6.3

Suppose we have determined that the intrinsic growth rate for a population is 0.3 and occurs when the population size equals 300. Suppose that the carrying capacity of the population occurs at $L = 1000$. Then we want a growth rate graph similar to the one in Figure 6.13. We will use a parabola to model this growth rate. Its vertex is at the point $(300, 0.3)$, and it goes through the point $(1000, 0)$.

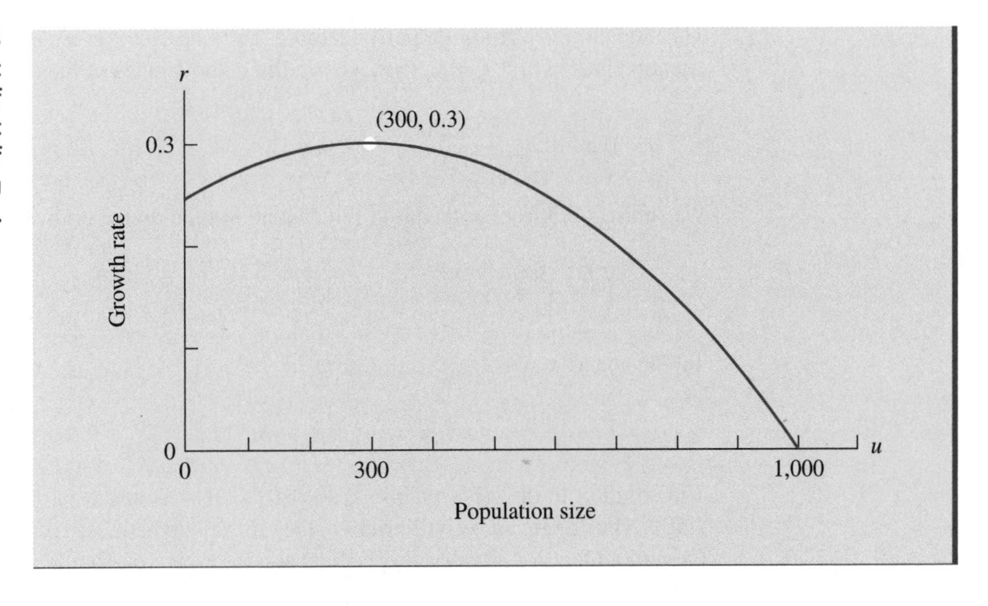

FIGURE 6.13 Intrinsic growth rate of $b = 0.3$ occurs at population size of $m = 300$. Carrying capacity is $L = 1000$.

Using the equation $r - k = a(u - h)^2$, with $h = 300$ and $k = 0.3$, gives

$$r - 0.3 = a(u - 300)^2$$

Since we know the parabola goes through the point $(u, r) = (1000, 0)$, substitution gives

$$0 - 0.3 = a(1000 - 300)^2$$

or $\quad a = -\dfrac{0.3}{700^2}$

Thus, after simplifying, our growth rate is given by the function

$$r = 0.3 - 0.3\left(\frac{u - 300}{700}\right)^2 \quad \text{or} \quad r = 0.3 - 0.3\left(\frac{u(n-1) - 300}{700}\right)^2$$

Substitution into the dynamical system $u(n) = u(n-1) + ru(n-1)$ gives

$$u(n) = u(n-1) + \left(0.3 - 0.3\left(\frac{u(n-1) - 300}{700}\right)^2\right)u(n-1)$$

Using fractions allows us to find exact equilibrium values fairly easily. To find the equilibrium values for this dynamical system, we solve the equation

$$E = E + \left(0.3 - 0.3\left(\frac{E - 300}{700}\right)^2\right)E$$

Subtracting E from both sides gives

$$0 = \left(0.3 - 0.3\left(\frac{E - 300}{700}\right)^2\right)E$$

This equation is already partially factored, and one factor is E. Thus, one solution to this equation is $E = 0$. If $E \neq 0$, then we can divide both sides of the equation by E to get

$$0 = 0.3 - 0.3\left(\frac{E - 300}{700}\right)^2$$

We subtract 0.3 from both sides of the equation then divide both sides by -0.3 to get

$$1 = \left(\frac{E - 300}{700}\right)^2$$

Taking square roots of both sides gives

$$1 = \frac{E - 300}{700} \quad \text{or} \quad -1 = \frac{E - 300}{700}$$

The solution to the left equation is $E = 1000$. The solution to the right equation is $E = -400$. Thus there are 3 equilibrium values for this dynamical system, -400, 0, and 1000. Even though $E = -400$ has no physical significance, you should be able to determine that it is a stable equilibrium. As before, $E = 0$ is an unstable equilibrium and $E = 1000$ is a stable equilibrium.

It is quite difficult to find the maximum interval of stability for the equilibrium value 1000. From the web analysis of Figure 6.14, it is clear that if $0 < u(0) < 1757$, then $u(n)$ goes to 1000. If the initial population is a little larger than 1757, say $u(0) = 2000$, then $u(n)$ goes toward -400. If the initial population is even larger, say $u(0) = 2100$, then $u(n)$ goes toward 1000 again. There are actually many intervals for which if $u(0)$ is in any one of them, then $u(n)$ goes toward $E = 1000$. The same is true for the equilibrium value $E = -400$. It would be quite difficult to find the union of all of these "intervals of stability."

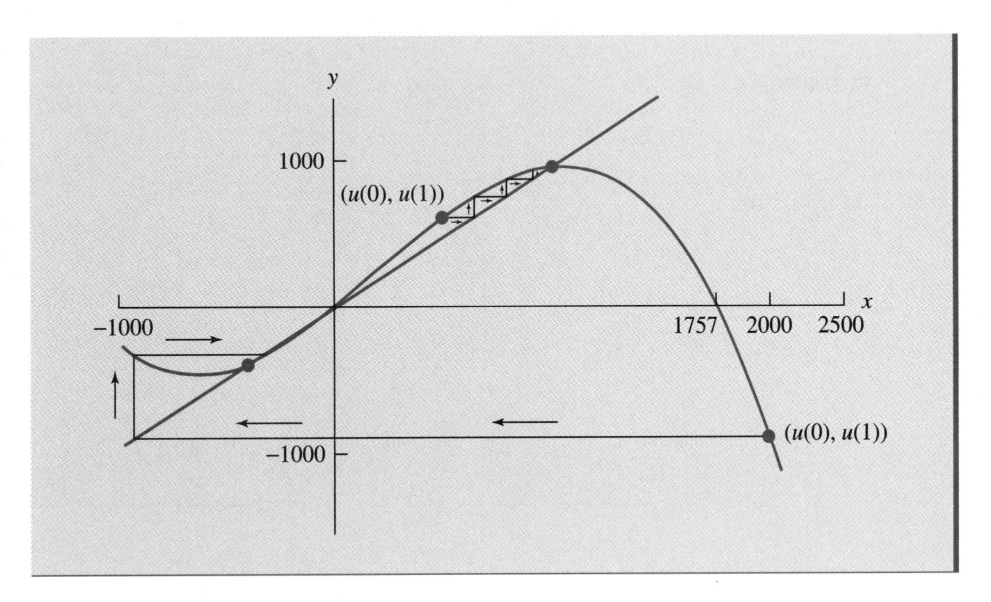

FIGURE 6.14 Web analysis shows $E = 1000$ and $E = -400$ are stable. But when $u(0) = 2000$, then $u(n)$ goes to -400, not 1000.

You should find that, if $u(0)$ is in the interval of stability for $E = 1000$, then

$$\frac{u(n) - 1000}{u(n-1) - 1000} \approx 0.15$$

for moderately large values of n. This means that the population size gets 85% closer to 1000 each year.

The values for $u(n)$ approach 1000 so rapidly that if you use a moderately large value for n when computing the ratio, your calculator or computer may give you an error because it has rounded the denominator to 0. ■

Whenever the growth rate function $r(u)$ increases, then decreases to 0 at the carrying capacity, such as the growth rate in Figure 6.13, the model is said to be a **depensation model.** In later sections, we will see that when a species population size is modeled by a depensation model, certain approaches to harvesting and hunting can have devastating effects on the population.

For some populations, the growth rate may actually be negative when the population size is small. This could be due to a birth rate that is lower than the death rate because it is so difficult for individuals to find a mate. Such models are called **critical depensation models** and have growth rate functions with shapes similar to Figure 6.15. Figure 6.15 indicates that if the population size drops below M, then the population will have a negative growth rate and will continue to decrease in size until it becomes extinct. For critical depensation models, M is called the **minimum viable population.** The value L is still the carrying capacity, since that is where the growth rate function decreases to 0. Examples of critical depensation will be explored in the problems.

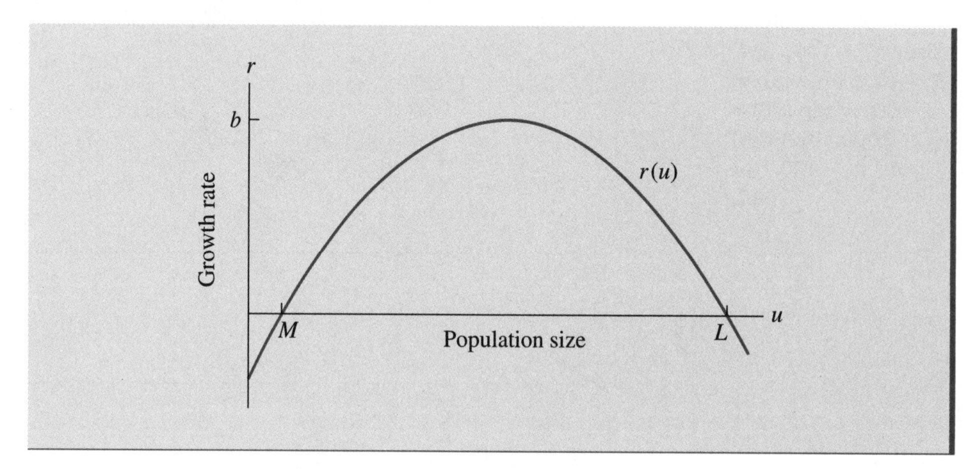

FIGURE 6.15 The growth rate is negative when $u < M$. This is a critical depensation model.

DEFINITION

The value M is called a **minimum viable population** if, when the population size drops below M, then the population eventually becomes extinct.

This section has taken a limited view about how populations grow. This view has allowed us to develop some understanding of the dynamics involved in population growth. To develop a better understanding of population growth, we could combine the approach in this section with the study of population age structure that was discussed in Section 2.4. Among other factors that affect population growth and should therefore be part of a comprehensive model are seasonal effects, multispecies and ecosystem interactions, and spatial effects such as the diffusion of a species within an ecosystem. Models are currently being developed and analyzed that incorporate many of these effects. One problem with these models is that it is difficult to obtain good empirical data with which to test the models, but this should not detract from their use in helping us understand and manage our resources.

6.3 Problems

1. Suppose we have a species in which the growth rate decreases and the intrinsic growth rate is $b = 0.4$, which occurs at $u = 0$. Also suppose that the carrying capacity for this species is 500. Use a parabola with vertex at $(0, 0.4)$ and that goes through the point $(500, 0)$ for the growth rate. A graph of this problem can be seen in Figure 6.16. Find the corresponding dynamical system. Find the equilibrium values for this dynamical system. Determine the stability of the positive equilibrium population size. If the positive equilibrium is stable, find the rate at which the size of the population approaches equilibrium.

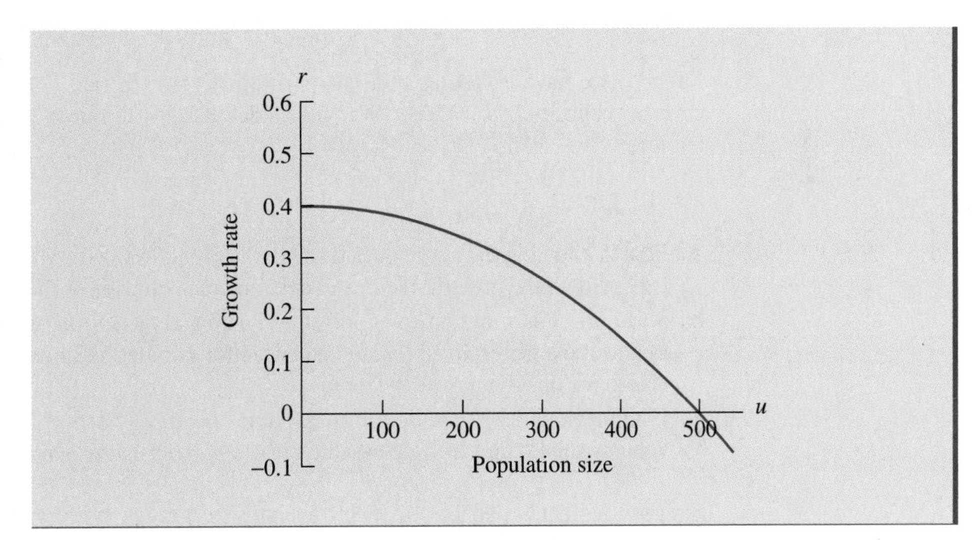

FIGURE 6.16 Graph of growth rate function for Problem 1.

2. Suppose we have a species in which the growth rate decreases and the intrinsic growth rate is $b = 0.7$, which occurs at $u = 0$. Also suppose that the carrying capacity for this species is 2000. Use a parabola with vertex at $(0, 0.7)$ and that goes through the point $(2000, 0)$ for the growth rate. Find the corresponding dynamical system. Find the equilibrium values for this dynamical system. Determine the stability of the positive equilibrium population size. If the positive equilibrium is stable, find the rate at which the size of the population approaches equilibrium.

3. Suppose we have a species in which the growth rate decreases and the intrinsic growth rate is $b = 1.1$, which occurs at $u = 0$. Also suppose that the carrying capacity for this species is 1000. Use a parabola with vertex at $(0, 1.1)$ and that goes through the point $(1000, 0)$ for the growth rate. Find the corresponding dynamical system. Find the equilibrium values for this dynamical system. Determine the stability of the positive equilibrium population size. If the positive equilibrium is stable, find the rate at which the size of the population approaches equilibrium.

4. Suppose we have a species that has an intrinsic growth rate $b = 0.2$ when $u = 0$ and a carrying capacity of $L = 1000$. We will model the growth rate for this population with the cubic

$$r = 0.2 + au^3$$

a. Find a, knowing the curve goes through the point $(1000, 0)$. Use this to find an equation for r. Graph the growth rate r, and describe the behavior of the growth rate.

b. Substitute this function for r into the dynamical system $u(n) = u(n-1) + ru(n-1)$ to develop a generalized logistic equation for population growth. Find the equilibrium values for this dynamical system.

c. Find the stability of the equilibrium values for this system. If the positive equilibrium value is stable, find the rate at which $u(n)$ approaches this equilibrium population.

5. Suppose we have a species that has an intrinsic growth rate $b = 0.3$ when $u = 0$ and a carrying capacity of $L = 2000$. We will model the growth rate for this population with the cubic

$$r = 0.3 + au^3$$

a. Find a, knowing the curve goes through the point $(2000, 0)$. Use this to find an equation for r. Graph the growth rate r, and describe the behavior of the growth rate.

b. Substitute this function for r into the dynamical system $u(n) = u(n-1) + ru(n-1)$ to develop a generalized logistic equation for population growth. Find the equilibrium values for this dynamical system.

c. Find the stability of the equilibrium values for this system. If the positive equilibrium value is stable, find the rate at which $u(n)$ approaches this equilibrium population.

6. Suppose we have a species that has an intrinsic growth rate $b = 0.2$ when $u = 0$ and a carrying capacity of $L = 100$. We will model the growth rate for this population with the equation

$$r = 0.2 + au^4$$

a. Find a, knowing the curve goes through the point $(100, 0)$. Use this to find an equation for r. Graph the growth rate r, and describe the behavior of the growth rate.

b. Substitute this function for r into the dynamical system $u(n) = u(n-1) + ru(n-1)$ to develop a generalized logistic equation for population growth. Find the equilibrium values for this dynamical system.

c. Find the stability of the equilibrium values for this system. If the positive equilibrium value is stable, find the rate at which $u(n)$ approaches this equilibrium population.

7. Suppose we have a species that has an intrinsic growth rate $b = 0.4$ when $u = 0$ and a carrying capacity of $L = 200$. We will model the growth rate for this population with the equation

$$r = 0.4 + au^4$$

a. Find a, knowing the curve goes through the point $(200, 0)$. Use this to find an equation for r. Graph the growth rate r and describe the behavior of the growth rate.

b. Substitute this function for r into the dynamical system $u(n) = u(n-1) + ru(n-1)$ to develop a generalized logistic equation for population growth. Find the equilibrium values for this dynamical system.

c. Find the stability of the equilibrium values for this system. If the positive equilibrium value is stable, find the rate at which $u(n)$ approaches this equilibrium population.

8. Suppose we have a species in which the growth rate increases until the population size reaches 200, then decreases until the population size reaches the carrying capacity of 1000. The intrinsic growth rate is $b = 0.3$, which occurs at $u = 200$. Use a parabola with vertex at $(200, 0.3)$ and that goes through the point $(1000, 0)$ for the growth rate. Graph your function for the growth rate. Find the corresponding dynamical system. Find the equilibrium values for this dynamical system. Determine the stability of the positive equilibrium population size. If the positive equilibrium is stable, find the rate at which the size of the population approaches equilibrium.

9. Suppose we have a species in which the growth rate increases until the population size reaches 500, then decreases until the population size reaches the carrying capacity of 2000. The intrinsic growth rate is $b = 0.7$, which occurs at $u = 500$. Use a parabola with vertex at $(500, 0.7)$ and that goes through the point $(2000, 0)$ for the growth rate. Graph your function for the growth rate. Find the corresponding dynamical system. Find the equilibrium values for this dynamical system. Determine the stability of the positive equilibrium population size. If the positive equilibrium is stable, find the rate at which the size of the population approaches equilibrium.

10. Suppose we have a species in which the growth rate increases until the population size reaches 1200, then decreases until the population size reaches the carrying capacity of 2200. The intrinsic growth rate is $b = 0.3$, which occurs at $u = 1200$.

a. Find the equation for a parabola with vertex at $(u, r) = (1200, 0.3)$ and that goes through the point $(2200, 0)$, and use it for the growth rate. Graph this parabola using an appropriate viewing window, and use the graph to help find the range of population sizes for which this population has a negative growth rate.

b. Develop the corresponding dynamical system. Find the 3 equilibrium values for this dynamical system. Determine the stability of each of the equilibrium population sizes. If an equilibrium value is stable, find the rate at which the size of the population approaches equilibrium. What is the minimum viable population size?

c. Draw cobwebs near each of the stable equilibrium values. You will need to use a different viewing window for each equilibrium value in order to see the cobwebs.

11. Suppose we have a species in which the growth rate increases until the population size reaches 800, then decreases until the population size reaches the carrying capacity of 1300. The intrinsic growth rate is $b = 0.1$ which occurs at $u = 800$.

a. Find the equation for a parabola with vertex at $(u, r) = (800, 0.1)$ and that goes through the point $(1300, 0)$, and use it for the growth rate. Graph this parabola using an appropriate viewing window, and use the graph to help find the range of population sizes for which this population has a negative growth rate.

b. Develop the corresponding dynamical system. Find the 3 equilibrium values for this dynamical system. Determine the stability of each of the equilibrium population sizes. If an equilibrium value is stable, find the rate at which the size of the population approaches equilibrium. What is the minimum viable population size?

c. Draw cobwebs near each of the stable equilibrium values. You will need to use a different viewing window for each equilibrium value in order to see the cobwebs.

12. Suppose we have a species in which the growth rate increases until the population size reaches 1000, then decreases until the population size reaches the carrying capacity of 2000. The intrinsic growth rate is $b = 0.2$, which occurs at $u = 1000$.

a. Find the equation for a parabola with vertex at $(u, r) = (1000, 0.2)$ and that goes through the point $(2000, 0)$, and use it for the growth rate. Graph this parabola using an appropriate viewing window. What does the graph imply about the growth rate of the population when the size of the population is small?

b. Develop the corresponding dynamical system. Find the 2 equilibrium values for this dynamical system. Determine the stability of each of the equilibrium population sizes. (Be careful when determining the stability of the equilibrium value $E = 0$. Use positive and negative initial values even though negative initial values have no physical meaning.) If an equilibrium value is stable, find the rate at which the size of the population approaches equilibrium. What is the minimum viable population size?

6.4 Graphical Approach to Harvesting

Historically, the world has not protected its renewable resources; fish, trees, water, and so on. For example, in the early 1900s, it is estimated that there were approximately 400,000 fin whales, 200,000 blue whales, and 125,000 humpback whales worldwide. Currently, the estimates are 80,000 fin whales, 9,000 blue whales and 3000 humpback whales. The bar graph of this data in Figure 6.17 is quite striking. Most of this decline in whale populations is the result of unregulated overfishing. Our future depends on reasonable management of our renewable resources.

In this section, we will model the management of a fish population. Models similar to these have been applied to the fishing industry for years in an attempt to understand how to preserve a population of fish while maintaining a high yield of fish from year to year. The results obtained can be applied to any resource whose growth can be modeled by a generalized logistic equation.

In this section and the next, we are going to develop and use simple graphical techniques that will allow us to avoid working with the more complicated dynamical systems that result from harvesting models. These techniques allow us to easily determine the equilibrium population sizes and their stability, given different approaches to harvesting.

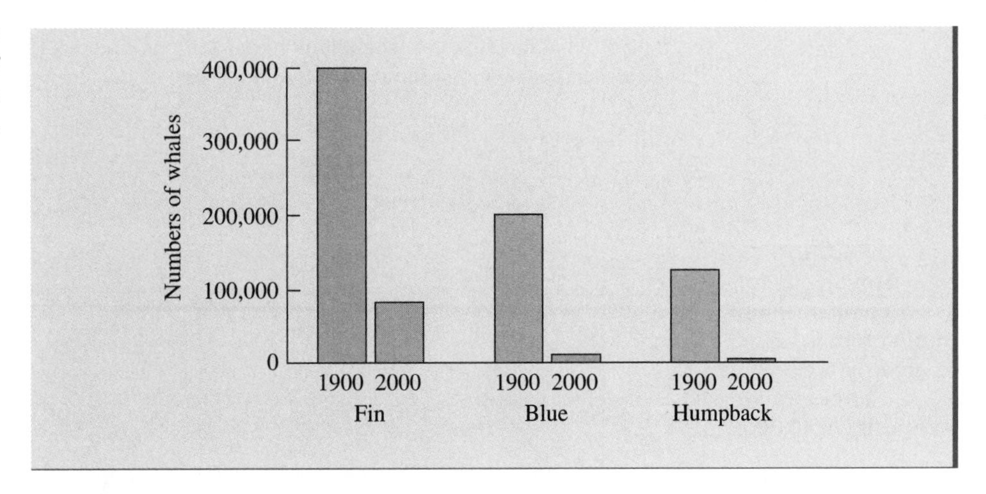

FIGURE 6.17 Estimates of whale population sizes.

Suppose we are modeling a population in which the growth rate decreases as the population size increases. Recall that this is a compensation model. Let's assume the intrinsic growth rate is 8% and the carrying capacity is 40,000. One such growth rate function is seen in Figure 6.18, where u represents the size of the population and r represents the growth rate given the population size u.

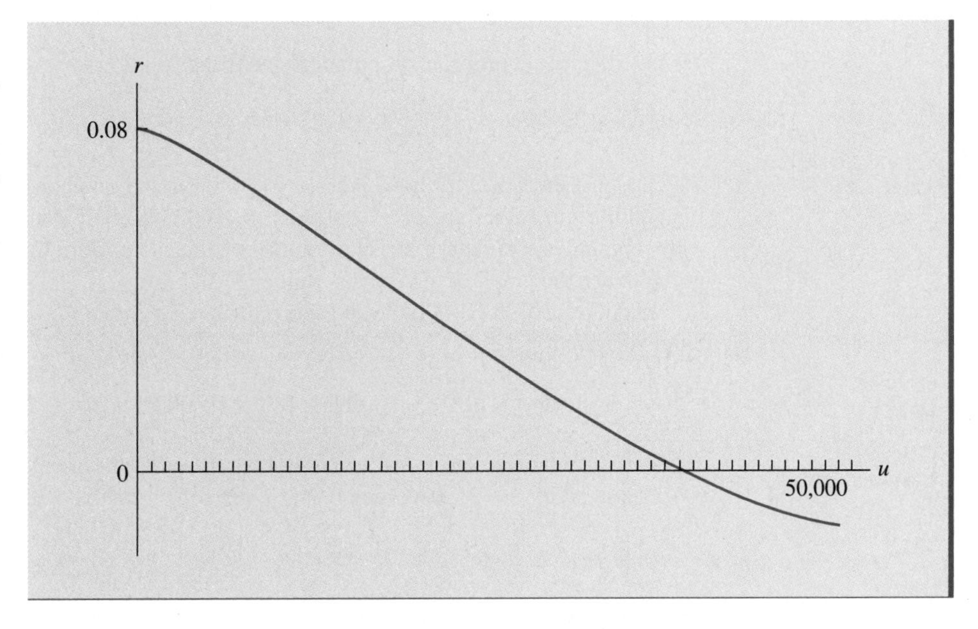

FIGURE 6.18 Growth rate, $r(u)$, for compensation model. Intrinsic growth rate is 8% and carrying capacity is 40,000.

The actual yearly growth in the population is given by the function

$$g(u) = r(u)\, u$$

A graph of what this function might look like is seen in Figure 6.19. The dynamical system that models such a population would be of the form

$$u(n) = u(n-1) + g(u(n-1))$$

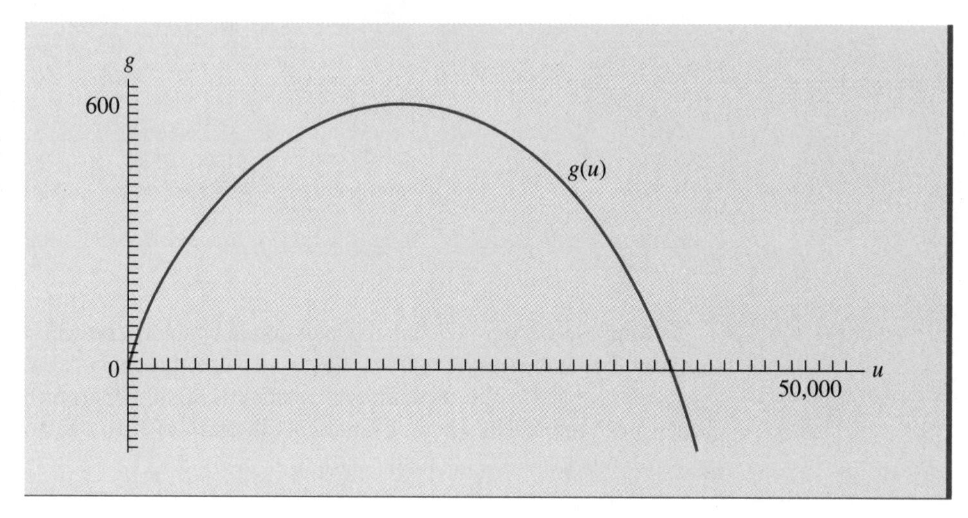

FIGURE 6.19 Growth $g(u)$ for compensation model with intrinsic growth rate of 8% and carrying capacity of 40,000.

We note that the equilibrium population size is the solution to

$$E = E + g(E)$$

which is just the values, E, at which $g(E) = 0$. From the graph in Figure 6.19, we see that the equilibrium values are $E = 0$ and $E = 40,000$. We could generate tables of values or use web analysis to determine the stability of these equilibrium values. Instead, we are going to use the graph of $g(u)$ to determine the stability.

Suppose $u(0) = 10,000$. From Figure 6.19, we see that $g(u(0)) \approx 480$. From the dynamical system,

$$u(1) = u(0) + g(u(0)) \approx 10,000 + 480 > 10,000 = u(0)$$

In short, the next population size is larger than the previous population size.

If instead $u(0) = 35,000$, then from Figure 6.19, $g(u(0)) \approx 320$. Again we have

$$u(1) > u(0)$$

since the growth was positive. In fact, whenever $0 < u(0) < 40,000$, we see that $g(u(0)) > 0$. This means that $u(1) = u(0) + g(u(0)) > u(0)$; that is, the next population size is larger than the previous one. We can denote this on the graph of Figure 6.19 by drawing arrows going to the right on the segment of the graph where $0 < u < 40,000$. This is seen in Figure 6.20.

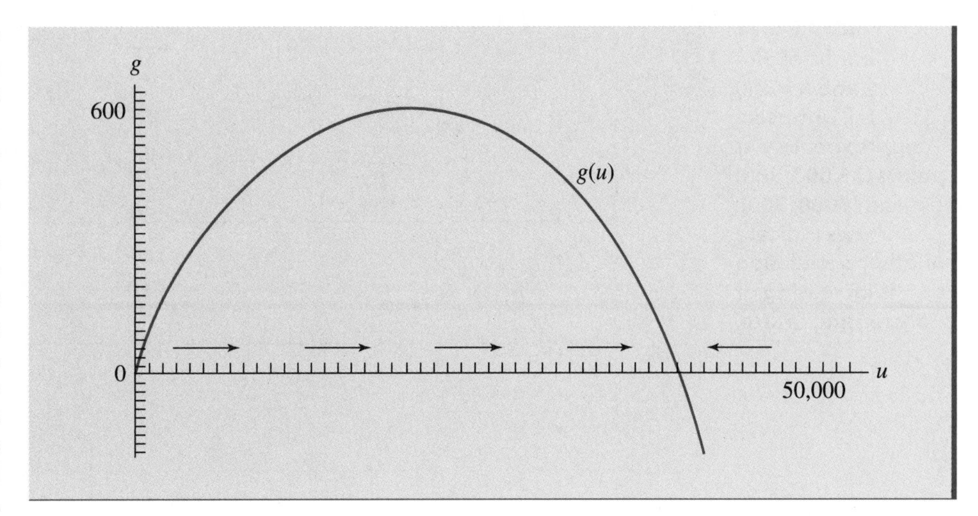

FIGURE 6.20 **If population size is between 0 and 40,000, then growth is positive and population is increasing. If population size is over 40,000, the growth is negative and population is decreasing. Equilibrium population sizes are 40,000 and 0, the horizontal intercepts.**

Suppose that $40,000 < u(0)$. From the graph of Figure 6.20, we see that $g(u(0)) < 0$; that is, there is negative growth. This means that $u(1) = u(0) + g(u(0)) < u(0)$; that is, the next population size is smaller than the previous population size. We can denote this on the graph by drawing arrows going to the left on the segment of the graph where $40,000 < u$. This is also seen in Figure 6.20. From the figure 6.20, we see that if $u(0) > 0$, then $u(1)$ is closer to 40,000 than $u(0)$. This same logic implies that if $u(n-1) > 0$, then $u(n)$ is closer to 40,000 than $u(n-1)$. This means that $E = 40,000$ is a stable equilibrium. Also, for small population sizes, $u(n)$ is farther from 0 than $u(n-1)$. This implies that $E = 0$ is an unstable equilibrium.

Notice how easy it is to determine the equilibrium values and their stability from the graph of the function g. We don't need to generate time graphs or webs. We are going to use this form of graphical analysis to analyze different approaches to harvesting.

Suppose that each year, we harvest 300 of the species whose growth function is given in Figure 6.20. Let's assume that the harvest is at the end of the year, so it doesn't effect the growth rate. The idea behind this graphical analysis is to graph the growth function $g(u)$ and the harvest function, $h = 300$ in this case, on the same axis, which is seen in Figure 6.21. We then use these graphs to locate population sizes for which the population is growing and population sizes for which the population is decreasing.

The equilibrium population sizes are when the growth g equals the harvest h, which are the u-coordinates of the points at which their graphs intersect. The intersection points are seen to be about (5000, 300) and (35,000, 300). Thus, the equilibrium population sizes are $E = 5000$ and $E = 35,000$.

Suppose that the population size is $u = 10,000$. From the graphs in Figure 6.21, we see that $g(u) \approx 480$ and the harvest is 300. Thus, next year's population size will be approximately

$$u_{\text{new}} = u_{\text{old}} + g - h \approx 10,000 + 480 - 300 = 10,180 > u_{\text{old}}$$

The same will be true if the population size is any value between 5000 and 35,000; that is, the next year's population size will be larger.

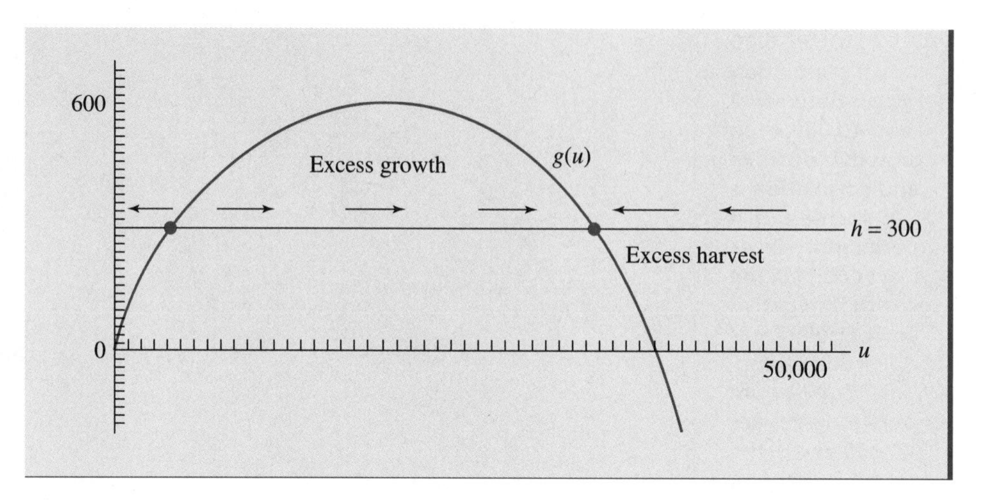

FIGURE 6.21 Graphs of $g(u)$ and $h = 300$ intersect approximately at points (35,000, 300) and (5000, 300). Arrows indicate whether population is increasing or decreasing in size.

On the other hand, if $u = 3000$ then $g \approx 220$, so

$$u_{\text{new}} \approx 3000 + 220 - 300 = 2920 < u_{\text{old}}$$

In summary, if $5000 < u < 35,000$, then the growth is greater than the harvest, so the population will increase in size the next year. This is denoted by the arrows going to the right in Figure 6.21. On the other hand, if $u < 5000$ or $u > 35,000$, then the growth is smaller than the harvest and the population will decrease in size in the next year. This is denoted by the arrows going to the left in Figure 6.21. This means that $E = 35,000$ is a stable equilibrium population size and $E = 5000$ is an unstable equilibrium population size.

DEFINITION

A **sustainable yield** is an amount that can be harvested every year, for an indefinite period of time.

In analyzing Figure 6.21, we have determined that if we harvest 300 per year from this population, than the population will stabilize at the equilibrium population of 35,000, assuming the initial population size was large enough. The harvest of 300 per year is called a **sustainable yield.**

We have also found that there is an unstable equilibrium population size of 5000. This population size is the **minimum viable population** given this harvest strategy. If the population drops below 5000, it will die out unless the harvesting is curtailed.

Suppose the yearly harvest depends on the population size, and is given by the function $h(u)$. Suppose the function $g(u)$ gives the growth for the population. We graph the functions $h(u)$ and $g(u)$ on the same axes, as seen in Figure 6.22. The equilibrium population sizes are where $g(E) = h(E)$. This corresponds to the u-values where the curves $g(u)$ and $h(u)$ intersect. This means that $E = M$ and $E = P$ are equilibrium population sizes.

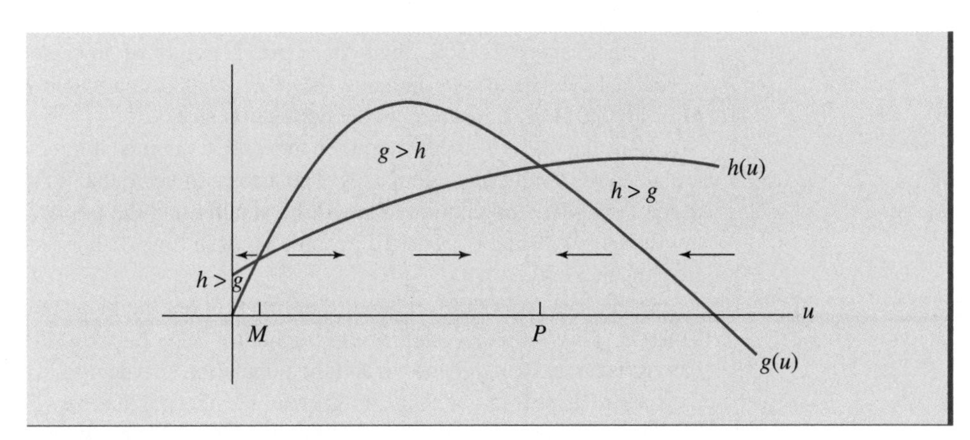

FIGURE 6.22 Growth and harvest functions intersect at $u = M$, the minimum viable population size, and $u = P$, the stable equilibrium population size. Arrows indicate increasing and decreasing populations.

To study the stability of these equilibrium values, note that if $M < u < P$, then the growth is larger than the harvest so

$$u_{\text{new}} > u_{\text{old}}$$

and u_{new} is closer to P and further from M than u_{old}. If $u > P$ then the growth is less than the harvest, so the population size will decrease toward P. Thus, P is stable.

Similarly, if $u < M$ then $g < h$, so the population size will decrease, going away from M. Thus, M is unstable.

We are going to use this type of logic to analyze different harvesting strategies. Let's reconsider the population modeled by the growth function given in Figure 6.19. We now want to determine what happens with different constant harvests. In Figure 6.23, we graph the growth function along with 3 possible harvest functions, $h = 500$, $h = 600$, and $h = 700$.

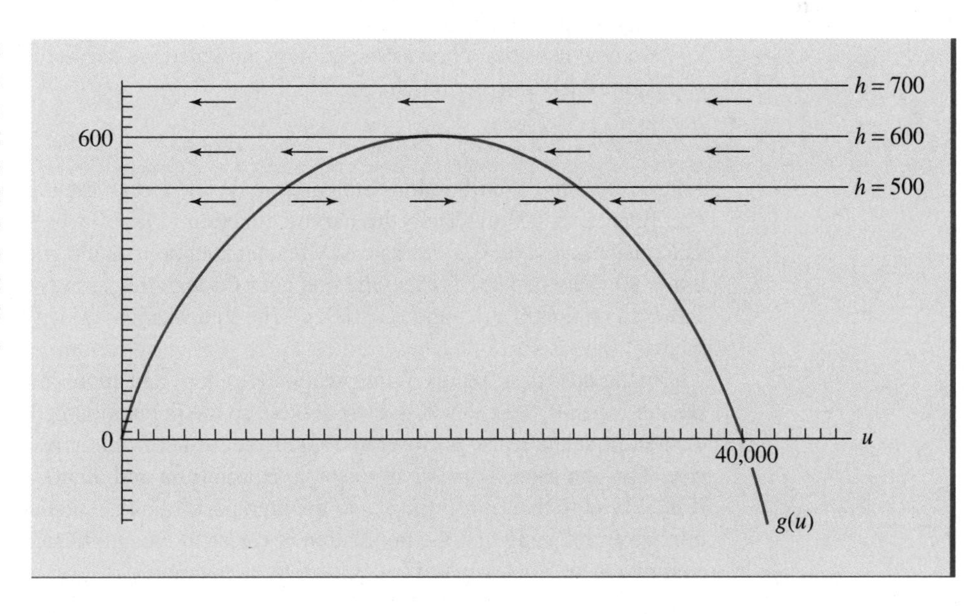

FIGURE 6.23 Growth function along with 3 different constant harvests. Arrows below each harvest function indicate relationship between current and next population size.

If the harvest is 500, then there are 2 points of intersection—$(11000, 500)$ and $(29000, 500)$. The arrows indicate that $P = 29,000$ is a stable population size and that $M = 11,000$ is the minimum viable population size.

If the harvest is 700 per year, then there are no points of intersection. This means there are no equilibrium population sizes. The arrows indicate that no matter what the value for u, the next year's population size will be smaller and the population will become extinct unless the harvesting is curtailed at some point in time. A harvest of 700 per year is not a sustainable yield.

If the harvest is 600 per year, then there is one point of intersection, $(20000, 600)$. Thus, $E = 20,000$ is an equilibrium population size. Suppose that $u < 20,000$. Then the harvest is larger than the growth and the population size decreases toward 0. If $u > 20,000$, the harvest is still larger than the growth and the population size decreases, but toward 20,000, not toward 0. Thus, $E = 20,000$ is a semistable equilibrium.

From the graph in Figure 6.23, we can see that if there is a constant harvest $h > 600$ per year, then the harvest line does not intersect the growth function and the population dies out.

If $h < 600$, then there are 2 points of intersection. One point of intersection corresponds to a stable equilibrium population size; the other corresponds to an unstable equilibrium, which is the minimum viable population size. The closer h is to 600, the closer the minimum viable population size is to the equilibrium population size. The closer these values are to each other, the more at risk the population. Something unexpected could drop the population size from the stable equilibrium to below the minimum viable population size, resulting in the eventual extinction of the population or curtailment of harvesting.

If $h = 600$, then there is only one semistable equilibrium population size. Since 600 is the largest harvest that results in an equilibrium population, then 600 is called the **maximum sustainable harvest.** But this harvest carries a great deal of risk. Over time, the population will stabilize at the equilibrium size, but a slight unexpected drop in the population size results in the population heading towards extinction. Semistable equilibria are normally not desirable.

We now consider a harvesting strategy in which we harvest a fixed proportion of the population each time period, that is,

$$h = pu$$

where p is some fixed fraction. Suppose we decide to harvest 1% of the population each year, that is, $p = 0.01$. Then the harvest function is $h = 0.01u$. This is a line that goes through the origin $(0, 0)$. We can easily find another point that it goes through. For example, if $u = 50,000$, then $h = 500$, so the line goes through the point $(50000, 500)$. Figure 6.24 shows the graph of $g(u)$ and $h = 0.01u$. The line was just drawn through the 2 indicated points.

In this case the 2 points of intersection are $(0, 0)$ and approximately $(35000, 350)$. We see that when $0 < u < 35,000$, $g(u) > h(u)$, so the population gets larger. This means that the new population size is closer to 35,000 and further from 0 than the previous population size. This indicates that 0 is an unstable equilibrium and 35,000 is a stable equilibrium. If $u > 35,000$, then $g(u) < h(u)$, so the new population is smaller than previous population, again meaning that the population is closer to the equilibrium value of 35,000. This confirms that 35,000 is a stable equilibrium population size.

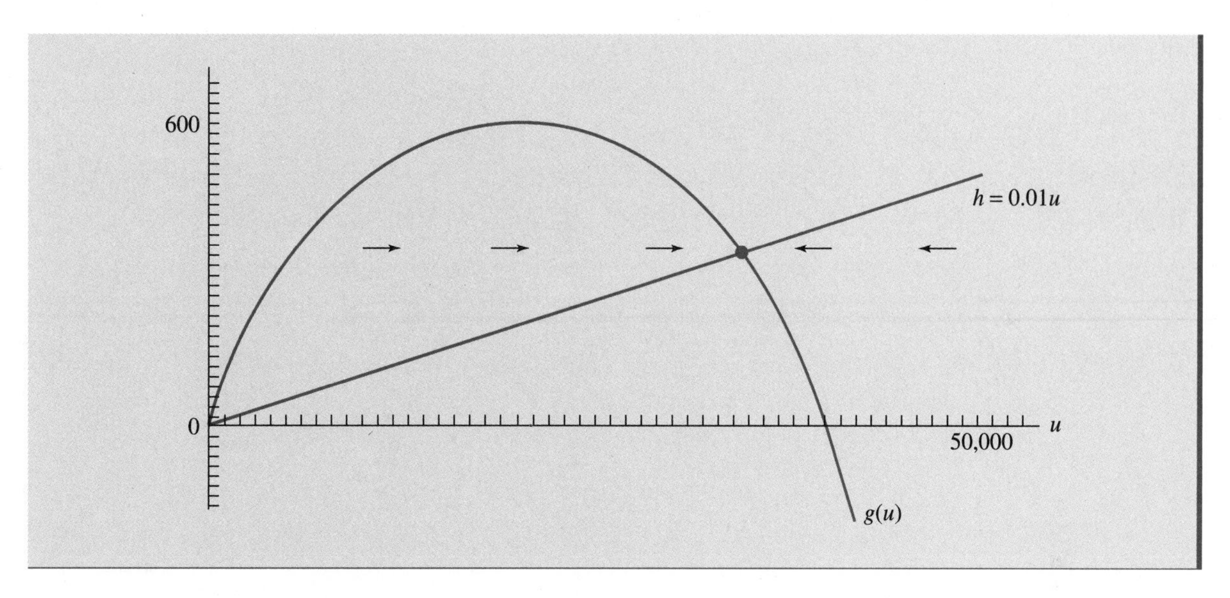

FIGURE 6.24 Comparison of proportional harvest to growth.

We have learned the following about this population model. If we harvest 1% per year, the population will stabilize at 35,000 over time. Since we are harvesting 1% per year, we will be harvesting about 350 per year over time. Thus, 350 per year is the sustainable yield, given this strategy. We also notice that there is no positive minimum viable population size. This means that this strategy is safe for this population.

In Figure 6.25, we have drawn the line that goes through the origin and the peak of the growth function $(20000, 600)$. The slope of this line is

$$\frac{600 - 0}{20,000 - 0} = 0.03$$

and the vertical intercept is 0. Thus, the equation for this line is

$$h = 0.03u$$

meaning we harvest 3% per year. In this case, the 2 points of intersection are $(0, 0)$ and $(20000, 600)$. Analysis similar to that done with Figure 6.24 indicates that 20,000 is the stable equilibrium population size for this harvesting strategy. Over time the population stabilizes at 20,000. Since we are harvesting 3%, we have a sustainable yield of 600 per year. This is the maximum sustainable harvest because it is the highest point on the growth function g. In this case, there is no positive minimum viable population. Thus, if we harvest 3% per year, we can obtain the maximum sustainable harvest of 600 without putting the population at risk.

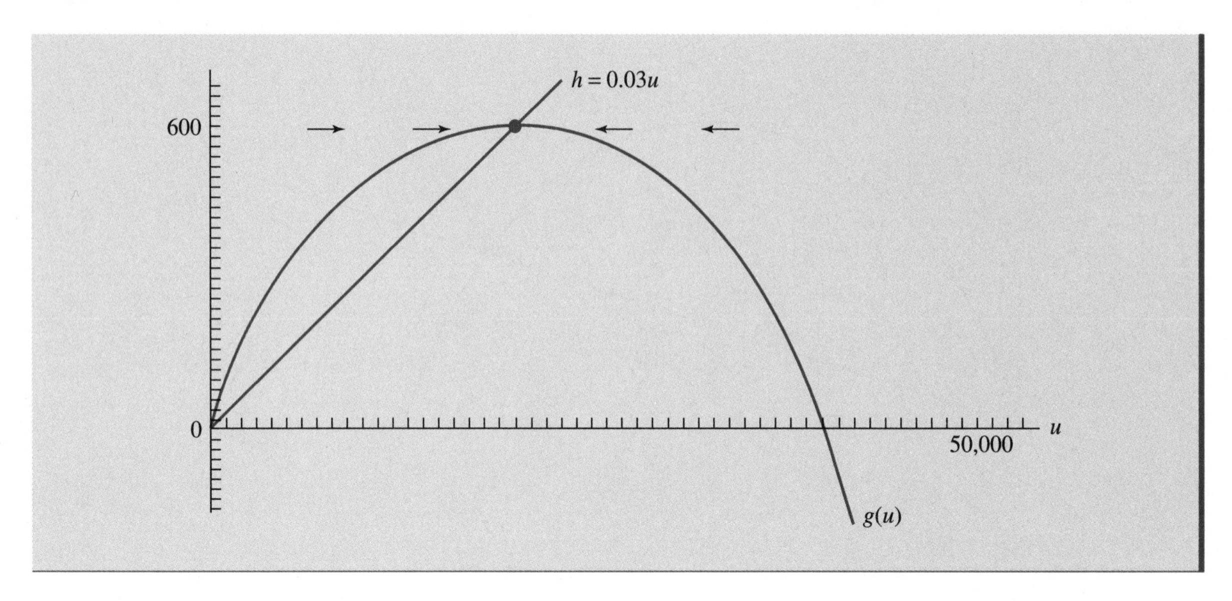

FIGURE 6.25 Growth $g(u)$ compared to harvest $h = 0.03u$; results in equilibrium at the peak of $g(u)$.

In Figure 6.26 is a comparison of the growth function to the harvest function $h = 0.05u$. In this case, the points of intersection are $(0, 0)$ and $(9000, 450)$. It takes more time, effort, and expense to harvest 5% instead of 3%. But since the population eventually stabilizes at 9000, the sustainable yield is only 450 per year instead of 600. Thus, we find that increasing the effort can result in an eventual decrease in the yearly harvest.

This result, increased fishing resulting in a lower sustainable harvest, has been observed many times in fishing. Before 1970, the annual Peruvian anchovy harvest was approximately 14 tons, 20% of the world's supply. A newly established fishing fleet increased the fishing effort. The result was an average yearly harvest of around 2 million tons per year from 1973 through 1989. The anchovy population is just now recovering from this overfishing.

If the percentage harvested exceeds the intrinsic growth rate for the population, then the harvest always exceeds the growth and the population dies out. From Figure 6.18, we know that 8% is the intrinsic growth rate for this population. Thus, if we harvest more than 8% per year, then the harvest line only intersects g at the origin and the population will die out. This also means that the size of the yearly harvest will slowly decrease to 0 over time. This is what was happening to the whales, which have a relatively small intrinsic growth rate.

The moral of the previous analysis is that, for some populations, we can get the same maximum sustainable harvest using a fixed harvest or a proportional harvest but the proportional harvest is safer.

In general, an expression for the growth function g cannot be found, but if we have some idea about the shape of the function g, we can make suggestions about appropriate harvesting strategies that protect the population.

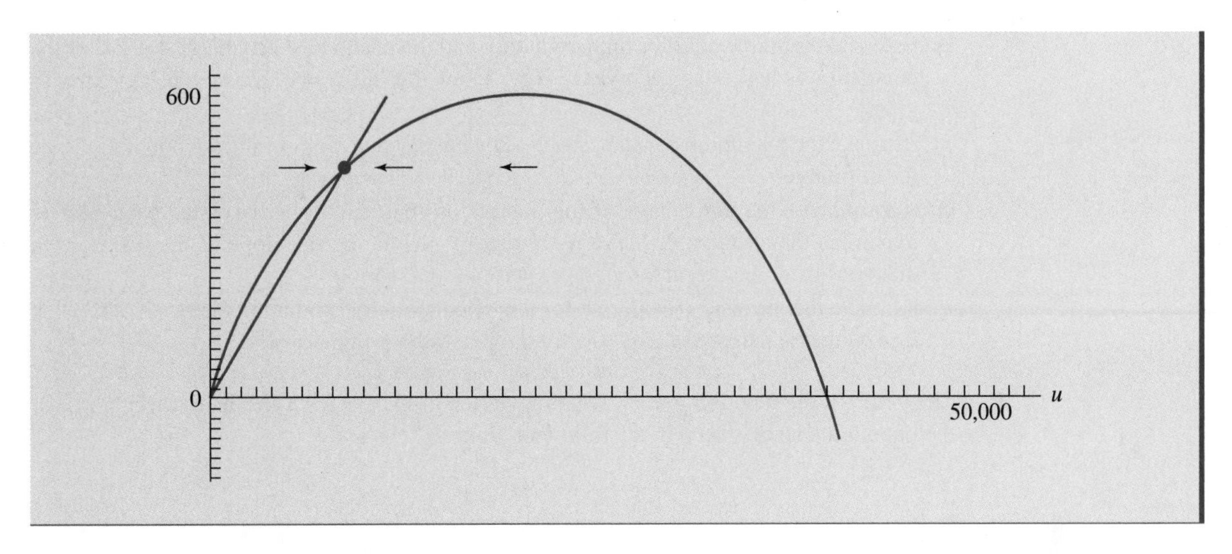

FIGURE 6.26 Harvest is $h = 0.05u$.

In the Problems, you will explore the results of constant and proportional rates of harvest for a variety of growth functions.

6.4 Problems

1. The graph of the function $g = ru$, which gives the growth of a species in terms of the population size, is seen in the following figure.

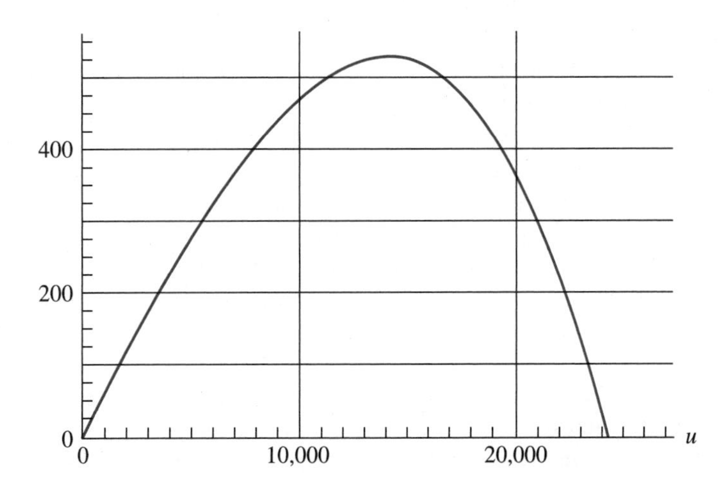

a. Estimate the stable equilibrium population and the minimum viable population if there is a constant yearly harvest of 200.

b. Estimate the stable equilibrium population and the eventual yearly harvest if 2% of the population is harvested each year. (*Hint:* Draw the line $h = 0.02u$ on the same axis as $g(u)$.

c. Estimate the maximum constant sustainable harvest and the equilibrium population size for this harvest.

d. Approximate the percentage of the population that should be harvested each year to maximize the sustainable harvest. Do this by estimating the slope of the line $h = pu$ that goes from the origin through the vertex of this graph.

e. Estimate the intrinsic growth rate for this population by estimating the slope of the line $h = pu$ that is tangent to the growth function at the origin.

2. The graph of the function $g = ru$, which gives the growth of a species in a year in terms of the population size, is seen in the following figure.

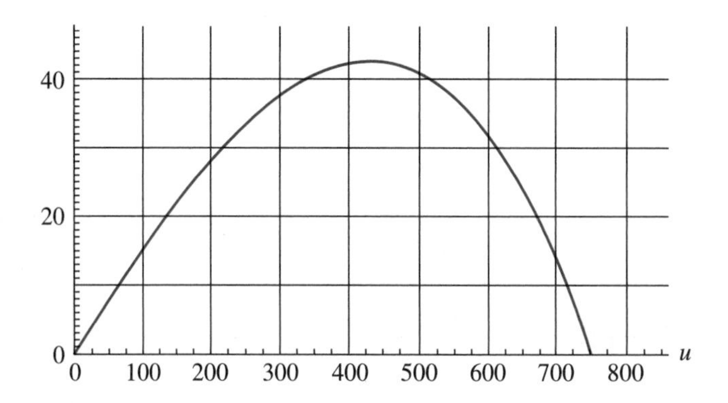

a. Estimate the stable equilibrium population and the minimum viable population if there is a constant yearly harvest of 30.

b. Estimate the stable equilibrium population and the eventual yearly harvest if 3% of the population is harvested each year.

c. Use this graph to estimate the maximum constant sustainable harvest and the equilibrium population size for this harvest.

d. Approximate the percent of the population that should be harvested each year to maximize the sustainable harvest by estimating the slope of the line $h = pu$ that goes from the origin through the vertex of this graph.

e. Estimate the intrinsic growth rate for this population by estimating the slope of the line $h = pu$ that is tangent to the growth function at the origin.

3. The graph of the function $g = ru$, which gives the growth of a species in a year in terms of the population size, is seen in the following figure.

a. Estimate the stable equilibrium population and the minimum viable population if there is a constant yearly harvest of 200.

b. Estimate the stable equilibrium population and the eventual yearly harvest if 20% of the population is harvested each year.

c. Use this graph to estimate the maximum sustainable harvest and the equilibrium population size.

d. Approximate the percentage of the population that should be harvested each year to maximize the sustainable harvest. For this level of effort, estimate the minimum viable population.

e. Estimate the proportion of the population harvested, p, that will result in a semistable equilibrium. This also equals the population's intrinsic growth rate.

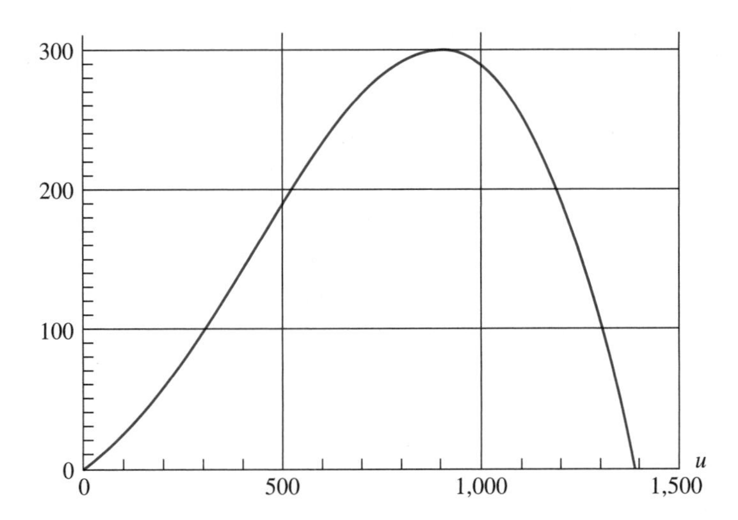

4. The graph of the function $g = ru$, which gives the growth of a species in a year in terms of the population size, is seen in the following figure. This is the growth function for a population that has negative growth for small populations; that is, a population whose model exhibits critical depensation.

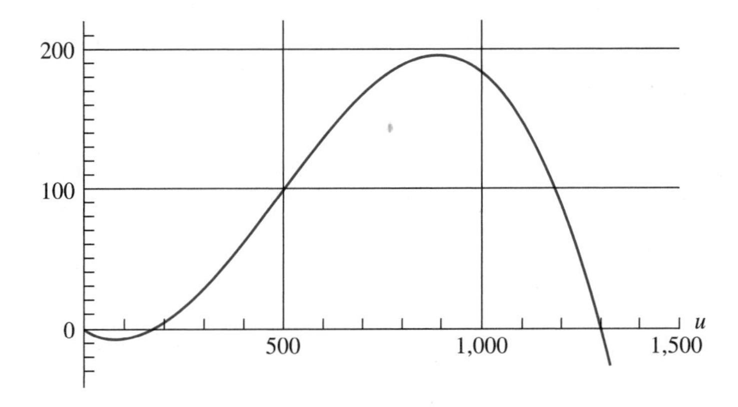

a. Estimate the stable equilibrium population and the minimum viable population if there is a constant yearly harvest of 100.

b. Estimate the stable equilibrium population and the eventual yearly harvest if 10% of the population is harvested each year.

c. Estimate the maximum sustainable harvest and the equilibrium population size.

d. Approximate the percentage of the population that should be harvested each year to maximize the sustainable harvest. For this level of effort, estimate the minimum viable population.

e. Estimate the proportion of the population harvested, p, that will result in a semistable equilibrium. This equals the intrinsic growth rate.

5. The graph of the function $g = ru$, which gives the growth of a species in a year in terms of the population size, is seen in the following figure.

a. The harvesting strategy is to harvest 100 plus an additional 0.5% of the population each year. Graph a linear function on the following graph that describes this harvesting policy. Develop a linear function $h = pu + b$ for this harvesting strategy; that is, find the values for p and b. What is the stable equilibrium population and the sustainable harvest? What is the minimum viable population?

b. The harvesting strategy is to harvest 100 plus an additional percentage of the population each year. About what percentage should be harvested to maximize the sustainable harvest? In this case, what is the minimum viable population size?

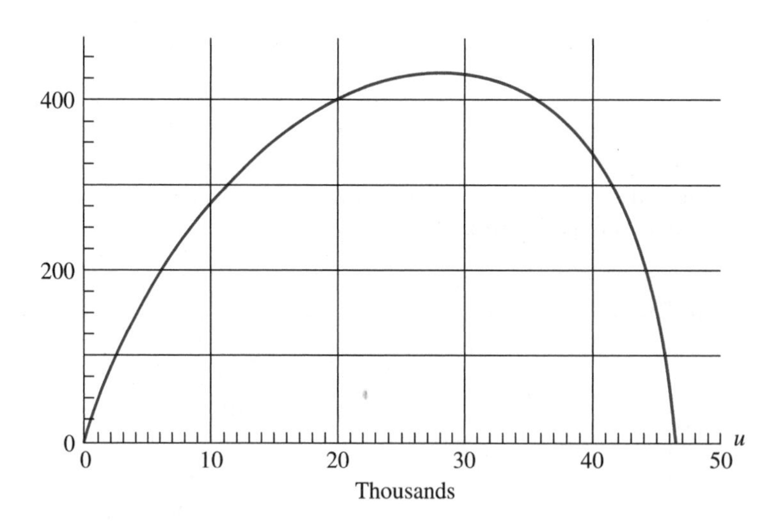

6. The graph of the function $g = ru$, which gives the growth of a species in a year in terms of the population size, is seen in the following figure.

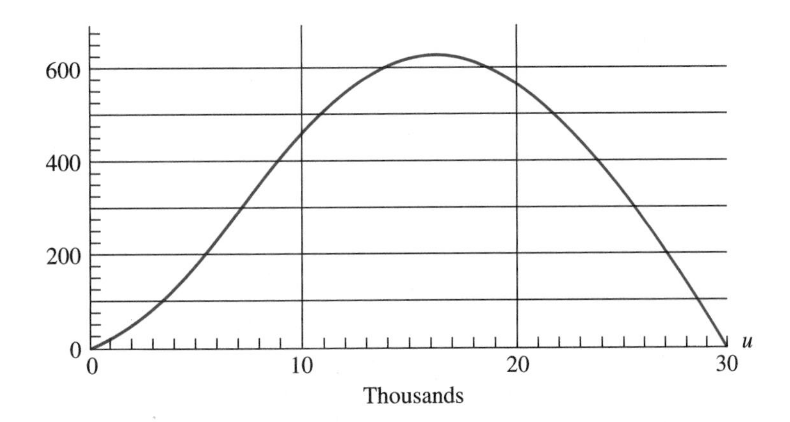

a. The harvesting strategy is to harvest 200 plus an additional 1% of the population each year. Graph a linear function on the following graph that describes this harvesting policy. Develop a linear function $h = pu + b$ for this harvesting strategy; that is, find the values for p and b. What is the stable equilibrium population and the sustainable harvest? What is the minimum viable population?

b. The harvesting strategy is to harvest 200 plus an additional percentage of the population each year. About what percentage should be harvested to maximize the sustainable harvest? In this case, what is the minimum viable population size?

7. The graph of the function $g = ru$, which gives the growth of a species in a year in terms of the population size, is seen in the following figure. The harvesting strategy is to harvest b plus an additional 3% of the population each year. About what value for b results in maximum sustainable harvest?

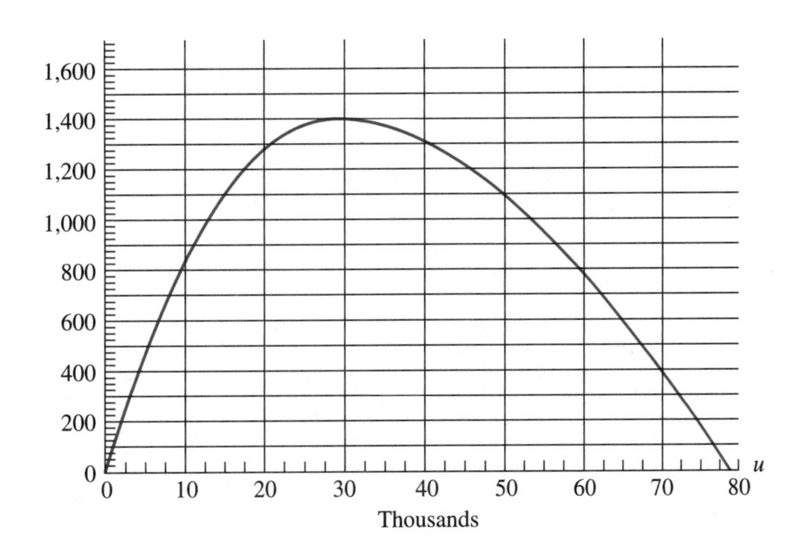

8. The graph of the function $g = ru$, which gives the growth of a species in a year in terms of the population size, is seen in the following figure. The harvesting strategy is to harvest b plus an additional 7% of the population each year. About what value for b results in maximum sustainable harvest?

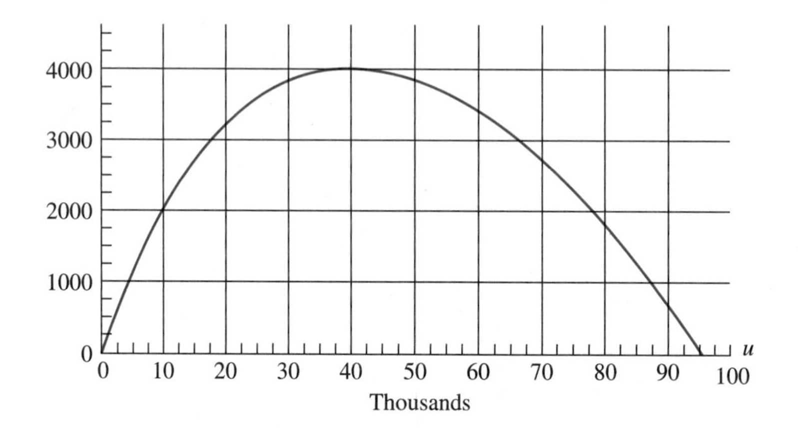

9. The graph of the function $g = ru$, which gives the growth of a species in a year in terms of the population size, is seen in the following figure.

a. The harvesting strategy is to harvest 3% of the population in excess of 10,000. For example, if the population size was 30,000, then the harvest would be 3% of 20,000 or 600. Develop a function $h = pu + b$ for this harvesting strategy. Graph this line on the figure, and use it to estimate the equilibrium population and the sustainable harvest.

b. The harvesting strategy is to harvest a fixed percentage of the population in excess of 10,000. What should that percentage be so that at the equilibrium population, the maximum sustainable harvest is obtained?

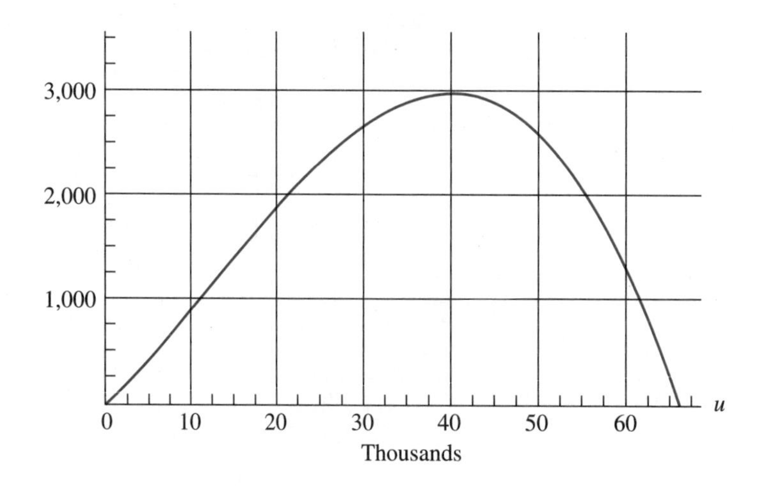

10. The graph of the function $g = ru$, which gives the growth of a species in a year in terms of the population size, is seen in the following figure.

a. The harvesting strategy is to harvest 2% of the population in excess of 20,000. Develop a function $h = pu + b$ for this harvesting strategy. Graph this line on the figure and use it to estimate the equilibrium population and the sustainable harvest.

b. The harvesting strategy is to harvest a fixed percentage of the population in excess of 20,000. What should that percentage be so that at the equilibrium population, the maximum sustainable harvest is obtained?

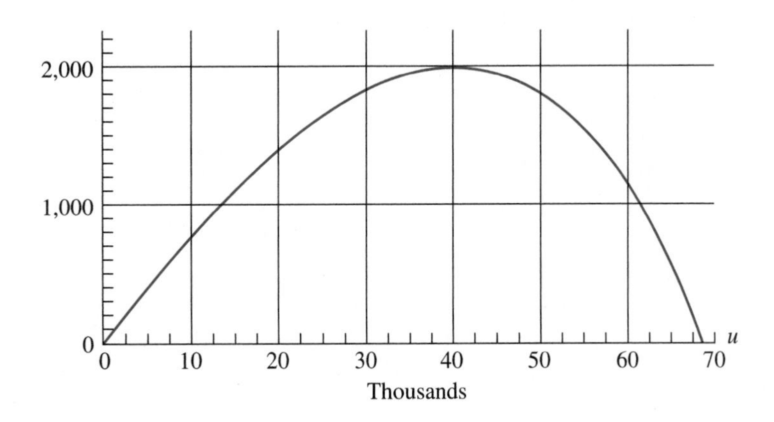

6.5 Analytic Approach to Harvesting

In this section, we are going to develop explicit expressions for the growth rate r and the growth function $g = ru$. We will then analyze the expressions, finding equilibrium population sizes for different harvesting strategies. The purpose is to determine what effect the shape of the growth rate function r has on harvesting strategies. Recall that in Example 6.1, we developed the function

$$r = 0.2 - 0.000025u$$

to model the linearly decreasing growth rate for a species with an intrinsic growth rate of 20% and a carrying capacity of 8000. This means that the growth function for this population is parabolic

$$g = 0.2u - 0.000025u^2$$

Let's suppose that at the end of each year, we harvest or remove 175 fish from this population, that is, $h = 175$. In the last section we learned that to find the equilibrium population, we only need to determine when the growth equals the harvest. This means we have to solve the equation $g = h$ or

$$0.2u - 0.000025u^2 = 175$$

Bringing all the terms to the same side gives

$$-0.000025u^2 + 0.2u - 175 = 0$$

Using the quadratic formula gives that there are 2 equilibrium values, $E = 1000$ and $E = 7000$. Alternatively, we can graph the functions

$$g = -0.000025u^2 + 0.2u \quad \text{and} \quad h = 175$$

using some form a technology, then find the points of intersection, which are seen in Figure 6.27. When doing this, we need to be careful in choosing an appropriate viewing window.

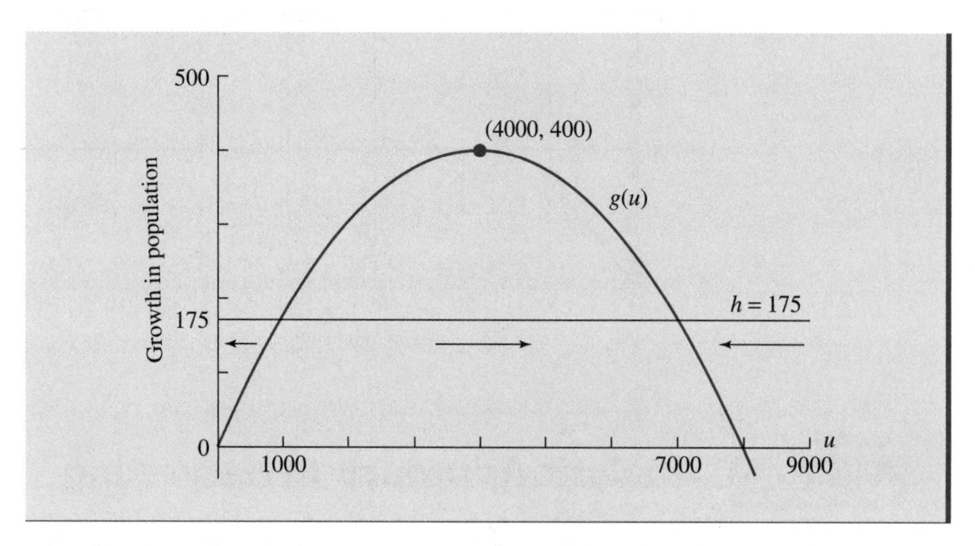

FIGURE 6.27 Graph of growth and harvest indicates $E = 7000$ is stable population and $E = 1000$ is unstable.

From the graph in Figure 6.27 and what we learned in Section 6.4, it is clear that $E = 1000$ is unstable and $E = 7000$ is stable. This means that if the initial population size is above 1000, we can harvest 175 of these fish each year and the population will stabilize at 7000. Thus, 175 is a sustainable yield, as long as the size of the population remains above 1000. The minimum viable population is 1000, given a harvest of 175 per year. If the population size drops below 1000, then a harvest of 175 per year will drive the population to extinction.

REMARK Alternatively, we could model this situation with the dynamical system

$$u(n) = u(n-1) + g(u(n-1)) - h$$

which after substitution becomes

$$u(n) = u(n-1) + 0.2u(n-1) - 0.000025u^2(n-1) - 175$$

If we let $u(0) = 8000$, then we find that $u(1) = 7825$, $u(2) = 7684$, $u(10) = 7170$, $u(20) = 7033$, and $u(50) = 7000$. It appears that the population size has now stabilized at the equilibrium population of 7000. We could make a time graph to analyze this dynamical system, or we could make webs. But the approach of comparing g to h, as in Figure 6.27, is easier and gives the same information.

To find the maximum sustainable yield, we only need to find the vertex of the parabola $g = 0.2u - 0.000025u^2$.

Let's review some facts about parabolas. Suppose we have a parabola $y = f(x)$. One method for finding the vertex of a parabola is to use the fact that a parabola is symmetric about the vertical line that goes through its vertex. This means that if we find the 2 points at which the parabola goes through the x-axis (assuming the parabola actually goes through the x-axis), then the x-coordinate of the vertex is halfway between those 2 points. For example, consider the parabola

$$y = x^2 - 4x - 12$$

We find the x-intercepts by solving

$$0 = x^2 - 4x - 12$$

Factoring gives

$$0 = (x - 6)(x + 2)$$

so the x-intercepts are at $x = 6$ and $x = -2$. The x-coordinate of the vertex is halfway between these values, at $x = 2$. The y-value of the vertex is found by substitution into the equation for the parabola

$$y = 2^2 - 4(2) - 12 = -16$$

A sketch of this parabola is seen in Figure 6.28.

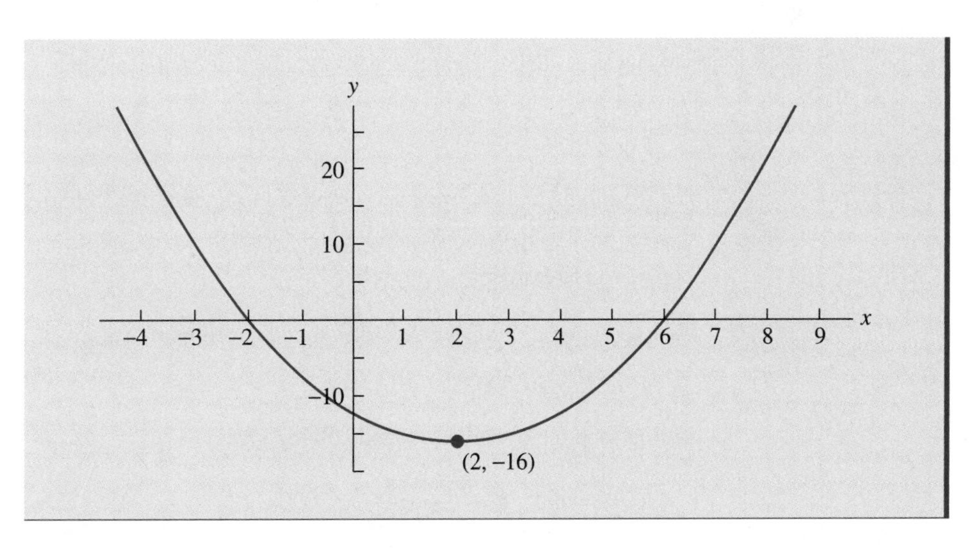

FIGURE 6.28 **Vertex of parabola is halfway between x-intercepts.**

Consider the growth function $g = 0.2u - 0.000025u^2$. We solve the equation $0 = 0.2u - 0.000025u^2$ to get the horizontal intercepts, $u = 0$ and $u = 8000$. This means that the vertex of this parabola is at $u = 4000$. The g-value of the vertex is $g = 0.2(4000) - 0.000025(4000)^2 = 400$, which is shown in Figure 6.27.

If the harvest is above 400, the horizontal line does not intersect the graph of g. In this case, no matter what the population size, the harvest is larger than the growth each year, and the population will begin to die out.

In this problem, the maximum sustainable harvest is at $h = 400$. At this level, the horizontal line is tangent to the vertex of the parabola. The intersection point is $(4000, 400)$ meaning that if the harvest is 400 per year and the population size is at 4000, then the population size will remain at 4000. But for any other population size, the growth is less than the harvest and the population size will decrease. If the population size is above 4000, it will decrease toward 4000. If the population size is below 4000, it will decrease toward extinction. Thus, 4000 is a semistable equilibrium population size.

To review, suppose we have the function $g = ru$ that gives the growth of the population as a function of population size. Suppose we have a fixed yearly harvest of size h. If the horizontal line of height h intersects the graph of g, then h is a sustainable harvest and the points of intersection give the equilibrium populations. The largest value for h at which the horizontal line intersects g is the maximum sustainable harvest, but this constant harvest puts the population at risk because the equilibrium value is semistable.

We are now going to assume that the harvest is given by the equation

$$h = pu$$

where u is the population size and p is the fixed proportion of the fish that are caught. The value for p depends on the effort put into fishing. This is called the **catch per unit effort** hypothesis.

Let's see what proportional harvesting means for the species with growth function

$$g = 0.2u - 0.000025u^2$$

which was studied in Example 6.1. Assume that each year's harvest is 2.5% of the population, that is,

$$h = 0.025u$$

The function g and the line h are graphed in Figure 6.29. To find the points of intersection, we find the points where $g = h$; that is, we solve

$$0.2u - 0.000025u^2 = 0.025u$$

Subtracting $0.025u$ from both sides, then factoring out a u, gives

$$u(0.175 - 0.000025u) = 0$$

Setting each factor equal to 0 and solving gives

$$u = 0 \quad \text{and} \quad u = \frac{0.175}{0.000025} = 7000$$

The points at which the line and the curve intersect are therefore $(0, 0)$ and $(7000, 175)$. The point $(7000, 175)$ gives both the positive equilibrium population size and the sustainable harvest using this harvesting strategy.

FIGURE 6.29 Growth and proportional harvest indicate 7000 is stable population.

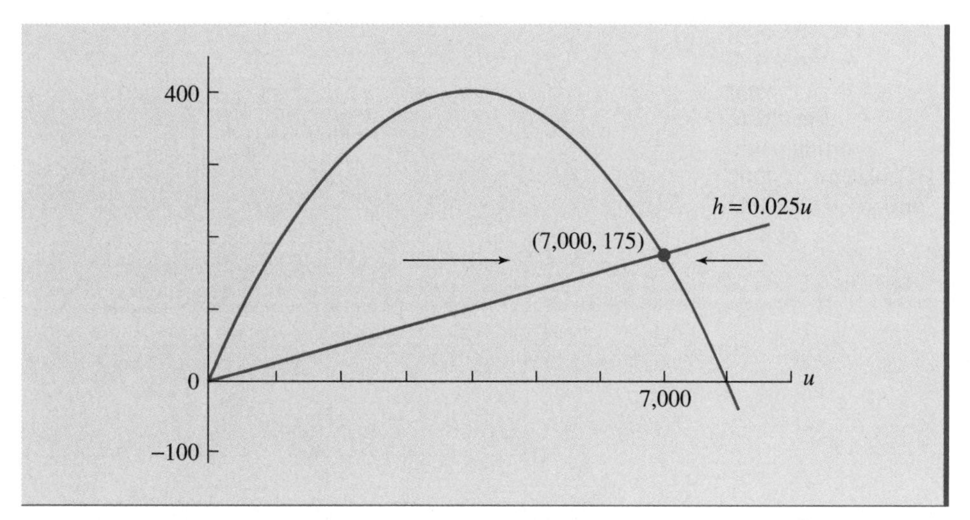

From Figure 6.29, we see that if the population size is below 7000, then the harvest is less than the growth in population that year and so the population size will increase. If the population size is above 7000, then the harvest exceeds the growth and the population size will decrease that year. Thus it appears that 7000 is a stable equilibrium value, as we have already concluded. An advantage of proportional harvest over fixed harvest in this case is that there is no minimum viable population. The species is safe.

For this species, if the effort is increased slightly, then the slope of the line $h = pu$ is increased slightly. As you can see from Figure 6.29, if you increase the slope of the line slightly, the point of intersection only moves a little. This means that small changes in effort result in small changes in the stable equilibrium population and small changes in the sustainable harvest.

We now show how this graphical analysis helps us easily determine the value for p that results in the maximum sustainable harvest. All we need do is find the slope, p, that results in the line $h = pu$ going through the vertex of the parabola. We know that the line goes through the point $(0, 0)$. We want it to go through the vertex, which is at $(4000, 400)$. The slope of the line must be

$$p = \frac{400 - 0}{4000 - 0} = 0.1$$

An effort that results in 10% of the fish being harvested each year gives a harvest equation of

$$h = 0.1u$$

The graphs of g and the line h are seen in Figure 6.30. The point of intersection is the vertex of the parabola $(4000, 400)$ and gives both the equilibrium population and the maximum sustainable harvest.

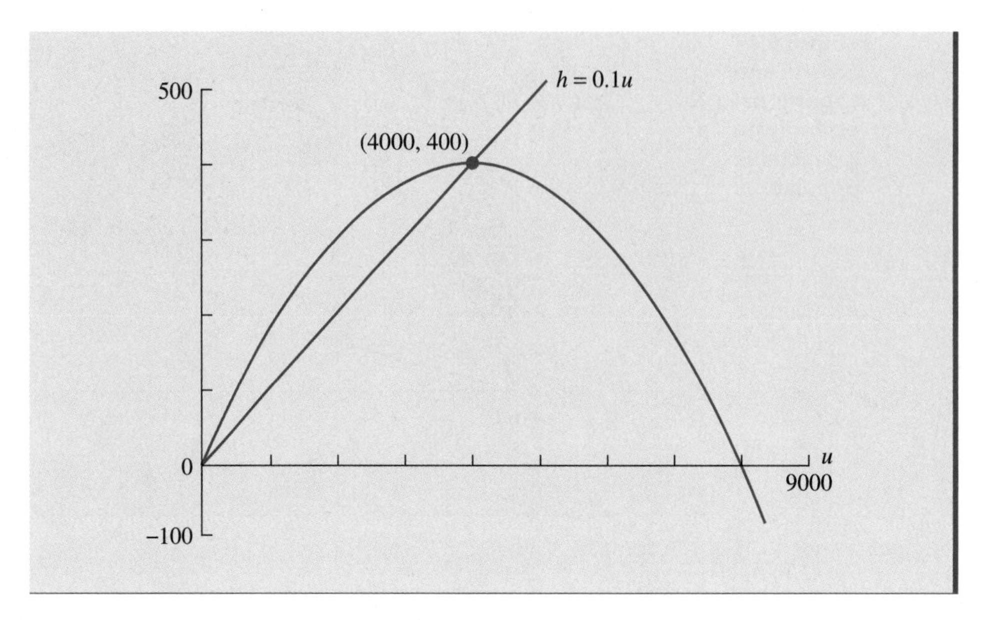

FIGURE 6.30 A harvest of 10% each year results in equilibrium population of 4000 and yearly harvest of 400.

Suppose we increase our effort to $p = 0.15$, meaning that 15% of the fish are harvested each year. The points of intersection are $(0, 0)$ and $(2000, 300)$, which can be seen in Figure 6.31. This means that the sustainable harvest is 300 per year.

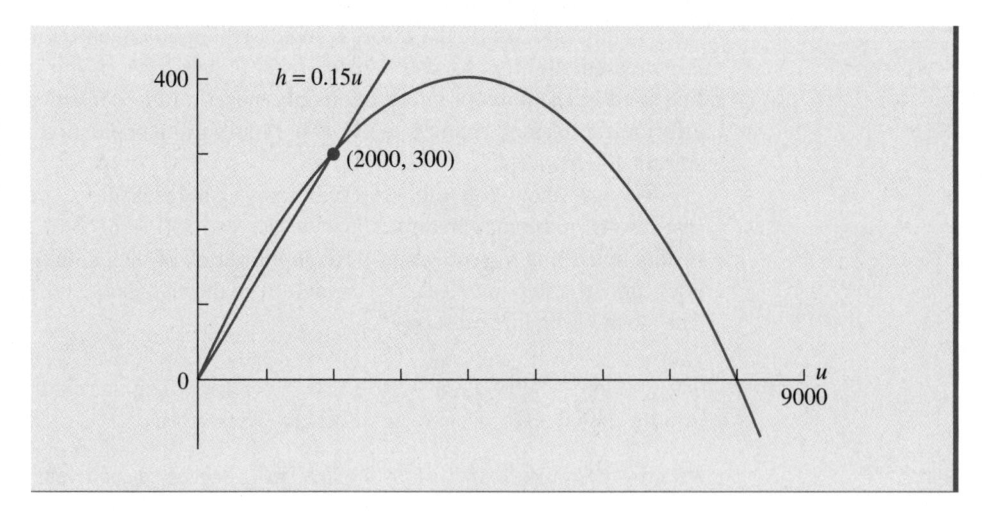

FIGURE 6.31 Harvest of 15% per year results in equilibrium population of 2000 and yearly harvest of 300.

Increasing the effort from 10% caught to 15% caught lowers the actual sustainable harvest. It will cost more money in boats and labor to increase the effort, but the result is fewer fish. Thus, it would seem unwise to increase the effort beyond where the maximum sustainable harvest occurs, but in practice, it appears that the effort often exceeds the optimal effort.

It would seem that the proportional harvest is far superior to the constant harvest, at least in terms of the safety of the population being harvested. But problems can occur using the proportional-effort harvest also. If the proportion harvested is too large, then there is no positive equilibrium value, as seen in Figure 6.32, in which 30% of the species is harvested each year.

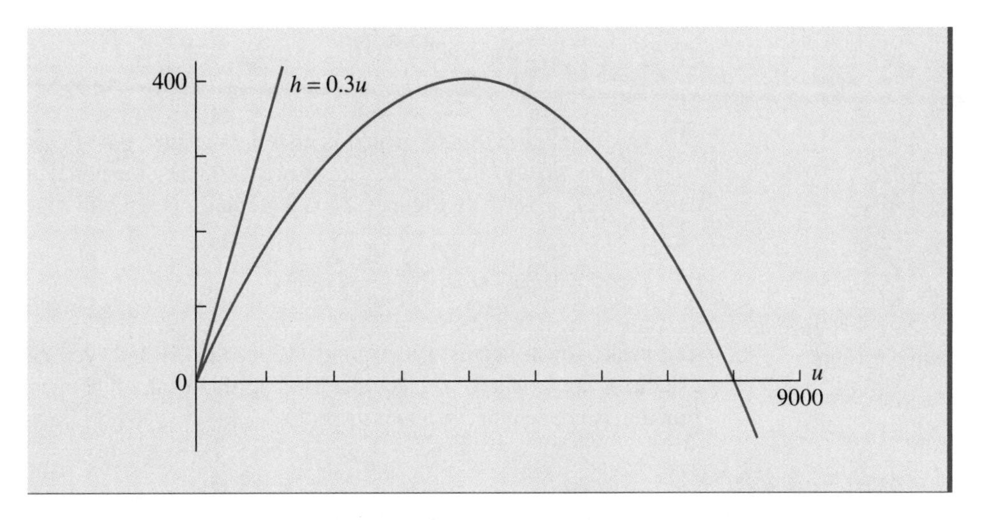

FIGURE 6.32 **A harvest of 30% results in no positive equilibrium and the population goes toward extinction.**

Let's find the values for p that result in no positive equilibrium. To do this, we first set the rate of harvest equal to the growth,

$$0.2u - 0.000025u^2 = pu$$

and solve for u. Bringing all the terms to the right and factoring gives

$$0 = (0.000025u + p - 0.2)u$$

Setting each factor equal to 0 gives the equilibrium values

$$E = 0 \quad \text{and} \quad E = \frac{0.2 - p}{0.000025}$$

Our problem is to determine the values for p for which there are no positive equilibrium values. There are no positive equilibrium values when

$$\frac{0.2 - p}{0.000025} \leq 0$$

Multiplying both sides by 0.000025, then adding p to both sides gives that when

$$p \geq 0.2$$

there is no positive equilibrium. This means that when the rate of harvest, p, exceeds the intrinsic growth rate 0.2, or 20%, then the population goes toward extinction.

We note that when $p = 0.2$, then the line $h = 0.2u$ is tangent to the parabola at the origin.

Example 6.4

In this example, we will compare constant harvest and proportional harvest for a species that satisfies a depensation model. We will find disadvantages to both approaches.

Consider a population whose growth rate is given by

$$r = 0.3 - 0.3\left(\frac{u - 900}{900}\right)^2$$

The maximum growth rate, which is also the intrinsic growth rate, is $b = 0.3$ and occurs when the population size is 900. The growth is 0 when the population size is at the carrying capacity of $L = 1800$. In Figure 6.33 is a graph of the growth function

$$g = ru = 0.3u - 0.3u\left(\frac{u - 900}{900}\right)^2$$

The peak growth occurs at the point $(1200, 320)$, which you can find using a graphing calculator, a computer, or calculus. Finding this peak algebraically is difficult, since the function is not symmetric about its roots.

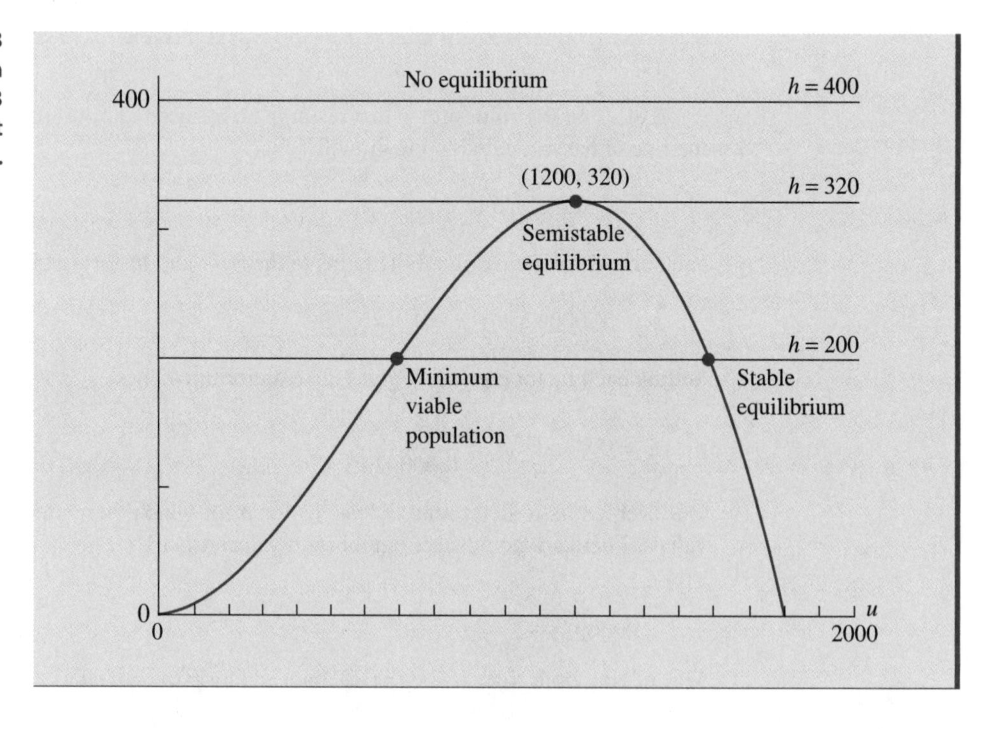

FIGURE 6.33 Depensation model with 3 different constant rates of harvest.

Constant Rate of Harvest

Suppose we are harvesting a constant amount each year. To determine the equilibrium population for this harvest, we draw a horizontal line on the graph of Figure 6.33at the height of this harvest. The equilibrium values will be the u-values where the line intersects the curve. There are 3 possible results.

1. If the constant rate of harvest is less than 320, the line intersects the curve at 2 points. The point on the left gives the minimum viable population, and the point on the right gives the stable population size. This can be seen by the line $h = 200$ in Figure 6.33. The points of intersection are found using a graphing tool.
2. If the yearly harvest is greater than 320, there are no intersection points and the population will decrease to extinction. This can be seen by the line $h = 400$ in Figure 6.33.
3. If the yearly harvest equals 320, the horizontal line $h = 320$ intersects the peak of the curve. This means that we could have a constant maximum sustainable harvest of about 320 per year and a population size of 1200 will be semistable.

Proportional Harvest We now want to find the value p that results in the maximum sustainable harvest, where the harvest is given by the equation

$$h = pu$$

We want this line to go through the origin and through the point $(1200, 320)$. Thus, its slope is

$$p = \frac{320}{1200} = \frac{4}{15} \approx 0.267$$

which means a harvest of about 26.7% per year. The line $h = (4/15)u$ is seen in Figure 6.34, along with the graph of g.

In this case, there are 3 points of intersection—$(0, 0)$, $(1200, 320)$, and $(600, 160)$. These points can be found using a graphing tool or by solving

$$0.3u - 0.3u\left(\frac{u - 900}{900}\right)^2 = \frac{4}{15}u$$

algebraically. Clearly, one solution is $u = 0$. Dividing both sides by u gives

$$0.3 - 0.3\left(\frac{u - 900}{900}\right)^2 = \frac{4}{15}$$

We could multiply out all of these terms and use the quadratic formula, but there is an easier way. Subtract 0.3 from both sides, giving

$$-0.3\left(\frac{u - 900}{900}\right)^2 = -\frac{1}{30}$$

Divide both sides by -0.3, which gives

$$\left(\frac{u - 900}{900}\right)^2 = \frac{1}{9}$$

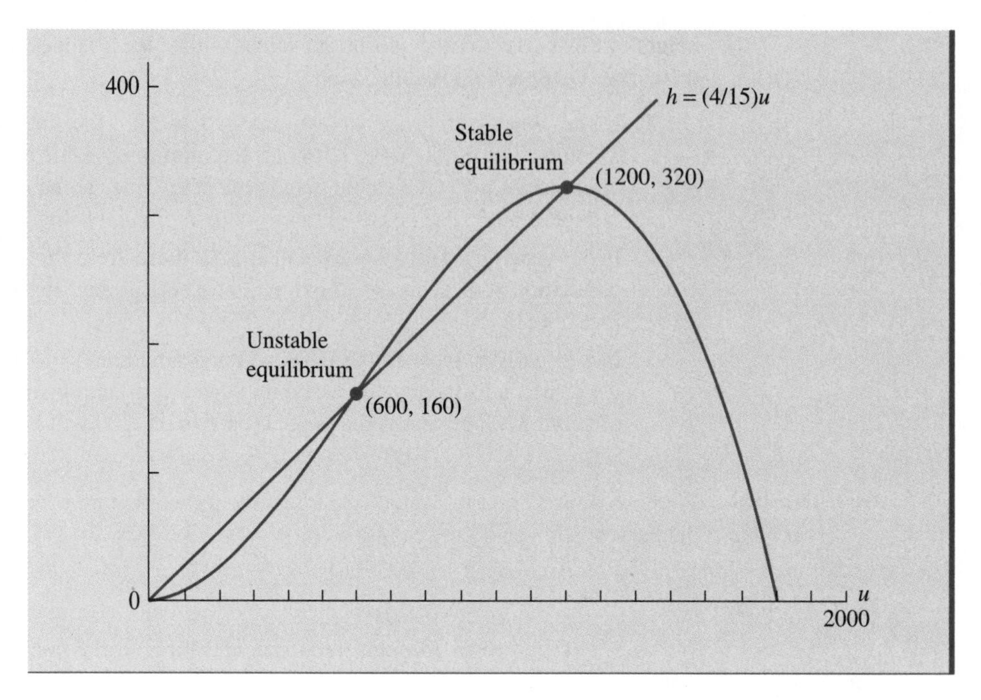

FIGURE 6.34 Harvesting 4/15 of the population results in a sustainable harvest of 320 with a stable population of 1200. The minimum viable population is 160.

Take the square root of both sides, remembering that the square root of the right side can be positive or negative. This gives

$$\frac{u-900}{900} = \frac{1}{3} \quad \text{or} \quad \frac{u-900}{900} = -\frac{1}{3}$$

The solutions to these 2 equations, which are the equilibrium population sizes, are $u = 1200$ and $u = 600$.

Figure 6.34 indicates that if the population size is between 600 and 1200, then the growth exceeds the harvest and the population size will increase. If the population size is above 1200, then the harvest exceeds the natural growth and the population decreases in size toward 1200. Thus, $E = 1200$ appears to be a stable equilibrium population and there is a maximum sustainable harvest of 320 of the species per year.

Similar analysis shows that $E = 600$ is an unstable equilibrium value and if the population drops below 600, then a harvest of about 26.7% per year will drive the population to extinction. So 600 is the minimum viable population size for this harvesting effort. Note that this also means that 0 is a stable equilibrium value for the population size.

For this population, even harvesting a fixed proportion of the population each year puts the population at risk.

Now observe that if the harvesting effort is increased, which is modeled by a line with steeper slope, the stable and unstable equilibrium values get closer together. How do we find the value of p for which there is only one positive equilibrium value, as seen in Figure 6.35? To find the value for p where the harvest line is tangent to the growth function, we solve $g = h$ or

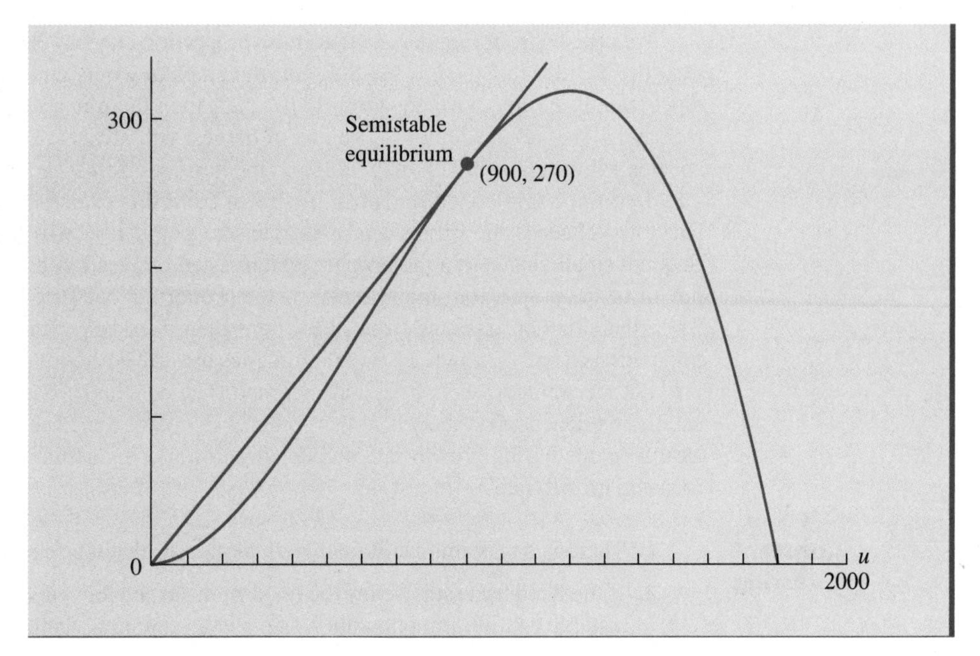

FIGURE 6.35 Harvesting 30% per year results in semistable equilibrium population of 900 and yearly harvest of 270.

$$0.3u - 0.3u\left(\frac{u-900}{900}\right)^2 = pu$$

We are looking for positive values of u, so we can divide both sides of the equation by u, giving

$$0.3 - 0.3\left(\frac{u-900}{900}\right)^2 = p$$

After subtracting 0.3 from both sides, then dividing both sides by -0.3, we get

$$\left(\frac{u-900}{900}\right)^2 = \frac{0.3-p}{0.3}$$

Taking square roots gives

$$\frac{u-900}{900} = \pm\sqrt{\frac{0.3-p}{0.3}}$$

There will be 2 solutions unless the term under the square root equals 0, in which case there is one solution. Thus, there is only one solution when

$$p = 0.3$$

that is, when the proportion harvested equals the intrinsic growth rate. In this case, $E = 900$ is a semistable equilibrium population.

We can see from Figure 6.35 that if the proportion exceeds 30% per year, the intrinsic growth rate, then the line $h = pu$ intersects the graph of g only at the origin. We can see this algebraically in the above equation. If $p > 0.3$, then there is a negative under the square root and there are no solutions. The rate of harvest will always exceed the growth, and the species will head toward extinction unless the harvest effort is reduced.

Two important conclusions can be drawn from this example. First, if the proportion harvested exceeds the intrinsic growth rate, the population will head toward extinction. Second, small changes in the proportion harvested can lead to critical changes in the behavior of the population. In particular, if the proportion harvested increases from 29% to 31%, the situation changes from one where there is a large stable positive equilibrium population to one in which the population goes toward extinction.

Let me summarize what we have learned. The conclusions are true for both compensation and depensation models, unless otherwise noted. Recall that compensation models assume a decreasing growth rate and depensation models assume an increasing, then decreasing growth rate.

Constant Harvesting

1. There is a maximum sustainable yield that results in a semistable population size.
2. If the fixed harvest is below the maximum sustainable yield, then there will be both a stable equilibrium population size and a minimum viable population size.

Figure 6.36 demonstrates these 2 conclusions. Both conclusions for constant harvesting are somewhat negative. In fact, agencies controlling fishing have estimated the maximum sustainable yield and have used this for the annual allowable harvest. It has now become clear that this is a dangerous approach, and the allowable harvest is now reduced to give a safety net. We know that this approach is also dangerous because of conclusion 2. The question is, How much of a reduction in constant harvest is "safe enough"?

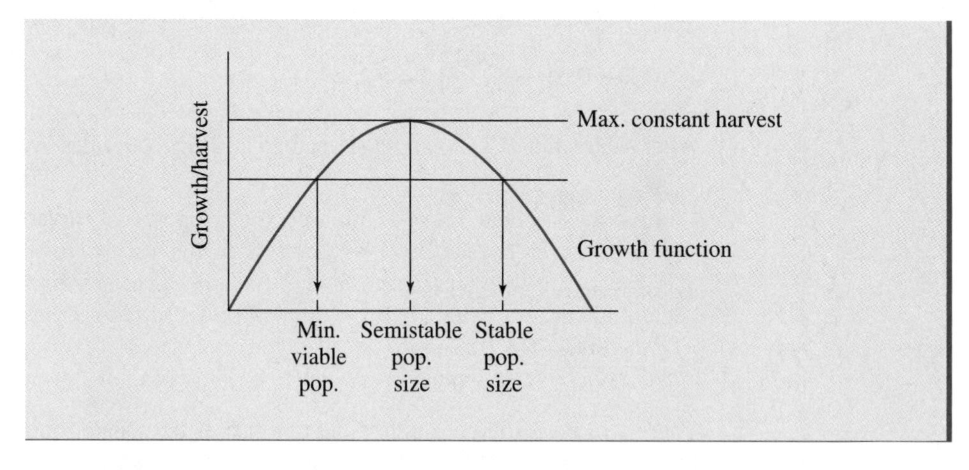

FIGURE 6.36 Demonstration of two conclusions concerning constant harvest.

Proportional Harvesting

1. To have a positive, stable equilibrium population, the proportion harvested must be less than the intrinsic growth rate; that is, $p < b$. (For a depensation model, $p = b$ results in a semistable positive equilibrium population, which also puts the population at risk of extinction.)

2. There is some value for p, the proportion harvested, that results in the maximum sustainable yield. For that value, there is a stable equilibrium population size.
3. If the species satisfies a compensation model, there is no minimum viable population when $p < b$.
4. If the species satisfies a depensation model, there may be a minimum viable population, even when p is small.

Figure 6.37 demonstrates these 4 conclusions. These graphs have 6 possible proportional harvest lines drawn and labeled. Conclusion 1 is seen from the harvesting lines 1 and 4. These 2 lines both have slopes equal to the intrinsic growth rate. Line 4 intersects growth function at point corresponding to a semistable equilibrium population. Conclusion 2 is demonstrated by lines 2 and 5 which intersect the growth function at a point whose vertical value corresponds to the maximum sustainable harvest and whose horizontal value gives the stable equilibrium population size. Conclusion 3 is demonstrated by line 3. The intersection of this line and the growth function corresponds to a stable equilibrium population size. There is no minimum viable population size. Conclusion 4 is demonstrated by line 6. Even though this line has a small slope, there is still a minimum viable population size corresponding to the lower point of intersection of this line with the growth function.

FIGURE 6.37 Graphs demonstrate 4 conclusions concerning proportional harvesting.

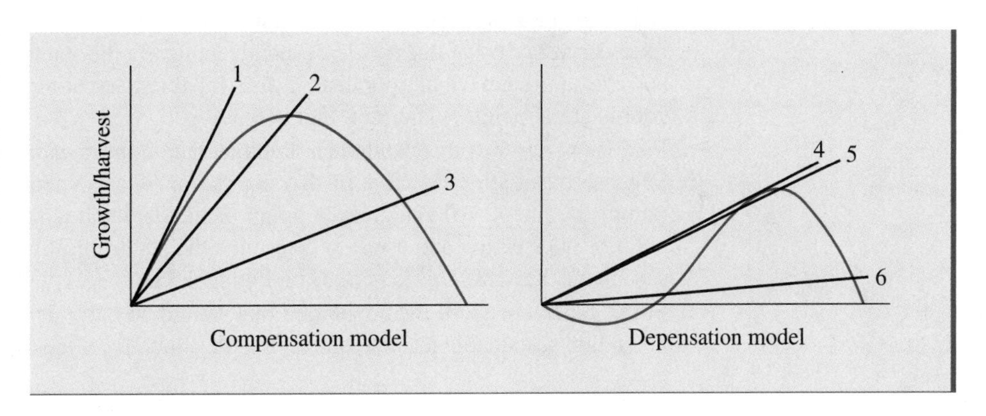

Conclusions 1, 2, and 3 are positive aspects of proportional harvesting. Notice that conclusion 3 only holds for some species. Conclusion 4 is negative for certain species. In particular, if a species satisfies a depensation model, there can be a repeated improvement in the harvesting strategy while the population still goes toward extinction. This can be caused by an initially large value for p, so that the population decreases in size. The effort is then reduced, but the population size is still below the minimum viable population corresponding to the new p-value. Thus, the population continues to decline. Again, p is reduced, but again the population is below the new minimum viable population. This behavior can be repeated, with the population continuing its decline toward extinction.

In practice, we cannot develop expressions for r and g because it is too difficult to obtain accurate information about the population, especially considering the other factors that can influence a species growth rate. Then you might ask, Why did we analyze species algebraically? The reason is the following. The algebraic analysis shows that when r is decreasing (a compensation model), then the growth function g is shaped like a parabola. The parabolic shape of g is why condition 3 holds. When the species satisfies a depensation model, meaning r is shaped like a parabola, then the growth function g has a "cubic" shape, in that it changes from "bowed," or concave, up to "bowed," or concave, down. In Figure 6.34, the function g is concave up for $u < 600$ and concave down for $u > 600$. The point where the shape changes, (600, 160) in Figure 6.34, is called a point of inflection. A little calculus makes it easy to find the point of inflection. This change in basic shape of g is why there is a minimum viable population for some values of p, which is why condition 4 holds.

When working the exercises, notice if r satisfies a compensation model or a depensation model. Then notice whether condition 3 or condition 4 holds.

6.5 Problems

1. Consider a species that satisfies the logistic equation; that is, the growth rate decreases linearly. The intrinsic growth rate is $b = 0.4$, and the carrying capacity is 1000.

a. Find an equation for the growth rate of this species.

b. Find a parabola that gives the actual growth g of this species in terms of the population size. The graph of this parabola is seen in Figure 6.38. Suppose that each year there is a fixed harvest of 64 of this species. Find algebraically the stable equilibrium population and the minimum viable population size for this harvesting strategy, that is, the u^2 coordinate of points A and B on Figure 6.38.

c. Find the vertex v of this parabola to find the maximum sustainable harvest.

d. Suppose that each year, 12% of this population is harvested. One point of intersection of the line $h = 0.12u$ and the graph of the parabola is the origin. Find the other point of intersection and use it to determine the stable equilibrium population and the sustainable harvest. This is point C on Figure 6.38.

e. Find the percentage of the population that should be harvested each year to obtain the maximum sustainable harvest, that is, find the slope p of line 1 in Figure 6.38.

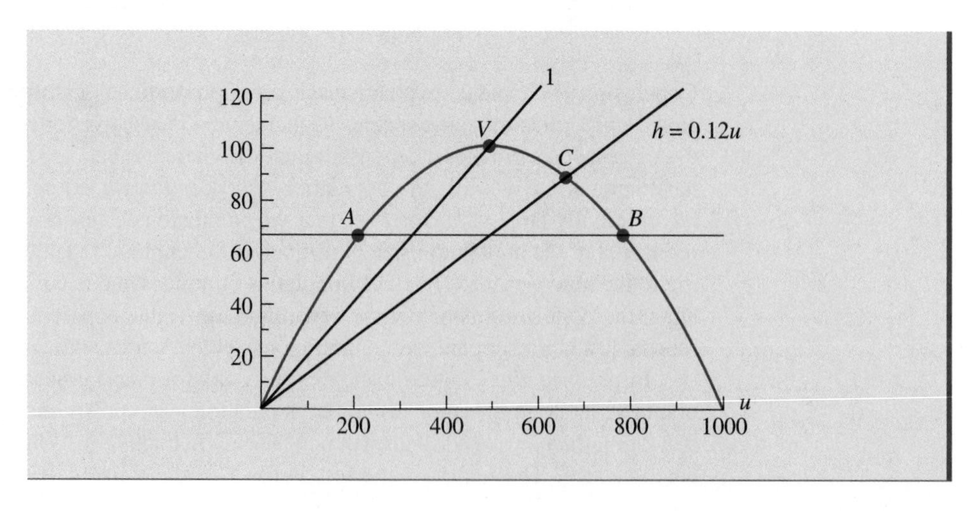

FIGURE 6.38 Graphs of growth and harvest function for Problem 1.

2. Consider a species that satisfies the logistic equation; that is, the growth rate decreases linearly. The intrinsic growth rate is $b = 0.3$, and the carrying capacity is 5000.

a. Find an equation for the growth rate of this species.

b. Find a parabola that gives the actual growth g of this species in terms of the population size. Suppose that each year there is a fixed harvest of 360 of this species. Find the stable equilibrium population and the minimum viable population size for this harvesting strategy.

c. Find the vertex of this parabola to find the maximum sustainable harvest.

d. Suppose that each year, 24% of this population is harvested. Find the point of intersection of the curve and line in the first quadrant and use it to determine the stable equilibrium population and the sustainable harvest.

e. Find the percentage of the population that should be harvested each year to obtain the maximum sustainable harvest. This results in a line $h = pu$ that goes through the vertex of the parabola.

3. Consider a species that satisfies the logistic equation, that is, the growth rate decreases linearly. The intrinsic growth rate is $b = 0.1$, and the carrying capacity is 10,000.

a. Find an equation for the growth rate of this species.

b. Find a parabola that gives the actual growth g of this species in terms of the population size. Suppose that each year there is a fixed harvest of 160 of this species. Find the stable equilibrium population and the minimum viable population size for this harvesting strategy.

c. Find the vertex of this parabola to find the maximum sustainable harvest.

d. Suppose that each year, 3.8% of this population is harvested. Find the point of intersection of the line $h = 0.038u$ and the graph of the parabola in which u is positive. Use this point to determine the stable equilibrium population and the sustainable harvest.

e. Find the percentage of the population that should be harvested each year to obtain the maximum sustainable harvest.

4. Consider a species that satisfies the logistic equation; that is, the growth rate decreases linearly. The intrinsic growth rate is $b = 0.8$, and the carrying capacity is 8000.

a. Find an equation for the growth rate of this species.

b. Find a parabola that gives the actual growth g of this species in terms of the population size. Suppose that each year there is a fixed harvest of 1311 of this species. Find the stable equilibrium population and the minimum viable population size for this harvesting strategy.

c. Find the vertex of this parabola to find the maximum sustainable harvest.

d. Suppose that each year, 47% of this population is harvested. Find the point of intersection of the line $h = 0.47u$ and the graph of the parabola in which u is positive. Use this point to determine the stable equilibrium population and the sustainable harvest.

e. Find the percentage of the population that should be harvested each year to obtain the maximum sustainable harvest.

5. Suppose that the growth rate for a species is given by the equation

$$r = 0.4 - 0.4\frac{u^2}{1000^2}$$

This is a population with an intrinsic growth rate of 0.4, a carrying capacity of 1000, and a nonlinear, decreasing growth rate.

a. Graph the growth function $g = ru$, which is a cubic. Estimate the maximum possible growth, which is also the maximum sustainable yield. Estimate the population size for that harvest.

b. Estimate the value for p, the fraction of the population that should be harvested to yield the maximum sustainable harvest.

6. Suppose that the growth rate for a species is given by the equation

$$r = 0.1 - 0.1\frac{u^2}{5000^2}$$

This is a population with an intrinsic growth rate of 0.1, a carrying capacity of 5000, and a nonlinear, decreasing growth rate.

a. Graph the cubic polynomial $g = ru$. Estimate the maximum possible growth, which is also the maximum sustainable yield. Estimate the semistable population size for that harvest.

b. Estimate the value for p, the fraction of the population that should be harvested to yield the maximum sustainable harvest.

7. Suppose that the growth rate for a species is given by the equation

$$r = 0.1 - 0.1\frac{(u - 500)^2}{2000^2}$$

This is a depensation model for a population with an intrinsic growth rate of 0.1 occurring at a population size of 500 and a carrying capacity of 2500.

a. Graph the growth function $g = ru$. Estimate the maximum possible growth, which is also the maximum sustainable yield. Estimate the semistable population size for that harvest.

b. Estimate the value for p, the fraction of the population that should be harvested to yield the maximum sustainable harvest. What is the minimum viable population?

c. Find, algebraically, the semistable equilibrium population size and the corresponding harvest when the proportion harvested equals the intrinsic growth rate; that is, $p = 0.1$.

8. Suppose that the growth rate for a species is given by the equation

$$r = 0.2 - 0.2\frac{(u - 1500)^2}{2000^2}$$

This is a depensation model for a population with an intrinsic growth rate of 0.2 occurring at a population size of 1500 and a carrying capacity of 3500.

a. Graph the growth function $g = ru$. Estimate the maximum possible growth, which is also the maximum sustainable yield. Estimate the semistable population size for that harvest.

b. Estimate the value for p, the fraction of the population that should be harvested to yield the maximum sustainable harvest. Estimate the minimum viable population.

c. Find, algebraically, the semistable equilibrium population size and the corresponding harvest when the proportion harvested equals the intrinsic growth rate; that is, $p = 0.2$.

9. Suppose that the growth rate for a species is given by the equation

$$r = 0.2 - 0.2\frac{(u-2500)^2}{2000^2}$$

This is a critical depensation model of a population with an intrinsic growth rate of 0.2 occurring at a population size of 2500 and a carrying capacity of 4500.

a. Graph the growth function $g = ru$. Estimate the maximum possible growth, which is also the maximum sustainable yield. Estimate the semistable population size for that harvest.

b. Estimate the value for p, the fraction of the population that should be harvested to yield the maximum sustainable harvest. Find the minimum viable population.

c. Find, algebraically, the semistable equilibrium population size and the corresponding harvest when the proportion harvested equals the intrinsic growth rate; that is, $p = 0.2$.

Project 1. Assume that a population has a linearly decreasing growth rate

$$r = b - \frac{b}{L}u$$

and assume a proportional harvest, $h = pu$. Show that the value for p that results in the maximum sustainable yield is half the intrinsic growth rate, that is,

$$p = \frac{b}{2}$$

Also show that the equilibrium population size in this case is half the carrying capacity; that is,

$$E = \frac{L}{2}$$

(In practice, agencies controlling harvesting have estimated the intrinsic growth rate for a population. They have used this estimate to control harvesting. The assumption is that the population will stabilize at half its carrying capacity, allowing us to estimate the carrying capacity. But using depensation models in which r decreases like a parabola shows that these assumptions may not be valid.)

6.6 Economics of Harvesting

In Sections 6.4 and 6.5, we were concerned with maximizing the sustainable harvest. This we will call the **social maximum** since it results in society having the largest possible supply of the species over an indefinite period of time. The social maximum is not always the goal in harvesting.

A second goal we will consider is **maximizing profit.** This is more complicated than it might initially seem. Specifically, we could be interested in maximizing the profit for the entire fishing industry or for one particular company or individual.

Let's consider maximizing profit from the point of view of a country. Fish caught and sold furnishes a major source of hard currency for many developing countries. This currency is often more important for the people of the country than is the food furnished by the fish. Suppose a country has control over the entire fishing industry. This country would then try to control the harvesting of the fish so as to maximize the profit for the entire industry over a sustained period of time. Thus, the country should develop long-term goals.

Let's consider maximizing profit from the point of view of the individual. An individual person is not interested in the country's profit. This person is interested in maximizing current profit to provide for his or her family. Thus, this person has short-term goals.

Let's reconsider the species of fish from Example 6.1, in which the intrinsic growth rate was $b = 0.2$, the carrying capacity was $L = 8000$, and the growth rate was linearly decreasing

$$r = 0.2 - 0.000025u$$

This resulted in the growth function

$$g = 0.2u - 0.000025u^2$$

We will assume that a fixed proportion of the fish is harvested each time period

$$h = pu$$

Recall that we found the equilibrium population size by solving the equation $h = g$ or

$$pu = 0.2u - 0.000025u^2$$

which simplifies to

$$p = 0.2 - 0.000025u$$

Solving for the equilibrium population in terms of the fraction harvested gives

$$u = 8000 - 40{,}000p$$

Since the harvest is $h = pu$, we have, by substitution, that the sustainable harvest given in terms of the fraction harvested is

$$h = 8000p - 40{,}000p^2$$

This means that the harvest is a parabolic function of the fraction harvested. As we did in Section 6.5, we can determine the fraction that should be harvested to maximize the sustainable harvest. To do this, we find the vertex of this parabola, which is $(0.1, 400)$ as seen in Figure 6.39. So the maximum sustainable harvest is 400 fish, which is obtained by harvesting 10% each year. We now take the economics of the situation into account.

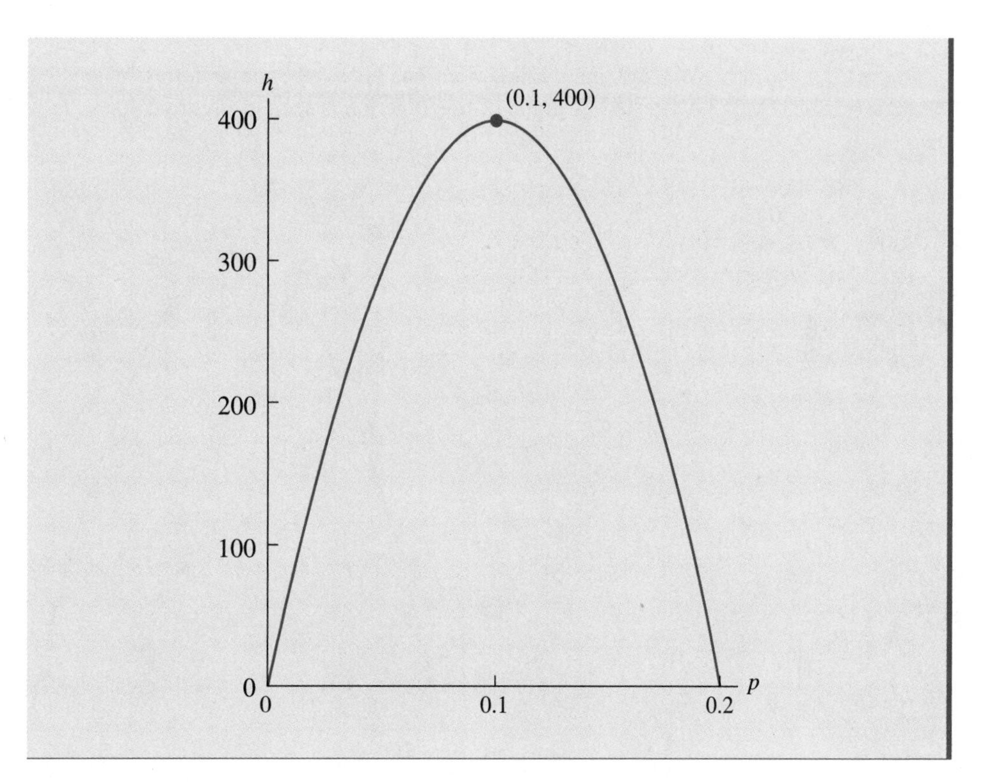

FIGURE 6.39 **Harvest as function of fraction harvested.**

First, we assume that there is some fixed price at which we can sell the fish, say \$2 per fish. This assumption might seem a little artificial. If there are more fish for sale, we might assume that the price offered per fish would go down. On the other hand, there are many different types of fish available for world consumption, and if the price of 1 fish increases, other fish may replace it in the marketplace. We are also looking at this problem from the point of view of one country. This one country may have only a small portion of the world market for this fish. So considering these aspects, assuming a fixed price per fish, no matter what the harvest, is not that unreasonable.

To make our pricing model more realistic, we should give the harvest, not in number of fish, but in weight. So we assume that $h = 400$ means 400 kg of fish, not 400 fish. The price would then be per kilogram, not per fish. This is more realistic, since the fish might come in all different sizes.

We now have a revenue function

$$\text{revenue} = \text{price per kg} \times \text{kg of fish}$$

which we write as

$$R = 2h$$

Thus, the revenue function is

$$R = 2h = 16{,}000p - 80{,}000p^2$$

We now have a function that gives revenue in terms of the fraction of the fish that are harvested. A graph of this function is seen in Figure 6.40.

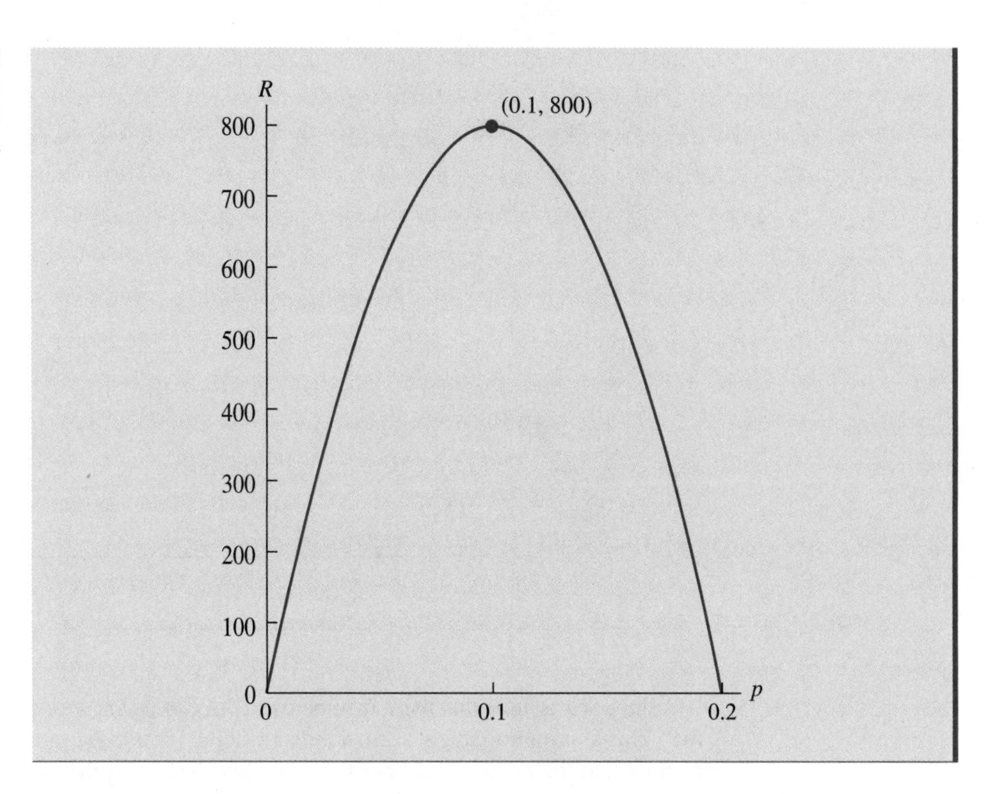

FIGURE 6.40 Revenue in terms of fraction harvested.

Notice that the maximum revenue occurs at the same value of p as the maximum harvest. This makes sense, since the revenue is just a multiple of the harvest. Thus, maximizing the sustainable harvest is the same as maximizing the sustainable revenue.

We will now consider profit. Recall that

$$\text{profit} = \text{revenue} - \text{cost}$$

Thus, we need a function that describes the cost for harvesting some amount of fish. We will assume that the cost is proportional to the fraction of fish caught, that is

$$C = cp$$

for some constant c. This should make sense. To double the fraction of fish caught, we would have to double the amount of time spent harvesting. This would double the labor costs and might double the number of boats used. While a proportional cost function is not exact, it is a good first approximation for our study.

Let's assume that $c = 1600$. This would mean that

$$C = 1600p$$

This means that the cost for harvesting 10% , that is, $p = 0.1$, is $C = \$160$. The cost for harvesting 20% would be $C = \$320$, and the cost for harvesting 30% would be $C = \$480$.

We now have that the profit function is

$$P = (16{,}000p - 80{,}000p^2) - 1600p = 14{,}400p - 80{,}000p^2$$

The profit function is also a parabola. The maximum profit occurs at the vertex of this parabola, which is at $(0.09, 648)$, as seen in Figure 6.41. This means that to achieve the maximum profit, the harvesting should be 9%, not 10%.

FIGURE 6.41 **Profit as function of fraction harvested.**

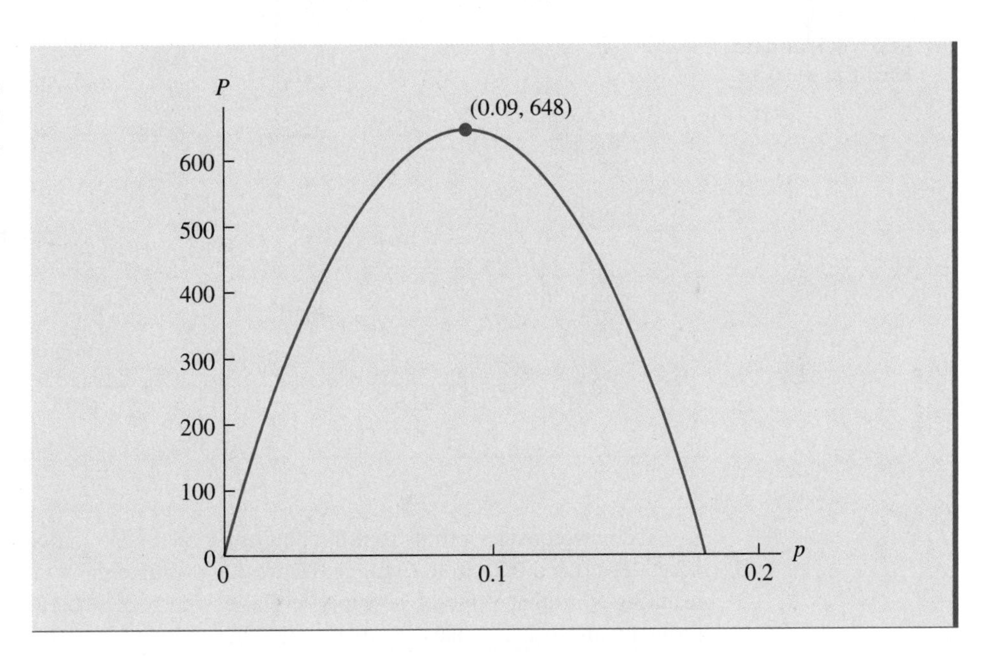

Recall that the equilibrium population was given by the equation

$$u = 8000 - 40{,}000p$$

and the sustainable harvest was

$$h = 8000 - 40{,}000p^2$$

Using $p = 0.09$ gives a sustainable harvest of 396 instead of the maximum sustainable harvest of 400. On the other hand, the equilibrium population has increased from 4000 to

4400. In general, maximizing total profit results in a slightly lower sustainable harvest but a higher equilibrium population. Maximizing profit actually benefits the species.

Figure 6.42 shows graphs of the revenue function and the cost function. Maximum profit occurs where there is a maximum distance between the revenue and costs functions, which is at $p = 0.09$. Notice that the two functions intersect at $p = 0.18$. This point can be found by finding where $R = C$ or solving

$$16,000p - 80,000p^2 = 1600p$$

This means that harvesting 18% each time period results in no profit.

FIGURE 6.42 Revenue and cost functions. Maximum profit is at $p = 0.09$, maximum revenue at $p = 0.1$, and no profit at $p = 0.18$.

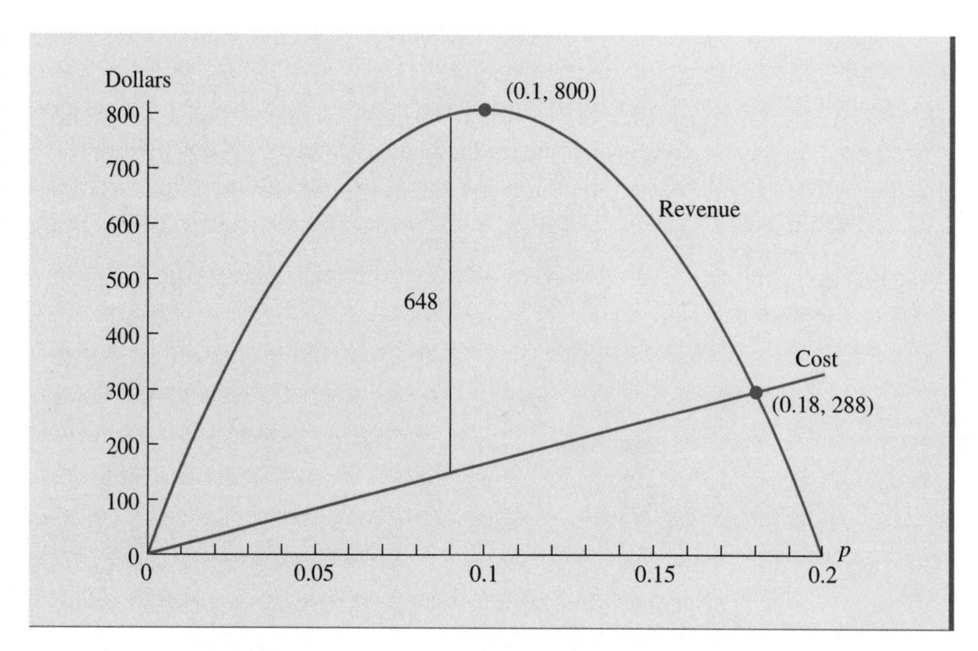

We now consider profit from the point of view of the individual. Pick any value for p between 0 and 0.2. From Figure 6.42, we can estimate the total revenue and total profit resulting from that value of p. Suppose that currently, $p < 0.18$, meaning that there is a profit for those individuals involved in fishing. Also assume that there are unemployed individuals. They observe that people are making a profit in fishing. If these unemployed have access to a boat, they will also begin fishing, viewing a small profit as better than no income at all. Thus, the value for p will increase.

An industry in which anyone can participate is called **open access.** The fishing industry has traditionally been open access. From the previous argument, we see that whenever the value for p is less than the value that results in no profit, $p = 0.18$ in this case, then people enter the industry and the value for p increases, since more people are fishing. This continues to happen until the value for p is at or near the no profit value. In our example, open access results in $p = 0.18$. The result is that no one that is harvesting the fish is making a profit, or if they are, it is very small.

Recall that the sustainable harvest is $h = 8000p - 40{,}000p^2$, which is 144 when $p = 0.18$. This is far below the maximum sustainable harvest of 400. The equilibrium population is $u = 8000 - 40{,}000p$, which is 800 in this case. This is also far below the equilibrium population of 4000 that occurs when harvesting the maximum sustainable harvest.

In summary, open access policies may result in (1) the equilibrium population size being very low, (2) the sustainable harvest being much smaller than it could be, and (3) individuals involved in fishing receiving very little profits.

There have been many occurrences of fishing stock collapses around the world. For example, in 1970, there were 12 million tons of Peruvian anchovy caught. The landings declined to about 10 million tons in 1971 and 4 million tons in 1972. From 1973 through 1989, the yearly catch was mostly under 2 million tons per year, with a couple of exceptional years when a little over 3 million tons were caught. While other factors, such as ocean temperature, may have contributed, it is believed that overfishing because of open access policies contributed to this collapse in Peruvian anchovies and to the collapse of other fish stocks.

A number of countries have had some success with an approach of using Individual Transferable Quotas or ITQs. Essentially, the government determines the optimal value for p, the fraction of the fish that can be caught. It conducts studies that indicate the amount of fish present that year. Combining these estimates, the government determines how many fish should be caught that year and issues permits allowing each fisherman to catch a certain part of that quota. Using this approach, it guarantees those in this fishing industry a decent profit each year, and a sustainable profit. There are many objections to this policy. Those who are not given permits feel that it is unfair to keep them out of the market. It also can lead to corruption, from people trying to payoff someone to obtain permits and to fishermen who catch more than their quota. There are many advantages also, from more stable employment and more optimal use of fishing boats.

Example 6.5

Suppose we are studying a population of fish that has an intrinsic growth rate of $b = 0.5$, a carrying capacity of $L = 500$ tons, and its growth rate is linearly decreasing. Let's assume we are harvesting a constant proportion of the species, $h = pu$. Assume that the fish can be sold for \$100 per ton and the cost function is $C = 20{,}000p$. Let's answer the following questions.

1. What fraction p should be harvested to achieve the maximum sustainable revenue? In this case, what is the sustainable revenue, the sustainable harvest and the equilibrium population size?
2. What fraction p should be harvested to achieve the maximum sustainable profit? In this case, what is the sustainable profit, the sustainable harvest and the equilibrium population size?
3. What fraction p should be harvested to achieve no profit, which is the result of an open access policy? In this case, what will be the sustainable harvest and the equilibrium population size?

To answer these questions, we first develop the growth rate function

$$r = b - \frac{b}{L}u = 0.5 - \frac{0.5}{500}u = 0.5 - 0.001u$$

This means that the growth function is

$$g = 0.5u - 0.001u^2$$

We now equate harvest and growth, giving $pu = 0.5u - 0.001u^2$, which simplifies to

$$p = 0.5 - 0.001u \quad \text{or} \quad u = 500 - 1000p$$

After substitution, the harvest function is

$$h = pu = p(500 - 1000p) = 500p - 1000p^2$$

The revenue is harvest times price, or

$$R = 100(500p - 1000p^2) = 50{,}000p - 100{,}000p^2$$

To answer question 1, we need to find the vertex of the parabola given by the revenue function. The intercepts are found by solving

$$R = 50{,}000p - 100{,}000p^2 = 0$$

which gives $p = 0$ and $p = 0.5$. The vertex is halfway between, at $p = 0.25$. Substitution of 0.25 for p into the function for R gives $R = \$6250$. Substitution into the harvest function $h = 500p - 1000p^2$ gives a sustainable harvest of $h = 62.5$ tons. Substitution into the population equation $u = 500 - 1000p$ gives an equilibrium population of $u = 250$ tons.

To answer question 2, we need to find the profit function, which is

$$P = R - C = 50{,}000p - 100{,}000p^2 - 20{,}000p$$

which simplifies to

$$P = 30{,}000p - 100{,}000p^2$$

The horizontal intercepts for this parabola are $p = 0$ and $p = 0.3$. Thus, the vertex is at $p = 0.15$. Substitution of $p = 0.15$ into the profit function gives a maximum profit of $P = \$2250$. Substitution into the harvest function gives a sustainable harvest of $h = 52.5$ tons. Substitution into the u equation gives that the equilibrium population is $u = 350$ tons.

To answer question 3, we need to find where the revenue function and cost function are equal, that is, solve $R = C$. This gives

$$50{,}000p - 100{,}000p^2 = 20{,}000p$$

which has solutions $p = 0$ and $p = 0.3$. Thus, a harvest of 30% per year results in no profit, an equilibrium harvest of $h = 60$ tons and an equilibrium population of $u = 200$ tons. The solution $p = 0$ also gives no profit because there is no harvest. We were looking for the positive solution.

In this example, no profit results in a higher sustainable harvest than maximum profit. ■

6.6 Problems

1. Suppose we are studying a population of fish that has an intrinsic growth rate of $b = 0.8$, a carrying capacity of $L = 1000$ tons, and its growth rate is linearly decreasing. Let's assume we are harvesting a constant proportion of the species, $h = pu$. Assume that the fish can be sold for \$50 per ton and the cost function is $C = 6250p$. Let's answer the following questions.

a. What fraction p should be harvested to achieve the maximum sustainable revenue? This is the p-coordinate of the vertex of the harvest parabola seen in Figure 6.43. In this case, what is the sustainable revenue, the sustainable harvest, and the equilibrium population size?

b. What fraction p should be harvested to achieve the maximum sustainable profit? This is the p-coordinate of the vertex V of the profit parabola seen in Figure 6.43. In this case, what is the sustainable profit, the sustainable harvest, and the equilibrium population size?

c. What fraction p should be harvested to achieve no profit, which is the result of an open access policy? This is the p-coordinate of the horizontal intercept C of the profit parabola seen in Figure 6.43. In this case, what will be the sustainable harvest and the equilibrium population size?

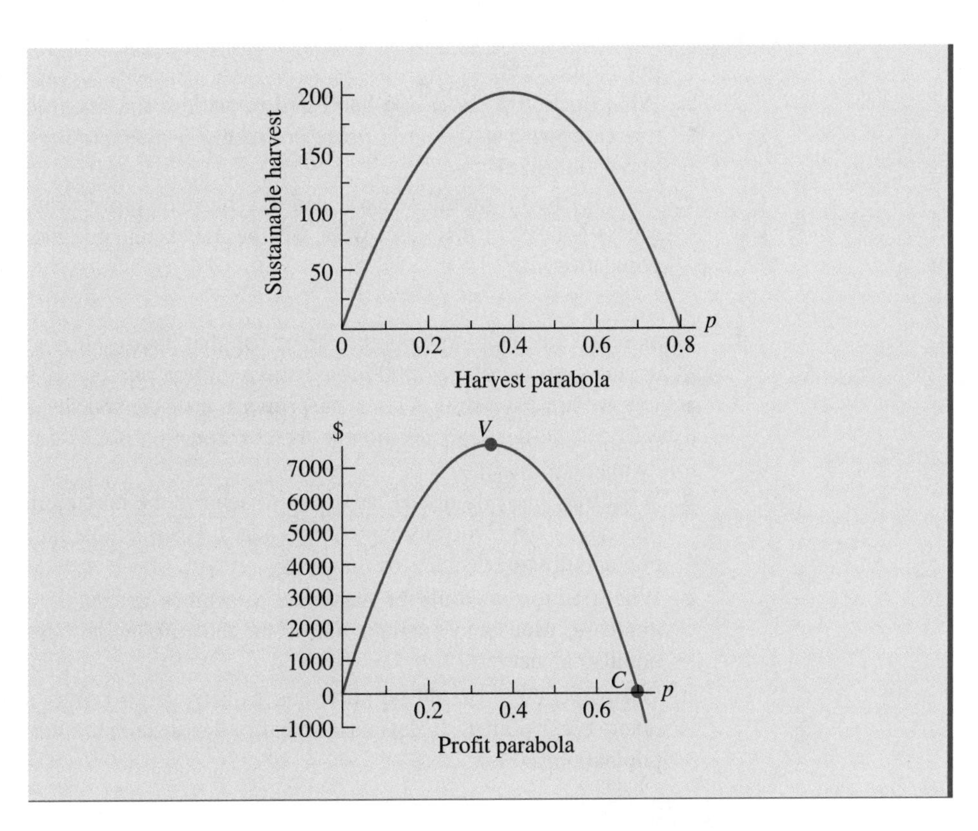

FIGURE 6.43 Graphs of harvest and profit functions for Problem 1.

2. Suppose we are studying a population of fish that has an intrinsic growth rate of $b = 0.2$, a carrying capacity of $L = 10,000$ tons, and growth rate that is linearly decreasing. Let's assume we are harvesting a constant proportion of the species, $h = pu$. Assume that the fish can be sold for \$250 per ton and the cost function is $C = 625,000p$. Let's answer the following questions.

a. What fraction p should be harvested to achieve the maximum sustainable revenue? In this case, what is the sustainable revenue, the sustainable harvest, and the equilibrium population size?

b. What fraction p should be harvested to achieve the maximum sustainable profit? In this case, what is the sustainable profit, the sustainable harvest, and the equilibrium population size?

c. What fraction p should be harvested to achieve no profit, which is the result of an open access policy? In this case, what will be the sustainable harvest and the equilibrium population size?

3. Suppose we are studying a population of fish that has an intrinsic growth rate of $b = 0.6$, a carrying capacity of $L = 200$ tons, and a growth rate that is linearly decreasing. Let's assume we are harvesting a constant proportion of the species, $h = pu$. Assume that the fish can be sold for \$80 per ton and the cost function is $C = 4000p$. Let's answer the following questions.

a. What fraction p should be harvested to achieve the maximum sustainable revenue? In this case, what is the sustainable revenue, the sustainable harvest, and the equilibrium population size?

b. What fraction p should be harvested to achieve the maximum sustainable profit? In this case, what is the sustainable profit, the sustainable harvest, and the equilibrium population size?

c. What fraction p should be harvested to achieve no profit, which is the result of an open access policy? In this case, what will be the sustainable harvest and the equilibrium population size?

4. Suppose we are studying a population of fish that has an intrinsic growth rate of $b = 1.2$, a carrying capacity of $L = 3000$ tons, and a growth rate that is linearly decreasing. Let's assume we are harvesting a constant proportion of the species, $h = pu$. Assume that the fish can be sold for \$250 per ton and the cost function is $C = 187,500p$. Let's answer the following questions.

a. What fraction p should be harvested to achieve the maximum sustainable revenue? In this case, what is the sustainable revenue, the sustainable harvest, and the equilibrium population size?

b. What fraction p should be harvested to achieve the maximum sustainable profit? In this case, what is the sustainable profit, the sustainable harvest, and the equilibrium population size?

c. What fraction p should be harvested to achieve no profit, which is the result of an open access policy? In this case, what is the sustainable harvest and the equilibrium population size?

5. Suppose we are studying a population of fish where

$$r = 0.81 - \frac{0.81}{100^2}u^2 = 0.81 - 0.000081u^2$$

Let's assume we are harvesting a constant proportion of the species, $h = pu$. Assume that the fish can be sold for \$90 per ton and the cost function is $C = 2000p$. Let's answer the following questions.

a. Find a function for u in terms of p by setting $g = h$ and solving for u.

b. What fraction p should be harvested to achieve the maximum sustainable revenue? In this case, what is the sustainable revenue, the sustainable harvest, and the equilibrium population size? (*Hint:* You will need a graphing calculator or computer to finish this, and the next 2 parts.)

c. What fraction p should be harvested to achieve the maximum sustainable profit? In this case, what is the sustainable profit, the sustainable harvest, and the equilibrium population size?

d. What fraction p should be harvested to achieve no profit, which is the result of an open access policy? In this case, what will be the sustainable harvest and the equilibrium population size?

6. Suppose we are studying a population of fish where

$$r = 0.25 - \frac{0.25}{1000^2}u^2 = 0.25 - 0.00000025u^2$$

Let's assume we are harvesting a constant proportion of the species, $h = pu$. Assume that the fish can be sold for \$200 per ton and the cost function is $C = 15{,}000p$. Let's answer the following questions.

a. Find a function for u in terms of p by setting $g = h$ and solving for u. Use this to find a function for h in terms of p.

b. What fraction p should be harvested to achieve the maximum sustainable revenue? In this case, what is the sustainable revenue, the sustainable harvest, and the equilibrium population size? (*Hint:* You will need a graphing calculator or computer to finish this and the next 2 parts.)

c. What fraction p should be harvested to achieve the maximum sustainable profit? In this case, what is the sustainable profit, the sustainable harvest, and the equilibrium population size?

d. What fraction p should be harvested to achieve no profit, which is the result of an open access policy? In this case, what is the sustainable harvest and the equilibrium population size?

7. Suppose we are studying a population of fish where

$$r = 0.64 - 0.64\left(\frac{u - 500}{1000}\right)^2$$

Let's assume we are harvesting a constant proportion of the species, $h = pu$. Assume that the fish can be sold for \$100 per ton and the cost function is $C = 70{,}000p$. Let's answer the following questions.

a. Find a function for the stable equilibrium u in terms of p by setting $g = h$ and solving for u. (To find the stable equilibrium, take positive square roots.) Use this to find a function for h in terms of p.

b. What fraction p should be harvested to achieve the maximum sustainable revenue? In this case, what is the sustainable revenue, the sustainable harvest and the equilibrium population size? (*Hint:* You will need a graphing calculator or computer to finish this and the next 2 parts.)

c. What fraction p should be harvested to achieve the maximum sustainable profit? In this case, what is the sustainable profit, the sustainable harvest, and the equilibrium population size?

d. What fraction p should be harvested to achieve no profit, which is the result of an open access policy? In this case, what is the sustainable harvest and the equilibrium population size?

8. Suppose we are studying a population of fish where

$$r = 0.25 - 0.25\left(\frac{u - 100}{500}\right)^2$$

Let's assume we are harvesting a constant proportion of the species, $h = pu$. Assume that the fish can be sold for \$150 per ton and the cost function is $C = 50{,}000p$. Let's answer the following questions.

a. Find a function for the stable equilibrium u in terms of p by setting $g = h$ and solving for u. (To find the stable equilibrium, take positive square roots.) Use this to find a function for h in terms of p.

b. What fraction p should be harvested to achieve the maximum sustainable revenue? In this case, what is the sustainable revenue, the sustainable harvest, and the equilibrium population size? (*Hint:* You will need a graphing calculator or computer to finish this and the next 2 parts.)

c. What fraction p should be harvested to achieve the maximum sustainable profit? In this case, what is the sustainable profit, the sustainable harvest, and the equilibrium population size?

d. What fraction p should be harvested to achieve no profit, which is the result of an open access policy? In this case, what will be the sustainable harvest and the equilibrium population size?

7 Genetics

7.1 Introduction to Population Genetics

This chapter begins a study of genetics. This study will use much of the mathematics you have learned in this book. Throughout this chapter, you will be finding equilibrium values, determining their stability, and estimating the rate at which solutions converge to equilibrium.

You may already be familiar with the basics of genetics, such as that many traits are determined by a pair of genes, one inherited from the mother and one inherited from the father. For simplicity, let's assume the gene comes in two forms, B and G, which are called **alleles.** This means that an individual can have one of the four combinations, GG, GB, BG, and BB, where the first letter represents an allele inherited from the mother and the second letter represents an allele inherited from the father. The particular combination of alleles determines the **genotype** of the individual. We consider the 3 genotypes: individuals with the combination GG, which we call G-homozygotes; individuals with the combination BB, which we call B-homozygotes; and individuals with either of the combinations GB and BG, which we denote by "BG" and which are called heterozygotes.

Often, G-homozygotes exhibit one trait, B-homozygotes exhibit a second trait, and heterozygotes exhibit a third trait. In some cases the heterozygotes exhibit the same trait as one of the homozygotes, say the B-homozygotes. In this case, B would be called a dominant allele and G would be called a recessive allele. The particular traits determine the **phenotype** of the individual. By our assumptions, we always have 3 genotypes, but we have only 2 phenotypes if B or G is dominant, seen in Figure 7.1. Examples of different phenotypes might be hair color, eye color, and the ability to curl your tongue.

In Section 7.2, we introduce some of the basics of genetics. In Sections 7.3 and 7.4, we are going to study 2 factors that have a dynamic influence on the genetic makeup of a population over time, mutation and selection.

Mutation is the spontaneous changing of one allele into another allele. While the probability of any one allele mutating to another is small, over an extended period of time mutation can have an appreciable effect on the genetic makeup of a population. In Section 7.3,

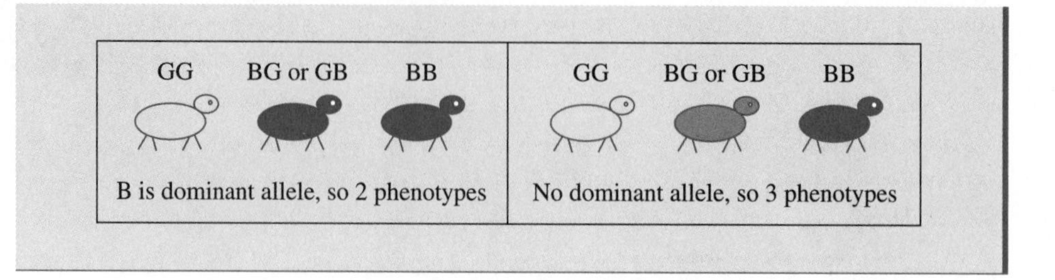

FIGURE 7.1 Difference in number of phenotypes depending on alleles.

we develop affine dynamical systems that model the change in a trait over time as a result of mutation. We will find explicit expressions for some of the functions that satisfy these dynamical systems.

We will examine the effects of selection in Section 7.4. Selection refers to the natural selection of one phenotype over another, because one phenotype is more likely to survive existing environmental conditions than another. This can cause a species to have a different genetic makeup in one geographic area than in another. Modeling selection leads to the development of nonlinear dynamical systems. The analysis of these systems leads to a number of surprising results.

7.2 Basics of Genetics

To get an understanding of the process of passing traits from one generation to the next, you are going to simulate a genetic situation in which two-thirds of the alleles in the population are of the form B and one-third are G. One way to do this is to get a collection of 2 objects that are identical except one is a different color from the other. For example, you might have a collection of blue and green beads. Put twice as many blue beads into a cup as green beads, say 12 blue and 6 green beads. This cup represents an initial adult population in which two-thirds of the alleles in the population are of one type, called type B, and one-third of the alleles are of the other type, G. The beads in this cup represent the genetic makeup of generation 1. We will let $g(n)$ and $b(n)$ represent the fraction of the alleles in generation n that are G and B, respectively. For this example, we have that $g(1) = \frac{1}{3}$ and $b(1) = \frac{2}{3}$.

We do not consider the person that an allele comes from. We are assuming that all of the alleles of all the women are in one giant cup and all of the alleles of all of the men are in a second cup. We are assuming that there are twice as many B-alleles as G-alleles in each cup. A child randomly draws one allele from each cup, the result determining the genetic makeup of that child.

To simulate a child receiving an allele from its mother, you draw one bead from the cup at random. Thus, the probability the child receives a B from its mother is two-thirds. Once the bead has been drawn and its color recorded, you return the bead to the cup, so there are always twice as many B-alleles in the cup as G-alleles.

You can do this simulation in other ways. You could generate a random integer from 1 to 3 on a calculator or computer. If 1 or 2 occurs, the child receives a B from its mother, but if a 3 occurs, the child receives a G from its mother. Or you could roll a die and if 1 through 4 occurs, you get a B, but if 5 or 6 occurs, you get a G.

Now that you know how to randomly generate an allele for the child, you will simulate the birth of the first child to generation 2. To simulate the first birth, generate a random B or G, where the probability of B is $\frac{2}{3}$ and the probability of G is $\frac{1}{3}$. Record the result. (If you are drawing beads from a cup, replace the bead in the cup so that there are still 2 B beads for every one G bead.) Generate another B or G, and record it. This represents the allele the child receives from the father. At this point, you have recorded either GG, GB, BG, or BB. These are the alleles of the first child.

Repeat this process for a total of 18 children. These are the children of generation 2. At this point, we will assume that all of the children survive to adulthood. So these are also the adults of generation 2. I obtained the results seen in Figure 7.2. Your results are likely to differ from mine. Since there were 18 children, the total number of B- and G-alleles drawn was 36.

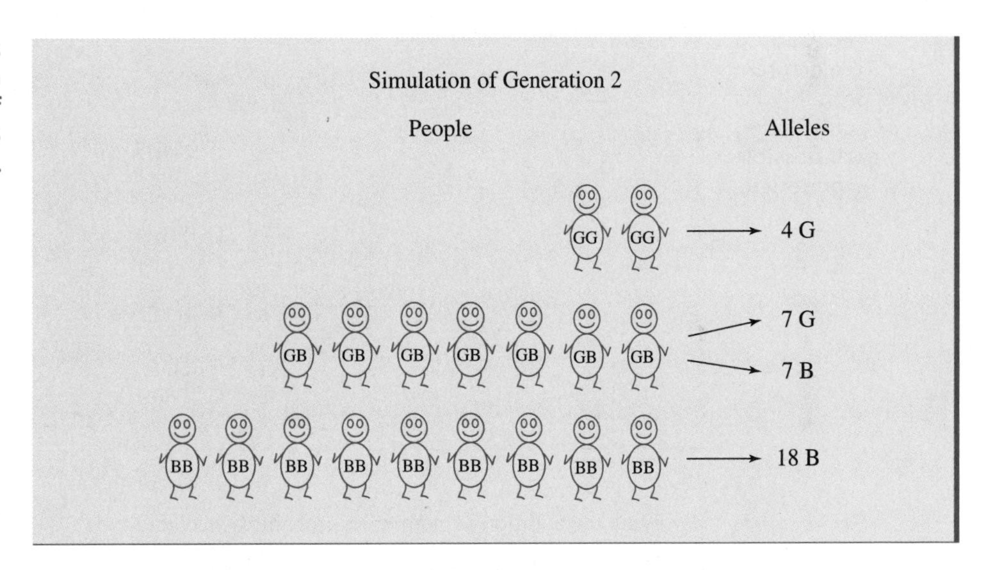

FIGURE 7.2 Simulation results in a total of 11 G-alleles and 25 B-alleles.

We will compare the actual results of the simulation with a model that predicts the results of the simulation, but first, we need to review some probability. Recall the **multiplication principle for probability** from Section 4.3. It states that the probability of A happening on the first step of a process then B happening on the second step of the process (with the 2 events being independent) is the product of the 2 probabilities, that is,

$$P(AB) = P(A)P(B)$$

Also recall the **addition principle for probability,** which states that

$$P(C \text{ or } D) = P(C) + P(D)$$

assuming C and D cannot both occur.

We now apply the multiplication and addition principles for probability to our study of genetics. Assume that the fraction of G-alleles in a population is g. By this I mean that the fraction of G-alleles among females is g and the fraction of G-alleles among males is also g. Assume that the fraction of B-alleles is b among both females and males. The first step in the process of determining the genetic makeup of a child is to draw an allele from a female. The second step is to draw an allele from a male. In this case, there are 4 possible results, BB, GG, BG, and GB. It is clear that the result of the first draw does not effect the result of the second draw. This means we have independent events and can use the multiplication principle to compute the probability of each possible result. The probability of each result occurring is then

$$\text{P(GG)} = \text{P(G)P(G)} = g^2 \quad \text{P(BB)} = b^2 \quad \text{P(GB)} = gb \quad \text{and} \quad \text{P(BG)} = bg$$

These probabilities can be seen visually from the tree diagram in Figure 7.3. The tree diagram is just a visualization of the multiplication principle for probability.

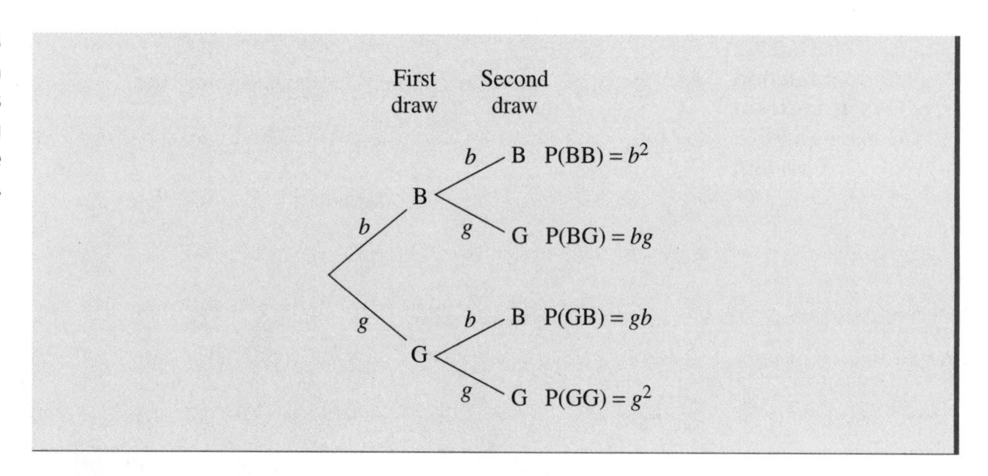

FIGURE 7.3 Tree diagram gives probabilities of a child being each possible genotype.

Applying the addition principle for probability, we get that

$$\text{P(“BG”)} = \text{P(BG or GB)} = \text{P(BG)} + \text{P(GB)} = gb + bg = 2gb$$

We now apply these results to our example in which two-thirds of the alleles are B and one-third are G in generation 1. This means that $b = \frac{2}{3}$ and $g = \frac{1}{3}$, and the likelihood of a child being each genotype is

$$\text{P(BB)} = b^2 = (\tfrac{2}{3})^2 = \tfrac{4}{9}$$

$$\text{P(“BG”)} = 2bg = 2(\tfrac{2}{3})(\tfrac{1}{3}) = \tfrac{4}{9}$$

$$\text{P(GG)} = g^2 = (\tfrac{1}{3})^2 = \tfrac{1}{9}$$

Recall that in our simulation, 18 children were born. If 18 children are born, then we expect the total number of BB-children to be

$$18\text{P(BB)} = 18(\tfrac{4}{9}) = 8.$$

Similarly, we expect the total number of GB and BG children to be

$$18\text{P}(\text{"BG"}) = 18(\tfrac{4}{9}) = 8$$

and the total number of GG-children to be

$$18\text{P}(\text{GG}) = 18(\tfrac{1}{9}) = 2$$

Figure 7.4 is similar to Figure 7.2 except that it shows the **predicted** results for generation 2 instead of the **actual** results. You should compare the predicted results to the actual results you obtained in your simulation.

FIGURE 7.4 **Predicted makeup results in a total of 12 G-alleles and 24 B-alleles.**

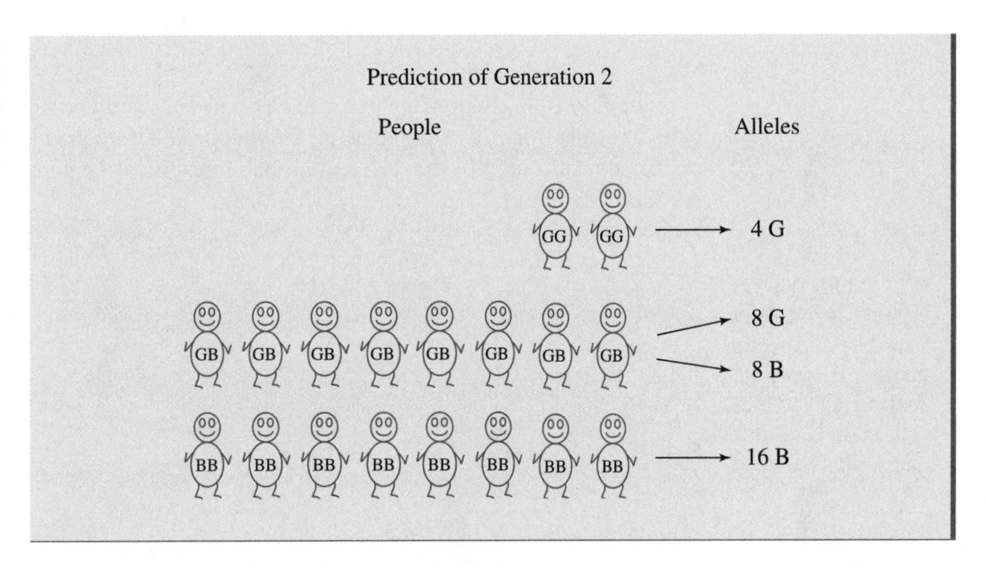

Make sure you understand the difference between the simulated results and the predicted results. If you flip 20 fair coins, there is a good chance you will obtain a result other than 10 heads and 10 tails. But you "expect" to get 10 heads and 10 tails in the 20 flips in the sense that, if you flip 20 coins repeatedly, the average number of heads will be about 10. When using dynamical systems, we are often predicting the "average" result but expect real results to differ somewhat from our predictions.

Example 7.1

Suppose that 80% of the alleles are G and 20% are B in generation 1. We define $g(n)$ and $b(n)$ as the fraction of the alleles in generation n that are green and blue, respectively. In this case, we have that $g(1) = 0.8$ and $b(1) = 0.2$. You could simulate this using a cup with 8 green beads and 2 blue beads as in Figure 7.5, or generate random integers from 1 to 10 with 1 to 8 being green and 9 or 10 being blue. This means that

$$\text{P}(\text{BB}) = (0.2)^2 = 0.04$$

$$\text{P}(\text{"BG"}) = 2(0.2)(0.8) = 0.32$$

$$\text{P}(\text{GG}) = (0.8)^2 = 0.64$$

If you are not sure how these values were computed, refer back to the tree diagram in Figure 7.3.

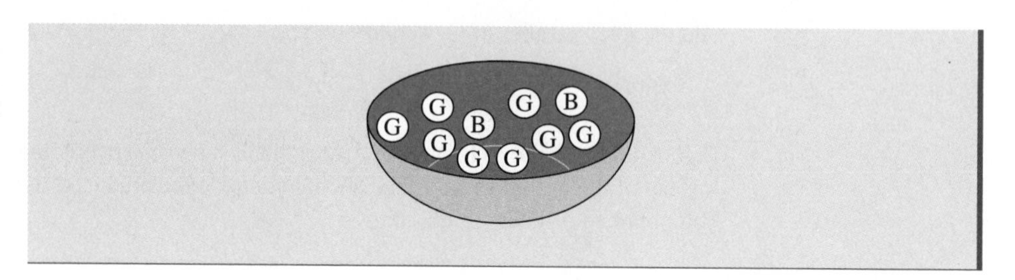

FIGURE 7.5 Bowl represents genetic makeup of generation 1.

Suppose 1000 children are going to be born. This could be simulated by drawing 1000 pairs of beads from the cup, replacing the first bead before drawing the second bead. The predicted results are just the different probabilities times the number of children born and are "seen" in Figure 7.6.

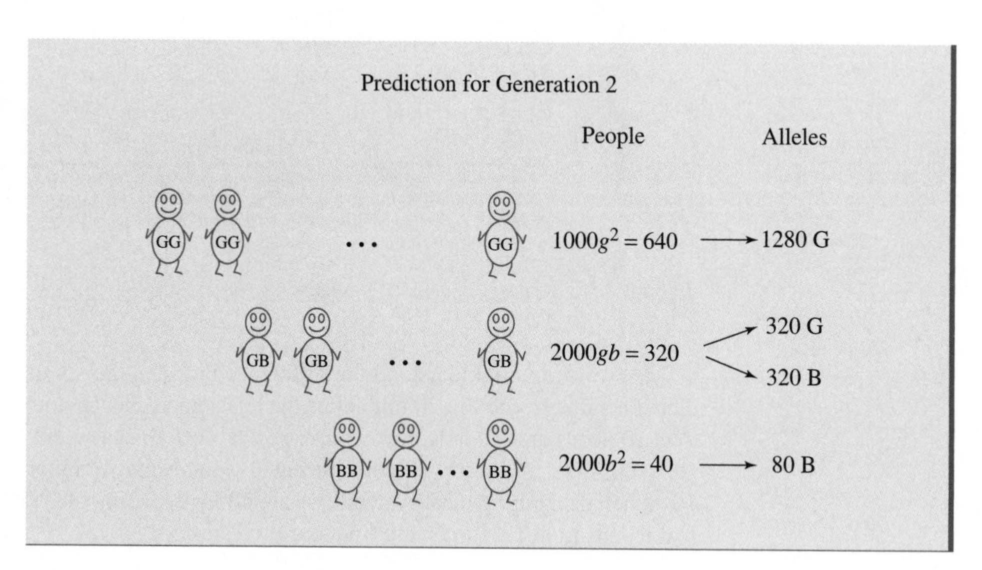

FIGURE 7.6 Predicted number of people of each genotype, resulting in a total of 1600 G-alleles and 400 B-alleles.

The 640 GGs result in 1280 G-alleles and the 320 GBs result in 320 G-alleles for a total of 1600 G-alleles. Similarly, there are 400 B-alleles. This means that in generation 2, 1600 out of the 2000 alleles are G, so

$$g(2) = \frac{1600}{2000} = 0.8 \quad \text{and} \quad b(2) = \frac{400}{2000} = 0.2$$

Note that the fraction of alleles of each type did not change from one generation to the next. That is because there were no other factors acting on the population. ■

Assume generation $n-1$ consists of 1000 adults. We know that the fraction of alleles that are G in generation $n-1$ is denoted by $g(n-1)$. We also know the fraction of the alleles that are B is

$$b(n-1) = 1 - g(n-1)$$

since the beads that aren't green must be blue. This can be modeled by a cup in which the fraction of beads that are green is $g(n-1)$ and the fraction that are blue is $b(n-1)$. To simulate 1000 children being born to generation n, we could draw green and blue beads from the cup, replacing beads after each draw, to get 1000 children. This means we would draw a total of 2000 beads, 2 beads (or alleles) for each child.

The tree diagram in Figure 7.3 gives that

$$P(GG) = g(n-1)g(n-1) = g^2(n-1)$$

$$P(\text{“BG”}) = P(BG) + P(GB) = 2g(n-1)b(n-1)$$

$$P(BB) = b^2(n-1)$$

where P(GG), P("BG"), and P(BB) represent the probabilities of a child being born with the allele combination GG, "BG," and BB, respectively, in generation n.

To compute the **number** of the 1000 children that we expect to be GG, we multiply the probability, $g^2(n-1)$, of a child being GG times 1000 births, giving $1000g^2(n-1)$. Similar computations give the expected number of "BG"- and BB-children. These results are summarized in Figure 7.7. The results seen in Figure 7.7 will be used extensively in the rest of this chapter. As seen in Figure 7.7, the $1000g^2(n-1)$ GG-people have $2000g^2(n-1)$ G-alleles, since each person has 2 G-alleles. The $2000b(n-1)g(n-1)$ heterozygotes each have 1 G-allele. Adding up the G-alleles from these 2 groups gives

$$\text{number of G-alleles} = 2000g^2(n-1) + 2000b(n-1)g(n-1)$$

Factoring $2000g(n-1)$ out of each term gives

$$\text{number of G-alleles} = 2000g(n-1)(g(n-1) + b(n-1))$$

FIGURE 7.7 **Number of people of each genotype born to generation n and resulting numbers of alleles.**

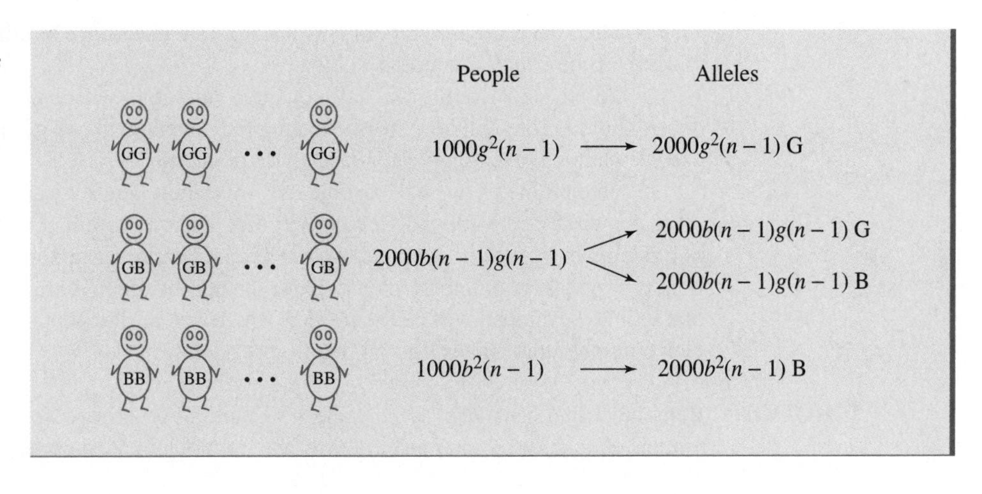

(Factoring early and often will simplify work in this chapter.) We know that $g(n-1)+b(n-1)=1$, since all of the alleles are G or B. This gives

$$\text{number of G-alleles} = 2000g(n-1)$$

This equation should make sense. Since $g(n-1)$ is the probability that any particular allele inherited by a child is G and there are 2000 alleles among the children, we should expect that the number of G-alleles is $2000g(n-1)$. We are going to use this logic extensively in Section 7.3. A similar calculation gives that

$$\text{number of B-alleles} = 2000b(n-1)$$

We can now compute $g(n)$ and $b(n)$, the fraction of green and blue alleles, respectively, among the adults of generation n. We find $g(n)$ by dividing the total number of green alleles, $2000g(n-1)$, by the total number of alleles, 2000, getting

$$g(n) = \frac{2000g(n-1)}{2000} = g(n-1)$$

Similarly, $b(n) = b(n-1)$. Assuming there are no other factors affecting the genetic makeup, the fractions of green and blue alleles remain constant. This means that from one generation to the next, the fraction of people of each genotype remains constant. Suppose the initial fraction of green and blue alleles are $g(0) = g$ and $b(0) = b$. Then the fraction of children in each of the following generations that are born of each genotype will be

$$\begin{aligned} \text{Fraction born GG} &= 12mug^2 \\ \text{Fraction born GB or BG} &= 2bg \\ \text{Fraction born BB} &= b^2 \end{aligned}$$

This is called the Hardy–Weinberg law. It states that in absence of other factors, the genetic makeup of the population doesn't change.

In the next 2 sections, we will consider situations where the Hardy–Weinberg law doesn't hold. There will be factors effecting the genetic makeup of the population. We will predict what will eventually happen to these populations.

In some problems, we will assume that 1000 individuals are being born. In other problems, we will assume 900 individuals are born. There is nothing magical about the number of individuals we pick in our population. We could assume a population of 10,000 or 1 million, or even use an unknown such as T for the number being born. The dynamical systems that will be developed will be the same no matter what size population you pick. I usually pick a number that makes the computations easier.

CAUTION Remember that $g(n)$, $b(n)$, $g^2(n)$, and $b^2(n)$ are all **fractions.** To get the number of individuals of each type, you must multiply the fractions by the number of births.

7.2 Problems

1. Suppose that a population consists of 30 adults that are G-homozygotes, 70 heterozygotes, and 40 B-homozygotes. How many of the alleles of these 140 people are G? What fraction of the alleles are G? The following diagram might help.

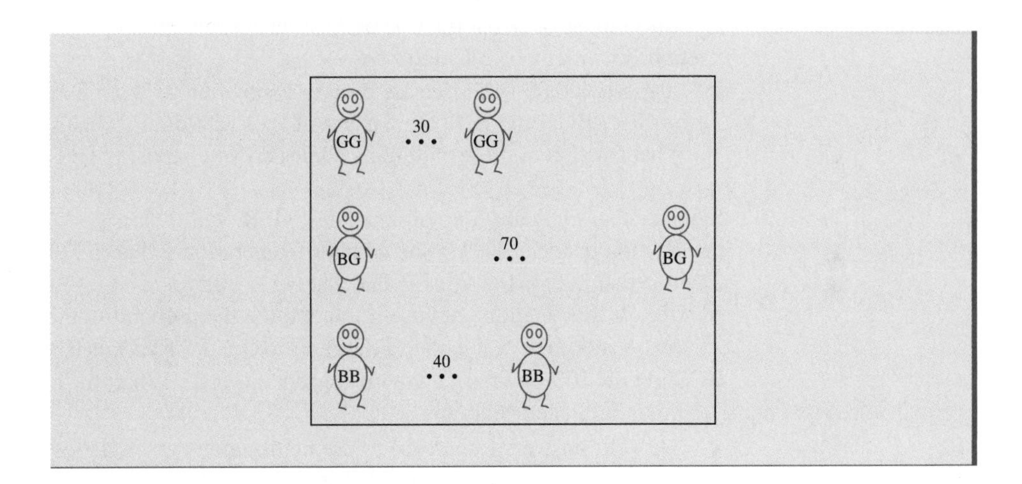

2. Suppose that a population consists of 55 adults that are G-homozygotes, 40 heterozygotes, and 17 B-homozygotes. How many of the alleles of these people are G? What fraction of the alleles are G?

3. Suppose that there are 3 possible alleles, G, B, and R. In this case there are 6 genotypes—GG, "BG", "RG", BB, "BR", and RR. Note that "BG" represents people that are BG or GB, and similarly for "RG" and "BR." Suppose that there are 10 GG-people, 20 "BG"-people, 40 "RG"-people, 25 BB-people, 70 "BR"-people, and 25 RR-people. How many of the alleles of these people are G? What fraction of the alleles are G?

4. Suppose that there are 3 possible alleles, G, B, and R. In this case there are 6 genotypes, GG, "BG", "RG", BB, "BR", and RR. Note that "BG" represents people that are BG or GB, and similarly for "RG" and "BR". Suppose that there are 55 GG-people, 80 "BG"-people, 120 "RG"-people, 65 BB-people, 20 "BR"-people, and 105 RR-people. How many of the alleles of these people are R? What fraction of the alleles are R?

5. Suppose that $g(1) = 0.3$ and $b(1) = 0.7$, meaning that 30% of the alleles in generation 1 are G while 70% are B.
   a. Suppose 100 children are born to generation 2. What percentage do you expect to be GG, "BG," and BB? How many of the 200 alleles among these children do you expect to be G? What fraction of the alleles among these 100 children do you expect to be G?

b. Suppose 3000 children are born to generation 2. What percentage do you expect to be GG, "BG," and BB? How many of these children's alleles do you expect to be G? What fraction of these children's alleles do you expect to be G, that is, predict $g(2)$?

6. Suppose that $g(1) = 0.37$ and $b(1) = 0.63$, meaning that 37% of the alleles in generation 1 are G while 63% are B.

a. Suppose 10,000 children are born to generation 2. What percentage do you expect to be GG, "BG," and BB? How many of the 20,000 alleles among these children do you expect to be G? What fraction of the alleles among these 10,000 children do you expect to be G.

b. Suppose 400,000 children are born to generation 2. What percentage do you expect to be GG, "BG," and BB? How many of these children's alleles do you expect to be G? What fraction of these children's alleles do you expect to be G, that is, predict $g(2)$?

7. Suppose that there are 3 possible alleles, G, B, and R. Let $g(n)$, $b(n)$, and $r(n)$ represent the fraction of the alleles in the adults of generation n that are G, B, and R, respectively. Assume that $g(1) = 0.2$, $b(1) = 0.3$, and $r(1) = 0.5$.

a. Draw a tree diagram and use it to compute the probability that a child will be each of the 6 genotypes, that is, find P(GG), P("BG"), P("RG"), P(BB), P("BR"), and P(RR).

b. Suppose 100 children are born to generation 2. What are the expected numbers of children of each genotype?

c. Use your answers to part (b) to predict the number of alleles of each type among the 100 children of generation 2. What fraction of the alleles are expected to be of each type?

8. Suppose that there are 3 possible alleles, G, B, and R. Let $g(n)$, $b(n)$, and $r(n)$ represent the fraction of the alleles in the adults of generation n that are G, B, and R, respectively. Assume that $g(1) = 0.1$, $b(1) = 0.6$, and $r(1) = 0.3$.

a. Draw a tree diagram and use it to compute the probability that a child will be each of the 6 genotypes.

b. Suppose 500 children are born to generation 2. What are the expected numbers of children of each genotype?

c. Use your answers to part (b) to predict the number of alleles of each type among the 500 children of generation 2. What fraction of the alleles are expected to be of each type?

7.3 Mutation

In this section, we will study the effect resulting from some alleles of one form spontaneously mutating to another form. Mutation is independent of the genotype of the person. Therefore, if we know how many alleles of each type there are in a population, we can predict how many will mutate.

It may help you understand mutation if you actually perform the following simulation. Let $g(n)$ and $b(n)$ represent the fraction of alleles in the reproductive population of generation n that are green and blue, respectively, after mutation has taken place for that generation. Let's assume that $g(1) = \frac{2}{3}$ and $b(1) = \frac{1}{3}$. This represents the initial genetic makeup of generation 1. One way to simulate this population is to put twice as many green beads as blue into a cup, say 12 green beads and 6 blue beads. Another method is to generate a random integer from 1 through 3, with 1 and 2 representing "green" and 3 representing "blue."

Let's assume 15 people are born to generation 2. This is a total of 30 alleles. We do not care what the genotypes of these people are. We only care what alleles they have. Therefore, we will generate 30 alleles by drawing one bead at a time from the cup, replacing the bead after each draw. My results, in the order they were obtained, are

GBBGGGBGGGBBGGGBBGGGBGBGGBBGGB

Thus, my simulated population has 18 G-alleles and 12 B-alleles. Mutation has not yet taken place, so we cannot yet compute $g(2)$ and $b(2)$.

We now assume that some of the G-alleles mutate to B-alleles. Let's assume the mutation rate is $\mu = 0.5$, meaning that half of the G-alleles mutate. This is unreasonably large, but it will make our computations easier. Later, we will consider more reasonable mutation rates. To simulate that half of the G-alleles mutate to B-alleles, I changed half of the G-alleles into B-alleles, which can be seen in Figure 7.8. This results in 9 G-alleles and 21 B-alleles, so I have that $g(2) = 9/30 = 0.3$ and $b(2) = 21/30 = 0.7$. This means that 30% of the alleles in generation 2 are G and 70% are B.

FIGURE 7.8 **Half of G-alleles become B-alleles.**

Simulation of mutation

Alleles before mutation	GGGGGGGGG	GGGGGGGGG	BBBBBBBBBBBB
		↓ Mutated	
Alleles after mutation	GGGGGGGGG	BBBBBBBBB	BBBBBBBBBBBB

In your simulation, if you have an odd number of G-alleles, you can flip a coin to decide if the last G-allele mutates to a B or not.

We have just done a simulation to estimate $g(2)$ and $b(2)$. We are now going to **predict** $g(2)$ and $b(2)$ using some probability. Recall that $g(1) = \frac{2}{3}$ and $b(1) = \frac{1}{3}$. We assumed that generation 2 had 30 alleles. Since $g(1) = \frac{2}{3}$, we expect two-thirds of these alleles to be G, for a total of 20 G-alleles. Similarly, we expect one-third of the 30 alleles to be B, for a total of 10.

To model mutation we replace half of the G-alleles with B-alleles, which is seen in Figure 7.9. From this, we see that of the 30 alleles, 10 are G and 20 are B. Thus,

$$g(2) = \frac{6}{18} = \frac{1}{3} \quad \text{and} \quad b(2) = \frac{12}{18} = \frac{2}{3}$$

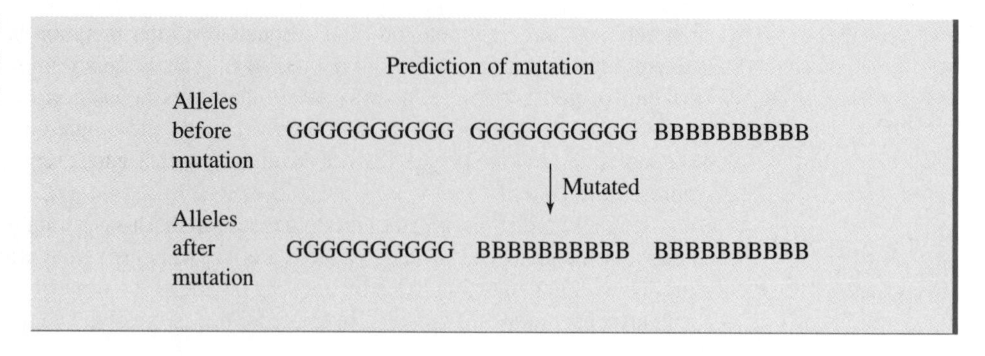

FIGURE 7.9 Half of G-alleles become B-alleles.

Let's now develop a dynamical system to model the genetic makeup of each population, assuming that half of the G-alleles each generation mutate to B-alleles. The proportions of green and blue alleles of generation $n-1$ are represented by $g(n-1)$ and $b(n-1)$, respectively. Figure 7.10 outlines the process that we are going to follow. There is no selective advantage of one genotype over another, so we only need to compute the number of alleles, not the number of each genotype. For simplicity, let's assume that 500 individuals are born, for a total of 1000 alleles. We could pick any number of children to be born to generation n. The result would be the same.

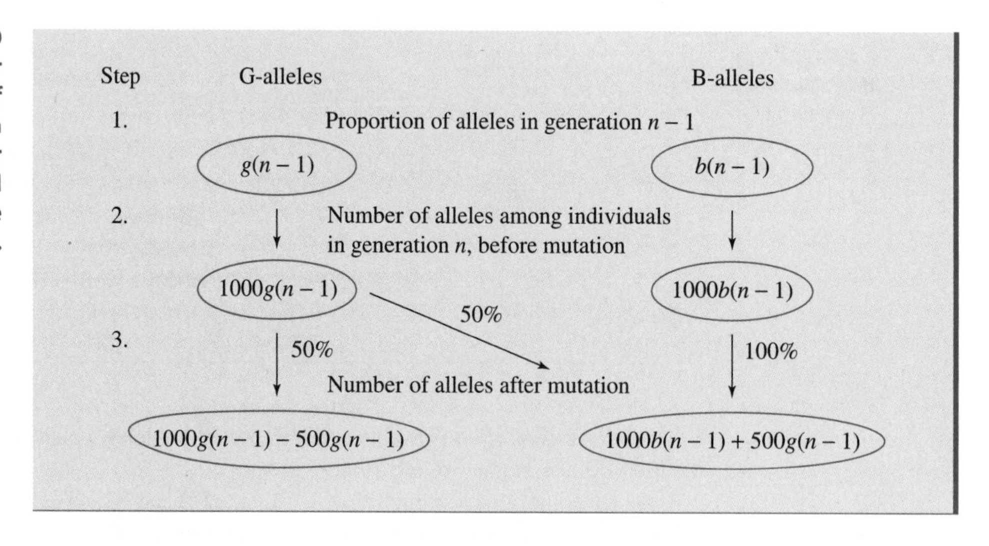

FIGURE 7.10 Steps for finding number of G- and B-alleles in generation *n* after mutation. To find *g*(*n*) and *b*(*n*), divide by the total, 1000.

If $g(n-1) = \frac{3}{4}$ and $b(n-1) = \frac{1}{4}$, meaning that three-fourths of the alleles among the adults of generation $n-1$ are G and 1/4 are B, then, from Step 2 Figure 7.10, the 500 children of generation n will have $1000(\frac{3}{4}) = 750$ G-alleles and $1000(\frac{1}{4}) = 250$ B-alleles. Note that the total number of alleles is $750 + 250 = 1000$, which makes sense because there are 1000 alleles. From Step 3, we assume that 375 of the 750 G-alleles become B-alleles. This means that 375 G-alleles remain and we have $250 + 375 = 625$ B-alleles. To find $g(n)$, we divide the number of G-alleles, 375, by the total number of alleles, 1000, to get

$$g(n) = \frac{375}{1000} = 0.375$$

Similarly,

$$b(n) = \frac{625}{1000} = 0.675$$

Let's repeat the steps without using a speci c value for $g(n-1)$ and $b(n-1)$. From the steps in Figure 7.10, we see that, after mutation, the number of G- and B-alleles in generation n is $500g(n-1)$ and $1000b(n-1)+500g(n-1)$, respectively. This means that

$$g(n) = \frac{\text{number of G-alleles}}{\text{total number of alleles}} = \frac{500g(n-1)}{1000} = 0.5g(n-1)$$

Similarly, the fraction of alleles that are B is

$$b(n) = \frac{1000b(n-1)+500g(n-1)}{1000} = b(n-1)+0.5g(n-1)$$

This gives the 2 dynamical systems

$$g(n) = 0.5g(n-1)$$

and

$$b(n) = b(n-1)+0.5g(n-1)$$

You should repeat the previous steps with values other than 1000 for the original number of alleles. You should get the same 2 dynamical systems, regardless of the original number of alleles.

Using the fact that $b(n-1)+g(n-1)=1$, or that $g(n-1)=1-b(n-1)$, allows us to rewrite the second dynamical system as $b(n)=b(n-1)+0.5(1-b(n-1))$, or

$$b(n) = 0.5b(n-1)+0.5$$

after simplifying. From Section 3.3, we know that the function that satisfies the dynamical system $g(n)=0.5g(n-1)$ can be written explicitly as

$$g(n) = c_g(0.5)^n$$

where c_g can be determined from $g(1)$. From this form of the function, we know that 0 is a stable equilibrium value and that $g(n)$ goes to 0 exponentially at a rate of 0.5, meaning it gets 50% closer to 0 each generation.

From Section 3.5, we know that the function that satisfies the dynamical system $b(n) = 0.5b(n-1)+0.5$ can be written explicitly as

$$b(n) = c_b(0.5)^n+1$$

where c_b can be determined from $b(1)$. In this case, $E = 1$ is a stable equilibrium value. The rate at which $b(n)$ goes to equilibrium is 0.5, which means that $b(n)$ gets 50% closer to 1 each generation. This tells us that over time, "most" of the alleles will be B.

Example 7.2

In this example, we will assume a more realistic mutation rate, that is, that $\mu = 0.0001$ of the G-alleles mutate to B-alleles each generation. We will develop dynamical systems for $g(n)$ and $b(n)$ and will use explicit functions that satisfy these dynamical systems to determine the half-life of the G-alleles, that is, the number of generations it will take for the proportion of G-alleles to be cut in half. Figure 7.11 outlines the process we will follow.

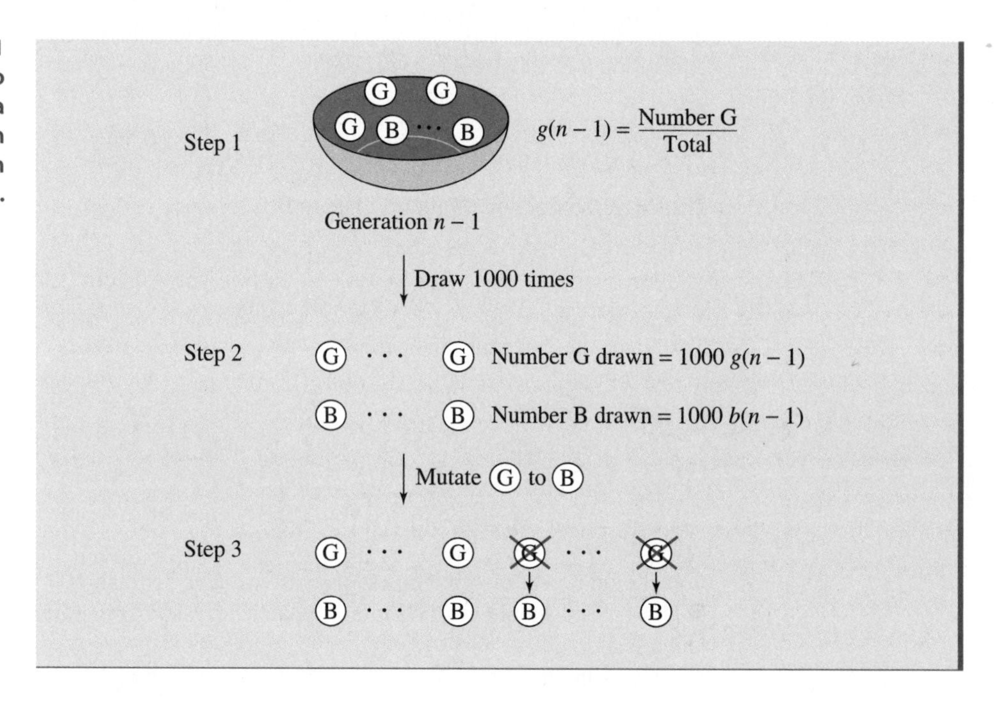

FIGURE 7.11 Steps to developing a dynamical system to model situation in Example 7.2.

Assume that the proportion of G-and B-alleles in generation $n-1$, after mutation has already occurred, is $g(n-1)$ and $b(n-1)$, respectively. Assume that 500 children are born to generation n, which gives a total of 1000 alleles. From step 2 of Figure 7.10, we expect

$$1000g(n-1) \text{ and } 1000b(n-1)$$

of alleles to be G and B, respectively.

We assume that 0.0001 of the G-alleles will mutate to B-alleles, which is

$$(0.0001)1000g(n-1) = 0.1g(n-1)$$

alleles. From Step 3 of Figure 7.10, we subtract $0.1g(n-1)$ from the total G-alleles and add it to the total B-alleles, leaving

$$1000g(n-1) - 0.1g(n-1) = 999.9g(n-1)$$

G-alleles and

$$1000b(n-1)+0.1g(n-1)$$

B-alleles in generation n after mutation. Dividing the number of G-alleles by the total number of alleles, 1000, gives the dynamical system

$$g(n)=\frac{999.9g(n-1)}{1000}=0.9999g(n-1)$$

Similarly, dividing the number of B-alleles by 1000 gives

$$b(n)=\frac{1000b(n-1)+0.1g(n-1)}{1000}=b(n-1)+0.0001g(n-1)$$

After substituting $1-b(n-1)$ for $g(n-1)$ and simplifying, this last dynamical system becomes

$$b(n)=0.9999b(n-1)+0.0001$$

The explicit form of the function that satisfies $g(n)=0.9999g(n-1)$ is

$$g(n)=c(0.9999)^n$$

Suppose we know that $g(1)=0.4$. Substitution into the function gives that

$$0.4=c(0.9999)$$

so

$$c=\frac{0.4}{0.9999}$$

The function is then

$$g(n)=\frac{0.4(0.9999)^n}{0.9999}=0.4(0.9999)^{n-1}$$

Suppose we want to know the "half-life" of the G-alleles. This means we are finding n when $g(n)=0.2$. Substitution of the expression for $g(n)$ gives

$$\frac{0.4(0.9999)^n}{0.9999}=0.2$$

Divide both sides by 0.4 and multiply both sides by 0.9999 to get

$$0.9999^n=0.49995$$

Taking logarithms of both sides, then bringing the exponent out in front gives

$$n\log(0.9999)=\log(0.49995)$$

Dividing both sides by log(0.9999) gives

$$n = \frac{\log(0.49995)}{\log(0.9999)} \approx 6932$$

generations. We say that the half-life of the G-alleles is about 7000 generations. The mutation rate can only be estimated, so using any greater degree of accuracy for the half-life would be misleading.

The same half-life would have been found using any other number for $g(1)$, where $0 < g(1) \leq 1$. ■

Note that the length of time for a generation varies from one species to another. For humans, a generation is about 20 years, so the half-life of this G-allele would be about 140,000 years. For bacteria, a generation could be 2 hours. Thus, for bacteria, the half-life would be about 1.5 years. The bar graph in Figure 7.12 dramatizes this difference. The bar for the bacteria is too small to see. This implies that mutation alone has little impact on human evolution in modern times, but that it can have a dramatic influence of the genetic makeup of bacteria in your lifetime.

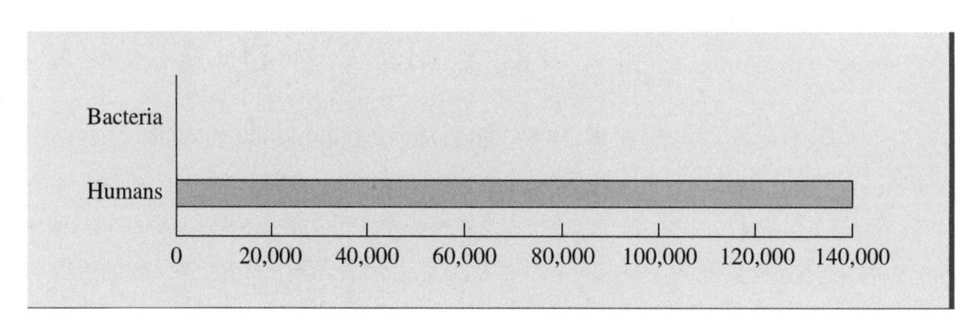

FIGURE 7.12 Half-life for a trait assuming mutation rate of 0.01%.

The half-life of the G-allele depends on the mutation rate, $\mu = 0.0001$ in this case, as well as length of time of a generation of the species. In the next section, we will see that selection can result in a much quicker evolution in the genetic makeup of a species.

The mutation of one allele to another indicates that an allele is dying out in a population. In the exercises, you will explore how mutation can lead to genetic diversity if alleles mutate in both directions, such as some of the G-alleles mutating to B-alleles and some of the B-alleles mutate to G-alleles.

REMARK Once you have values for $g(n-1)$ and $b(n-1)$, you can use the tree diagram in Figure 7.3 to estimate the fraction of the individuals of generation n that will be born with each genotype. These fractions will be as follows. The fraction that are:

GG is	$g^2(n-1)$
BG or GB is	$2g(n-1)b(n-1)$
BB is	$b^2(n-1)$

7.3 Problems

1. Suppose that $g(1) = 0.3$ and $b(1) = 0.7$, meaning that 30% of the alleles in generation 1 after mutation are G and 70% are B. Assume that 5000 individuals are born into generation 2.

a. Find the number of G-and B-alleles in generation 2 before mutation. Remember that the total should equal 10,000 alleles.

b. Suppose that 2% of the G-alleles mutate to B-alleles. Find the number of G- and B-alleles among the 10,000 alleles in generation 2 after mutation. Use these numbers to find $g(2)$ and $b(2)$.

c. Pick your own number for the number of individuals born into generation 2. Use these to find the number of G-and B-alleles, before mutation. Using the mutation rate in part (b), find $g(2)$ and $b(2)$ for this population of children. How does this result compare to the answer to part (b).

2. Suppose that $g(1) = 0.6$ and $b(1) = 0.4$, meaning that 60% of the alleles in generation 1 are G and 40% are B. Assume that 2000 are born to generation 2.

a. Find the number of G- and B-alleles in generation 2, before mutation.

b. Suppose that 5% of the B-alleles mutate to G-alleles. Find the number of G- and B-alleles in generation 2 after mutation. Use these numbers to find $g(2)$ and $b(2)$.

c. Pick your own number for the number born to generation 2. Use this to find the number of G-and B-alleles before and after mutation. Find $g(2)$ and $b(2)$. How does this result compare to the answer to part (b)?

3. Suppose that $g(1) = 0.1$ and $b(1) = 0.9$. Assume that 1000 are born to generation 2.

a. Find the number of G- and B-alleles in generation 2, before mutation.

b. Suppose that 20% of the G-alleles mutate to B-alleles and 30% of the B-alleles mutate to G-alleles. The mutation occurs simultaneously; that is, an allele cannot mutate from G to B and then back to G. Find the number of G- and B-alleles in generation 2 after mutation. Use these numbers to find $g(2)$ and $b(2)$.

c. Pick your own number for the number born to generation 2. Use this to find the number of G- and B-alleles before and after mutation. Find $g(2)$ and $b(2)$. How does this result compare to the answer to part (b)?

4. Suppose that $g(1) = 0.5$ and $b(1) = 0.5$. Assume that 10,000 are born to generation 2.

a. Find the number of G- and B-alleles in generation 2 before mutation.

b. Suppose that 5% of the G-alleles mutate to B-alleles and 10% of the B-alleles mutate to G-alleles. The mutation occurs simultaneously; that is, an allele cannot mutate from G to B and then back to G. Find the number of G- and B-alleles in generation 2 after mutation. Use these numbers to find $g(2)$ and $b(2)$.

c. Pick your own number for the number born to generation 2. Use this to find the number of G- and B-alleles before and after mutation. Find $g(2)$ and $b(2)$. How does this result compare to the answer to part (b)?

5. Suppose that 10% of the G-alleles mutate to B-alleles. Let $g(n-1)$ and $b(n-1)$ be the proportion of G- and B-alleles in generation $n-1$ after mutation has occurred. Assume that 500 individuals are born to generation n.

a. Develop a dynamical system for $g(n)$ and another for $b(n)$.

b. Assuming that $g(1) = 0.8$ and $b(1) = 0.2$, find expressions for the functions $g(n)$ and $b(n)$.

c. Use the expressions found in part (b) or the dynamical systems to find $g(10)$ and $b(10)$. At what rate is $g(n)$ going to 0?

d. What is the half-life of the G-alleles in numbers of generations?

6. Suppose that 5% of the G-alleles mutate to B-alleles. Assume that 500 individuals are born to generation n.

a. Develop a dynamical system for $g(n)$ and another for $b(n)$.

b. Assuming that $g(1) = 0.6$ and $b(1) = 0.4$, find expressions for the functions $g(n)$ and $b(n)$.

c. Find $g(100)$ and $b(100)$. At what rate is $g(n)$ going to 0?

d. What is the half-life of the G-alleles in numbers of generations?

7. Suppose that 1% of the B-alleles mutate to G-alleles. Assume that 500 individuals are born to generation n.

a. Develop a dynamical system for $g(n)$ and another for $b(n)$.

b. With $g(1) = 0.3$ and $b(1) = 0.7$, find an expression for the functions $b(n)$ and $g(n)$.

c. Find $b(100)$. At what rate is $b(n)$ going to 0?

d. What is the half-life of the B-alleles in numbers of generations?

8. Suppose that the fraction of the B-alleles that mutate to G-alleles is 0.0005. Assume that 500 individuals are born to generation n.

a. Develop a dynamical system for $g(n)$ and another for $b(n)$.

b. Assuming that $g(1) = 0.06$ and $b(1) = 0.94$, find an expression for the function for $b(n)$.

c. Find $b(1000)$. At what rate is $b(n)$ going to 0?

d. What is the half-life of the B-alleles in numbers of generations?

9. Suppose that 20% of the G-alleles mutate to B-alleles and 30% of the B-alleles mutate to G-alleles. The mutation is simultaneous; that is, an allele cannot mutate from G to B and back to G. Assume that 500 individuals are born to generation n.

a. Develop a dynamical system for $g(n)$. The following graphical outline of the steps, which is similar to Figure 7.10, may help. Use the fact that $b(n-1) = 1 - g(n-1)$ to write this dynamical system only in terms of $g(n-1)$.

b. What is the equilibrium value E_g for $g(n)$? At what rate does $g(n)$ go to E_g?

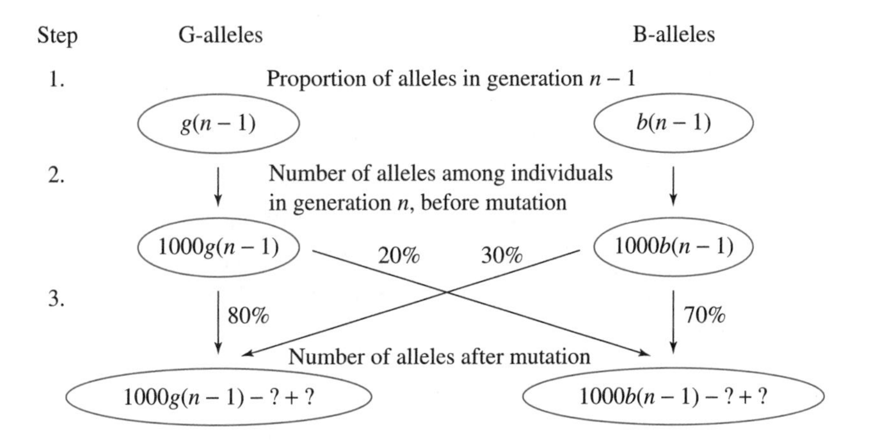

c. In part (b) you found that over time, the fraction of the G-alleles among the adults of this population will stabilize at the value E_g. This means that for "large" n-values, $g(n-1) \approx g(n) \approx E_g$. This also implies that $b(n-1) \approx b(n) \approx E_b = 1 - E_g$. Use this to predict the fraction of the individuals born with each genotype each generation, that is, what fraction of the individuals will be born GG, "BG", and BB? Refer to the tree diagram in Figure 7.3 and the Remark at the end of this section for help.

10. Suppose that 10% of the G-alleles mutate to B-alleles and 20% of the B-alleles mutate to G-alleles. Assume that 500 individuals are born to generation n.

a. Develop a dynamical system for $g(n)$. Use the fact that $b(n-1) = 1 - g(n-1)$ to write this dynamical system only in terms of $g(n-1)$.

b. What is the equilibrium value E_g for $g(n)$? At what rate does $g(n)$ go to E_g?

c. In part (b) you found that over time, the fraction of the G-alleles after mutation will stabilize at the value E_g. This also implies that the fraction of alleles that are B-alleles will stabilize at $E_b = 1 - E_g$. Use this to predict the fraction of the individuals born with each genotype each generation, that is, what fraction of the individuals will be born GG, "BG," and BB?

11. Suppose that 2% of the G-alleles mutate to B-alleles and 1% of the B-alleles mutate to G-alleles.

a. Develop a dynamical system for $g(n)$ in terms of $g(n-1)$ only.

b. What is the equilibrium value for $g(n)$? At what rate does $g(n)$ go to equilibrium?

c. For large values of n, predict the fraction born with each genotype—GG, "BG," and BB.

12. Suppose that 0.1% of the G-alleles mutate to B-alleles and 0.4% of the B-alleles mutate to G-alleles.

a. Develop a dynamical system for $g(n)$ in terms of $g(n-1)$ only.

b. What is the equilibrium value for $g(n)$? At what rate does $g(n)$ go to equilibrium?

c. For large values of n, predict the fraction born with each genotype—GG, "BG," and BB.

13. Suppose that the fraction of the G-alleles that mutate to B-alleles is p and the fraction of the B-alleles that mutate to G-alleles is q.

a. Develop a dynamical system for $g(n)$ in terms of $g(n-1)$, p, and q.

b. What is the equilibrium value for $g(n)$ in terms of p and q? Given that $b(n) = 1 - g(n)$, what is the equilibrium value for $b(n)$?

14. Suppose there are 3 types of alleles, G, B, and R. Let $g(n-1)$, $b(n-1)$, and $r(n-1)$ be the proportions of G-, B-, and R-alleles in generation $n-1$, after mutation. Suppose that 10% of the G-alleles mutate to B-alleles and 30% of the G-alleles mutate to R-alleles.

a. Develop a dynamical system of 3 equations for $g(n)$, $b(n)$, and $r(n)$.

b. Assuming that $g(1) = 0.6$, and $b(1) = 0.3$, and $r(1) = 0.1$, use the dynamical system to find $g(30)$, $b(30)$, and $r(30)$.

c. Assuming that $g(1) = 0.4$, $b(1) = 0.3$, and $r(1) = 0.3$, use the dynamical system to find $g(30)$, $b(30)$, and $r(30)$. (This dynamical system of 3 equations has infinitely many equilibrium points. The particular one that a solution goes to depends on the initial values.)

15. Suppose there are 3 types of alleles, G, B, and R. Let $g(n-1)$, $b(n-1)$, and $r(n-1)$ be the proportions of G-, B-, and R-alleles in generation $n-1$ after mutation. Suppose that 3% of the G-alleles mutate to B-alleles among the adults and that 2% of the G-alleles mutate to R-alleles.

a. Develop a dynamical system of 3 equations for $g(n)$, $b(n)$, and $r(n)$.

b. Assuming that $g(1) = 0.6$ and $b(1) = 0.2$ and $r(1) = 0.2$, use the dynamical system to find $g(100)$, $b(100)$, and $r(100)$.

c. Assuming that $g(1) = 0.4$ and $b(1) = 0.2$ and $r(1) = 0.4$, use the dynamical system to find $g(100)$, $b(100)$, and $r(100)$. (This dynamical system of 3 equations has infinitely many equilibrium points. The particular one that a solution goes to depends on the initial values.)

16. Suppose there are 3 types of alleles, G, B, and R. Suppose that 30% of the G-alleles mutate to B-alleles, that 20% of the B-alleles mutate to R-alleles, and that 28% of the R-alleles mutate to G-alleles.

a. Develop a dynamical system of 3 equations for $g(n)$, $b(n)$, and $r(n)$.

b. Assuming that $g(1) = 0.3$ and $b(1) = 0.3$ and $r(1) = 0.4$, use the dynamical system to find $g(30)$, $b(30)$, and $r(30)$.

c. Assuming that $g(1) = 0.4$ and $b(1) = 0.2$ and $r(1) = 0.4$, use the dynamical system to find $g(30)$, $b(30)$, and $r(30)$.

d. This dynamical system of 3 equations has one stable equilibrium point in which $g + b + r = 1$. See if you can find it.

17. Suppose there are 3 types of alleles—G, B, and R. Suppose that 10% of the G-alleles mutate to B-alleles, 30% of the B-alleles mutate to R-alleles, and 5% of the R-alleles mutate to G-alleles.

a. Develop a dynamical system of 3 equations for $g(n)$, $b(n)$, and $r(n)$.

b. Assuming that $g(1) = 0.3$, $b(1) = 0.3$, and $r(1) = 0.4$, use the dynamical system to find $g(30)$, $b(30)$, and $r(30)$.

c. Pick your own initial values for $g(1)$, $b(1)$, and $r(1)$, making sure that each value is between 0 and 1, and the total of the values equals 1. Use the dynamical system to find $g(30)$, $b(30)$, and $r(30)$.

d. This dynamical system of 3 equations has 1 stable equilibrium point in which $g+b+r=1$. See if you can find it.

7.4 Selection

In this section, we are going to investigate how natural selection can effect the genetic makeup of a population. This means that not all of the children survive to adulthood. In fact, the survival rate may be different for each of the 3 genotypes, GG, "BG" and BB.

Let's consider the allele that causes sickle cell anemia. Sickle cell anemia is a genetic disease caused by a "defective" allele of a certain gene. Each person has two alleles that determine if that person has sickle cell anemia or not. We will denote the "normal" allele by A. In reality, there are a group of sickle cell diseases caused by a number of different alleles. To simplify matters we assume there is only one such allele, denoted S.

When both of a person's alleles are S, the person has sickle cell anemia, a disease that was, until recently, almost always fatal. When one allele is S and one is A, the person has sickle cell trait, which usually results in nonfatal health problems. When both of the alleles are A, then the person does not have any health problems related to sickle cell.

To better understand the prevalence of the sickle cell allele in many different populations, including people of African, Mediterranean, Arabian, and East Indian ancestry, it is necessary to consider malaria. Malaria is a parasitic disease that is spread by the Anopheles mosquitoes. There are about 2 million deaths from malaria each year, making it one of the world's deadliest diseases. Most of the fatal cases of malaria are in African children under the age of 5. In some areas, up to 40% of the children die of malaria. Forty percent of the world's population is at risk of contracting malaria.

In individuals with two A-alleles, the malaria parasite can infect the red blood cells. The bursting of these infected cells can cause kidney and liver failure or block blood vessels to vital organs; children under the age of 5 have a high risk of death if this occurs. The red blood cells of the AS or SA-individuals (which we will henceforth call "AS") are resistant to the malaria parasite. Thus, whether the sickle cell allele is considered defective depends on the level of risk of contracting malaria. In the United States, the sickle cell allele would be considered defective, since a person with even one of the alleles will have health problems with no benefit. But if a person lives in an area inhabited by mosquitoes that carry the malaria parasite, then the sickle cell allele can be considered positive in the sense that one sickle cell allele and one normal allele creates a condition in the blood cells that gives some protection from the malaria parasite while not giving the person serious health problems related to sickle cell anemia. This relationship is represented in Figure 7.13.

The proportion of A-alleles and S-alleles in generation n is denoted as $a(n)$ and $s(n)$, respectively. Remember that $a(n)+s(n)=1$. Assume that

1. none of the SS-children survive to adulthood because they have sickle cell disease,
2. only one-third of the AA-children survive to adulthood, the rest perishing from malaria, and

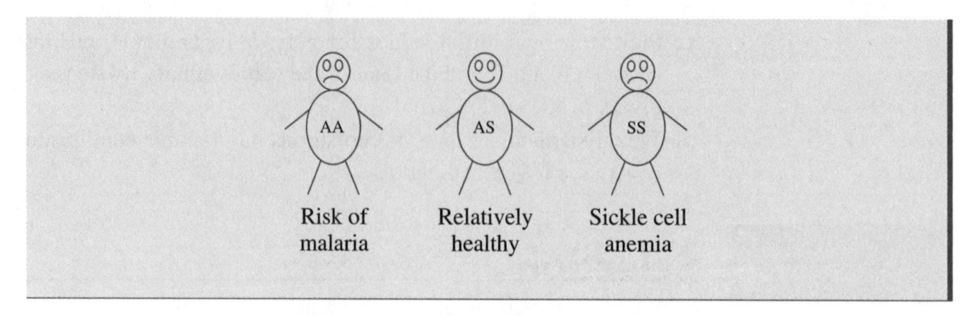

FIGURE 7.13 Different phenotypes using normal A and sickle cell S alleles.

3. all of the "AS"-children survive to adulthood, since they do not have sickle cell disease and are resistant to malaria.

To model the change in A and S-alleles over time, we must consider what fraction of children are born with each genotype. This is because there will be different survival rates for each of the genotypes. We now outline the process of developing a dynamical system for $s(n)$. To make the process easier conceptually, we assume that 900 children are born to generation n (so we can evenly divide by 3 later).

Step 1. **Determining the number of survivors of each genotype**
From the tree diagram of Figure 7.3, we know that

the fraction born AA is	$a^2(n-1)$
the fraction born AS or SA is	$2a(n-1)s(n-1)$
and the fraction born SS is	$s^2(n-1)$

Since there are going to be 900 children, then we multiply each fraction by 900 to get the number of children of each genotype. This is seen in Figure 7.14.

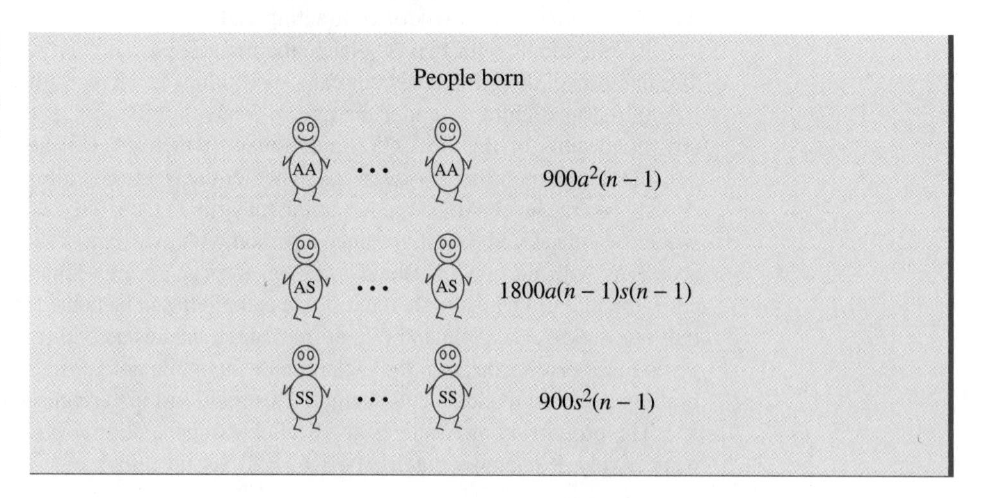

FIGURE 7.14 Number of people born with each genotype.

The genetic pool for the next generation will consist of those 900 people who survive both sickle cell anemia and malaria to reach adulthood. This consists of one-third of the AAs, all of the "AS"s, and none of the SSs. The survivors are seen in Figure 7.15, along with the number of A and S alleles that these survivors have.

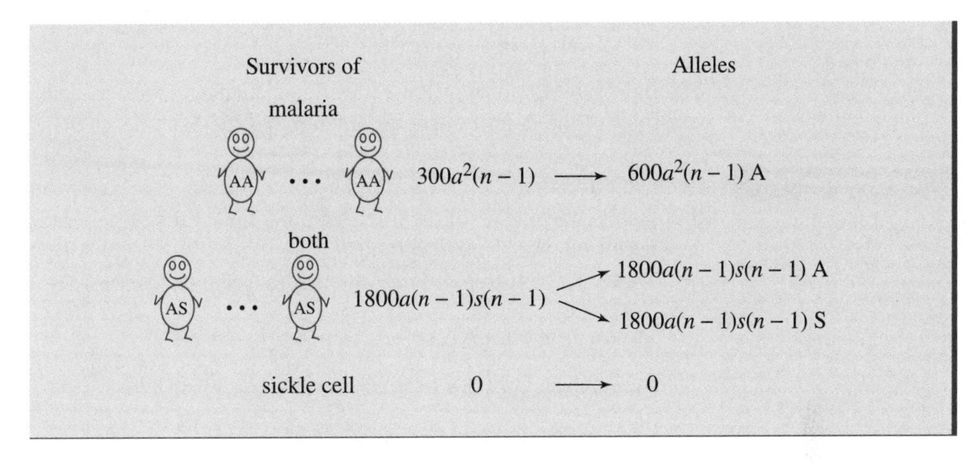

FIGURE 7.15 Number of individuals surviving both malaria and sickle cell anemia, along with number of A- and S-alleles among these individuals.

Step 2. Determining number of each kind of allele among the survivors

From Figure 7.15, we see that the total number of A-alleles among the AA-survivors and the "AS"-survivors is

$$600a^2(n-1)+1800a(n-1)s(n-1)$$

Similarly, we see that the total number of S-alleles, which all come from the "AS"-survivors, is

$$1800a(n-1)s(n-1)$$

The total number of A-and S-alleles is

$$600a^2(n-1)+3600a(n-1)s(n-1)$$

Step 3. Computing the ratios

We can now find $s(n)$ by finding the ratio of S-alleles to the total number of alleles. This is

$$s(n)=\frac{1800a(n-1)s(n-1)}{600a^2(n-1)+3600a(n-1)s(n-1)}$$

The denominator can be factored into

$$600a(n-1)(a(n-1)+6s(n-1))$$

This gives that

$$s(n)=\frac{1800a(n-1)s(n-1)}{600a(n-1)[a(n-1)+6s(n-1)]}=\frac{3s(n-1)}{a(n-1)+6s(n-1)}$$

after canceling.

Recall that $a(n-1) = 1 - s(n-1)$, since all of the alleles are either A or S. After making this substitution and collecting terms, we get that the nonlinear dynamical system

$$s(n) = \frac{3s(n-1)}{1+5s(n-1)}$$

describes how the proportion of S-alleles changes from one generation to the next when one-third of the A-homozygotes and none of the S-homozygotes survive to adulthood.

Example 7.3

Let's go through the steps for finding $s(2)$ in the specific case where $a(1) = 0.8$ and $s(1) = 0.2$, that is, when 80% of the alleles among the adults of generation 1 are normal and 20% are sickle cell. We assume 900 births.

We begin with step 1. From the tree diagram of Figure 7.3, we know that

the fraction born AA is $a^2(1) = 0.64$

the fraction born AS or SA is $2a(1)s(1) = 0.32$

and the fraction born SS is $s^2(1) = 0.04$

Since there are 900 births, we multiply each fraction by 900 to get the number of children of each genotype. This is seen in Figure 7.16.

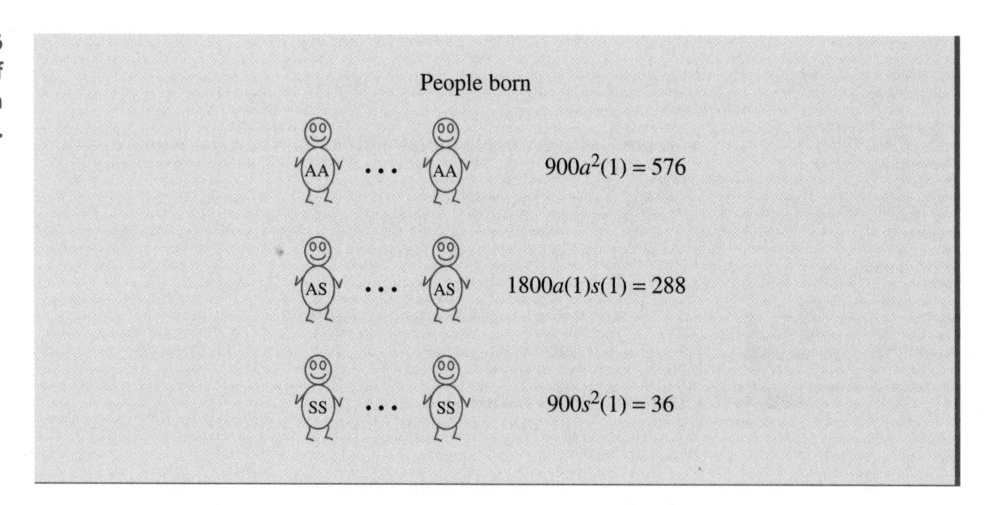

FIGURE 7.16 Number of people born with each genotype.

The genetic pool for the next generation will consist of those 900 people who survive both sickle cell anemia and malaria to reach adulthood. This consists of one-third of the AAs, all of the "AS"s, and none of the SSs. The survivors are seen in Figure 7.17, along with the number of A and S alleles that these survivors have.

Step 2. Determining number of each kind of allele among the survivors

From Figure 7.17, we see that the total number of A-alleles among the AA-survivors and the "AS"-survivors is

$$600a^2(1) + 1800a(1)s(1) = 384 + 288 = 672$$

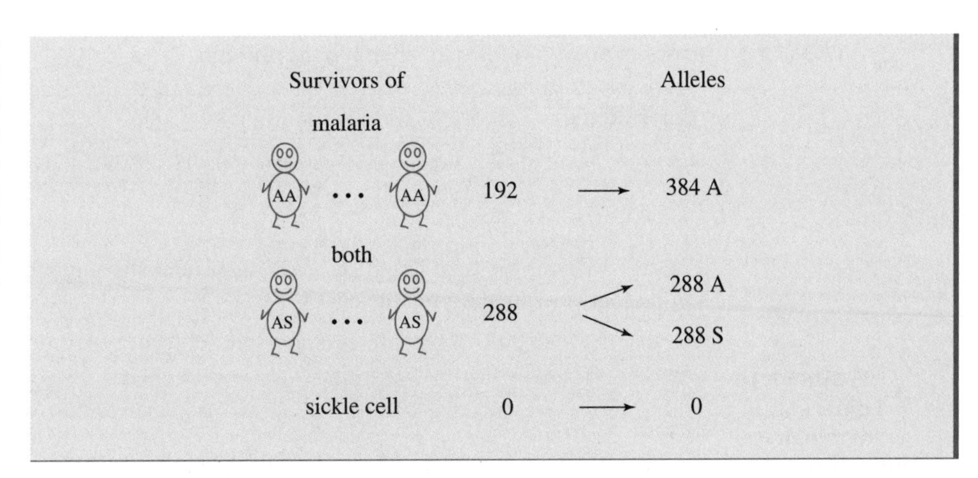

FIGURE 7.17 Number of individuals surviving both malaria and sickle cell anemia, along with number of A- and S-alleles among these individuals.

Similarly, we see that the total number of S-alleles, which all come from the "AS"-survivors, is

$$1800a(1)s(1) = 288$$

The total number of A- and S-alleles is

$$600a^2(1) + 3600a(1)s(1) = 384 + 576 = 960$$

Step 3. **Computing the ratios**

We can now find $s(2)$ by finding the ratio of S-alleles to the total number of alleles. This is

$$s(2) = \frac{1800a(1)s(1)}{600a^2(1) + 3600a(1)s(1)} = \frac{288}{960} = 0.3$$

There was no need to factor the denominator, since we knew the actual numbers.

Despite the fact that the S-homozygotes exhibit a lethal trait, the actual proportion of S-alleles in the population of adults increased from 20% to 30% of the total in just one generation. ■

Let's now examine the consequences of the dynamical system

$$s(n) = \frac{3s(n-1)}{1 + 5s(n-1)}$$

Table 7.1 gives my results when letting $s(1) = 0.2$ (and therefore $a(1) = 0.8$). Note that $s(n)$ seems to stabilize at 0.4, meaning that after several generations, the percentage of S-alleles in this population will be about 40% and, consequently, the percentage of A-alleles will be about 60%. You should try other starting values for $s(1)$ between 0 and 1. You will find that $s(n)$ always stabilizes at 0.4. Thus, 0.4 seems to be a stable equilibrium value for this dynamical system.

TABLE 7.1 Proportion of A-alleles in several generations.

generation n =	1	2	3	4	5	6	7	8	20
$s(n)$ =	0.2	0.3	0.36	0.386	0.395	0.398	0.399	0.400	0.400

Figure 7.18 gives the time graph $(n, s(n))$ for several starting values. All of the functions $s(n)$ have a horizontal asymptote of 0.4.

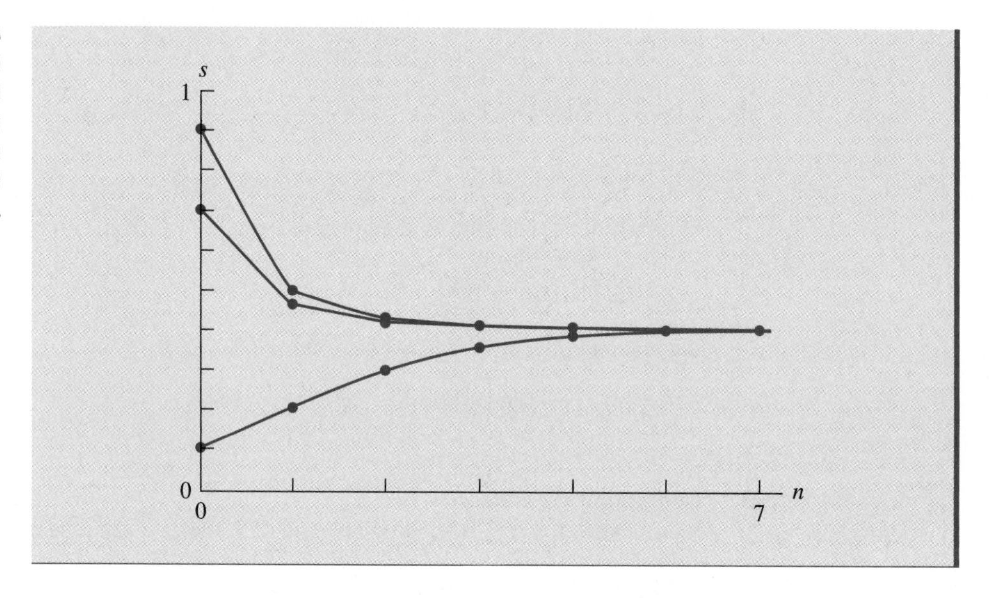

FIGURE 7.18 Points $(n, s(n))$ for dynamical system modeling sickle cell/malaria relationship. $s(n)$ goes to 0.4.

To find the equilibrium values, we solve

$$E_s = \frac{3E_s}{1+5E_s}$$

Multiplying both sides of the equation by $1+5E_s$ gives

$$E_s(1+5E_s) = 3E_s$$

Collecting all terms on the left side of the equation gives

$$5E_s^2 - 2E_s = 0$$

This equation can be factored as

$$E_s(5E_s - 2) = 0$$

The 2 solutions, which are found by setting each factor equal to 0, are $E_s = 0.4$ and $E_s = 0$.

From the results of Table 7.1 and Figure 7.18, it appears that 0.4 is a stable equilibrium value and 0 is an unstable equilibrium value. Further evidence of this is the web seen in Figure 7.19. The web in which $s(1) < 0$ has no physical meaning but was drawn to demonstrate that $E_s = 0$ is unstable.

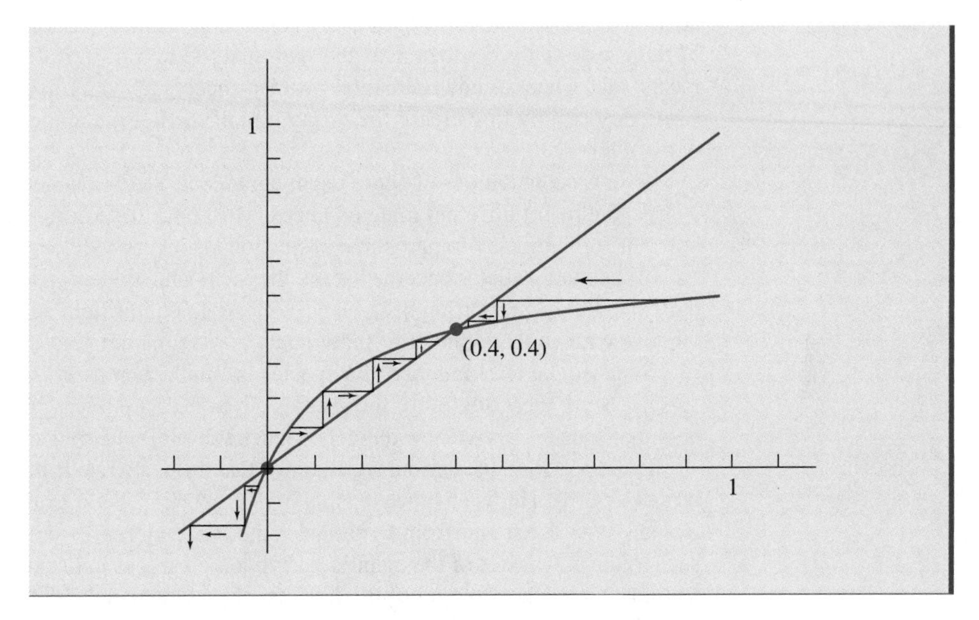

FIGURE 7.19 Webs for sickle cell/malaria dynamical system show $E_S = 0.4$ is a stable equilibrium and $E_S = 0$ is an unstable equilibrium.

Knowing that the equilibrium for $s(n)$ is $E_s = 0.4$ and, consequently, the equilibrium value for $a(n)$ is $E_a = 0.6$ allows us to predict the fraction of the population born with each genotype. Recall that

the fraction born AA is	$a^2(n-1)$
the fraction born AS or SA is	$2a(n-1)s(n-1)$
and the fraction born SS is	$s^2(n-1)$

Since $a(n-1) \approx E_a = 0.6$ and $s(n-1) \approx E_s = 0.4$ means that

the fraction born AA is	0.36
the fraction born AS or SA is	0.48
and the fraction born SS is	0.16

This means that about 16% of this population will be born with sickle cell anemia each generation. It also means that nearly half of the population will not get either disease. But the AA and "AS" individuals will appear to be the same, so parents won't know if their children are in the 36% that are susceptible to malaria or the 48% that have some resistance to malaria.

We can find that

$$\frac{s(n) - 0.4}{s(n-1) - 0.4} \approx 0.33$$

by making that computation for moderately large values of n. This means that $s(n)$ gets about 67% closer to equilibrium every generation. This is very rapid convergence to equilibrium, especially compared to the extremely slow rate resulting from mutation. This means that when an environmental problem occurs, such as malaria, then selection can work very rapidly to increase the proportion of the helpful allele in the population to the optimal level.

Over time, mutations of genes occur at random. So for a large population, there will be a large number of different mutated genes. Most have little or no effect on the population, but some of them may impart some protection against new diseases or environmental hazards. When these alleles become useful, they will tend to increase quickly in prevalence (in our example the proportion became 67% closer to equilibrium each generation) to a point where the maximum number of individuals survive; nature tends to optimize.

Suppose the total number of a species is small, such as in the case of an endangered species. Since there are few of the species alive, there are fewer mutated genes available to help that species survive new dangers. Genetic diversity helps a species survive.

In the previous discussion, we assumed that the death rate from malaria for AA-children was two-thirds. This is an unreasonably high proportion, but it made the computations easy. Actually, the death rate from malaria for the AA-children is dependent on many factors, such as the prevalence of the malaria-carrying mosquitoes in the region, the amount of protection people have from mosquito bites (such as screened windows and bed nets), and the availability of medical care in the event of a case of malaria. In the exercises, you will learn how to estimate the risk of malaria for a particular area by knowing the fraction of children born with sickle cell anemia. The fraction born with sickle cell anemia is E_s^2. Thus, if we know the fraction of the children born with sickle cell anemia, we can easily estimate the equilibrium value for $s(n)$.

It is important to note that the equilibrium genetic makeup is actually the makeup that ensures that the largest percentage of children survive both malaria and sickle cell anemia. Suppose that 900 children were born to a population in which $a(1) = 0.6$ and $s(1) = 0.4$. Since the fraction of the children that are AA is about 0.36, this means that about 324 of the children will be AA. Similarly, about 48% of the 900 children will be ìASî and about 16% will be SS. Our model predicts that the 144 SS-children will die of sickle cell anemia, that only one-third of the 324 AA-children will survive malaria, and that all of the 432 ìASî-children will survive. Thus, we expect that 540 of the 900 children will survive to adulthood. In Table 7.2 are the predictions of the number of 900 children that will survive to adulthood given other initial proportions for the S-allele. You should check these numbers on your own.

TABLE 7.2 Predicted number of 900 children to survive to adulthood for different genetic makeup of parents.

$s(1) =$	0.8	0.6	0.4	0.2
number of 900 children surviving to adulthood in generation 2	300	480	540	480

In the problems, you will actually show that the equilibrium value for $s(n)$ gives the optimal outcome for the population.

As discussed, one of the reasons that the sickle cell allele is relatively prominent in some human populations is that in areas where the malaria parasite thrives, the presence of the sickle cell allele results in the survival of more of the children. It is believed that this type of relationship exists for other diseases and genetic traits. For example, there is some evidence that people with just one of the alleles that causes cystic fibrosis have an increased chance of surviving cholera.

Let me turn this discussion to a related subject. Negative eugenics, a flawed movement begun in the 1800s, is the use of sterilization to eliminate harmful genetic traits from human society. It was mainly used to try to rid society of mentally retarded and "mentally defective" individuals. Laws were passed in a number of states allowing individuals to be sterilized, beginning in 1907 in Indiana. In 1927, the United States Supreme Court upheld the Virginia sterilization bill by an 8–1 vote in *Buck v. Bell.* In this case, the Supreme Court allowed the sterilization of Carrie Buck. In writing for the majority, Oliver Wendell Holmes wrote the now famous phrase, "three generations of imbeciles are enough." It is estimated that by 1935, about 20,000 forced sterilizations had been performed in the United States alone. Nazi Germany had performed nearly 400,000 sterilizations by the end of World War II, including 4000 for blindness and deafness. Sweden was the first nation to establish an institute on racial biology. It passed its first sterilization law in 1934 and expanded it in 1941. During the 1940s and 1950s, Sweden sterilized about 2000 people per year. Their program was not stopped until 1974.[1,2]

The main assumption behind negative eugenics is that by sterilizing individuals with an "undesirable" recessive trait, that trait will die out of the population. Sterilizing such individuals is the same as having them not survive to adulthood, from a genetic point of view. This is equivalent to assuming that the S-allele is lethal to the S-homozygotes (or the SS-individuals are sterilized), but that all of the AA- and "AS"-children survive. The assumption was that under such conditions, the proportion of the harmful allele would quickly go to 0. As you will discover in the exercises, this is not the case. In fact, the decline of such an allele is so slow, its prevalence in the population appears to remain constant from one generation to the next. This tragic movement doesn't work.

In Project 3, you will investigate a situation that is similar to negative eugenics, which is lethal recessive traits. Again, this would be a situation in which all of the AA- and "AS"-children survive, but all of the SS-children die of the related genetic disease. Often, the reason that lethal recessive traits occur is that normal alleles mutate to the defective alleles. Mutation rates are difficult to estimate directly. Using your understanding of the theory, you will learn how to approximate the mutation rate for lethal traits, given the equilibrium genetic makeup of the population. It is actually easy to determine the equilibrium genetic makeup by obtaining information on the fraction of children born with the genetic trait.

One disease for which this applies is galactosemia. Infants with this disease cannot metabolize lactose and galactose, 2 sugars found in milk. This means too much galactose will build up in the blood, resulting in, among other things, liver failure. Galactosemia

[1] *The Flamingoís Smile: Re ections in Natural History* by Stephen Jay Gould, WW Norton and Company, Inc., 1985

[2] *Washington Post,* August 29, 1997

was first discovered in 1908. Historically, it is likely that nearly 100% of infants with galactosemia died. This means that occurrence of galactosemia reached an equilibrium in human populations. This equilibrium is approximately 1 infant with galactosemia out of 30,000 births. Fortunately, effective treatments have been developed for galactosemia. These treatments, including complete removal of milk from the diet, need to be started within the first 10 days of life. It happens that galactosemia is easy to diagnose through genetic screening and most cases are discovered within the first week.

There are actually a number of different alleles that can cause variations of galactosemia. In Project 3, you are finding the joint mutation rate for all of these. There are studies that actually break down the cases of galactosemia by allele. We could use these studies to estimate the mutation rate for each possible allele.

7.4 Problems

1. Assume that there are 9000 births, that 90% of the AAs survive malaria, and that none of the SSs survive sickle cell anemia.

a. Suppose $s(1) = 0.7$. How many children are expected to be born AA, "AS," and SS?
b. How many of each genotype are expected to survive to adulthood?
c. How many alleles are there expected to be among the surviving adults? How many of them are expected to be A? What should $s(2)$ be?

2. Assume that there are 30,000 births, that 60% of the AA's survive malaria, and that none of the SSs survive sickle cell anemia.

a. Suppose $s(1) = 0.3$. How many children are expected to be born AA, "AS," and SS?
b. How many of each genotype are expected to survive to adulthood?
c. How many alleles are there expected to be among the surviving adults? How many of them are expected to be A? What should $s(2)$ be?

3. Assume that there are 10,000 births, that 40% of the AAs survive malaria, and that none of the SSs survives sickle cell anemia.

a. Suppose $s(1) = 0.8$. Predict $s(2)$. Do this by computing the estimated number of children that are of each genotype, the number of adults of each genotype, and finally, the number of A- and S-alleles among these surviving adults.
b. Repeat part (a) using $s(1) = 0.5$.
c. Repeat part (a) using $s(1) = 0.7$.

4. Assume that there are 10,000 births, that 20% of the AAs survive malaria, and that none of the SSs survive sickle cell anemia.

a. Suppose $s(1) = 0.2$. Predict $s(2)$.
b. Repeat part (a) using $s(1) = 0.5$.
c. Repeat part (a) using $s(1) = 0.7$.

5. Assume that 80% of the AAs survive malaria and that none of the SSs survive sickle cell anemia. Let $s(n-1)$ represent the fraction of alleles in generation $n-1$ that are S.

a. Develop a dynamical system for $s(n)$. To do this, assume that 1600 children are born to generation n. Develop expressions, in terms of $a(n-1)$ and $s(n-1)$, for the expected number of children of each genotype born to generation n. Use those expressions to develop expressions for the number of adults of each genotype. Finally, use those expressions to develop expressions for the total number of A- and S-alleles among the adults. The dynamical system for $s(n)$ is the ratio of the number of S-alleles to the total number of alleles. See Figures 7.14 and 7.15. Remember to substitute $1-s(n-1)$ for $a(n-1)$ in that dynamical system.

b. Find the equilibrium values for the dynamical system developed in part (a).

c. Draw a web for this dynamical system to determine the stability of the equilibrium values.

d. For the stable equilibrium value, determine the rate at which $s(n)$ approaches equilibrium. Use the rate to estimate what percent closer $s(n)$ is to equilibrium than $s(n-1)$.

6. Assume that 50% of the AAs survive malaria and none of the SSs survive sickle cell anemia. Let $a(n-1)$ represent the fraction of alleles in generation $n-1$ that are A.

a. Develop a dynamical system for $s(n)$.

b. Find the equilibrium values for the dynamical system developed in part (a).

c. Draw a web for this dynamical system to determine the stability of the equilibrium values.

d. For the stable equilibrium value, determine the rate at which $s(n)$ approaches equilibrium. Use the rate to estimate what percentage closer $s(n)$ is to equilibrium than $s(n-1)$.

7. Assume that 50% of the AAs survive malaria and 50% of the SSs survive sickle cell anemia.

a. Suppose that 1000 children are born to generation 2 and $a(1)=0.3$. How many children are expected to be born AA, "AS," and SS?

b. How many of each genotype are expected to survive to adulthood?

c. How many alleles are there expected to be among the surviving adults? How many of them are expected to be A? What should $a(2)$ be?

d. Develop a dynamical system for $a(n)$. (Use a variation on Figure 7.15.) To do this, assume that 1000 children are born to generation n. Develop expressions in terms of $a(n-1)$ and $s(n-1)$ for the expected number of children of each genotype born to generation n. Use those expressions to develop expressions for the number of adults of each genotype. Finally, use those expressions to develop expressions for the total number of A- and S-alleles among the adults. The dynamical system for $a(n)$ is the ratio of the number of A-alleles to the total number of alleles. Remember to substitute $1-a(n-1)$ for $s(n-1)$ in that dynamical system.

e. Estimate the stable equilibrium value for this dynamical system by picking a value for $a(1)$ between 0 and 1 and then observing the value $a(n)$ approaches.

f. Find the equilibrium values for the dynamical system developed in part (d). To do this, multiply both sides of the equation by the denominator and bring all terms to the left. One equilibrium value will be 0, and a second equilibrium value will be 1. This should help you in factoring (or using the quadratic equation) to find the third equilibrium value.

g. Draw a web for this dynamical system to determine the stability of the equilibrium values.

h. For the stable equilibrium value, determine the rate at which $a(n)$ approaches equilibrium. Use the rate to estimate what percent closer $a(n)$ is to equilibrium than $a(n-1)$.

8. Assume that 70% of the AAs survive malaria and 30% of the SSs survive sickle cell anemia.

a. Suppose that 1000 children are born to generation 2 and $a(1) = 0.8$. How many children are expected to be born AA, "AS," and SS?

b. How many of each genotype are expected to survive to adulthood?

c. How many alleles are there expected to be among the surviving adults? How many of them are expected to be A? What should $a(2)$ be?

d. Develop a dynamical system for $a(n)$.

e. Estimate the stable equilibrium value for this dynamical system by picking a value for $a(1)$ between 0 and 1 and then observing the value $a(n)$ approaches.

f. Find the equilibrium values for the dynamical system developed in part (d).

g. Draw a web for this dynamical system to determine the stability of the equilibrium values.

h. For the stable equilibrium value, determine the rate at which $a(n)$ approaches equilibrium. Use the rate to estimate what percent closer $a(n)$ is to equilibrium than $a(n-1)$.

9. Assume that 50% of the AAs survive malaria and 25% of the SSs survive sickle cell anemia.

a. Develop a dynamical system for $a(n)$.

b. Estimate the stable equilibrium value for this dynamical system by picking a value for $a(1)$ between 0 and 1 and then observing the value $a(n)$ approaches.

c. Find the equilibrium values for the dynamical system developed in part (b).

d. Approximate the rate at which $a(n)$ approaches the stable equilibrium value. Use the rate to estimate what percent closer $a(n)$ is to equilibrium than $a(n-1)$.

e. Draw a web for this dynamical system to determine the stability of the equilibrium values.

10. Assume that 90% of the AAs survive malaria and 70% of the SSs survive sickle cell anemia.

a. Develop a dynamical system for $a(n)$.

b. Estimate the stable equilibrium value for this dynamical system by picking a value for $a(1)$ between 0 and 1 and then observing the value $a(n)$ approaches.

c. Find the equilibrium values for the dynamical system developed in part (b).

d. For the stable equilibrium value, determine the rate at which $a(n)$ approaches equilibrium. Use the rate to estimate what percent closer $a(n)$ is to equilibrium than $a(n-1)$.

e. Draw a web for this dynamical system to determine the stability of the equilibrium values.

11. Assume that 80% of the AAs survive malaria, and that none of the SSs survive sickle cell anemia. Suppose that 1000 children are born to generation n. Let x instead of $s(n-1)$ represent the fraction of the alleles in the adults of generation $n-1$ that are S. Let a represent the number of the 1000 children born to generation n that survive to adulthood. Find a quadratic for a in terms of x. Graph the function $a(x)$ for $0 \leq x \leq 1$. Find the value for x that results in the maximum number of adults in generation n. Do this by ndi ng the x-value of the vertex of the parabola. Compare that answer to the stable equilibrium value found in Problem 5.

12. Assume that 50% of the AAs survive malaria, and that none of the SSs survive sickle cell anemia. Suppose that 1000 children are born to generation n. Suppose that $s(n-1) = x$ of the alleles in the adults of generation $n-1$ are S. Find a quadratic for a in terms of x that gives the expected number of adults in generation n. Graph the function $a(x)$ for $0 \leq x \leq 1$. Find the value for x that results in the maximum number of adults in generation n. Do this by finding the x-value of the vertex of the parabola. Compare that answer to the stable equilibrium value found in Problem 6.

13. Assume that there are 1000 births, that 100% of the AAs survive malaria, and that none of the SSs survive sickle cell anemia. Assume that 9% of the A-alleles among the surviving children mutate to S-alleles in the adults.

a. Develop a dynamical system for $a(n-1)$. Do this by finding the expressions that go into the boxes in the following outline. The following outline follows the 3 steps for "selection" outlined in this section, followed by the steps for considering mutation, outlined in Section 7.3 (see figure on page 368).

b. Suppose that $a(1) = 0.8$ and $s(1) = 0.2$. Find $a(10)$ and $s(10)$.

c. Find the equilibrium values for the dynamical system.

d. Draw a web to determine the stability of the equilibrium values found in part c).

14. Assume that there are 1000 births, that 100% of the AAs survive malaria, and that none of the SSs survive sickle cell anemia. Assume that 25% of the A-alleles mutate to S-alleles.

a. Develop a dynamical system for $a(n-1)$.

b. Suppose that $a(1) = 0.6$ and $s(1) = 0.4$. Find $s(10)$ and $a(10)$.

c. Find the equilibrium values for the dynamical system.

d. Draw a web to determine the stability of the equilibrium values found in part (c).

Project 1. Assume that 100% of the AAs survive malaria and none of the SSs survive sickle cell anemia. Let $a(n-1)$ represent the fraction of alleles in generation $n-1$ that are A. (This situation models the situation created by negative eugenics.)

a. Develop a dynamical system for $a(n)$.

b. Find the equilibrium values for the dynamical system developed in part (a).

c. Draw a web for this dynamical system to determine the stability of the equilibrium values.

d. Suppose $a(1) = 0.5$. Find $a(2)$ through $a(5)$. Can you write down a formula for $a(n)$ in this case? For what value of n does $a(n) = 0.98$? For what value of n does $a(n) = 0.99$? Considering a generation is about 20 years, how long did it take to cut the proportion of S-alleles in half, from 2% to 1% of the total number of alleles?

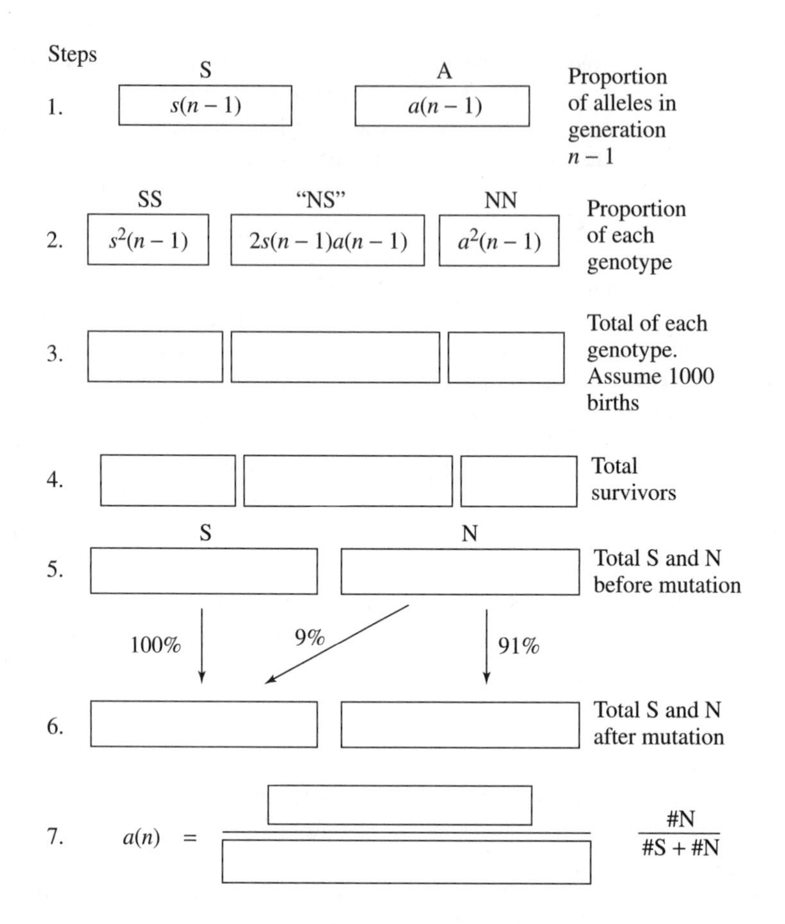

Project 2. Assume that the fraction of the AAs that survive malaria is m and none of the SSs survive sickle cell anemia. Let $a(n-1)$ represent the fraction of alleles in generation $n-1$ that are A.

a. Develop a dynamical system for $a(n)$. To do this, assume that 1000 children are born to generation n. Develop expressions, in terms of $a(n-1)$ and $s(n-1)$, for the expected number of children of each genotype born to generation n. Use those expressions to develop expressions for the number of adults of each genotype. Finally, use those expressions to develop expressions for the total number of A- and S-alleles among the adults. The dynamical system for $a(n)$ is the ratio of the number of A-alleles to the total number of alleles. Remember to substitute $1-a(n-1)$ for $s(n-1)$ in that dynamical system. The dynamical system will include ms.

b. Find the equilibrium values for the dynamical system developed in part (a). One equilibrium value will be given in terms of m.

c. Assume that $a(n)$ equals an equilibrium value other than 0 or 1. Assume that $s(n) = 1-a(n)$. Develop an expression for the proportion of the children of generation n that will have sickle cell anemia, that is, the proportion that is SS. This expression will involve an m.

d. Suppose in a particular region of Africa that 9% of the children of this population have sickle cell anemia. Set 0.09 equal to the expression found in part (c) and solve for m. This will give an estimate of the fraction of the AA-children that survive malaria in this area.

Project 3. Assume that there are 1000 births, that 100% of the AAs survive malaria, and that none of the SSs survive sickle cell anemia. Suppose $s(n-1)$ represents the fraction of S-alleles among the adults of generation $n-1$. Suppose the fraction of A-alleles that mutate to S-alleles among the adults of generation n is given by the fraction m.

a. Develop a dynamical system for $s(n)$ by (1) finding expressions for the number of A- and S-alleles among the surviving adults of generation n, (2) decreasing the number of A-alleles by the proportion m and adding the same number to the total of S-alleles, and (3) using the new totals of A- and S-alleles to compute $s(n)$, the number of S-alleles divided by the total. Remember that you can substitute $1-s(n-1)$ for $a(n-1)$.

b. Find the positive equilibrium values for the dynamical system found in part (d). This equilibrium value will be given in terms of the fraction m.

c. Assume that $s(n)$ equals the equilibrium value found in part (b) and that $a(n) = 1 - s(n)$. Write an expression for the proportion of the children in the next generation that will be born with the genotype SS.

d. The fraction of children born with galactosemia is between 0.0001 and 0.00002. Set these decimals equal to the proportion found in part (c) and use that to find a range for the mutation rate from normal alleles to alleles causing galactosemia.

Answers to Odd-Numbered Problems

Chapter 1

Section 1.2

1. **a.** $u(n) = u(n-1) + 17$
b. $u(n) = u(n-1) - 3.2$
c. $u(n) = u(n-1) + 7.4$

3. **a.** n is the number of games; $p(n)$ is the total pay for playing n games; The domain is $n = 0.1, \ldots.$
b. $p(n) = 30{,}000 + 5000n$; $p(120) = 30{,}000 + 5000(120) = 630{,}000$ dollars; solve $30{,}000 + 5000n = 1{,}000{,}000$, giving $n = 194$ games.
c.

Number of games played	0	1	2	3	4
Money made	30,000	35,000	40,000	45,000	50,000

d.

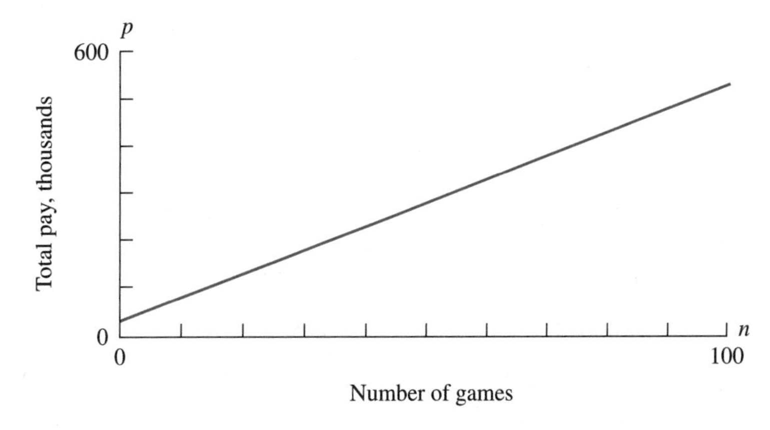

e. The slope is 5000, which is salary per game, and the y-intercept is 30,000, which is the bonus.

f. 5000 → salary

g. $p(n) = p(n-1) + 5000$

5. a. n is the number of meters above ground; $t(n)$ is the temperature, in degrees Centigrade, at a height of n meters; the domain is $0 \leq n \leq 12{,}000$.

b. $t(n) = 28 - 0.01n$; $t(1000) = 28 - 0.01(1000) = 18$; $0 = 28 - 0.01n$ gives that the temperature is 0°C at a height of $n = 2800$ meters or 2.8 km.

c.

Number of kilometers above ground	0	1	2	3	4
Temperature in degrees Centigrade	28	18	8	−2	−12

d.

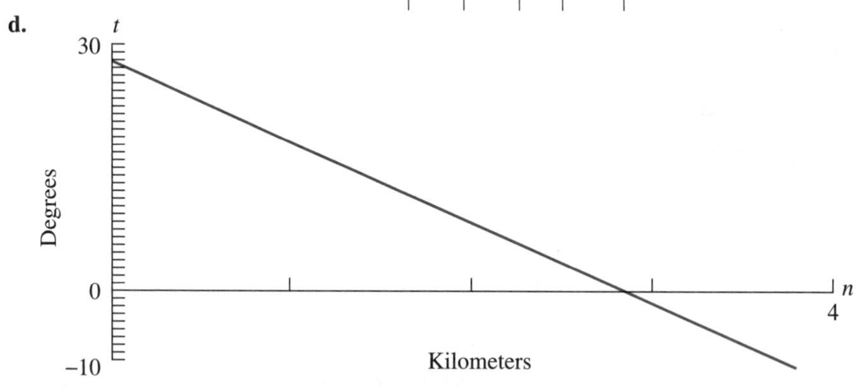

e. The slope is −0.01, which is the amount the temperature drops for a rise of 1 meter above ground. The y-intercept is 28, which is the ground temperature.

f. temperature → 0.01

g. $t(n) = t(n-1) - 0.01$

7. a. n is your speed in miles per hour, toward the sound. $f(n)$ is the frequency of the sound that you hear, in cycles per second, when traveling at speed of n miles per hour. The domain is unclear. It is any speed you can travel. It can also be negative if you travel away from the sound.

b. $f(n) = 256 + 256n/760$. Solving $271 = 256 + 256n/760$ gives that $n \approx 44.5$ miles per hour. Solving $512 = 256 + 256n/760$ gives that $n = 760$ miles per hour, or the speed of sound.

c.

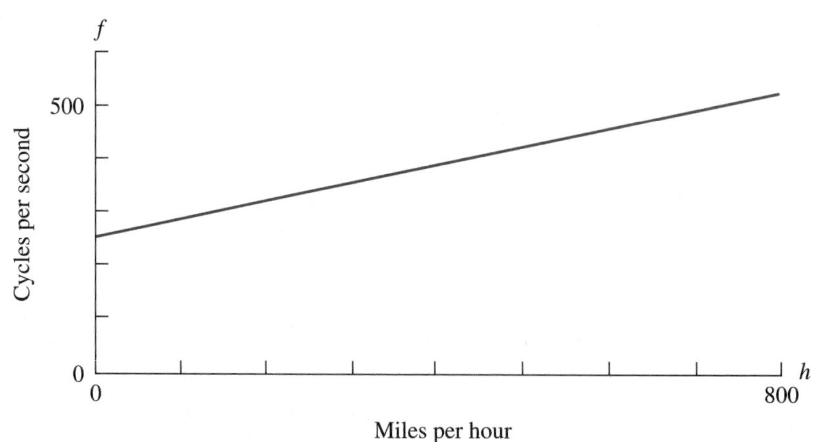

d. The slope is 256/760, or 0.34, which is the number of cycles per second the frequency increases as your speed increases toward the object by 1 mile per hour. The y-intercept is 256, which is the frequency of the sound when you are standing still.

e. 256/760 → frequency

f. $f(n) = f(n-1) + \frac{256}{760}$.

9. a. n is the depth, in centimeters, in saltwater. $p(n)$ is the pressure at n cm under saltwater.

b. $p(n) = 1030 + 1.026n$ g/cm^2; $p(500) = 1030 + 1.026(500) = 1543$ g/cm^2.

c. 1.026 → pressure

d. $p(n) = p(n-1) + 1.026$

Section 1.3

1. a. $u(1) = 10$; $u(2) = 14$; $u(3) = 18.4$; $u(4) = 23.24$

b. $u(1) = u(2) = u(3) = u(4) = -30$

3. a. $w(1) = 3.7$; $w(2) = 5.59$; $w(3) = 6.913$; $w(4) = 7.8391$

b. $w(0) = w(1) = 10$

5. a. $u(1) = 13$; $v(1) = 0$; $u(2) = 41$; $v(2) = 11$

b. $u(1) = 2$; $v(1) = 3$; $u(2) = 2$; $v(2) = 3$

7. a.

b. $a(n) = 1.0125a(n-1) - 1000$, with $a(0) = 50{,}000$

c. $a(4) = 48{,}472.64$

9. $p(n) = 1.01p(n-1) + 540{,}000$; with $p(1999) = 271{,}000{,}000$, then $p(2003) = 284{,}196{,}303$.

11. a.

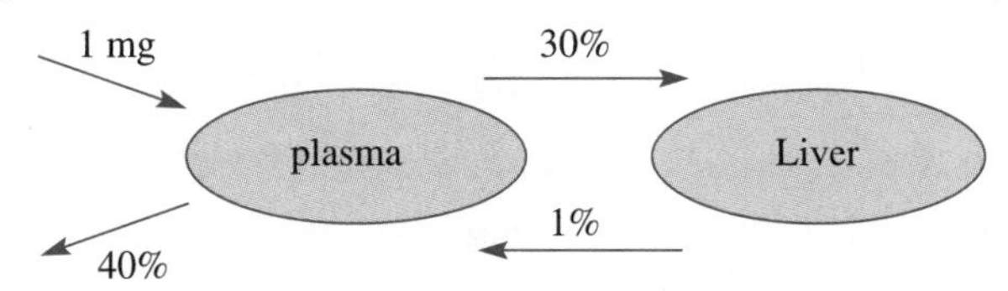

b. $p(n) = 0.3p(n-1) + 0.01l(n-1) + 1$, and $l(n) = 0.3p(n-1) + 0.99l(n-1)$

c. $p(1) = 3.3$, $l(1) = 80.7$, $p(2) = 2.8$, and $l(2) = 80.9$.

13. a.

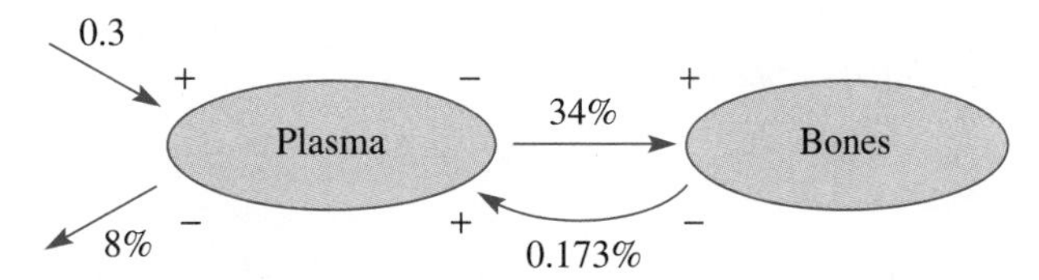

b. $p(n) = 0.58p(n-1) + 0.00173b(n-1) + 0.3$, and $b(n) = 0.99827b(n-1) + 0.34p(n-1)$. $p(3) = 0.575$, and $b(3) = 0.263$ mg, to 3-decimal places.

15. $c(n) = c(n-1) - 200$, and

$$w(n) = \left(1 - \frac{130}{3600}\right)w(n-1) + \frac{c(n-1) - 200}{3600}$$

Section 1.4

1. a. $u(30) = 604.52$
b. $u(43) = 2160.5$

3. a. $w(30) = 9.9998$; $w(50) = 10$ (actually 9.999999838)
b.

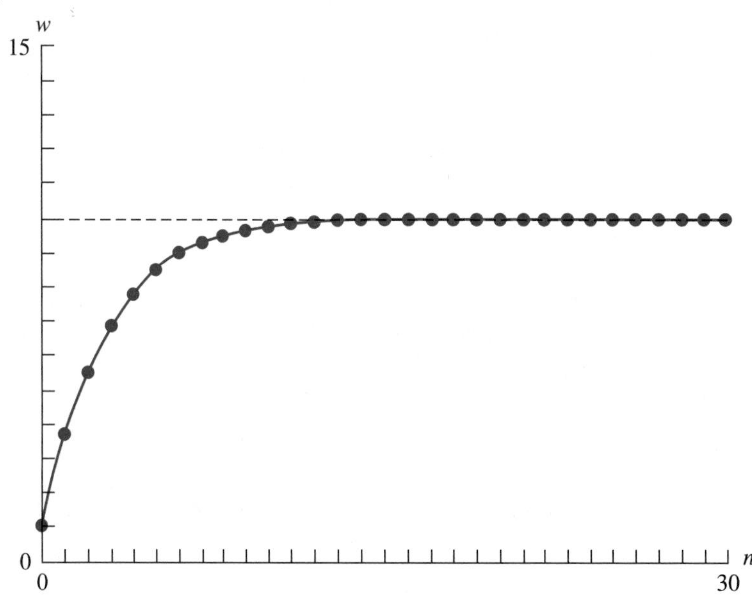

5. Any value where $5.203284887 < x < 5.2167741036$

7. Any value where $20.51107378 < x < 20.63715259$

9. a. With $p(n) = 0.58p(n-1) + 0.00173b(n-1) + 0.3$ and $b(n) = 0.99827b(n-1) + 0.34p(n-1)$, then $p(365) = 1.05$ and $b(365) = 82.76$ mg.
b. $p(197) = 0.9006$.

11. $w(n) = 0.97w(n-1) + x - 0.25$ with $w(0) = 160$ and $144 < w(52) < 146$. Then $4.4461 < x < 4.5218$. This is approximately 16,200 calories per week or 2314 calories per day.

13. $l(750) = 5885.8$ inches

Section 1.5

1. $u(0) = 250{,}000$, and $u(n) = (1 + 0.081/12)u(n-1) - p$
a. The monthly payment is \$2106.69, with the last payment being \$2104.20.
b. The monthly payment is \$1851.87 with the last payment being \$1850.71.

3. $u(n) = (1 + (0.081/12))u(n-1) - 1100$

a. You can borrow about \$131,000 to the nearest thousand. \$130,000 is acceptable. (The actual answer is \$130,537.)

b. You can borrow about \$148,000 to the nearest thousand. \$149,000 is acceptable. (The actual answer is \$148,498.)

5. $u(0) = 150{,}000$, $u(10) = 250{,}000$ and $u(n) = (1+R)u(n-1)$. $R = 0.052$ or 5.2% to the nearest 0.1%.

7. $a(0) = 2000$, $a(5) = 3500$, and $a(n) = (1+I)u(n-1) - 50$. Then $I = 0.139$ or 13.9%

9. a. $c(n) = (1+I)c(n-1) + 100$, with $c(0) = 200$ and $c(13) = 2000$. $I = 0.040$, or 4% interest.

b. $I = -0.062$, meaning your comics lost 6.2% of their value each year.

11. $u(n) = 1.08u(n-1) - 40{,}000$ with $u(30) \approx 0$ gives $u(0) = \$451{,}000$ to nearest thousand, given that $u(30) > 0$. Added to the initial \$40,000 withdrawal, you need 491,000. (Actual answer is \$490,311.)

13. 7.2%

15. a. $a(40) = \$30{,}691.42$.

b. $a(79) = -44.00$ meaning you withdrew \$44 too much. So the last withdrawal must be $\$1000 - \$44 = \$956.00$. The account is depleted after 79 quarters or 19.75 years.

c. You can withdraw between \$2840.58 and \$2841.46 each quarter.

17. a. $a(n) = 1.07a(n-1) - 1.03w(n-1)$ and $w(n) = 1.03w(n-1)$. $w(0) = 40{,}000$ and $a(0) = x - 40{,}000$.

b. Find that $a(0) = 689{,}000$ so $x = \$729{,}000$ in your account when you retire.

Chapter 2

Section 2.2

1. a. $E = 6$

b. $E = 4$

c. $E = 2$

d. no equilibrium

e. $E = 5$

f. $E = -50$

g. Every real number is an equilibrium value.

h. $E = 4$

3. $b = -0.6$

5. $a = -0.6$

7.

$$w(n) = w(n-1) - \frac{130}{3600}w(n-1) + \frac{16{,}000}{3600}$$

The equilibrium value is $E = 16{,}000/130 \approx 123.077$ pounds $w(69) = 122.04$, and $w(70) = 122.08$, so 70 weeks.

9. $a(n) = a(n-1) - ra(n-1) + 40$. Substituting 140 in for $a(n)$ and $a(n-1)$ gives $r = 0.286$, so the body eliminates about 28.6% of the medicine every day.

11. $a(n) = 0.6a(n-1) + b$. Substituting the equilibrium value of 45 for $a(n)$ and $a(n-1)$ gives $b = 18$ mg.

13. $(E, F) = (32, -6)$

15. $(E, F) = (50, -10)$

17. no equilibrium point

19. $(E, F) = (5, -6)$

21. $a = -1.71, b = 1.59$.

23. $r = 0.8, s = 14/11 \approx 1.27$.

25. **a.** $p(n) = 0.3p(n-1) + 0.01l(n-1) + 1$ and $l(n) = 0.3p(n-1) + 0.99l(n-1)$
b. $(p, l) = (2.5, 75)$

27. **a.**

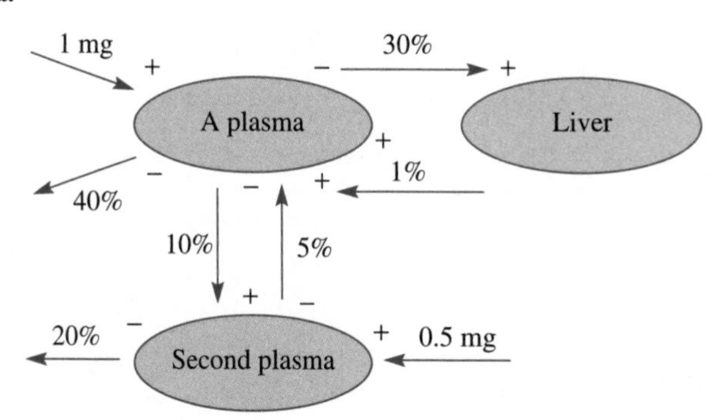

$p(n) = 0.2p(n-1) + 0.01l(n-1) + 0.05s(n-1) + 1$, $l(n) = 0.3p(n-1) + 0.99l(n-1)$, and $s(n) = 0.1p(n-1) + 0.75\,s(n-1) + 0.5$
b. $p = 55/24 \approx 2.29$, $l = 68.75$, and $s = 35/12 \approx 2.92$
c. 0.2 mg of second chemical each day. $l = 90$ mg, and $s = 2$ mg

Section 2.3

1. **a.** stable
b. unstable
c. unstable
d. stable
e. stable

3. $b = -0.6$; equilibrium is stable

5. $a = -0.6$; equilibrium is stable

7. $u(n) = 1.005u(n-1) - 200$; equilibrium is $E = 40{,}000$, which is unstable. If $u(0) < \$40{,}000$, you will eventually deplete your account. If $u(0) > \$40{,}000$, the account will continue to grow.

9. $(E, F) = (32, -6)$ is stable

11. $(E, F) = (50, -10)$ is stable.

13. $a = -1.71$; $b = 1.59$. The equilibrium point is stable.

15. $r = 0.8$; $s = 14/11 \approx 1.27$. The equilibrium point is unstable.

Section 2.4

1. $$\frac{u(n) - 50}{u(n-1) - 50} = 0.7$$

for any n. Each value is 30% closer to 50 or 70% as far away from 50.

3. **a.** Equilibrium is 400
b. It is stable
c. $$\frac{u(n) - 400}{u(n-1) - 400} = 0.85$$

for any n, so each value is 15% closer to 400 than the previous value.

5. **a.** Equilibrium is 2
b. It is unstable
c. $$\frac{u(n) - 2}{u(n-1) - 2} = -2.5$$

for any n, so each value is 150% further from 2 than the previous value.
d. Values alternate between positive and negative.

7. Each value is 1.07 less than the previous one so the values decrease linearly with a slope of -1.07.

9. Each value is 0.3 times the previous value, or is 30% as large as the previous value, or is 70% smaller than the previous value.

11. Each value is 0.97 times as far from the equilibrium value $E = 15$ as the previous value, or is 97% as far from equilibrium as the previous value, or is 3% closer to equilibrium than the previous value.

13. **a.** about 0.5% further
b. about 0.5% further
c. In part (a), the amount in the account is decreasing toward 0, while in part (b), the amount in the account is increasing toward infinity.

15. **a.** 1.25
b. 1.75

17. **a.** $(E, F) = (-40, 2)$ which is unstable
b. $(u(n) + 40)/(u(n-1) + 40) \approx 1.5$ so $u(n)$ is getting about 50% further from -40 each time. (For some choices of initial values, you might get $(u(n) + 40)/(u(n-1) + 40) \approx 1.2$, but we choose the larger value, 1.5.)
c. $(v(n) - 2)/(v(n-1) - 2) \approx 1.5$, so $v(n)$ is getting about 50% further from 2 each time.
d. $u(n) + 40 \approx 1.5(u(n-1) + 40)$, or $u(n) \approx 1.5u(n-1) + 20$

19. **a.** $(E, F) = (30, 7)$, which is unstable.
b. $(u(n) - 30)/(u(n-1) - 30) \approx 1.1$, so $u(n)$ is getting about 10% further from 30 each time. (For some choices of initial values, you might get $(u(n) - 30)/(u(n-1) - 30) \approx 0.8$, but we choose the larger value, 1.1.)
c. $(v(n) - 7)/(v(n-1) - 7) \approx 0.8$, so $v(n)$ is getting about 20% closer to 7 each time. This is not a contradiction, since the points $(u(n), v(n))$ are still getting further from the point $(30, 7)$.
d. $u(n) - 30 \approx 1.1(u(n-1) - 30)$, or $u(n) \approx 1.1u(n-1) - 3$

21. **a.** $(u, v) = (745, 30)$, which is stable.
b. $(u(n) - 745)/(u(n-1) - 745) \approx 0.9$, so $u(n)$ is getting about 10% closer to equilibrium each time. (For some choices of initial values, you might get $(u(n) - 745)/(u(n-1) - 745) \approx 0.6$, but we choose the larger value, 0.9.)
c. $(v(n) - 30)/(v(n-1) - 30) \approx 0.9$, so $v(n)$is getting about 10% closer to equilibrium each time.
d. $u(n) \approx 0.9u(n-1) + 74.5$

23. **a.** $(E, F) = (1.9, 1.625)$, which is stable.
b. $(u(n) - 1.9)/(u(n-1) - 1.9)$ varies. There is no fixed rate of convergence.
c. $(v(n) - 1.625)/(v(n-1) - 1.625)$ varies. There is no fixed rate of convergence.

25. **a.** $(u(n) - 0)/(u(n-1) - 0) \approx 0.94$, so $u(n)$ is getting about 6% closer to equilibrium each time. (For some choices of initial values, you might get $u(n)/u(n-1) \approx -0.14$, but we choose the larger value, 0.94.)
b. $(v(n) - 0)/(v(n-1) - 0) \approx 0.94$, so $v(n)$ is getting about 6% closer to equilibrium each time. (For some choices of initial values, you might get $v(n)/v(n-1) \approx -0.14$, but we choose the larger value, 0.94.)
c. $u(n) \approx 0.94u(n-1)$

Section 2.5

1. $u(n)/(u(n) + v(n)) \approx 0.91$, and $v(n)/(u(n) + v(n)) \approx 0.09$, so $u(n)$ is about 91% and $v(n)$ is about 9%.

3. $u(n)/(u(n) + v(n)) \approx 1$, and $v(n)/(u(n) + v(n)) \approx 0$, so $u(n)$ is about 100% and $v(n)$ is about 0%.

5. $u(n)/(u(n) + v(n))$ varies. There is no stable distribution.

7. **a.** $p(n) = 0.3p(n-1) + 0.01l(n-1)$, and $l(n) = 0.3p(n-1) + 0.99l(n-1)$
b. $p(n)/p(n-1) \approx 0.994$, and $l(n)/l(n-1) \approx 0.994$
c. $p(n)/(p(n) + l(n)) \approx 0.014$, or 1.4% of the total. $l(n)/(p(n) + l(n)) \approx 0.986$, or 98.6% of the total.

Section 2.6

1. **a.** $E = 4$
b. neutral
c. period 2

3. **a.** The equilibrium point is $(5, 5)$ which is neutral.
b. period 4

5. **a.** The equilibrium point is $(-1, 2)$, which is neutral.
   b. period 6

7. **a.** The equilibrium point is $(0, 10)$
   b. stable
   c. The period is between 5 and 6, closer to 5.

9. **a.** The equilibrium point is $(3, 1)$
   b. neutral
   c. The period is between 5 and 6, closer to 6.

11. **a.** The equilibrium point is $(7, 11)$.
    b. unstable
    c. The period is between 10 and 11, closer to 11.

13. **a.** The equilibrium point is $(10, 0)$.
    b. neutral
    c. The period is between 5 and 6, closer to 5.

15. **a.** The equilibrium point is $(3, -5)$.
    b. unstable
    c. The period is between 15 and 16, closer to 15.

17. **a.** The equilibrium point is $(2, 0)$.
    b. neutral
    c. The period is a little over 3.

Chapter 3

Section 3.2

1. **a.** $u(n) = 0.3n + 15$
   b. $u(n) = 0.3n + 8.7$
   c. $u(n) = 0.3n + 1.1$

3. **a.** $u(n) = 1.5n - 12$
   b. $u(n) = 1.5n - 1.5$
   c. $u(n) = 1.5n + 138$

5. $u(n) = 10.36n + 7$

7. **a.** $u(n) = u(n-1) + 10$; $u(0) = 120$
   b. $u(n) = 10n + 120$
   c. $u(1) = 130$; $u(2) = 140$; $u(3) = 150$

9. **a.** $u(n) = u(n-1) + b$; $u(0) = 75$
   b. $u(n) = 7.7n + 75$, so you write nearly 8 pages per day.

11. a. $u(n) = 0.3n + 15.6$

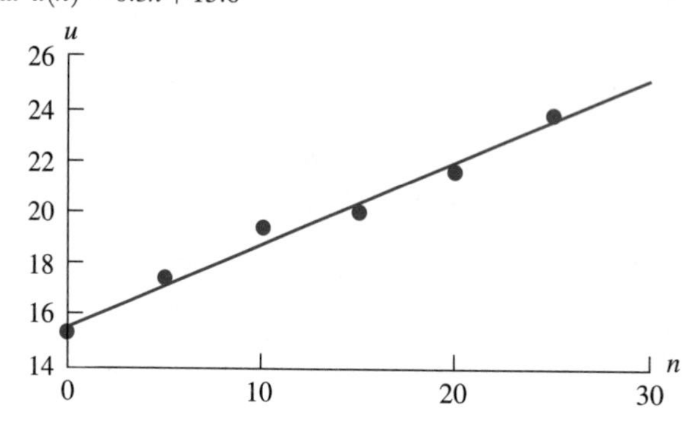

b. $u(n) = u(n-1) + 0.3$. About 0.3 quadrillion Btu of petroleum are being used each year.
c. $u(n) = 0.3166n + 15.643$

13. a. $u(n) \approx 3.5n + 85$

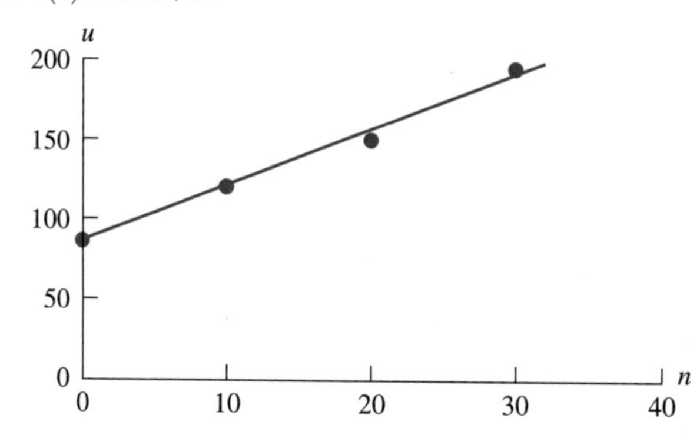

b. $u(n) = u(n-1) + 3.5$. The United States produces about 3.5 million tons more in waste each year.
c. $u(n) = 3.57n + 85.85$

Section 3.3

1. a. $u(n) = 12(1.3)^n$

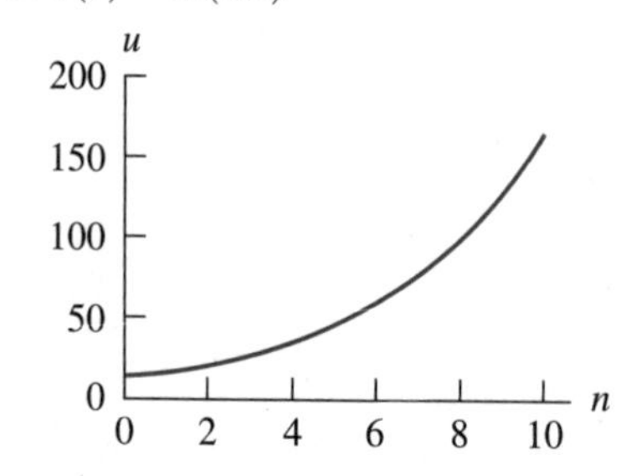

b. $u(n) = -20(1.3)^n$

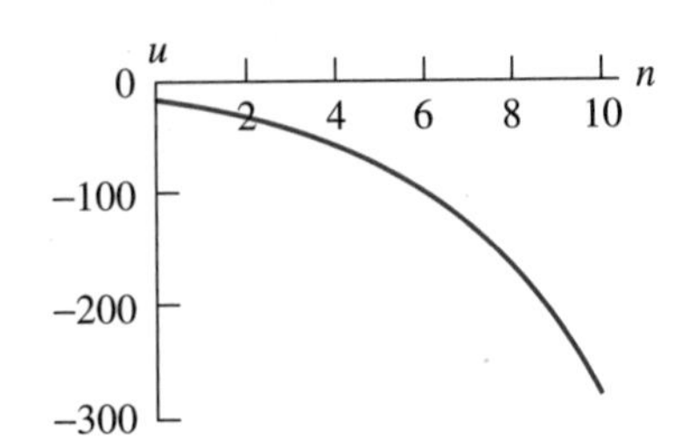

c. $u(n) = 0.059(1.3)^n = 3(1.3)^{n-15}$

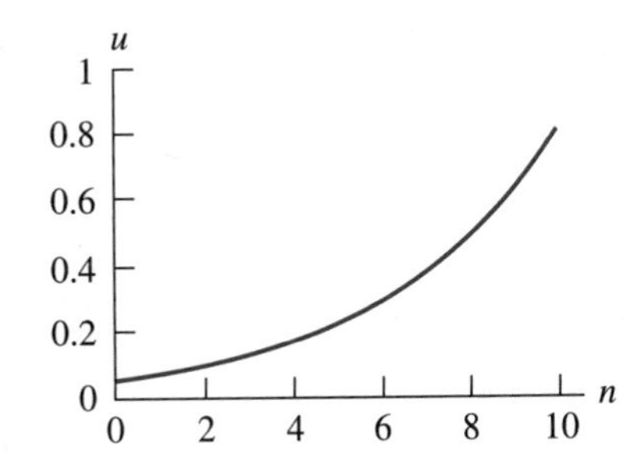

d. 30% is being added

3. a. $v(n) = 2(-0.3)^n$

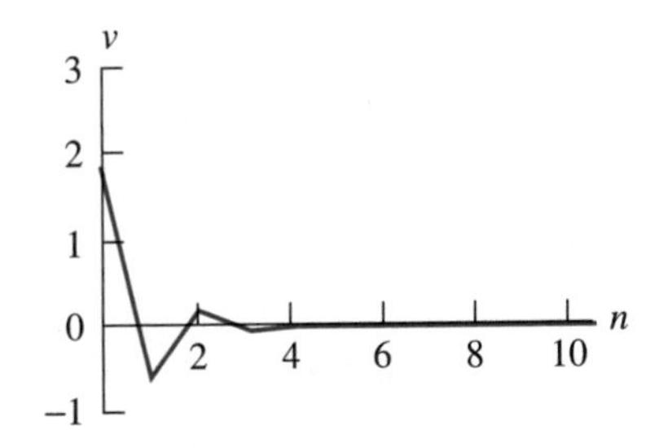

b. $v(n) = 30(-0.3)^n$

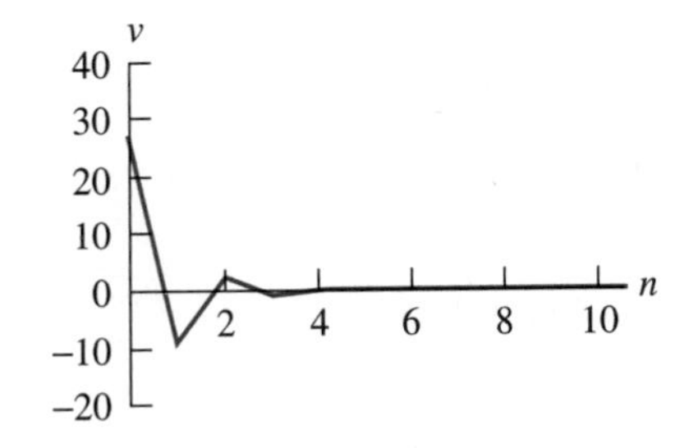

c. 130% is being subtracted each time.

5. **a.** $u(n) = 0.1(-4)^n$

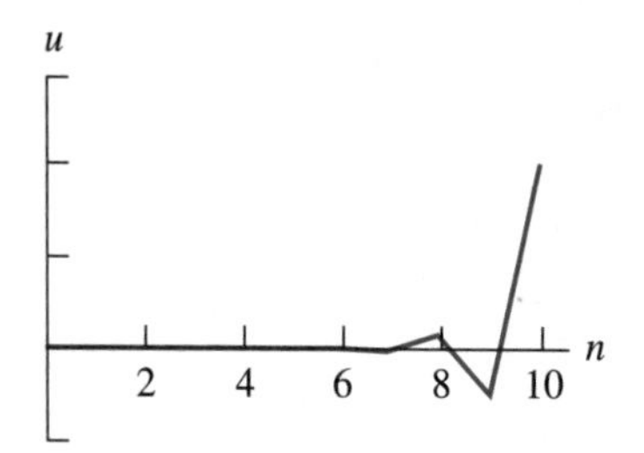

b. $u(n) = -0.04(-4)^n$

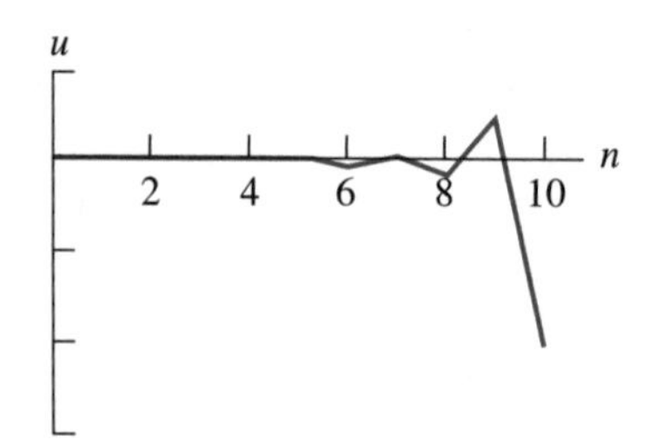

7. **a.** $u(n) = 2\left(3^{n/8}\right) \approx 2(1.15^n)$

b. $u(n) = 10\left(9^{n/12}\right) \approx 10(1.2^n)$

9. **a.** $u(n) = 0.92u(n-1)$

b. $u(n) = 5(0.92)^n$

c. $u(2000/60) = 5(0.92)^{2000/60} \approx 0.31$ grams.

11. **a.** $s(n) = (1+I)s(n-1)$

b. $s(n) = c(1+I)^n$; $s(0) = c = 2000$; $s(20) = 2000(1+I)^{20} = 7500$. $I = 0.06832$, so you earned about 6.8% per year.

Section 3.4

1. $n = 9.55$

3. $u(n) = 0.95u(n-1)$; half-life is 13.5 days.

5. $u(n) = 0.92u(n-1)$; $n = 8.3$ minutes.

7. $u(n) = 1.004u(n-1)$; $n = 173.6$ months, or about 14.5 years.

9. $u(n) = 0.99999u(n-1)$, or about 69,314 generations.

11. a. $m(n) = 0.0469n + 0.6417$

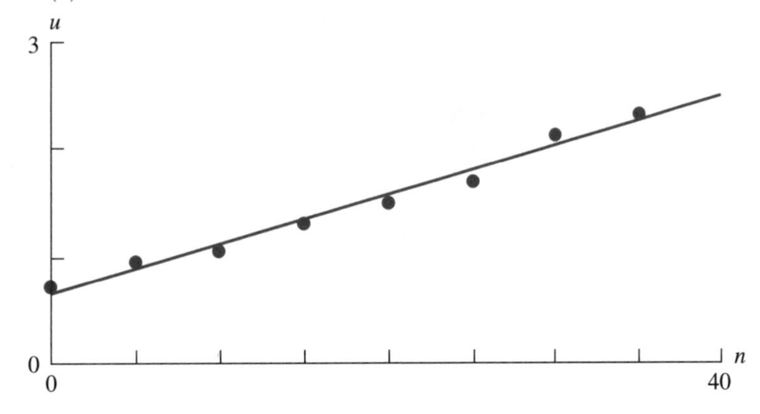

b. $m(n) = 0.754(1.034)^n$

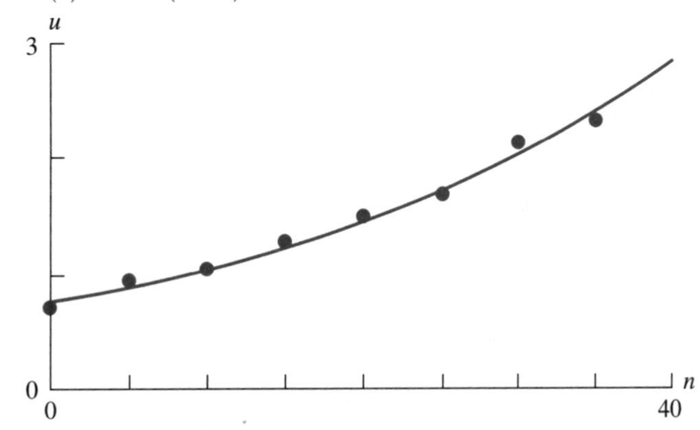

c. The exponential is a much better fit. $m(n)$ satisfies $m(n) = 1.034m(n-1)$, so the amount of travel goes up by about 3.4% per year.

13. a. $u(n) = 4.7(1.068)^n$

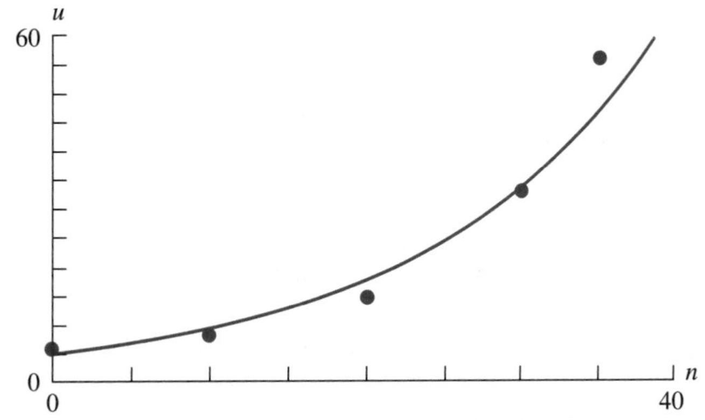

b. The ratios are not about the same. The ratios are increasing. An exponential function is not a good fit. All we can say is that the rate at which material is recycled is increasing.

15. The linear regression is $u(n) = 9.5n + 6.8$. The exponential regression is $u(n) = 11.9(1.36^n)$. The linear regression is clearly not good. The exponential function is pretty good.

17. a. $u(n) = 1.46n + 25.6$ which corresponds to $u(n) = u(n-1) + 1.46$. The population increases by 1.46 million per year. $u(55) \approx$ 105,900,000. The fit doesn't look bad, but there is a curved pattern to the data points that the line does not describe. See the following graph.

b. $u(n) = 30.105(1.026)^n$, which corresponds to $u(n) = 1.026u(n-1)$ The population increases by 2.6% per year. This is not a good fit either. This function has "more curve" than the data points. $u(55) \approx 123{,}500{,}000$. See the following graphs.

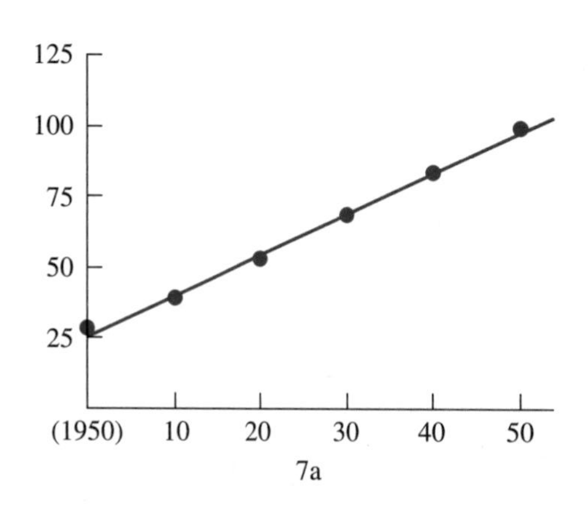

7a

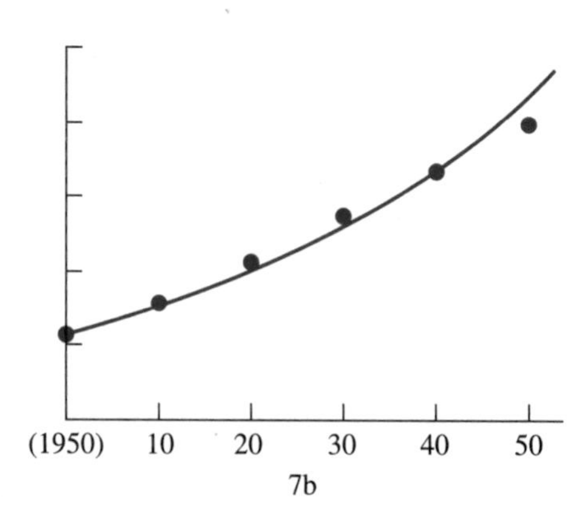

7b

c. $u(n) = 1.59n + 84.51$, which corresponds to $u(n) = u(n-1) + 1.59$. The population increases by 1.59 million per year. $u(15) \approx$ 108,300,000. The fit looks good. See the following graph.

d. $u(n) = 84.699(1.017)^n$, which corresponds to $u(n) = 1.017u(n-1)$. The population increases by 1.7% per year. This is not as good of a fit. The curve drops below the points in the middle. $u(15) \approx$ 109,700,000. See the following graphs.

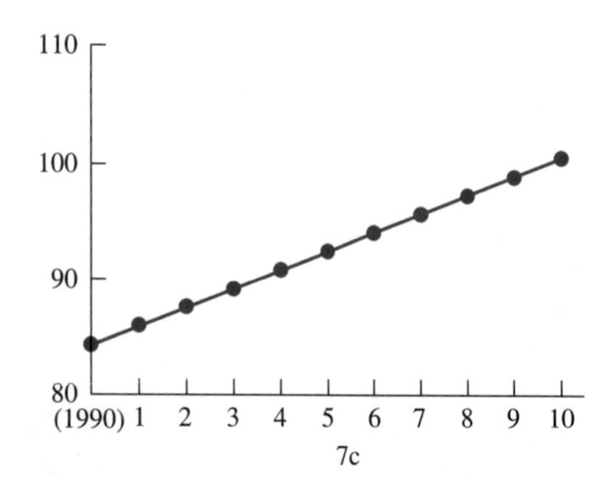

7c

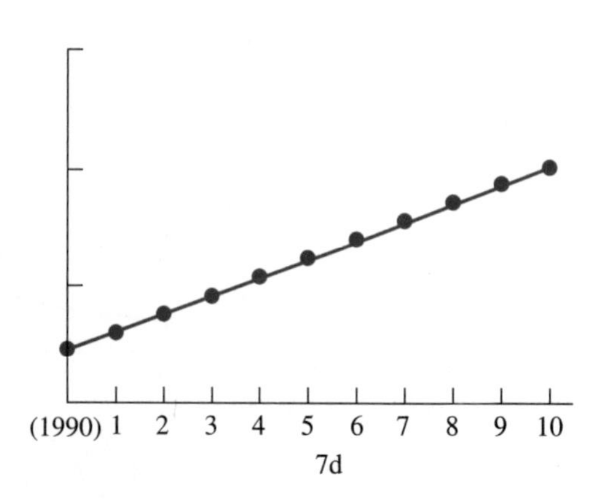

7d

e. The functions from parts (c) and (d) are better fits than those from parts (a) and (b). One reason is that a countries' growth rate can change over time. In parts (c) and (d), only recent population sizes are used, which would reflect current growth rates. In parts (a) and (b), older data is used which may not reflect current trends. It could be argued that the linear function from part (c) gives the best model, since this function seems to fit the data better. On the other hand, it could be argued that the exponential function from part (d) is better, since it gives more information about current growth rates.

Section 3.5

1. Using $u(0) = 20$

$n =$	0	1	2	3
using expression, $u(n) =$	20	17	14.6	12.68
using dynamical system, $u(n) =$	20	17	14.6	12.68

Expression is a solution

3. Using $u(0) = 6$

$n =$	0	1	2	3
using expression, $u(n) =$	6	5.2	5.04	5.008
using dynamical system, $u(n) =$	6	5.8	5.64	5.512

Expression is not a solution

5. **a.** $u(n) = 10(-0.5)^n + 8$
b. $u(n) = 32(-0.5)^n + 8$

7. **a.** $u(n) = -27(0.5)^n + 32$
b. $u(n) = 768(0.5)^n + 32$

9. $u(n) = 3(-2)^n + 1$

11. $u(n) = -16(-0.5)^n + 2$

13. $b = -11$

15. withdraw $11,745.96 each year.

Section 3.6

1. **b.** $u(n) = 33.4(0.927)^n$. The coefficient of determination is 0.896.
c. $u(n) = 28.38(0.913)^n + 5$. The coefficient of determination is 0.901.
d. $u(n) = 18.415(0.856)^n + 15$. The coefficient of determination is 0.920.
e. $u(n) = 10(0.5)^n + 25$. The coefficient of determination is 1.00.
f. Clearly the function in part 1 (e) is the best fit.

3. **b.** $u(n) = 8500(0.92)^n + 15{,}700$
c. $u(n) = 0.92u(n-1) + 1256$

5. a.

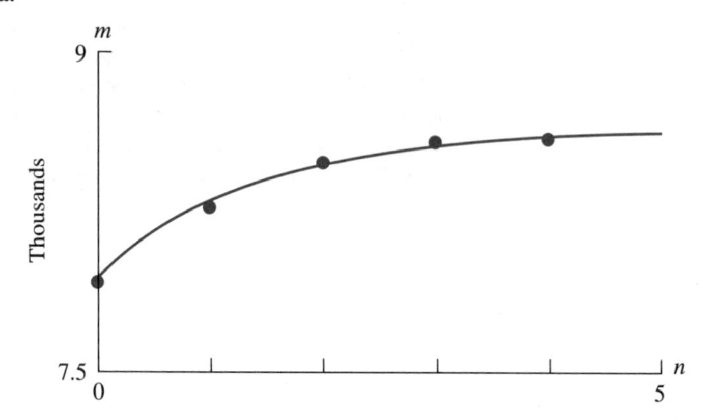

b. $m(n) = -718(0.5)^n + 8650$ has coefficient of determination 0.979. Its graph and points are seen in figure.

c. $m(n) = 0.5m(n-1) + 4325$. People are increasing driving to a point where they will drive about 8650 miles per year. They are getting about 50% closer to that total each year.

Chapter 4

Section 4.2

1. a. $u(n) = 4u(n-1) - 3u(n-2)$
b. $u(n) = 1.3u(n-1) - 5$
c. $u(n) = 4.2u(n-1) - 2u(n-2)$
d. $u(n) = u(n-1) + 2.3u(n-2) + 3$
e. $u(n) = 3u(n-1) + u(n-2)$
f. $u(n) = 7u(n-2) - u(n-3)$

3. a. $u(1) = 2$, $u(2) = 6$
b. $u(n) = 2u(n-1) + 2u(n-2)$
c. $u(10) = 18,272$
d. $u(10)/u(9) \approx u(16)/u(15) \approx u(32)/u(31) \approx 2.73$. There are about 2.73 times as many arrangements for each additional bead.

5. a. $u(1) = 1$, $u(2) = 3$, $u(3) = 6$
b. $u(n) = u(n-1) + 2u(n-2) + u(n-3)$
c. $u(4) = 13$, $u(5) = 28$, $u(6) = 60$, $u(7) = 129$

7. a. $u(1) = 2$, $u(2) = 4$, $u(3) = 8$, $u(4) = 16$
b. $u(n) = u(n-1) + 2u(n-2)$

Section 4.3

1. **a.** $P(Y_1) = 2/10$
   b. $P(Y_2) = 4/10$
   c. $P(Y_1 \text{ or } G_1) = P(Y_1) + P(G_1) = 7/10$
   d. $P(Y_1 Y_2) = P(Y_1)P(Y_2) = (2/10)(4/10) = 0.08$
   e. $P(YY \text{ or } GG \text{ or } RR) = P(YY) + P(GG) + P(RR) = P(Y)P(Y) + P(G)P(G) + P(R)P(R) =$ $(2/10)(4/10) + (5/10)(4/10) + (3/10)(2/10) = 0.34.$
   f. $P(YG \text{ or } YR \text{ or } RY \text{ or } GY) =$ $(2/10)(4/10) + (2/10)(2/10) + (3/10)(4/10) + (5/10)(4/10) = 0.44.$
   g. $P(RR \text{ or } RG \text{ or } RY \text{ or } YR \text{ or } GR) = 0.44$

3. **a.** $P(Y|H) = 0.2$
   b. $P(G|T) = 0.4$
   c. $P(HR) = P(H)P(R|H) = (0.5)(0.3) = 0.15$
   d. $P(Y) = P(HY \text{ or } TY) = P(H)P(Y|H) + P(T)P(Y|T) = 0.5(0.2) + 0.5(0.4) = 0.3$

5. **a.** $P(\text{no rain}) = 0.28$
   b. $P(R) = \frac{1}{3}(0.3) + \frac{2}{3}(0.6) = 0.5$ or 50% chance.

Section 4.4

1. You break even: You win \$8 two times and lose \$2 eight times.

3. **a.** $P(n) = \frac{20}{9}P(n-1) - \frac{11}{9}P(n-2)$
   b. $P(1) = 0.9048$
   c. 0.65 or 65% of the time you will go broke.

5. **a.**

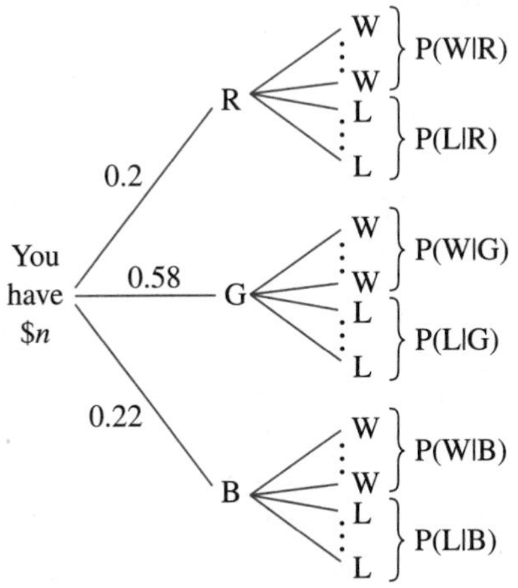

   You have \$n
   0.2 → R: W ... W } P(W|R); L ... L } P(L|R)
   0.58 → G: W ... W } P(W|G); L ... L } P(L|G)
   0.22 → B: W ... W } P(W|B); L ... L } P(L|B)

   b. $P(n) = 2.1P(n-1) - 1.1P(n-2)$
   c. $P(1) = 0.98254$
   d. $P(10) = 0.72$, or 72% of the time you lose.

Section 4.5

1. **a.** Solving $E = -0.4E + 0.45E + 2.85$ gives $E = 3$.
   b.

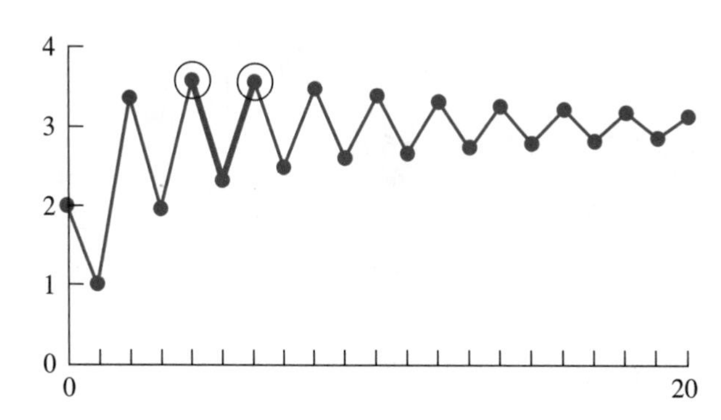

   c. The values are oscillating toward equilibrium, so the equilibrium value is stable.
   d. The solution oscillates, since it goes up and down. This means the period of 2.

3. **a.** Solving $E = 1.7E - 0.72E + 0.5$ gives $E = 25$.
   b.

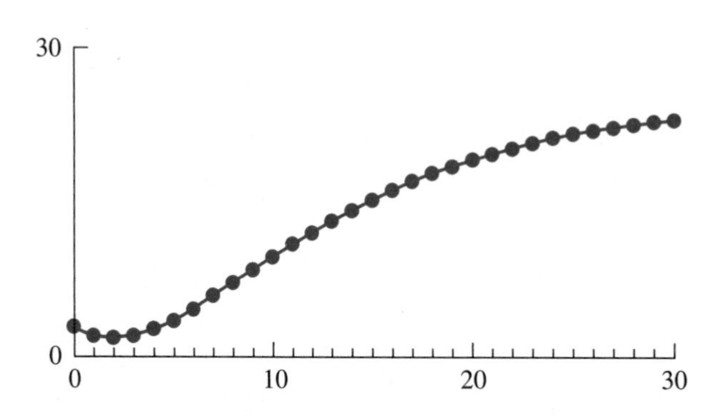

   c. The values are leveling off at $E = 25$, so the equilibrium is stable.
   d. Let $u(0) = 5$ and $u(1) = 10$. Then $(u(14) - 25)/(u(13) - 25) = 1.085$, $(u(24) - 25)/(u(23) - 25) = 0.925$, $(u(40) - 25)/(u(39) - 25) = 0.903$, and $(u(60) - 25)/(u(59) - 25) = 0.9003$. The rates appear to be approaching 0.9. Using other starting values gives the same result.
   e. Solving $E = 1.7E - 0.72E - 0.6$ gives $E = -30$, which is stable. Computing values such as $(u(60) + 30)/(u(59) + 30)$ results in a rate of 0.9, or 10% closer to equilibrium. The constant added to the end of the dynamical systems seems to effect the value of the equilibrium but not the stability or the rate.

5. **a.** Solving $E = E - 0.5E + 4$ gives $E = 8$.
   b. Let $u(0) = 25$ and $u(1) = 12$. The solution oscillates. It is difficult to determine the peaks from the graph. Using a table of values gives $u(2) = 6, \ldots, u(6) = 8.5$, $u(7) = 9$, and $u(8) = 8.75$, so the first peak is $u(7)$. Going down the table, the next peak is seen to be $u(15) = 8.0625$. Since $15 - 7 = 8$, the period appears to be 8. The next peak is $u(23) = 8.0039$. Since $23 - 15 = 8$, this seems to confirm the period is about 8.

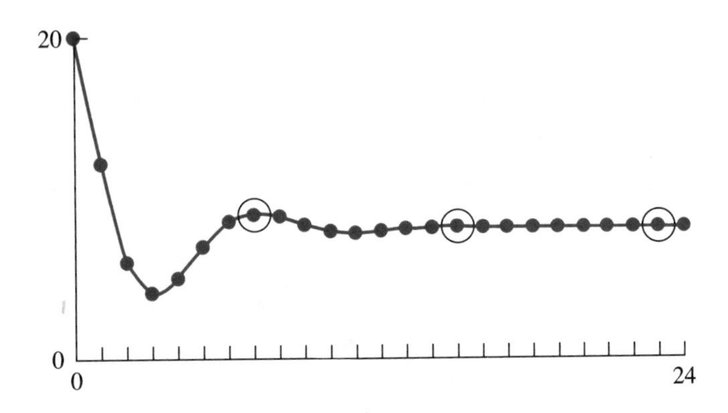

c. The values are approaching $E = 8$, so the equilibrium is stable.
d. no rate because it oscillates

7. a. Solving $E = E - E + 7$ gives $E = 7$.
b. Period is exactly 6, since the exact same 6 values keep repeating, $2, -4, 1, 12, 18, 13, \ldots$. The initial values have no affect on length of cycle, but they do affect the values of the 6 numbers that keep repeating.

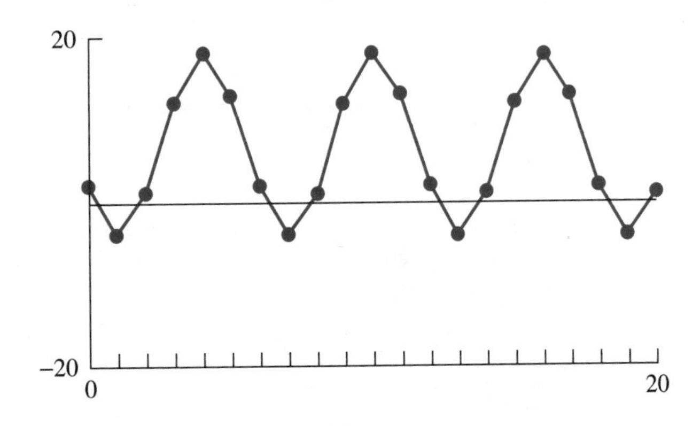

c. Neutral, since the size of the oscillations neither increase nor decreases. Initial values have no effect on stability.
d. The equilibrium value is $E = -3$. This value is neutral. The graphs again keep repeating 6 values. The constant at the end of this dynamical system seems to effect the value of the equilibrium but has no effect on length of period.

9. a. Solving $E = 1.8E - 0.8E + 3$ gives $0 = 3$, which has no solution. There is no equilibrium value

b. The graph does not oscillate, so there is no period.

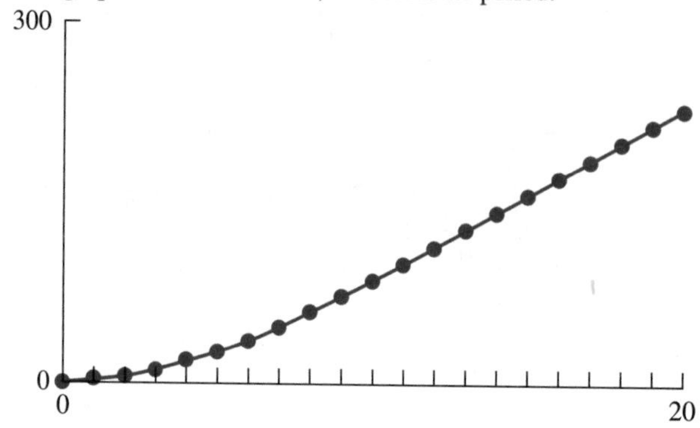

c. There is no equilibrium, so stability does not apply

d. The rate is about 1. If the difference between consecutive values is computed [e.g., $u(51) - u(50)$], it will be approximately 15. This means the solutions increase linearly, like $u(n) = 15n + c$.

11. a. Solving $E = 0.4E + 0.6E$ gives $E = E$, so every number is an equilibrium value

b. Period of oscillations is 2 since 2 line segments between peaks.

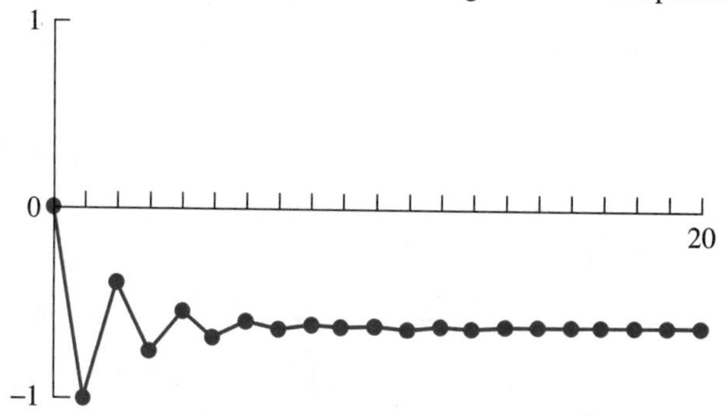

c. Solutions for different pairs of initial values oscillate to different equilibrium values. For any pair of initial values, $(u(0),\ u(1))$, it is difficult to predict which equilibrium value $u(n)$ will oscillate toward.

Section 4.6

1. a. $E = 10$

b. stable

c. The period is between 19 and 20. (The period is 19.5)

3. a. $E = 3\frac{1}{3}$

b. unstable

c. The period is between 11 and 12. (The period is 11.7.)

d. $t(1) \approx 4.8. t(8) \approx -4.7$

5. The equilibrium is neutral if $b = 2.5$ and is stable if $b < 2.5$.

7. a. $E = 2$ is stable.

b. $a = 0.8$. The economy becomes unstable.

9. a. $E = 2.4$ is stable with a period of about 8. (The period is 8.1.)
   b. $c = 2.5$. The economy is stable with a period of about 8.
   c. The value of c seems to have no effect on the stability or period.

Section 4.7

1. a. $g(n) = 1 + 0.15 = 1.15$:
   b. $g(n) = 1 - 0.3 = 0.7$
   c. $g(n) = 1.45$ when $t(n) = 2.5$ and $g(n) = 0.1$ when $t(n) = 7$.

3. a. $g(2) = 1.072$:
   b. $g(2) = 1.15$:
   c. $g(2) = 1.2$

5. a. $t(n) = \frac{20}{11}t(n-1) - \frac{12}{11}t(n-2) + \frac{15}{11}$. Equilibrium is unstable
   b. $t(n) = \frac{5}{3}t(n-1) - t(n-2) + \frac{5}{3}$. Equilibrium is neutral
   c. $t(n) = \frac{20}{13}t(n-1) - \frac{12}{13}t(n-2) + \frac{25}{13}$. Equilibrium is stable

7. $I = 80\%$

9. $I = 25\%$

11. a. $t(n) = 1.12t(n-1) - 0.72t(n-2) + 1.4$
    b. $E = 7/3$, which is stable.
    c. $I = 50\%$ and the economy is stable.

Section 4.8

1. a. $c_1 = 1$, and $c_2 = 2$
   b. $c_1 = -3$, and $c_2 = 0$

3. a. $u(n) = c_1 2^n + c_2 0.5^n - 6$
   b. $c_1 = 1$, and $c_2 = 2$

5. $u(n) = 3(2)^n + 2(-5)^n - 2$

7. $r \approx 0.77$; $\theta \approx 50°$; the period is approximately 7.2.

9. a. $r = 2$, $\theta = 60°$, and the period is 6.
   b. $u(n) = 2^n \cos(60n) + 2$, $c_1 = 1$, and $c_2 = 0$

Chapter 5

Section 5.2

1. $b = 12$ and $c = 3$

3. a. 9.26 grams
   b. 9.62 grams
   c. 9.80 grams
   d. 7.9, 8.2, and 8.4 ounces

5. **a.** The equilibrium is 14.7 grams, which is just over threshold for having an effect.
   b. The equilibrium is 33.6 grams, over the legal limit.
   c. The equilibrium is 373.8 grams, way over the level at which there is risk of death.
   d. No equilibrium, amount keeps increasing.

7. The amount eliminated increases as a increases, up to $a = 1.7$, then the amount eliminated decreases to 0. This is not a good model.

Section 5.3

1. Equilibrium values: $E = 0$ and $E = 5$. $E = 0$ is stable with a maximum interval of stability $(-10, 5)$; $u(n)$ goes toward 0 at the rate of 0.5, meaning it gets 50% closer each time. $E = 5$ is unstable.

3. Equilibrium values: $E = -1$ and $E = 3$. $E = -1$ is stable with maximum interval of stability $(-4\frac{1}{3}, 3)$ [acceptable is $(-4.3, 3)$]; $u(n)$ goes toward -1 at the rate of -0.2, meaning it gets 80% closer each time but alternates above and below -1. $E = 3$ is unstable.

5. Equilibrium values: $E = 1$ and $E = 5$. $E = 5$ is stable with a maximum interval of stability equal to $(1, 7.5)$. $u(n)$ goes toward 5 at a rate of -0.6, meaning it gets 40% closer each time but alternates above and below 5. $E = 1$ is unstable

7. The only equilibrium value is $E = 0.5$, and it is semistable from above.

9. Equilibrium values: $E = 1$ and $E = -4$. $E = 1$ is stable. $u(n)$ goes toward 1 at the rate of $-2/3$, or -0.67 to 2 decimal place accuracy. This mean $u(n)$ gets about 33% closer each time. $E = -4$ is unstable.

11. There are three equilibrium values: $E = 0$ and $E = 3$, which are both stable, and $E = 1$, which is unstable. The rate at which $u(n)$ approaches $E = 0$ is 0.4. The rate at which $u(n)$ approaches $E = 3$ is -0.2. [Determining the maximum interval of stability is very difficult for this problem. There are an infinite number of disjoint intervals within which, $u(n)$ goes to $E = 0$ if $u(0)$ is in the interval. The same is true for $E = 3$. The union of these intervals is approximately the interval $(-2.156886, 4.8663926)$. This interval also includes some values of $u(0)$ for which $u(n)$ goes to the unstable equilibrium value $E = 1$.]

13. There are two equilibrium values, $E = 0$ and $E = 2$. $E = 0$ is stable and $u(n)$ approaches $E = 0$ at the rate of 0.2. $E = 2$ is semistable from above.

Section 5.4

1.

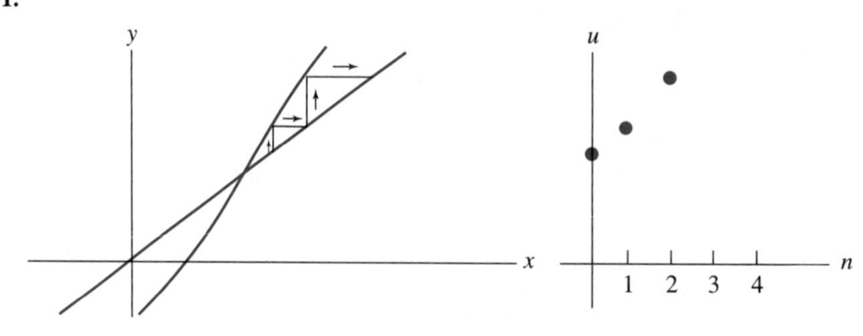

3.

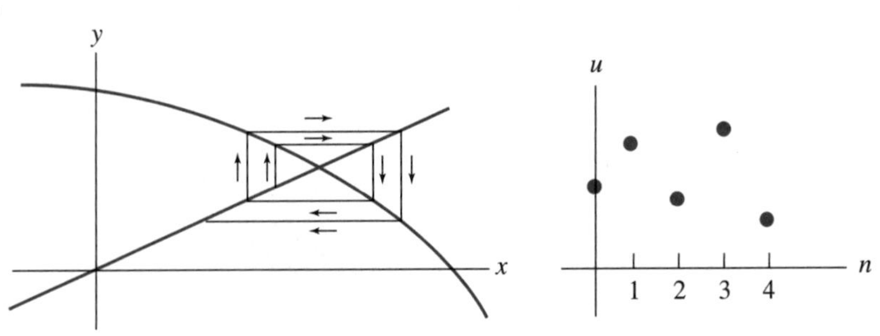

5. The equilibrium value is unstable.

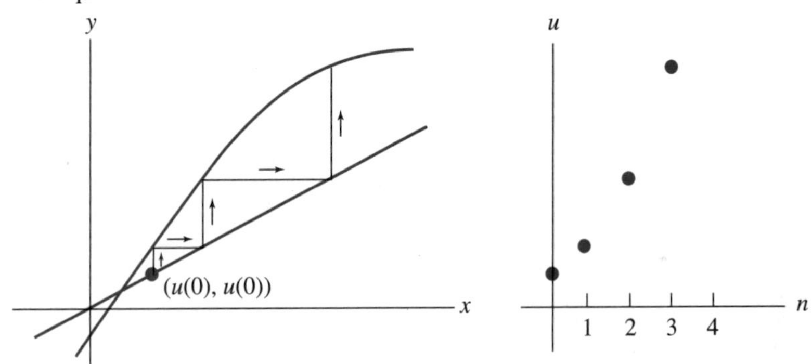

7. Stability is unclear from this web.

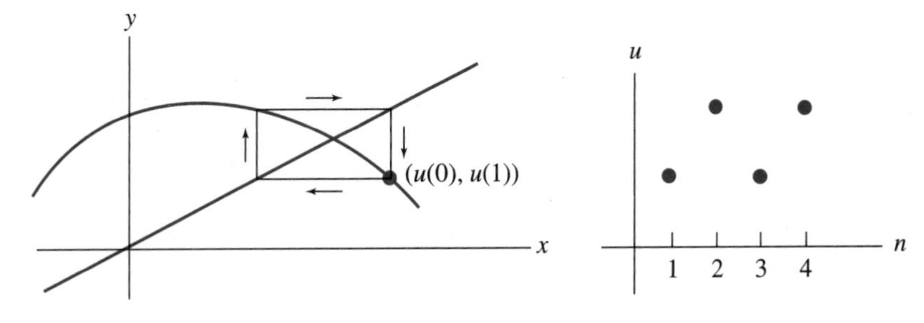

9. a. $E = 0$ is stable and $E = 5$ is unstable.

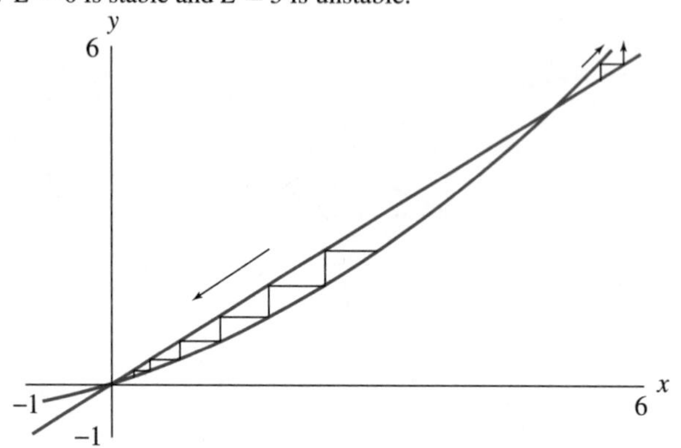

b. The maximum interval of stability is $(-10, 5)$.

11. a.

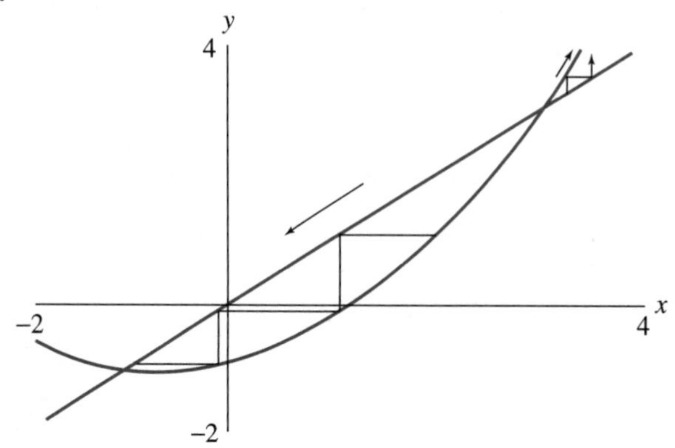

b. The maximum interval of stability is $(-4\frac{1}{3}, 3)$.

13. a.

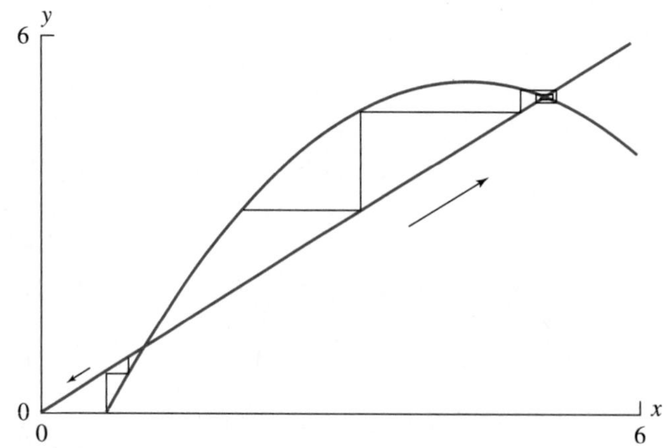

b. The maximum interval of stability is $(1, 7.5)$.

15. a.

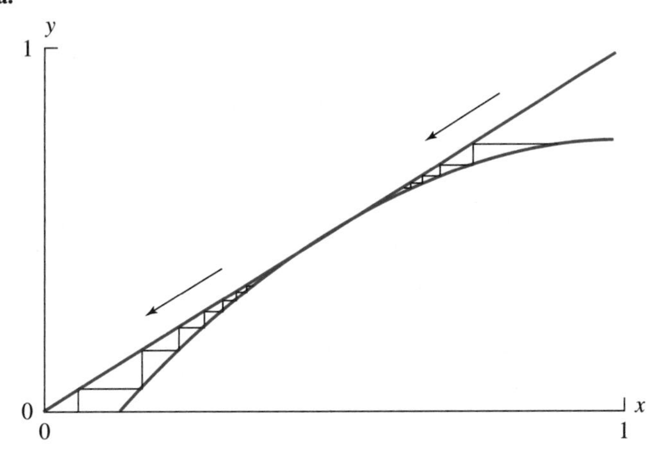

b. The maximum interval of stability is $(0.5, 1.5)$.

17. a.

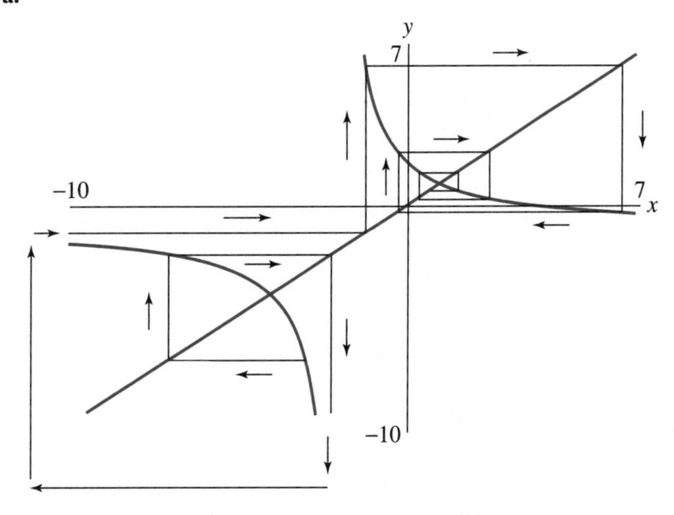

b. $u(n)$ circles around -4 going away, then circles inward toward 1.

c. Both webs result in a zero in the denominator.

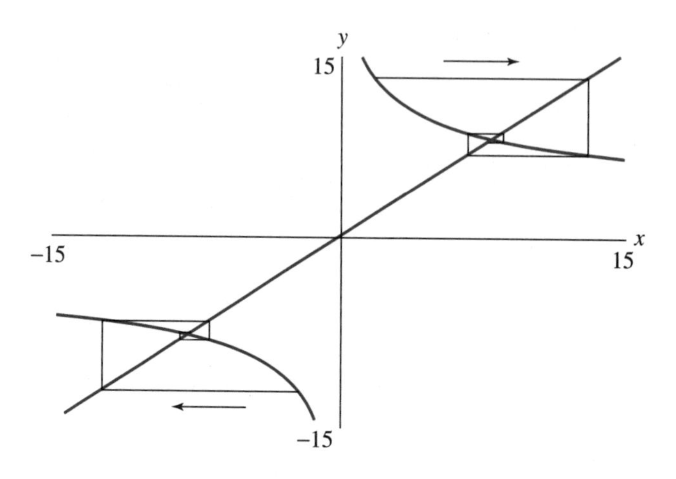

19. a.

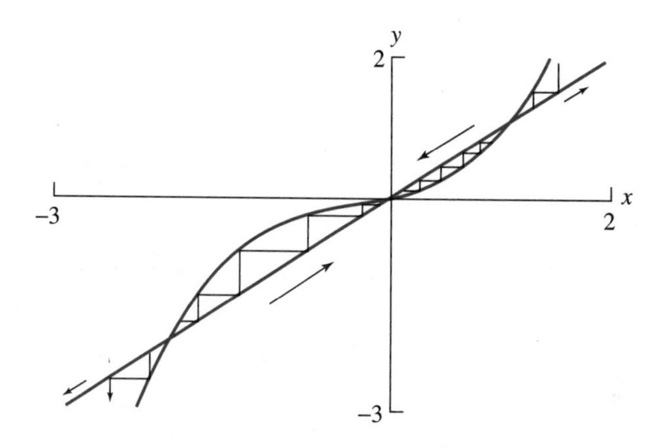

b. If $-2 < u(0) < 1$, then $u(n)$ goes to 0. If $u(0) < -2$, $u(n)$ goes to negative infinity. If $u(0) > 1$, $u(n)$ goes to infinity. If $u(0)$ equals an equilibrium value, it stays there.

21. a.

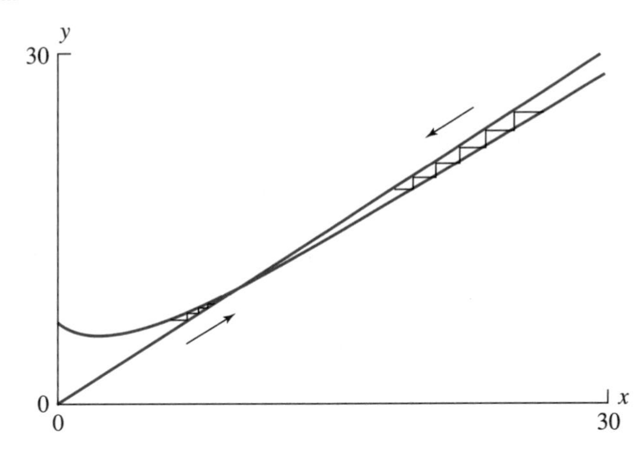

b.

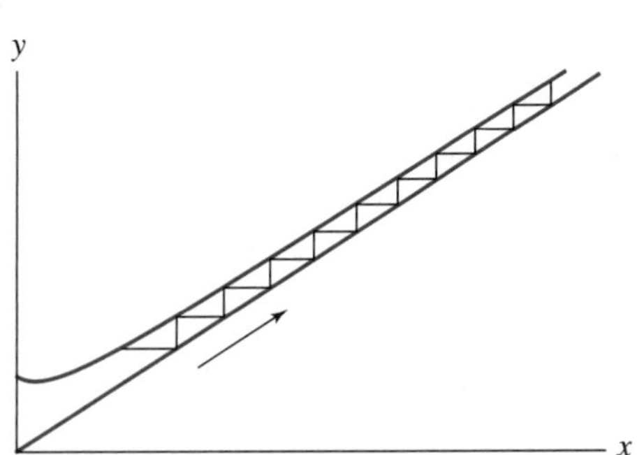

Chapter 6

Section 6.1

1. **a.** 0.9; **1b.** 0.9; **1c.** 0.9; L seems to have little or no effect on the rate.

3. **a.** 0.7; **3b.** 0.7; **3c.** 0.7; L seems to have little or no effect on the rate.

5. The rate is -0.4. This agrees with $1-1.4$.

7. $b=2$

9. $(0,6000)$; if $u(0)=6000$, then $u(1)=0$.

11. The maximum interval of stability is $(0,s)$, where

$$s=\frac{L(1+b)}{b}=\frac{L}{b}+L$$

When $b=1$, then $s=2L$.

Section 6.3

1. $r = 0.4 - 0.0000016u^2(n-1)$. The dynamical system is

$$u(n) = 1.4u(n-1) - 0.0000016u^3(n-1)$$

It has the equilibrium values $x = 0, -500$, and 500. $L = 500$ is stable and $u(n)$ approaches 500 at the rate of 0.2, meaning it gets 80% closer each year.

3. $r = 1.1 - 0.0000011u^2(n-1)$. The dynamical system is

$$u(n) = 2.1u(n-1) - 0.0000011u^3(n-1)$$

It has the equilibrium values $x = 0, -1000$, and 1000. $L = 1000$ is unstable.

5. **a.** The growth rate is

$$r = 0.3 - 0.3\frac{u^3}{2000^3}$$

The growth rate decreases.

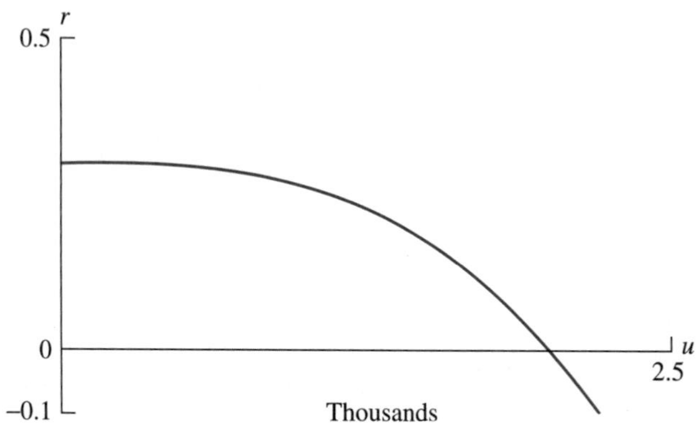

b.

$$u(n) = 1.3u(n-1) - \frac{0.3}{2000^3}u^4(n-1)$$

The equilibrium values are 0 and 2000.

c. Here, 0 is unstable, and 2000 is stable. $u(n)$ approaches 2000 at a rate of 0.1, meaning it gets 90% closer each year.

7. **a.** The growth rate is

$$r = 0.4 - 0.4\frac{u^4}{200^4}$$

The growth rate decreases.

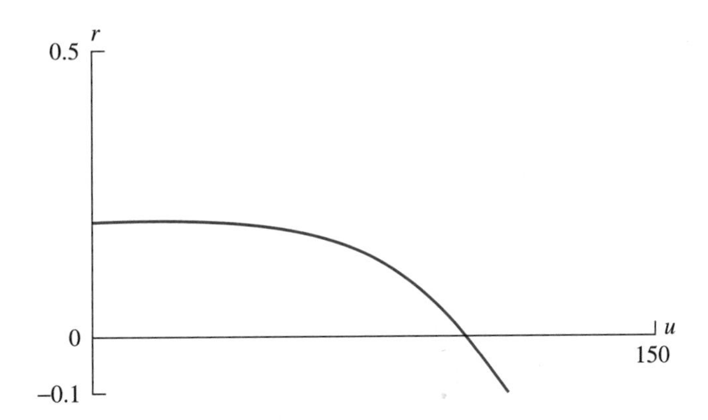

b. $u(n) = 1.4u(n-1) - \dfrac{0.4}{200^4}u^5(n-1)$

The equilibrium values are -200, 0, and 200.

c. -200 is stable, 0 is unstable, and 200 is stable. $u(n)$ approaches 200 at a rate of -0.6, meaning it gets 40% closer each year and the population size oscillates above and below 200.

9.

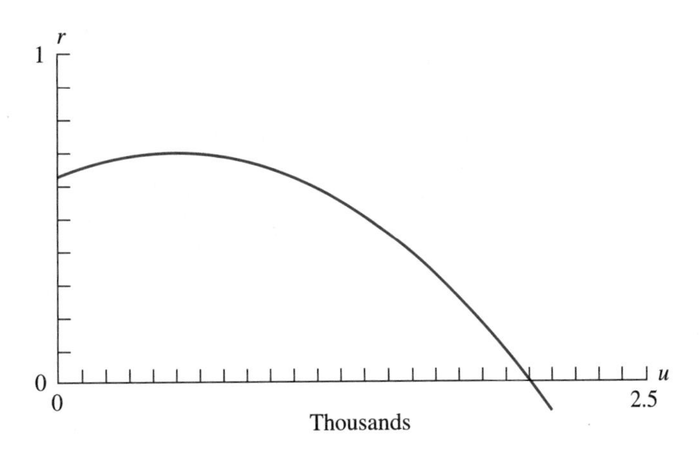

The growth rate is

$$r = 0.7 - 0.7\frac{(u-500)^2}{1500^2}$$

The dynamical system is

$$u(n) = 1.7u(n-1) - \frac{0.7u(n-1)(u(n-1)-500)^2}{1500^2}$$

The equilibrium values are -1000, 0, and 2000; -1000 is stable, 0 is unstable and 2000 is stable. The rate at which $u(n)$ approaches 2000 is -0.87, meaning that $u(n)$ gets 13% closer to 2000 each year, but oscillates above and below equilibrium.

11. a.

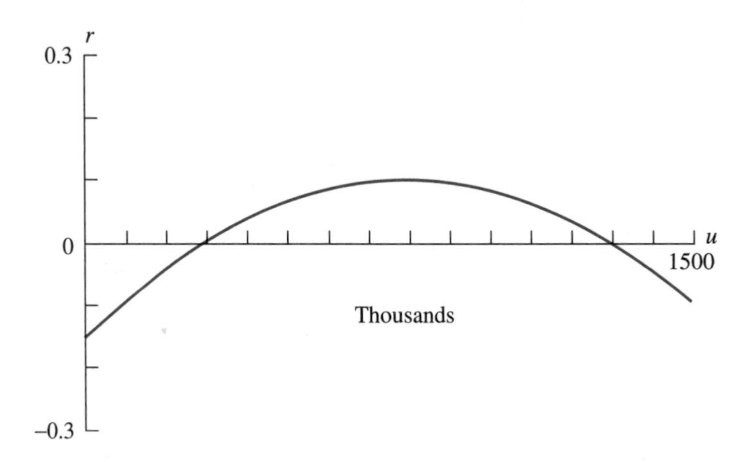

The growth rate is

$$r = 0.1 - 0.1\frac{(u-800)^2}{500^2}$$

Negative growth rate if population size is below 300 or over 1300.

b. The dynamical system is

$$u(n) = 1.1u(n-1) - \frac{0.1u(n-1)(u(n-1)-800)^2}{500^2}$$

The equilibrium values are 300, 0, and 1300; 300 is unstable, 0 is stable and 1300 is stable. The rate at which $u(n)$ approaches 1300 is 0.48, meaning that $u(n)$ gets 52% closer to 1300 each year. The rate at which $u(n)$ approaches 0 is 0.844, meaning the population decreases toward extinction by 15.6% each year. The minimum viable population size is 300.

c.

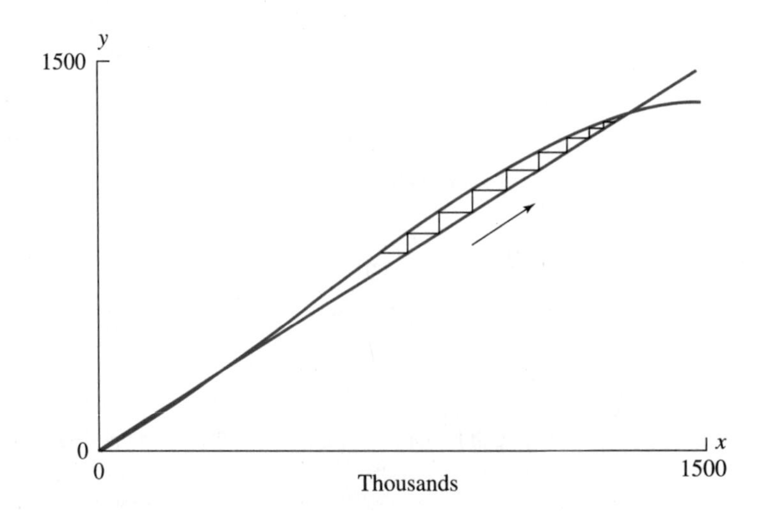

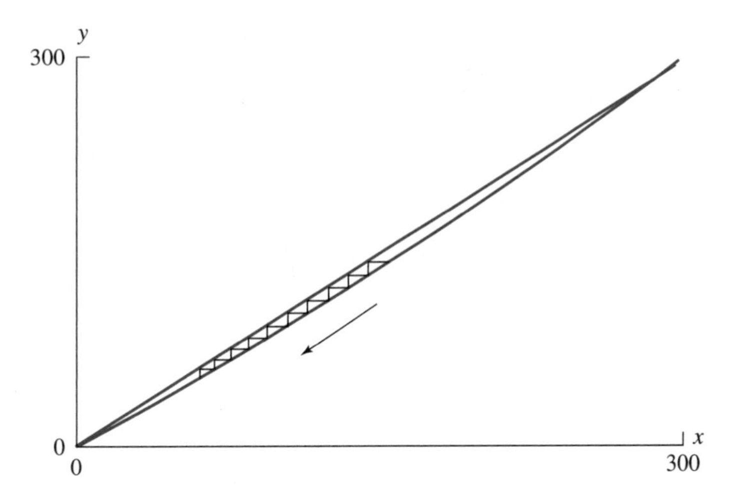

Section 6.4

1. **a.** stable equilibrium of about 22,000 and minimum viable population of about 3500
   b. stable equilibrium about 19,000 and sustainable harvest about 380
   c. maximum sustainable harvest of about 525 and equilibrium population of about 14,000
   d. about 3.75%
   e. about 6%

3. **a.** stable equilibrium of about 1200 and minimum viable population of about 530
   b. stable equilibrium about 1150 and sustainable harvest about 230
   c. maximum sustainable harvest of about 300 and equilibrium population of about 900
   d. About 33%; the minimum viable population is about 300.
   e. about 40%

5. **a.** $h = 0.005u + 100$; stable equilibrium of about 41,000, sustainable harvest of about 305, and minimum viable population of about 2500.
   b. Harvest of about 1.2% of population. Minimum viable population of 4000.

7. $b \approx 500$.

9. **a.** $h = 0.03u - 300$. The stable equilibrium is about 59,000, and the sustainable harvest is about 1470.
   b. 10%

Section 6.5

1. **a.** $r = 0.4 - 0.0004u$
   b. $g = 0.4u - 0.0004u^2$. The minimum viable population size is 200, and 800 is a stable population.
   c. The vertex is at $(500, 100)$. The maximum sustainable harvest is 100, and the semistable population size is 500.
   d. The intersection point is $(700, 84)$, so the sustainable harvest is 84 and the stable population is 700.
   e. $p = 100/500 = 0.2$ or 20%.

3. **a.** $r = 0.1 - 0.00001u$

b. $g = 0.1u - 0.00001u^2$. The minimum viable population size is 2000, and 8000 is a stable population.
c. The vertex is at $(5000, 250)$. The maximum sustainable harvest is 250, and the semistable population size is 5000.
d. The intersection point is $(6200, 235.6)$, so the sustainable harvest is about 235.6 and the stable population is 6200.
e. $p = 250/5000 = 0.05$, or 5%

5. a. A maximum harvest of about 154 results in a semistable population size of about 577.

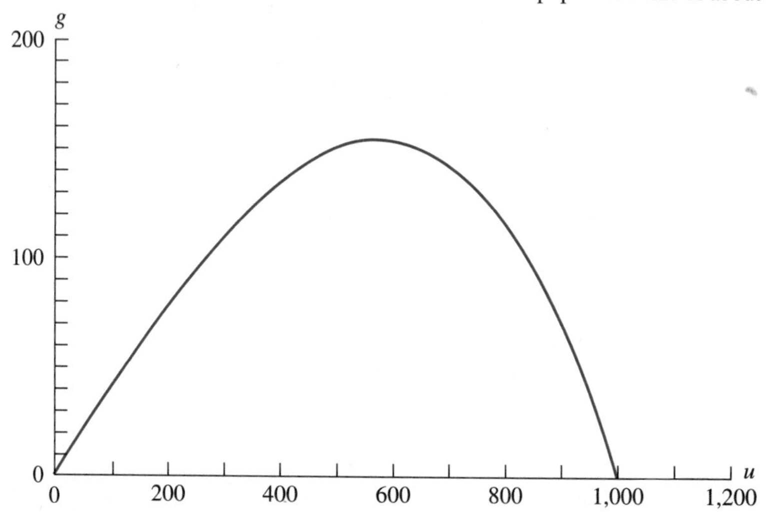

b. The slope should be about $p = 154/577 = 0.267$, or about 27% harvested each year.

7. a. A maximum harvest of about 112.5 results in a semistable population size of about 1500.
b. The slope should be about $p = 0.075$ or about 7.5% harvested each year. Minimum viable population is 0.

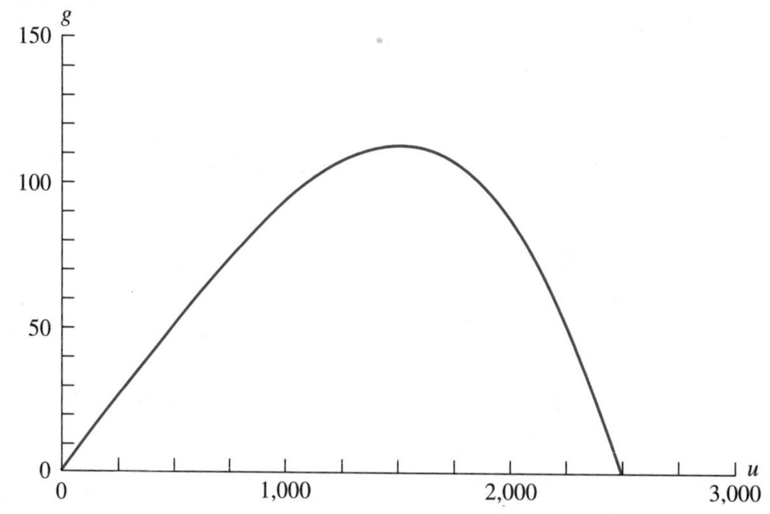

c. Equilibrium population size is 500 and harvest is about 50.

9. a. A maximum harvest of about 564 results in a semistable population size of about 3091.

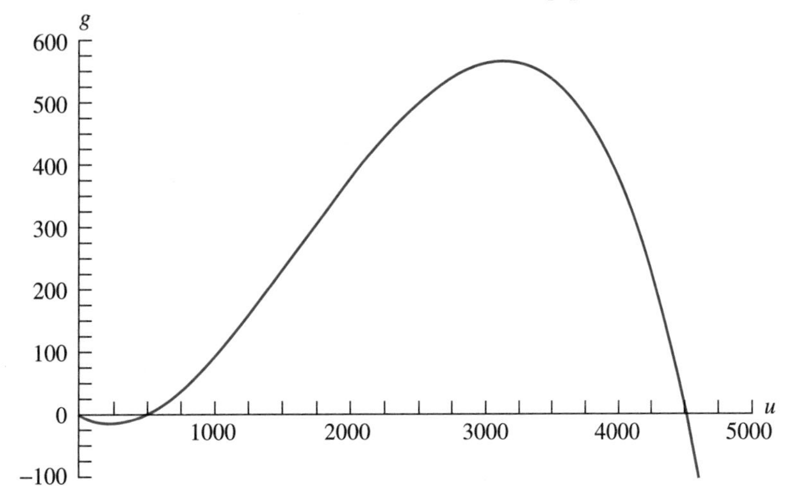

b. The slope should be about $p = 0.18$, or about 18% harvested each year. The minimum viable population size is about 1909.

c. The population size is 2500, and the harvest is 500.

Section 6.6

1. a. $p = 0.4$; revenue is \$10,000; $h = 200$; and $u = 500$
b. $p = 0.35$; profit is \$7656.25; $h = 196.875$; and $u = 562.5$
c. $p = 0.7$; $h = 87.5$; $u = 125$

3. a. $p = 0.3$; revenue is \$2400; $h = 30$; and $u = 100$
b. $p = 0.225$; profit is \$1350; $h = 28.125$; and $u = 125$
c. $p = 0.45$; $h = 22.5$; $u = 50$

5. a. $u = 100\sqrt{0.81 - p}/0.9$, and $h = 100p\sqrt{0.81 - p}/0.9$
b. $p = 0.54$; revenue is \$2806; $h = 31.177$; and $u = 57.7$
c. $p = 0.46$; profit is \$1801; $h = 30.238$; and $u = 65.7$
d. $p = 0.77, h = 17.111, u = 22.22$

7. a. $u = 1250\sqrt{0.64 - p} + 500$, and $h = 1250p\sqrt{0.64 - p} + 500p$
b. $p = 0.5193$; revenue is \$48,517; $h = 485.17$; and $u = 934$
c. $p = 0.3714$; profit is \$16,633; $h = 426$; and $u = 1148$
d. $p = 0.6144$; $h = 430$; $u = 700$

Chapter 7

Section 7.2

1. 130 G-alleles, fraction is $130/280 = 13/28 \approx 0.46$

3. 80 G-alleles, fraction is $80/380 = 4/19 \approx 0.21$

5. a. 9% GG, 42% "BG," and 49% BB; 60 G-alleles; fraction of G-alleles is 0.3.
b. 9% GG, 42% "BG," and 49% BB; 1800 G-alleles; $g(2) = 0.3$.

7. a.

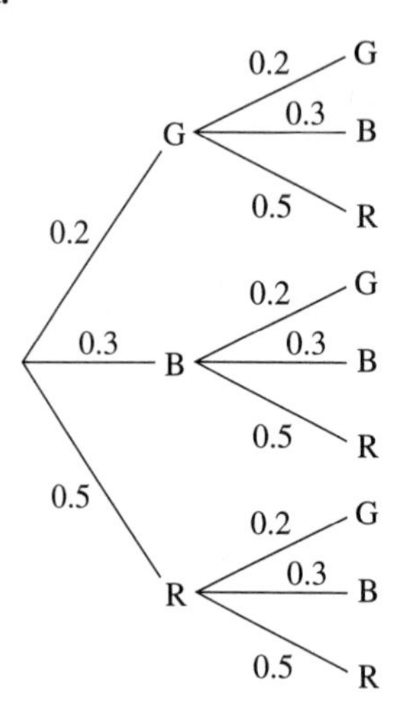

P(GG) = 0.04; P("BG") = 0.12; P(BB) = 0.09; P("RG") = 0.2; P("BR") = 0.3; P(RR) = 0.25
b. 4 GG, 12 "BG", 9 BB, 20 "RG," 30 "BR," and 25 RR
c. 40 G, 60 B, and 100 R; fraction G is 0.2; fraction B is 0.3; fraction R is 0.5.

Section 7.3

1. a. 3000 G and 7000 B
b. 2940 G and 7060 B; $g(2) = 0.294$ and $b(2) = 0.706$
c. $g(2) = 0.294$ and $b(2) = 0.706$; same as in part (b).

3. a. 200 G and 1800 B
b. 700 G and 1300 B, $g(2) = 0.35$ and $b(2) = 0.65$
c. $g(2) = 0.35$, and $b(2) = 0.65$; same as in part b.

5. a. $g(n) = 0.9g(n-1)$, and $b(n) = 0.9b(n-1) + 0.1$
b. $g(n) = 0.9^{n-1}(0.8)$, and $b(n) = 0.9^{n-1}(-0.8) + 1$
c. $g(10) \approx 0.31$ and $b(10) \approx 0.69$. Rate for $g(n)$ is 0.9 or 10% closer to 0.
d. Half-life is approximately 6.6 generations

7. a. $g(n) = 0.99g(n-1) + 0.01$ and $b(n) = 0.99b(n-1)$
b. $g(n) = 0.99^{n-1}(-0.7) + 1$ and $b(n) = 0.99^{n-1}(0.7)$
c. $b(100) \approx 0.26$. Rate for $b(n)$ is 0.99 or 1% closer to 0.
d. Half-life is approximately 69 generations

9. a. $g(n) = 0.5g(n-1) + 0.3$
b. $E = 0.6$ and $g(n)$ gets 50% closer to E each generation
c. Fraction GG is 0.36, "BG" is 0.48, BB is 0.16

11. a. $g(n) = 0.97g(n-1) + 0.01$
b. $E = \frac{1}{3}$ and $g(n)$ gets 3% closer to E each generation
c. Fraction GG is $\frac{1}{9}$, "BG" is $\frac{4}{9}$, BB is $\frac{4}{9}$

13. a. $g(n) = (1-p-q)g(n-1)+q$
b. $E_g = \frac{q}{p+q}$, and $E_b = \frac{p}{p+q}$

15. a. $g(n) = 0.95g(n-1)$, $b(n) = 0.03g(n-1)+b(n-1)$, $r(n) = 0.02g(n-1)+r(n-1)$
b. $g(100) \approx 0.004$, $b(100) \approx 0.56$, $r(100) \approx 0.44$
c. $g(100) \approx 0.0025$, $b(100) \approx 0.44$, $r(100) \approx 0.56$

17. a. $g(n) = 0.9g(n-1)+0.05r(n-1)$, $b(n) = 0.1g(n-1)+0.7b(n-1)$, $r(n) = 0.3b(n-1)+0.95r(n-1)$
b. $g(30) \approx 0.3$, $b(30) \approx 0.1$, $r(30) \approx 0.6$
c. $g(30) \approx 0.3$, $b(30) \approx 0.1$, $r(30) \approx 0.6$
d. $E_g = 0.3$, $E_b = 0.1$, $E_r = 0.6$. Solve $0.1E_g = 0.05E_r$, $0.1E_g = 0.3E_b$, and $E_g+E_b+E_r = 1$.

Section 7.4

1. a. 810 AA, 3780 "AS", 4410 SS
b. 729 AA, 3780 "AS", 0 SS
c. 9018 total alleles, 5238 A-alleles, $s(2) \approx 0.42$

3. a. $s(2) \approx 0.48$
b. $s(2) \approx 0.42$
c. $s(2) \approx 0.46$

5. a.

$$s(n) = \frac{5s(n-1)}{4+6s(n-1)}$$

b. $E = 0$, $E = \frac{1}{6}$
c. 0 is unstable and $\frac{1}{6}$ is stable.

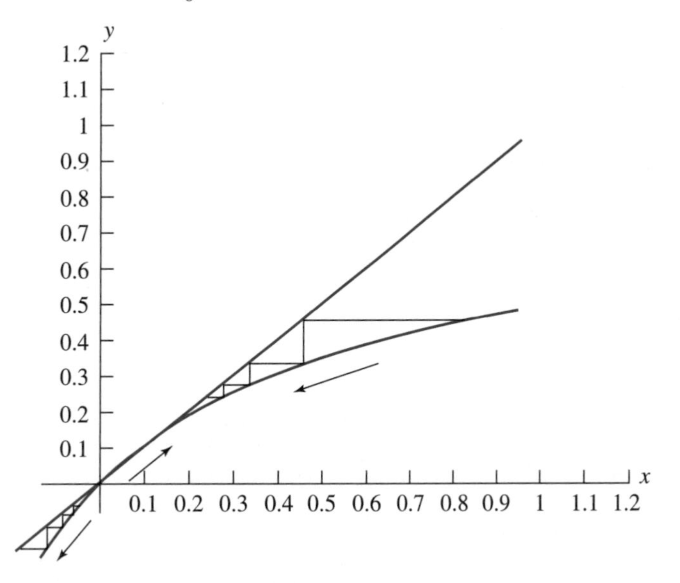

d. Rate is 0.8, so 20% closer to $\frac{1}{6}$.

7. a. 90 AA, 420 "AS," 490 SS
b. 45 AA, 420 "AS," 245 SS
c. 1420 total alleles, 510 A-alleles, $a(2) \approx 0.36$
d.

$$a(n) = \frac{a(n-1) - 0.5a^2(n-1)}{0.5 + a(n-1) - a^2(n-1)}$$

e. $E \approx 0.5$
f. $E = 0; E = 0.5; E = 1$
g. 0 and 1 are unstable; 0.5 is stable.

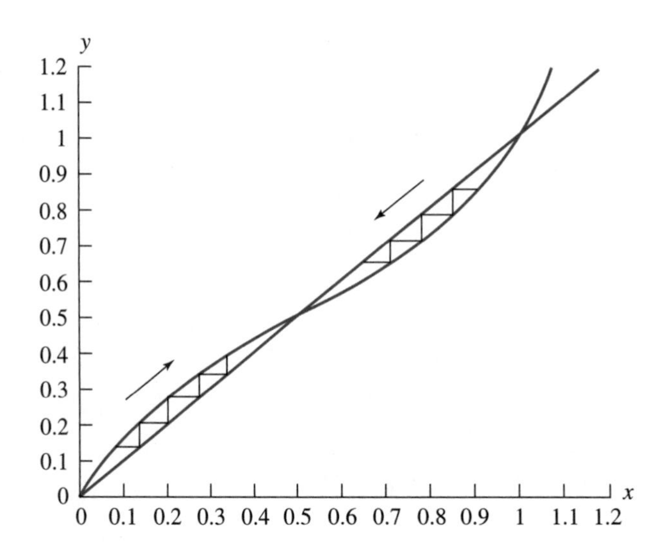

h. The rate is $\frac{2}{3}$, so 33% closer to 0.5.

9. a.

$$a(n) = \frac{a(n-1) - 0.5a^2(n-1)}{0.25 + 1.5a(n-1) - 1.25a^2(n-1)}$$

b. $E \approx 0.6$
c. $E = 0; E = 0.6; E = 1$
d. The rate is approximately 0.57, meaning getting 43% closer to 0.6.
e. Both 0 and 1 are unstable; 0.6 is stable.

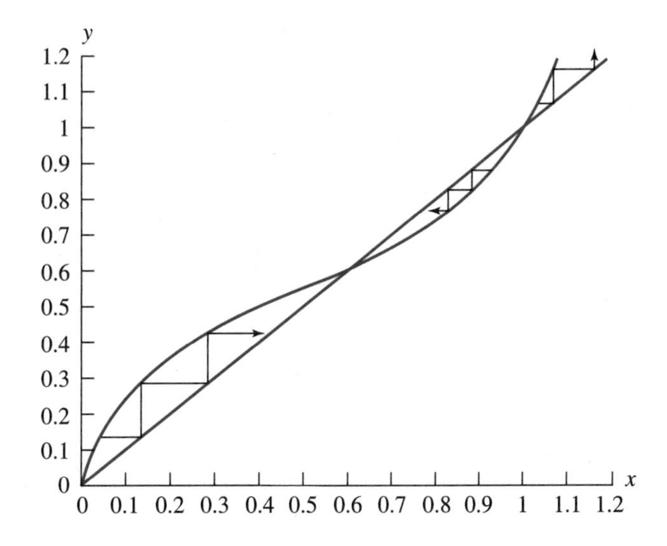

11. $a = 800 + 400x - 1200x^2$. Vertex when $x = \frac{1}{6}$ of the alleles are S. This is same as the equilibrium value in Problem 5.

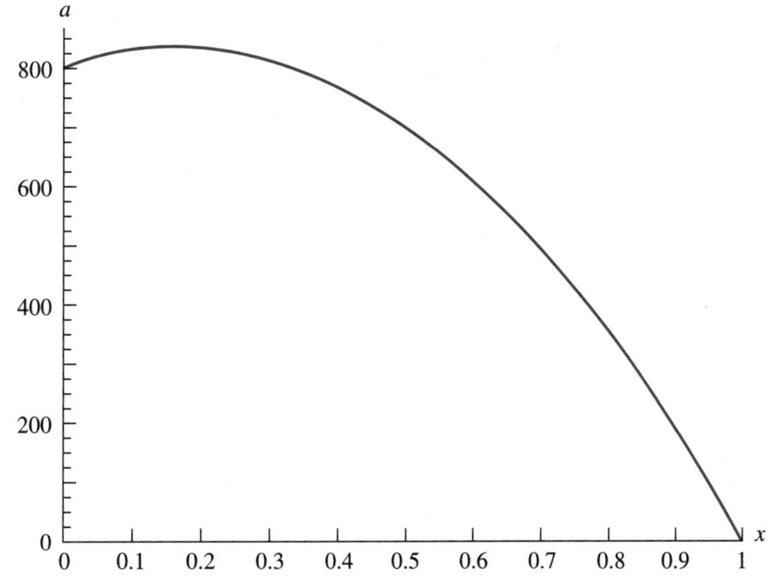

13. a.

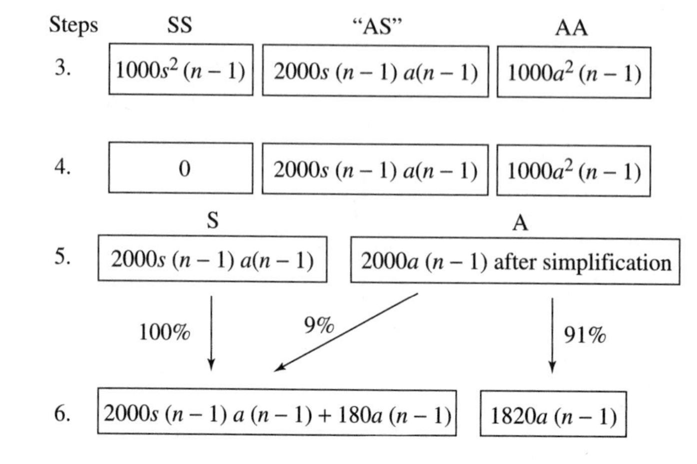

7. $$u(n) = \frac{1820a(n-1)}{2000a(n-1)[s(n-1)+1]} = \frac{0.91}{2-a(n-1)}$$

b. $a(10) = 0.70046$ and $s(10) = 0.29954$
c. $E = 0.7$ and $E = 1.3$
d. From web, 0.7 is stable and 1.3 is unstable.

Index